Birds of the Seychelles

Adrian Skerrett and Ian Bullock
Illustrated by Tony Disley

PRINCETON UNIVERSITY PRESS

PRINCETON, NEW JERSEY

Published in the United States, Canada, and the Philippine Islands by
Princeton University Press, 41 William Street, Princeton, New Jersey 08540

In the United Kingdom and European Union, published by Christopher Helm (Publishers) Ltd,
a subsidiary of A & C Black (Publishers) Ltd, 35 Bedford Row, London WC1R 4JH

Library of Congress Catalog Card Number 00-109195

ISBN 0-691-08863-2

This book has been composed in Optima

www.birds.princeton.edu

Printed in Singapore

10 9 8 7 6 5 4 3 2 1

Princeton Field Guides

Rooted in field experience and scientific study, Princeton's guides to animals and plants are the authority for professional scientists and amateur naturalists alike. **Princeton Field Guides** present this information in a compact format carefully designed for easy use in the field. The guides illustrate every species in color and provide detailed information on identification, distribution, and biology.

Coral Reef Fishes, by Ewald Lieske and Robert Myers
Birds of Kenya and Northern Tanzania: Field Guide Edition, by Dale A. Zimmerman, Donald A. Turner, and David J. Pearson
Birds of India, Pakistan, Nepal, Bangladesh, Bhutan, Sri Lanka, and the Maldives, by Richard Grimmett, Carol Inskipp, and Tim Inskipp
Birds of Australia, by Ken Simpson and Nicolas Day
Birds of Europe, by Killian Mullarney, Lars Svensson, Dan Zetterström, and Peter J. Grant
Birds of Nepal, by Richard Grimmett, Carol Inskipp, and Tim Inskipp
Birds of the Seychelles, by Adrian Skerrett and Ian Bullock

Princeton University Press
www.birds.princeton.edu

CONTENTS

Introduction	4
Acknowledgements	5
Style and layout of the book	6
The islands	9
Important birdwatching sites	17
The climate	19
Abbreviations used in the text	19
Seychelles Bird Records Committee	20
Plants mentioned in the text	21
Glossary of terms	23
Origins of the birds of Seychelles	25
Checklist of the birds of Seychelles	27
Colour plates	32
Species accounts	139
Bibliography	310
Index	318

INTRODUCTION

This volume describes and illustrates every species that has been recorded in Seychelles up to mid-2000 and some species of uncertain status reported from the country. More than 250 species are described and illustrated. The text is aimed at the serious observer, rather than the casual one interested only in common and obvious species, but it is at the same time of a non-technical style and therefore accessible to all bird-watchers. For breeding birds, the text goes beyond that which would be expected in a standard field guide with sections on breeding biology, origins, threats, and conservation. It is hoped this will enhance inter-est in these species and underline their importance.

Seychelles lies approximately between latitude 4°S and 10°S and longitude 46°E and 54°E in the west-ern Indian Ocean. The landmass is small though the islands are spread over a considerably larger area of ocean. The Exclusive Economic Zone of Seychelles (see map page 12) covers approximately 1,374,000 km^2, of which only 455 km^2 is land. Roughly one-third of this landmass is the island of Mahé and a fur-ther third the atoll of Aldabra. The number of islands in Seychelles depends partly on definition: it is often given as 115 but the 1992 Constitution of the Republic of Seychelles lists 155. These can be divided into two distinct biogeographical regions, the granitic islands (or granitics) and the outer islands.

The major attractions of Seychelles to the birdwatcher are threefold: the unique landbirds, the impor-tant seabird colonies, and the host of migrants and vagrants which turn up due to the location of the islands. The granitic islands of Seychelles are the world's oldest oceanic islands. Not surprisingly, unique endemic landbirds have evolved here, their origins reflecting the fact the islands are situated at the cross-roads between Asia and Africa. Most of the coralline outer islands do not have endemics, with the excep-tion of the Aldabra group, which is of greater antiquity than the islands of the Amirantes, Alphonse group and Farquhar group. The second attraction of Seychelles for birdwatchers is the enormous seabird colonies, where birds breed mainly during the southeast monsoon. Visits to the nature reserve islands of Cousin and Aride and to the privately protected island of Bird can be particularly exciting at this time. Some seabirds breed at the opposite time of year, or throughout the year, or have two seasons, with one species, Bridled Tern, having a unique eight-month cycle. At any time of year there is always something going on in the seabird colonies. The third attraction of birdwatching in Seychelles arises not from the uniqueness of its islands, but from the uniqueness of the Indian Ocean. Unlike other oceans, the Indian Ocean is enclosed on its northern limits by a gigantic landmass. Across Eurasia – the world's greatest land-mass – several billion birds migrate to and from Africa each year. Many of these cover huge distances even before they reach Africa, with migrations of 10,000 kilometres or more, twice per annum. Some birds from eastern Asia – including falcons, warblers, wheatears and others – will cross the Indian Ocean on non-stop flights equivalent in distance to the Saharan crossing performed by birds breeding further west. Not surprisingly, given the tendency of individual birds to stray from the main route, many of these birds reach Seychelles.

Birdwatching from October to December can be a particularly exciting time to see rare and unexpect-ed visitors to Seychelles. The spring return migration produces fewer rarities, perhaps partly due to the absence of newly fledged individuals with their higher propensity to deviate from traditional routes. However, some species are actually more likely to be seen during spring (indeed some only at this time) with evidence pointing to the existence of a return migration over the Indian Ocean in certain cases, par-ticularly in the Aldabra group.

ACKNOWLEDGEMENTS

We are grateful for the support, encouragement and contribution of many people. We must firstly acknowledge our colleagues on the Seychelles Bird Records Committee (SBRC), both past and present, Michael Betts, Phil Chantler, Chris Feare, James Ferguson-Lees, David Fisher, Ron Gerlach, Peter Harrison, Robert Lucking, John Marchant, Tony Palliser, John Phillips, Robert Prys-Jones, Steve Rooke, Bob Scott, Ian Sinclair and Don Turner. Their contributions continue to transform our knowledge of the birds of Seychelles. We also thank Maurice Loustau-Lalanne of Air Seychelles and Seychelles Islands Foundation, Glenny Savy and Justin Moustache of Islands Development Company, Guy Savy of Bird Island, Peter Hitchins and Fred Keeley of Cousine Island and David Rowat and Glynis Sanders of Seychelles Underwater Centre for logistical assistance. Michael Betts, Chris Feare, Ron Gerlach, Janet Hunter, Steve Parr and Gérard Rocamora provided helpful comments on the text and/or plates. Robert Prys-Jones of The Natural History Museum, Tring, UK and Tony Parker of Liverpool Museum, UK assisted with access to museum specimens and scientific papers. Judith Skerrett typed the first version (of many versions!) of the manuscript.

In addition to all the above, we would like to thank all the birdwatchers on the front line who have ever submitted records of their sightings in Seychelles and other information: Francis Alcinder, Carl Anderson, Marinette Assary, Phillip Baccus, Carl Axel-Bauer, Lisa Ballance, Mitch Bergeson, Brian Betsy, Darryl Birch, John Bowler, Steven Bright, Charlie Brown, Graeme Buchanan, Alan Burger, Glynn Burridge, James Cadbury, Peter and Harriet Carty, Gill Castle, Roselle Chapman, Lindsay Chong-Seng, John Collie, Perley Constance, Richard Coomber, Laura Davis, Sarah Davis, Roy Dennis, Tony Diamond, Jean Dogley, Charlotte van Duÿl, Nick Dymond, Stephen Eccles, Rodney Fanchette, Mike and Liz Fraser, Cliff Frith, Gill Gerlach, Justin Gerlach, Gemma Gessy, Liam Gibbon, John Gooders, A and G Goodwin, Phil Gregory, Adam Gretton, Eelko Groenendaal, Mark Hallett, Jacques Harter, Ian Hartley, Andrew Henderson, José Hetimier, Mike Hill, Lars Holmberg, Janet Hunter, Nigel Hunter, Sharon Johnson, Louise and Elvet Jones, Lars Jonsson, Tony Jupiter, Richard Kelly, Michael Knoll, Ken Kraaijeveld, Ahtee Labonte, Victorin Laboudallon, Florine Ladouceur, E Langley Roberts, Jack Laual, Andrea Lawrence, John Lawton, Tasso Leventis, Anna Liljevik, PB and MA Lockwood, Michael Savery, Stella Le Maitre, Philippa and Robert Lidstone-Scott, John Lines, Vicki Lucking, Michael Maher, KK Malmstrom, Michele Martin, Rod Martins, Pat Matyot, Mickey Mason, Bernard Master, Neil McCulloch, Don Merton, Carole Middleton, Robert Mileto, Simon Moffat, Marieke Moller-Holtkamp, Geoff Morgan, Jeanne Mortimer, Kathryn Murray, Tony Murray, John Nevill, Peter Njoroge, Malcolm Nicoll, George Norah, K Norris, Frank Oatman, Bill Oddie, Lars Olsson, Debbie Pain, Edwin Palmer, Karen Passmore, Viv Phillips, Murielle Pierre, Bob Pitman, Alan and Judith Prowse, John Quinn, Ulrich Radomski, Ray Rainbolt, John Raines, Mike Rands, Selby Rémie, David Richardson, Serge Robert, FJ Roberts, Jeremy Roberts, Peter Roberts, S Roebuck, Eric and Lesley Roest, DE Sargeant, Michael Savery, Guy Savy, Bruce Schoeberl, Ann Scott, Andrew Shaw, Humphrey Sitters, Femmie Smit, Bo Soderstrom, France Sophola, Joel Souyave, Jim Stevenson, Terry Stevenson, Margaret Stewart, Paul Sweet, Colin Taylor, Robert Threadgold, Regis Tiatousse, Ralph and Brenda Todd, Andrew Upton, Thor Veen, Ross and Janet Wanless, Ben Warren, Jeff Watson, Bernard and Elizabeth Watts, Mike Whiteside, Wolfgang Wiltschko, James Wolstencroft, Gordon Wright, Neil Young and Roy Youpa.

STYLE AND LAYOUT OF THE BOOK

The Illustrations

There are around 1,000 illustrations on 53 colour plates, depicting most variations of plumage likely to be seen in Seychelles. These have been produced from a variety of sources, including field sketches of live birds in Seychelles and (for migrants) Europe, published and unpublished photographs and museum specimens in collections held at Liverpool Museum, UK and The Natural History Museum (Tring), UK. Birds in similar poses on the same plate are drawn to the same scale as each other for purposes of comparison. For migratory species, the emphasis is on non-breeding plumage, though breeding plumage is also illustrated where this differs to a significant degree. Racial differences have also been illustrated where relevant. Opposite each plate is a text summarising key plumage features. Size represents length unless otherwise stated, with wingspan preferred for pelagic seabirds, birds of prey, swifts and swallows.

English names

We have followed the most frequently used common names in recent works notably, *Birds of Africa*, *Handbook of Birds of the World*, etc. in most cases. Exceptions have been made only where there is a well-established and perfectly reasonable name in Seychelles (such as Fairy Tern *Gygis alba*, a name that befits this species). In all likelihood, alternative bird names will survive, and for those who use them or simply find them entertaining, these bird names are included under 'Other name(s)'.

Creole names

Creole names are given for all species recorded in Seychelles. Being able to put a name to a bird in one's own language is a first step in taking an interest in ornithology. In recognition of this, the authors developed during the course of production of this book the concept of giving Creole names to all species recorded in the country. Traditionally, Creole names have been given to a minority of birds recorded in the country and very few of these names have been unique to just one species. A list of names has been produced (Skerrett and Matyot 2000) and these are given in the text.

French names

French is the third official language of Seychelles and, in recognition, French names are given for all species.

Taxonomy

In general, our taxonomic treatment is the same as found in other recent volumes. Races are treated more fully than in some guides, even where differences are minor or not easily recognisable in the field. In the case of endemics and other breeding species this is in part to draw attention to the importance of their conservation. In the case of migrants and vagrants it is largely to draw attention to the need to gain a better understanding of where these birds have come from. Consideration of race is also important, as it is possible some races will eventually be recognised as full species (e.g. races of Yellow Wagtail *Motacilla flava*).

Description

This section commences with the length and wingspan of each species followed by a brief summary of the principal characteristics of each species, useful in establishing its identity. This is followed by a more detailed description including differences according to season, age and sex.

Voice

This is given for all species and, where known, reference is made to the likelihood that non-breeding visitors to Seychelles may call, be silent, or perhaps give a subdued version of calls or songs heard at the breeding grounds.

Behaviour

An important part of identification, this section includes notes on habitat, social behaviour and characteristics of flight, stance and movement, food and method of feeding.

Range

This section deals with the world distribution of each species with an emphasis on occurrence in regions closest to Seychelles. Reference is made to different races with particularly those known to occur, or likely to occur, in Seychelles. Notes on racial characteristics are given even where, in a field situation, it may be difficult or impossible to identify races. This serves as a reminder of the importance of collecting as much detail, including a specimen if possible should one be chanced upon, and reporting this to Seychelles Bird Records Committee and/or The Nature Protection Trust of Seychelles (see SBRC below).

Status

This section relates to the species' status in Seychelles. Several titles are used within this section as follows:

Endemic: a species or race confined to Seychelles.

Most authorities recognise 12 endemic species: Seychelles Kestrel *Falco araea*, Seychelles Blue Pigeon *Alectroenas pulcherrima*, Seychelles Scops-owl *Otus insularis*, Seychelles Swiftlet *Aerodramus elaphrus*, Seychelles Bulbul *Hypsipetes crassirostris*, Seychelles Magpie-robin *Copsychus sechellarum*, Seychelles Warbler *Acrocephalus sechellensis*, Seychelles Paradise Flycatcher *Terpsiphone corvina*, Seychelles Sunbird *Nectarinia dussumieri*, Seychelles White-eye *Zosterops modesta*, Seychelles Fody *Foudia sechellarum* and Aldabra Drongo *Dicrurus aldabranus*.

In addition, Sinclair and Langrand (1998) highlighted four other species, currently generally recognised only as races, that may justify full species status, raising the number of endemics to 16. These are Aldabra Rail *Dryolimnas (cuvieri) aldabranus*, Seychelles Black Parrot *Coracopsis (nigra) barklyi*, Abbott's Sunbird *Nectarinia (sovimanga) abbotti* (having two endemic races: *abbotti* and *buchenorum*) and Aldabra Fody *Foudia (eminentissima) aldabrana*. A fifth possible full species proposed by Sinclair and Langrand, Madagascar Sacred Ibis *Threskiornis (aethiopicus) bernieri*, occurs only in Seychelles (at Aldabra, race *abbotti*) and Madagascar. We follow this same treatment, with reasons given in the text, while noting that further genetic work may help to resolve the taxonomic questions relating to these five species.

Ten species are recognised as having a total of 13 endemic races: Green-backed Heron (two endemic races: *degens* and *crawfordi*), Cattle Egret (race: *seychellarum*), Madagascar Sacred Ibis (as mentioned above, race: *abbotti*), Madagascar Turtle Dove (races *rostrata* and *coppingeri*), Comoro Blue Pigeon (race: *minor*), Madagascar Coucal (race: *insularis*), Madagascar Nightjar (race: *aldabrensis*), Madagascar Bulbul (race: *rostratus*), Souimanga Sunbird (race: *aldabrensis*) and Madagascar White-eye (races *aldabrensis* and *menaiensis*). An endemic race of Audubon's Shearwater (*colstoni*) has been described from Aldabra but there is doubt concerning its validity. In addition, the form of turtle dove found on Cosmoledo may be an undescribed race.

Breeds: apart from the 26 endemic species or races, there are 41 non-endemics that breed in Seychelles (18 seabirds, 8 waterbirds and 15 landbirds). Landbirds in this group include 11-12 introduced species (the upper figure including Madagascar Fody, which may or may not be a natural coloniser). Excluding double counting of Cattle Egret (represented by both an endemic race and a non-endemic race) the total number of breeding species is 66.

Annual migrant: a migratory species that does not breed in Seychelles but occurs in Seychelles every year outside its normal breeding season (sometimes year round) in Seychelles. This list comprises 24 species: 16 waders, seven seabirds and one duck.

Vagrant: a species that does not occur each year. This makes up half of the species in the book: 127 in all. This includes all records accepted by SBRC up to March 2000, though two were accepted with the caveat 'origin unknown' implying possible ship-assisted passage or deliberate introduction (Mallard and Brown Fish Owl). We have not tried to break this category down by reference to the number of confirmed records. The collection of records in Seychelles is still relatively recent and any attempt at categorisation may become quickly dated. Any sighting of any of these species should be reported to SBRC.

Uncertain: species for which there have been reports of possible occurrence but there is a lack sufficient detail to clinch a first record for Seychelles. Their precise status remains unresolved pending more information. There are 29 such species (15 seabirds, 3 waterbirds, 5 waders and 6 landbirds).

Extinct: this comprises seven species, of which three were endemic species or races (Seychelles Parakeet, Seychelles Chestnut-flanked White-eye and Aldabra Warbler), two survive elsewhere, one was introduced and one is of unknown status (but was possibly a race of Purple Swamphen).

Details of islands and locations, past and present, are given for all categories. All islands and local names used in the text are shown on the maps (pages 000 to 000). Population figures and trends are also given for these species wherever these have been documented. Where applicable, the IUCN status is given.

Origins

This section is substituted for 'Range' in the case of endemic species or races and describes the relationship to other birds in other regions of the world.

Threats and Conservation

This section appears for all breeding species, except for introduced species. Given the special interest of species breeding in Seychelles to this book, the section serves to inform the interested reader of the vulnerability of such species and the steps that could be taken or are being taken to conserve them.

General

This section is added for some species to give miscellaneous points of interest.

Similar Species

This section draws attention to the most likely confusion species and the key features to help distinguish them. In addition to species of proven occurrence in Seychelles, particular attention is drawn to potential vagrants where these need to be considered. Apart from records submitted with no description, the major reason why species reported to SBRC have not been accepted or are accepted with a caveat is the failure by observers to consider potential vagrants. For example, in its first published report, SBRC accepted 11 records of Common Swift *Apus apus*, nine of them with a caveat concerning all-dark *Apus* swifts, even though none had ever been known to occur in Seychelles. Likewise, 27 of 37 records of Barn Swallow *Hirundo rustica* were accepted with a caveat concerning other possible similar species; 11 records of 'Cuckoo sp.' were accepted and other records carried similar caveats or qualifications. We have included brief details of 136 potential vagrants and have paid particular attention to long-distance migrants to East Africa and, to a lesser extent, southern India and Sri Lanka, which could possibly reach Seychelles. Line drawings, where useful, are added to assist identification or elimination of these possibilities and to encourage meticulous care. Some of these potential vagrants, we believe, will be recorded in Seychelles one day (indeed during the course of production of this book, several species were upgraded from this category to 'Vagrant' and a full description included). Others may never be recorded, but should, nevertheless, always be considered until a clearer picture emerges. Seychelles' unique location, its relative proximity to the migration routes of many Eurasian species which cross the Sahara, and the fact a number of Asian species have occurred in Seychelles and nowhere else in the entire Afro-Malagasy region, bear testimony to the potential for surprises. We encourage observers to consider all possibilities and not make assumptions, which can sometimes lead to rejection of records.

References

For brevity and to make text more readable, references are given at the end of each species account rather than within the text. Where acknowledgement is appropriate within the body of the text or where the reader may be interested to discover further details, a number in superscript refers to the reference section at the end of that particular species' account. Authors names and year of publication only are given, full details of which may be obtained from the Bibliography.

Bibliography

All references cited in the text are given in full in this section. Also included are other works consulted during production of the text.

THE ISLANDS

The granitic islands, Bird and Denis

The 40 granite islands of Seychelles (see map page 12) occupying the submerged Seychelles Bank are the world's only oceanic islands of continental rock, all others being coralline or volcanic. For geographical convenience, the islands of Silhouette and North are usually included in the granitic group, though they comprise much younger rocks of volcanic syenite. Bird and Denis, lying on the northern rim of the Seychelles Bank, are both very young coral cays which may also be lumped here by virtue of their location. The land area of the group is just over half the total for the whole of Seychelles at about 235 km².

Of some significance to the level of endemism in the fauna of the granitic group is not just the unique geology but the related fact they are the oldest ocean islands in the world. Seychelles has been isolated from continental regions for 65 million years. This was before mammals evolved. Hence there are no naturally occurring land mammals, though there are two endemic species of bat and the land mammals introduced by man.

The granite rocks are some 750 million years old. At one time, they formed part of the supercontinent of Pangaea, which united all the world's continents, the rocks of Seychelles being close to the point where Madagascar, Africa and India combined. About 200 million years ago, Pangaea divided in two: Laurasia (modern Eurasia and North America) to the north and Gondwanaland (South America, Africa, Antarctica, India and Australasia) to the south. These supercontinents sub-divided further and some 127 million years ago Madagascar, India and granitic Seychelles broke away from Africa as a single unit. This unit drifted across what is now the Indian Ocean, Madagascar breaking away about 82 million years ago. Finally, in the midst of an era famed as marking the extinction of the dinosaurs, Seychelles broke off from the western edge of India.

Granitic Seychelles was for the first few million years of its isolation a single landmass, covering 300,000 km², an area larger than Britain and Ireland combined. It was a huge single landmass as recently as 10,000 years ago during the last ice age when sea levels were lower than the present. Today, we can only speculate what birds might have occupied this land that no human eyes ever saw. Seychelles is fertile ground for investigation by those interested in ancient avifauna. As for the future, slowly but surely, granitic Seychelles is sinking under its own weight at a rate of about one metre per 10,000 years. At this rate, the tip of Morne Seychellois will dip beneath the waves in a mere nine million years (geologically speaking). However, that still leaves plenty of time for reclamation projects to keep pace or even expand the landmass.

Silhouette and North are of similar physical appearance to the granitic islands, having been exposed to similar weathering processes for millennia. However, they comprise much younger syenite of around 63 million years of age.

Bird and Denis are little more than sandbanks that may have emerged following a fall in sea levels caused by a change in local ocean currents only 2,000 years ago. Certainly, it is unlikely they existed as far back as 4,000 years ago.

Approximately 99.7% of the Seychelles population live in the granitic group, including 90% on Mahé and most of the balance on Praslin and La Digue. Silhouette has a population of about 200, with just a handful of people or none at all on the remaining islands.

There are three nature reserves important for birds: the islands of Aride and Cousin, and the Vev Reserve of La Digue. The islands of Booby, Mamelles, Beacon (Ile Sèche) and Ile aux Vaches Marines have protected status primarily due to their important seabird colonies. Additionally, there are two national parks important for protection of endemics: Morne Seychellois National Park on Mahé, and Praslin National Park, the latter including the World Heritage Site of Vallée de Mai. In addition, though they have no legal status, privately owned Cousine and North Island are run as nature reserves.

The Granitic Islands are as follows:

Island	Area (ha)	Distance from Victoria (km)
Mahé	15500	–
Praslin	2756	45
Silhouette	1995	30
La Digue	1010	50
Curieuse	286	52
Félicité	268	55
Frégate	219	55
St Anne	219	5
North	201	35
Cerf	127	5
Marianne	95	60
Grand Soeur	84	60
Thérèse	74	8
Aride	68	50

(Granitics – cont'd)

Island	Area (ha)	Distance from Victoria (km)
Conception	60	9
Petite Soeur	34	60
Cousin	29	44
Cousine	26	43
Long	21	7
Récif	20	34
Round, Praslin	19	46
Anonyme	10	9
Mamelles	9	6
Moyenne	9	18
Ile aux Vaches Marines	5	7
L'Islette	3	7
Beacon (Ile Sèche)	2	8
Cachée	2	7
Cocos	2	55
Round, Mahé	2	6
L'Ilot Frégate	1	50
Booby	<1	50
Chauve Souris, Mahé	<1	12
Chauve Souris, Praslin	<1	50
Ile La Fouche	<1	55
Hodoul	<1	<1
L'Ilot	<1	7
Rat	<1	10
Souris	<1	15
St Pierre, Praslin	<1	51
Zavé	<1	50
Harrison Rocks (Grand Rocher)	<1	9

Islands North of the Granitics:

Island	Area (ha)	Distance from Victoria (km)
Denis	143	95
Bird	101	105

The Outer Islands

West and southwest of Seychelles lie four groups of coral islands, referred to collectively as the outer islands.

The Amirantes is a linear archipelago of 26 islands similar to arc systems found elsewhere in the tropics (see map page 14). It lies on a shallow bank, which varies in depth from 11 m over parts of the rim, to 70 m in the centre. Less than 1 km below the surface, there is a basaltic base, but there are no exposed volcanic rocks. No island rises more than 3 m above sea level.

To the south of the Amirantes is the Alphonse group of three islands (see map page 14). This is sometimes included as part of the Amirantes, though it is separated from these islands by a deep trench.

South-southwest of the Alphonse group is the Farquhar group (see map page 14). This lies closer to Madagascar than it does to Alphonse. The island of Goëlettes in Farquhar atoll marks the most southerly extreme of Seychelles. Included in the Farquhar group is the raised coral platform island of St Pierre, which was the only point of land between the granitic islands and the Aldabra group just 4,000 years ago when sea levels were higher than at the present date.

The Aldabra group of raised coral islands lies in the extreme southwestern corner of Seychelles (see map page 15). The group includes the atolls of Aldabra, Cosmoledo and Astove together with the platform island of Assumption. Aldabra atoll is the world's largest raised coral atoll and a World Heritage Site. These raised coral islands have also submerged and re-emerged several times during their history. The most recent emergence was 125,000 years ago. Not surprisingly, given their antiquity, levels of endemism are high in the Aldabra group, in marked contrast with elsewhere in the outer islands.

There are also two isolated coral islands of little ornithological interest to the south of Mahé, Platte and Coëtivy.

The Outer Islands are as follows:

(a) Islands south of the Granitics

Island	Area (ha)	Distance from Victoria (km)
Coëtivy	931	290
Platte	54	140

(b) The Amirantes group

Island	Area (ha)	Distance from Victoria (km)
Desroches	394	230
Poivre Atoll	255	270
Alphonse	174	400
D'Arros	150	255
St Joseph Atoll	122	250
Marie Louise	53	310
Desnoeufs	35	325
African Banks	30	235
Rémire	27	245
St François	17	410
Boudeuse	1	530
Etoile	1	305
Bijoutier	<1	405

St Joseph Atoll comprises 14 islands: St Joseph, Ile aux Fouquets, Ressource, Petit Carcassaye, Grand Carcassaye, Benjamin, Bancs Ferrari, Chiens, Pélicans, Vars, Ile Paul, Banc de Sable, Banc aux Cocos and Ile aux Poules. Poivre Atoll comprises three islands (Poivre, Florentin and South Island) and African Banks two islands (African Banks and South Island; the latter was once a seabird island but is now covered at high tide). This gives a total of 29 islands in the Amirantes/Alphonse group.

(c) The Farquhar group

Island	Area (ha)	Distance from Victoria (km)
Farquhar Atoll	799	770
Providence Atoll	228	710
St Pierre	167	740

Farquhar Atoll comprises ten islands: Bancs de Sable, Déposés, Ile aux Goëlettes, Lapins, Ile du Milieu, North Manaha, South Manaha, Middle Manaha, North Island and South Island; and Providence Atoll two islands: Providence and Bancs Providence, giving a total of 13 islands in the group.

(d) The Aldabra group

Island	Area (ha)	Distance from Victoria (km)
Aldabra Atoll	15,380	1150
Assumption	1171	1140
Astove	661	1045
Cosmoledo Atoll	460	1045

Aldabra Atoll comprises 46 islands (Grand Terre, Picard, Polymnie, Malabar, Ile Michel, Ile Esprit, Ile aux Moustiques, Ilot Parc, Ilot Emile, Ilot Yangue, Ilot Magnan, Ile Lanier, Champignon des Os, Euphrate, Grand Mentor, Grand Ilot, Gros Ilot Gionnet, Gros Ilot Sésame, Heron Rock, Hide Island, Ile aux Aigrettes, Ile aux Cèdres, Iles Chalands, Ile Fangame, Ile Héron, Ile Michel, Ile Squacco, Ile Sylvestre, Ile Verte, Ilot Déder, Ilot du Sud, Ilot du Milieu, Ilot du Nord, Ilot Dubois, Ilot Macoa, Ilot Marquoix, Ilots Niçois, Ilot Salade, Middle Row Island, Noddy Rock, North Row Island, Petit Mentor, Petit Mentor Endans, Petits Ilots, Pink Rock and Table Ronde), and Cosmoledo 19 islands (Menai, Ile du Nord (West North), Ile Nord-Est (East North), Ile du Trou, Goëlettes, Grand Polyte, Petit Polyte, Grand Ile (Wizard), Pagode, Ile du Sud-Ouest (South), Ile aux Moustiques, Ile Baleine, Ile aux Chauve-Souris, Ile aux Macaques, Ile aux Rats, Ile du Nord-Ouest, Ile Observation, Ile Sud-Est and Ilot la Croix), giving a total of 67 islands in the group.

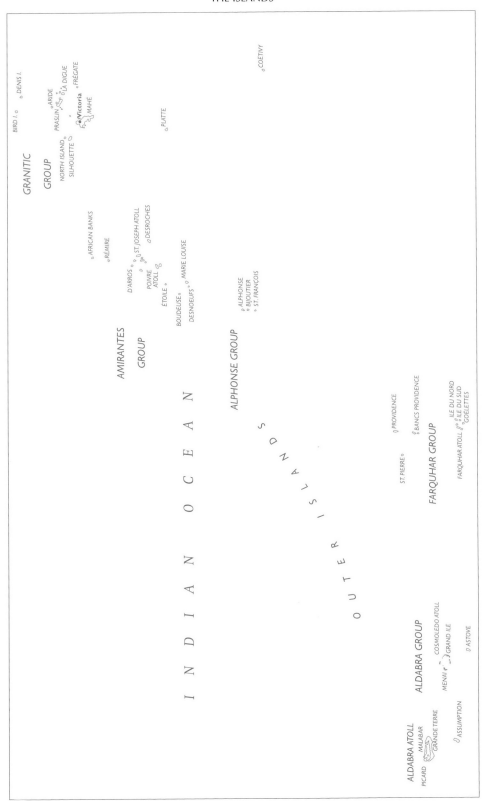

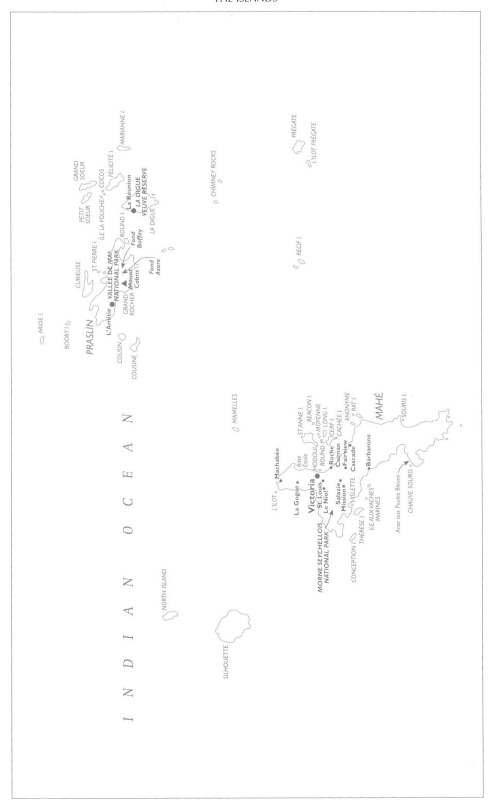

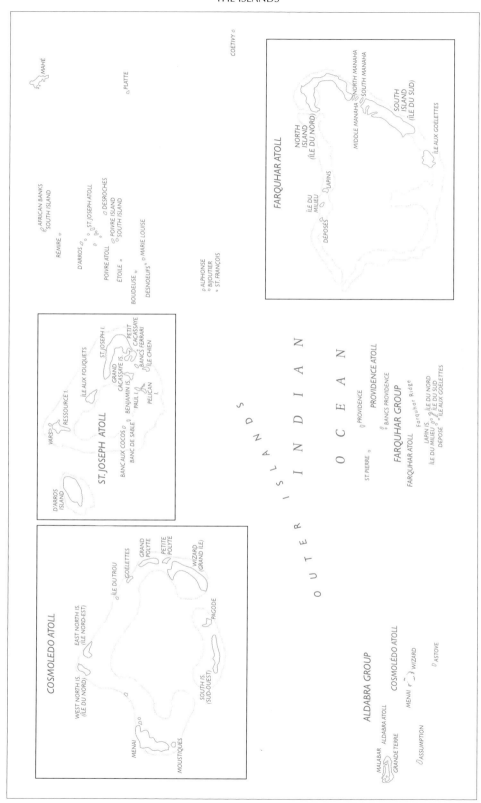

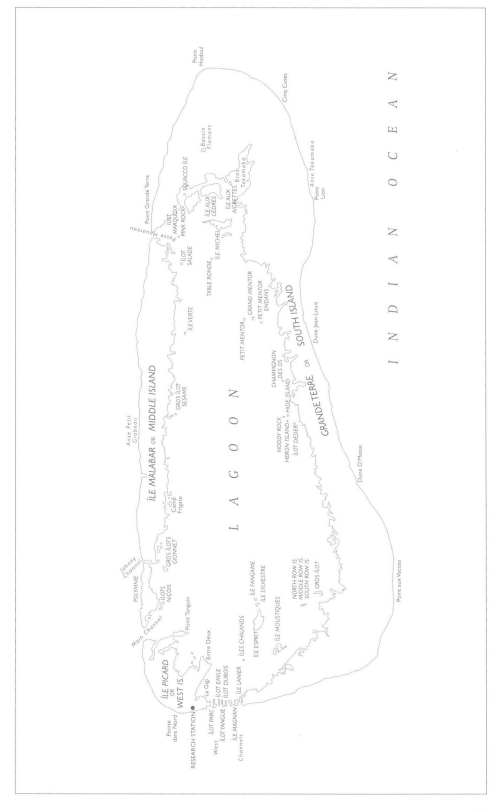

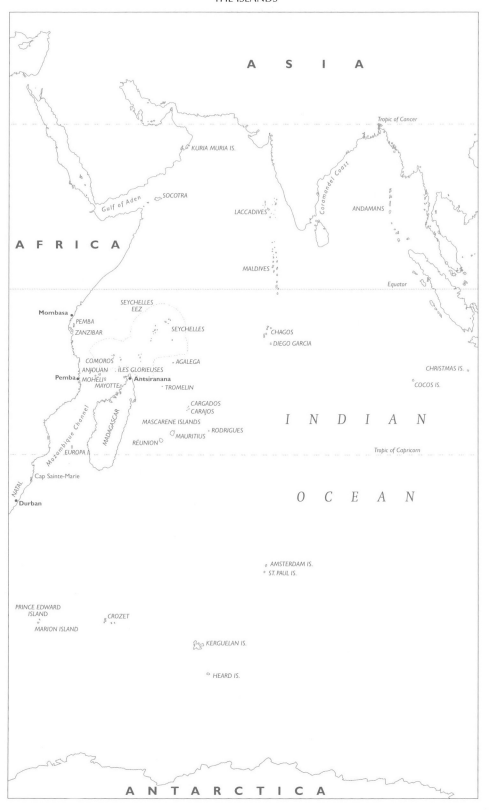

ASIA

AFRICA

Tropic of Cancer

KURIA MURIA IS.

Gulf of Aden
SOCOTRA

LACCADIVES

Coromandel Coast

ANDAMANS

MALDIVES

Equator

SEYCHELLES
EEZ

Mombasa
PEMBA
ZANZIBAR

SEYCHELLES

CHAGOS

DIEGO GARCIA

COMOROS
ANJOUAN   ÎLES GLORIEUSES
Pemba   MOHÉLI
MAYOTTE   Antsiranana
              TROMELIN

AGALEGA

CHRISTMAS IS.

COCOS IS.

CARGADOS
CARAJOS

MASCARENE ISLANDS
              RODRIGUES
RÉUNION   MAURITIUS

Mozambique Channel

MADAGASCAR

EUROPA I.

Cap Sainte-Marie

NATAL

Durban

I N D I A N

Tropic of Capricorn

O C E A N

AMSTERDAM IS.
ST. PAUL IS.

PRINCE EDWARD
ISLAND

MARION ISLAND

CROZET

KERGUELEN IS.

HEARD IS.

A N T A R C T I C A

IMPORTANT BIRDWATCHING SITES

1 Mahé

Seven endemic landbirds occur within Morne Seychellois National Park, which occupies the centre of northern Mahé. These include two of the most enigmatic and elusive species: Seychelles Scops-owl and Seychelles White-eye. The Park is the stronghold for Seychelles Scops-owl, confined to Mahé. Seychelles White-eye, found only on Mahé and neighbouring Conception (which is far more difficult to reach), also occur and sometimes breed. Seychelles Kestrel occur in fairly good numbers, Mahé holds most of the world population of Seychelles Kestrel, found from sea level to the highest hills. Other landbirds that can be seen with relative ease are Seychelles Blue Pigeon, Seychelles Swiftlet, Seychelles Bulbul and Seychelles Sunbird.

The coastal plateau areas of Mahé include important surviving wetland sites, though most are under considerable pressure from development. These include North-east Point marsh, Beau Vallon marsh, Roche Caiman Bird Sanctuary, Plantation Club marsh and Police Bay marsh. These areas are vital to the survival of Yellow Bittern in Seychelles. They have been the focal point for the recent natural colonisation of Seychelles by Black-crowned Night Heron. The east coast of Mahé includes mudflats close to Victoria, which are the best wader sites in Seychelles.

2 Praslin

Praslin National Park covers much of the centre of the southern half of Praslin. It includes within its boundaries the World Heritage Site of Vallée de Mai. For birds, the area is most famous as a stronghold for Seychelles Black Parrot, restricted to Praslin and Curieuse. Parrots may be seen throughout the island wherever suitable food plants exist, including in several hotel gardens at sea level. Five other endemic landbirds occur: Seychelles Kestrel, Seychelles Blue Pigeon, Seychelles Swiftlet, Seychelles Bulbul and Seychelles Sunbird. North Praslin and the grounds of Lemuria Resort include important wetland areas.

3 La Digue

La Digue is renowned as the home of Seychelles Paradise Flycatcher. The western plateau is the stronghold for this species, holding around 75% of known territories. Several pairs occur within the small Vev Reserve at La Passe. It is also one of the most important areas for Yellow Bittern, of special interest as it occurs only in Seychelles outside the Asian region. Apart from the flycatcher, four other endemics occur: Seychelles Blue Pigeon, Seychelles Swiftlet (breeding above the plateau), Seychelles Sunbird and Seychelles Bulbul.

4 Aride

The most northerly granite island, Aride, lies 10 km north of Praslin. Royal Society for Nature Conservation (RSNC) purchased the island in 1973, with funds provided by Christopher Cadbury. It has received legal protection as a nature reserve since 1979. It is the most important seabird island of the granitics in terms of number of species (ten) and number of birds (over one million). These include the world's largest colony of Lesser Noddy and the only surviving Roseate Tern colony in the granitics (there are two considerably smaller colonies in the Amirantes). Red-tailed Tropicbird breeds on Aride, the only nesting site north of the Aldabra group. Sooty Terns breed in large numbers, Aride being probably the last viable colony in the granitics (two considerably smaller colonies exist but receive no protection). Aride also hosts by far the largest population of Audubon's Shearwater in Seychelles (indeed probably more than the rest of the entire African/western Indian Ocean region combined). Other seabirds are Seychelles largest populations of Fairy Tern and Common Noddy, second largest population of Wedge-tailed Shearwater and smaller numbers of Bridled Tern. Landbirds include 80% of the world population of Seychelles Warbler and three other endemics. Of the four endemic landbirds, two have been translocated (Seychelles Warbler and Seychelles Magpie-robin) and two have recolonised by natural means (Seychelles Sunbird and Seychelles Blue Pigeon) following the regeneration of the native woodland. The numbers of roosting frigatebirds (predominantly Great Frigatebird but also Lesser Frigatebird) are higher on Aride than anywhere else in the granitics.

5 Cousin

Cousin is situated just 2 km west of the northwest tip of Praslin. Following a worldwide appeal, International Council for Bird Preservation (now BirdLife International) purchased the island in 1968, held in trust by RSNC. BirdLife International managed the island up to 1998 when the task was passed to BirdLife Seychelles. The island has been a nature reserve with legal protection since 1975. The purchase of Cousin probably saved Seychelles Warbler from extinction, this being its last surviving home at this time (it now also breeds on Aride and Cousine following successful transfers). Cousin also holds an important breeding population of Seychelles Magpie-robin, translocated to the island from Frégate. It is one of only four islands to hold Seychelles Fody. Two other endemics, Seychelles Sunbird and Seychelles Blue Pigeon, breed. Cousin is an important seabird island, with seven breeding species: Audubon's Shearwater, Wedge-tailed Shearwater, White-tailed Tropicbird, Bridled Tern, Common Noddy, Lesser Noddy and Fairy Tern.

6 Cousine

Privately owned Cousine lies 2 km west of Cousin and 4 km west of Praslin. It has no legal protection but is run as a reserve with resident conservationists. It hosts similar species to Cousin, including a breeding population of Seychelles Magpie-robin translocated from Frégate, Seychelles Warbler translocated from Cousin, and Seychelles Fody. The same seven seabird species found on Cousin all breed.

7 Frégate

Frégate remains a vitally important site for birds though it has no legal protection and is out of bounds to day-trippers. Until successful transfers to Cousin and Cousine this was the last breeding site for Seychelles Magpie-robin. It is one of only four islands where Seychelles Fody breed. Two other endemics occur: Seychelles Blue Pigeon and Seychelles Sunbird.

8 Bird

The sand cay of Bird Island takes its name from the huge colony of Sooty Tern that breeds on the island during the southeast monsoon. It also holds the largest population of Brown Noddy in Seychelles. There are no endemic landbirds, though the resident population of Madagascar Turtle Dove is interesting in showing characteristics of the Seychelles race *rostrata*. Due to its location at the northern edge of the Seychelles bank, Bird is the first landfall for many migratory birds and has a long list of vagrants, some of which have been recorded nowhere else in Seychelles.

9 The Amirantes

To the west of the granitic islands lie the Amirantes. There are no endemic landbirds but several islands host important seabird colonies. At one time there were probably seabird colonies on most islands but many have been wiped out due to direct human exploitation and introduced predators. Interest is now largely restricted to African Banks, St Joseph Atoll, Etoile, Boudeuse, Marie Louise, Desnoeufs and St François. The most important colonies contain Sooty Terns (on African Banks, Etoile and Desnoeufs), Black-naped Tern (on African Banks and St Joseph Atoll), Masked Booby (on Boudeuse), and Roseate Tern (on African Banks and Etoile).

10 Farquhar Group

South of the Amirantes and considerably closer to Madagascar than to Mahé, lies the Farquhar group. Again, there are no endemic landbirds but seabirds include a large colony of Sooty Tern (on Goëlettes) and Black-naped Tern (on Goëlettes and Bancs Providence).

11 Cosmoledo Atoll

Cosmoledo is the least visited atoll of Seychelles but the most important unprotected site for birds in Seychelles. It holds the nation's largest populations of Red-footed Booby and Masked Booby and the last few breeding pairs of Brown Booby. The colony of Sooty Tern is the largest in Seychelles. A few Great Frigatebird still breed, their numbers (like the boobies) heavily reduced by poachers. Bridled Tern breed in the only site south of the Amirantes, as do Brown Noddy, herons and egrets. Landbirds include Abbott's Sunbird, Madagascar Cisticola and two endemic races: Madagascar White-eye race *menaiensis*, and a race of Madagascar Turtle Dove, assumed to be race *coppingeri*, but is possibly an undescribed race.

12 Aldabra Atoll

The World Heritage Site of Aldabra lies in the southwestern corner of Seychelles. It is the world's largest raised coral atoll. It accounts for around one-third of the landmass of Seychelles, but is uninhabited except for the warden and staff of the research station. Apart from holding far more giant tortoises than anywhere else on earth, it is renowned for its birds. These include the last flightless bird of the Indian Ocean (Aldabra Rail) and the world's second largest frigatebird colony (with both Great Frigatebird and Lesser Frigatebird present). Other endemic landbirds include a further two full species (in addition to the rail): Aldabra Drongo and Aldabra Fody, and seven endemic races: Madagascar Turtle Dove, Madagascar Nightjar, Comoro Blue Pigeon, Madagascar Coucal, Madagascar Bulbul, Souimanga Sunbird and Madagascar White-eye. It is possible the endemic Aldabra Warbler, only discovered in 1967, may survive, although this species is generally thought to be extinct, with no sightings since 1983. Other seabirds include Audubon's Shearwater. The small population of Caspian Tern represents the only oceanic colony of this species in the world. The small population of Greater Flamingo sometimes breeds (possibly annually though this remains uncertain), this being the only atoll in the world where breeding occurs and the only oceanic site other than Galápagos. Madagascar Sacred Ibis breeds only here and on Madagascar.

THE CLIMATE

The climate of Seychelles is tropical: hot and humid all-year around with low annual and diurnal ranges of temperature. Though there are no seasons in the usual sense, the climate is influenced by two wind systems, each of which dominates for roughly half the year. These wind systems create a marked variation in wind direction, rainfall and humidity.

The southeast monsoon blows from May to October in the granitics. Further south in the Aldabra and Farquhar groups, it arrives a month or so later and may continue into December. This is a relatively dry wind system, particularly in the southern islands, which experience a seasonal drought (September-October rainfall at Aldabra averages 17 mm, compared to 300 mm at Mahé). Temperature averages 25-30°c at sea level during this period and humidity around 80%.

The northwest monsoon blows from November to April in the granitics and about one month later in the southern outer islands. The wind speed and direction is less constant at this time of year and temperatures are up to 3° higher. The islands lie outside the cyclone belt, which influences climate from approximately 15°S, but in exceptional years cyclones may hit the southern islands.

Between the two seasons, as the sun passes vertically overhead, winds may die away completely resulting in flat calm seas and periods of extreme humidity. As each wind season takes hold, wind speeds rise, particularly during southeast monsoon with very heavy seas around July to August. Heavy downpours may occur at any time of year particularly in the granitics where mean annual rainfall is 2400 mm compared to 1500 mm in the Amirantes and only 1000 mm at Aldabra. Rainfall reaches a peak around mid-December to mid-February.

The Indian Ocean can be affected by the El Niño phenomenon, which may result in torrential rain and high sea temperatures in Seychelles. In 1997/98 a particularly severe El Niño had a considerable impact, raising sea temperatures, raising tide levels and causing extreme coral bleaching. Inevitably such dramatic changes also have an impact on breeding seabird colonies. It also appeared to result in fewer migrants and vagrants being recorded in that season though precise data is lacking.

ABBREVIATIONS USED IN THE TEXT

c	circa, about
cm	centimetre
ed.	editor
e.g.	for example
et al.	and others
etc.	etcetera
g	grammes
ICBP	International Council for Bird Preservation
i.e.	that is
km	kilometre
m	metre
mm	millimetre
NPTS	Nature Protection Trust of Seychelles
pers. comm.	personal communication
RSNC	Royal Society for Nature Conservation
SBRC	Seychelles Bird Records Committee
SIF	Seychelles Islands Foundation
sp.	species

SEYCHELLES BIRD RECORDS COMMITTEE

The importance of the consideration of records of vagrants to Seychelles merits a special mention for the work of Seychelles Bird Records Committee (SBRC). SBRC was founded in 1992. It considers records of species of less than annual occurrence in Seychelles, which comprise about half of the species recorded in this volume.

New discoveries will continue, and gradually the status of many vagrants and migrants will become clearer. For the purposes of this book, the authors have generally avoided reference to the number of accepted records of a particular species. As collection of records is relatively recent, an emphasis on the number of records for most species would be misleading. Of the 110 species of less than annual occurrence accepted by SBRC up to 31 December 1995 and published in its first report, 44 species were represented in the record books by a single sighting and a further 21 by just two sightings (Skerrett 1996). As each year goes by, more records are received and the picture changes. Some species represented by a handful of sightings may one day be proved to be annual in occurrence. Clearly much more information is needed. This is something to which every birdwatcher visiting Seychelles, even for a short period, can contribute.

As the number of observations grows, it is likely that the status of some species will change. Seabirds such as Swinhoe's Storm-petrel *Oceanodroma monorhis* and White-faced Storm-petrel *Pelagodroma marina* may turn out to be annual visitors, even though there were no records for either species whatsoever accepted up to 31 December 1995. Migratory warblers, represented in SBRC's first report by just six records of five species, may not be such extreme rarities as this statistic would suggest. Indeed, it is interesting to note all but one of these warbler records were reported by professional ornithologists working on nature reserves in Seychelles; many more warblers may disappear unseen and unrecorded into the forests of Mahé, Praslin and Silhouette.

We urge every birdwatcher visiting Seychelles to report their observations to SBRC. Lists of all sightings, including commoner species are welcomed, while species of less than annual occurrence should be reported on a record form, accompanied whenever possible by original field notes, sketches, photographs or slides. Record forms are obtainable from the following address: Adrian Skerrett, Mahé Shipping Co. Ltd., PO Box 336, Seychelles or Hazeley Brook, Keele Road, Keele, Staffordshire, ST5 5AL, UK.

SBRC publish accepted records in *Birdwatch*, the quarterly journal of Nature Protection Trust of Seychelles. Subscription details are available from: Nature Protection Trust of Seychelles, PO Box 207, Seychelles.

Specimens of dead birds found in Seychelles, particularly of migrants, should be collected. Specimens of even more common migrants could yield information on matters such as the race and origin of birds occurring in Seychelles.

PLANTS MENTIONED IN THE TEXT

The following plants are mentioned in the text, where they are given their common name only (unless there is no common name in which case the scientific name is used). Often there is no English name and the common name used by English-speaking people is a Creole one or the scientific genus. Even where English names exist, Creole names are sometimes used in Seychelles by English and Creole speakers alike in preference to the English name and so are given in parentheses after the English name used in the text.

Acalypha claoxyloides A shrub of 1-2 m endemic to the Aldabra group.

Albizia *Paraserianthes falcataria* A tall, introduced tree of 10-40 m very common in the granitics, its spreading umbrella-like crown characteristic of river valleys.

Bilenbi *Averrhoa bilimbi* A small fruit tree up to 7-8 m, bearing small, sour green fruits directly on the trunk; introduced in the granitics.

Breadfruit (Friyapen) *Artocarpus altilis* An introduced tree up to 20 m, common around houses, especially in the granitics, bearing large, round green fruits.

Bwa dir *Canthium bibracteatum* A shrub of up to 8 m, indigenous to the granitics and Aldabra.

Bwa dou *Craterispermum microdon* A small tree up to 8 m, endemic to the granitics.

Bwa kafoul *Allophylus sechellensis* A shrub of up to 3 m, endemic to the granitics.

Bwa kalou *Memecylon elaeagni* A small tree, endemic to the granitics.

Bwa kuiyer *Tabernaemontana coffeoides* A shrub of 2-4 m, indigenous to the granitics, also present (but very rare) at Aldabra.

Bwa mozet *Mystroxylon aethiopicum* An indigenous shrub of 3-6 m found in the Aldabra group.

Bwa nwar *Albizia lebbeck* An introduced tree of up to 25 m found in low to mid-altitudes in the granitics.

Bwa rouz *Dillenia ferruginea* A tree of up to 20 m, endemic to the granitics, growing from sea level to high mountains.

Bwa siro *Premna serratifolia* An indigenous shrub of up to 6 m, found throughout the islands.

Bwa tabak *Tournefortia argentea* An indigenous shrub of up to 4 m, typical of the beach crest, found throughout Seychelles but particularly common in the outer islands.

Casuarina (Sed) *Casuarina equisetifolia* A tall (up to 40 m), fast-growing, pine-like coastal tree. It is uncertain whether this tree is a natural coloniser or introduced.

Cinnamon (Kannel) *Cinnamomum verum* An evergreen introduced spice tree of up to 15 m, common in the larger granitic islands, often invading high forest habitats.

Clove (Zirof) *Syzygium malaccense* An introduced spice tree of up to 20 m found in the granitics.

Coco de mer (Koko dmer) *Lodoicea maldivica* A tall, palm endemic to Praslin and Curieuse, the female tree bearing the world's largest nut.

Coconut (Koko) *Cocos nucifera* The symbol of the tropics, originally a beach crest plant but transplanted to all altitudes.

Dracaena *Dracaena reflexa* An indigenous shrub up to 15 m high, found on Aldabra and in the Amirantes.

Ficus *Ficus spp.* Small to medium-sized fruiting trees of up to 15 m including both indigenous and introduced species.

Guava (Gouyav) *Psidium spp.* An introduced, small fruit tree of up to 8 m found in the granitics.

Horne's pandanus (Vakwa Parasol) *Pandanus hornei* An endemic pandanus or screwpine of up to 18 m, favouring damper sites on Mahé, Praslin and Silhouette, the male tree bearing creamy white inflorescences, the female hard, orange fruits like pine cones.

Indian almond (Badanmyen) *Terminalia catappa* An indigenous tree, up to 20 m, found throughout Seychelles mainly in coastal areas. The green fruit, maturing to purple, contains an edible nut.

Jamaican vervain (Zepi ble kedera) *Stachytarpheta jamaicensis* An introduced plant of c50-100 cm, with pale blue flowers growing from a green, prostrate spike and dark green deeply serrated leaves.

Kalis *Tabebuia pallida* An introduced timber tree up to 15 m found in the granitics and some of the outer islands bearing pink tubular flowers.

Kapok *Ceiba pentandra* An introduced tree in the granitics of up to 20 m, bearing capsules filled with cotton like material.

Lantana (Vyeyfiy) *Lantana camara* An introduced, invasive shrub with spiny, wiry stems which bear umbrella-like clusters of small tubular flowers and small fruits similar to blackcurrants.

Latagnen lat *Verschaffeltia splendida* A tall endemic palm with stilt roots, up to 25 m, growing on steep hillsides and bearing white flowers and red fruits.

Lepeka *Passiflora suberosa* An introduced climbing plant found in the granitics and at Aldabra, bearing small flowers typical of the passion flower family and small fruits.

Mango (Mang) *Mangifera indica* An introduced fruit tree, up to 20 m, common in the granitics and other inhabited islands.

Manz tou *Asystasia gangetica* A small plant, c30 cm, with small white flowers, growing on poor soils.

Mourouk *Erythrina variegata* An indigenous tree of 5-25 m found in the granitics.

Palmis *Deckenia nobilis* A tall, endemic palm growing up to 30 m bearing inflorescences of small, creamy white flowers and small deep mauve fruits.

Pandanus tectorius An indigenous pandanus of the Aldabra group.

Pawpaw (Papay) *Carica papaya* A small introduced fruit tree of c4-8 m, common on all inhabited islands, bearing long fruits, green at first, orange when ripe.

Pemphis (Bwa-d-amann) *Pemphis acidula* An indigenous, dense shrub of up to 5 m, common on the outer islands, especially in the Aldabra group where it is the dominant plant over much of the landmass.

Pisonia (Bwa mapou) *Pisonia grandis* An indigenous tree of up to 10 m, particularly common on seabird islands where its sticky fruits adhere to feathers, sometimes killing birds.

Takamaka *Calophyllum inophyllum* An indigenous tree of up to 20 m with dense, dark green foliage, found mainly on the coastal plateaus of granitics and on some of the outer islands.

Vanilla *Vanilla planifolia* An introduced climbing plant bearing aromatic pods, both cultivated and found naturalised in the hills of the granitics.

Vouloutye *Scaevola sericea* An indigenous shrub, typical of the beach crest, growing 2-4 m, found throughout the islands.

Water hyacinth *Eichornia crassipes* An introduced water plant, which may choke freshwater marshes.

Water lettuce *Pistia stratiotes* An introduced water plant, which may choke freshwater marshes.

GLOSSARY OF TERMS

Adult: a bird that has achieved the fullest stage of plumage development.

Allopreening: the preening of one bird by another, usually a mate.

Alula: the 'thumb' of the wing (sometimes known as the 'bastard wing'), comprising a few short, stiff feathers at the **carpal** joint.

Asynchronous: not synchronous.

Australasia: the zoogeographical region surrounding Australia, New Zealand and New Guinea.

Autumn: refers to that season in the northern hemisphere.

Axillaries: the inner wing-lining feathers between the wing and body (the 'armpit').

Bare parts: the collective name for unfeathered areas, namely the bill, legs, feet, eyes and unfeathered skin.

Carpal: the forward pointing bend of the folded wing (the 'wrist').

Cere: the soft, fleshy covering at the base of the upper **mandible** in some birds (e.g. a **raptor**).

Conspecific: two or more races belonging to the same species.

Co-operative breeding: where a non-breeding bird, usually a relative, shares some or all nest duties with a breeding pair.

Coverts: small (contour) feathers which cover the quills of the tail and **flight feathers** and also the ear.

Crepuscular: active at dawn and dusk.

Crown: the top of the head.

Culmen: the top ridge of the bill.

Decurved: curved downward.

Dimorphic: having two distinct colour phases.

Diurnal: active by day.

Ear-coverts: feathers covering the external ear opening.

Ear-tufts: erectile head feathers (e.g. in some owls).

Eclipse: the dull post-breeding plumage as occurs for a short period in ducks.

Endemic: a species or race confined to a particular country or region.

Eye: for the purposes of this book includes the coloured iris.

Eye-ring: differently coloured feathers or skin circling the eye.

Eye-stripe: a coloured line appearing to run horizontally through the eye.

Facial disc: a circle of facial feathers radiating around the eye.

Flank: the side of the body below the folded wing.

Flavistic: an excess of yellow pigment in the plumage (e.g. in some fodies).

Flight feathers: the long **primaries** and **secondaries** of the wing.

Flycatch: the habit of launching forth in pursuit of aerial insect prey and returning to a perch or the ground.

Frontal shield: a flat, horn-like extension at the base of the upper **mandible** covering part of the forehead (e.g. in Common Moorhen).

Gape: the inside of the mouth.

Gonys: the outer ridge of the lower **mandible** that in some species (e.g. gulls) forms a distinct **gonydeal angle.**

Gonydeal angle: the angled ridge of the lower **mandible** of some species.

Graduated: refers to a tail in which the central feathers are longest, other feathers becoming progressively shorter moving out from the centre.

Greater coverts: See **Wing-coverts.**

Gular: referring to the throat.

Hawking: flying in pursuit of aerial prey.

Hepatic: a dark brownish red **morph** (e.g. in some cuckoos).

Hind-collar: a coloured band of feathers around the **hindneck.**

Hindneck: the back of the neck.

Immature: in this volume, all **plumages** prior to full **adult.**

Indigenous: a species or race native (but not necessarily **endemic**) to a particular country or region.

Iridescence: a metallic shiny gloss to the plumage.

Jizz: a combination of field characters which collectively identify a species.

Juvenile: a bird in its first full plumage after the nestling or downy stage.

Kleptoparasitism: where food collected by one bird is stolen by another.

Lesser coverts: See **Wing-coverts.**

Lore: the area between the bill and the eye.

Malar: the side of the throat immediately below the lower **mandible,** sometimes marked by a **malar-stripe.**

Malar-stripe: a coloured streak bordering the throat (e.g. in some pipits).

Mandible: in this volume, refers to either the upper half (the upper mandible) or the lower half (the lower mandible) of the bill, the plural being used for both together.

Mantle: the upper back.

Median coverts: See **Wing-coverts.**

Mirror: a white spot in the black wingtip of some gulls.

Monospecific: in reference to a genus, one with only one recognised species.

Monotypic: in reference to a species, one with no recognised subspecies.

Morph: a colour phase of a species with more than one plumage colour (see **Dimorphic** and **Polymorphic**).

Moult: the process by which old feathers are shed and replaced by newer ones.

Moustachial-stripe: a stripe running back from the base of the bill, sometimes above a second streak, the **submoustachial-stripe**.

Nape: the rear of the head between the **crown** and the **hindneck.**

Nocturnal: active at night.

Nominate race: the first race of a species to have been described (where two or more races exist).

Northwest monsoon: the season from November to April marked by higher rainfall and winds predominately from the northwest.

Passerine: a perching bird of the order *Passeriformes*.

Pectoral: pertaining to the breast.

Pelagic: pertaining to the open sea.

Pellet: a regurgitated compact mass of undigested food.

Plumage: a bird's feathers, used sometimes for a particular seasonal, sexual or age difference.

Polygyny: where a male mates with two or more females.

Polymorphic: having several distinct colour forms.

Primaries: the outer **flight feathers** (attached to the 'hand' and 'digits' of the wing).

Primary extension: the length by which the **primaries** extend beyond the **secondaries** in a folded wing.

Race: a population of a species that is geographically and morphologically distinct from other populations of the same species.

Raptor: a bird of prey.

Rump: the area between the lower back and the **uppertail-coverts.**

Scapulars: feathers between the **wing-coverts** and the **mantle**.

Secondaries: the inner **flight feathers** (attached to the 'forearm' of the wing).

Sedentary: making only small, local movements between seasons (i.e. not migratory).

Southeast monsoon: the season from May to October, marked by lower rainfall and humidity and southeasterly winds.

Species: a population or group of populations that do not normally interbreed with other such populations.

Speculum: a bright, often iridescent patch on the **secondaries** of many ducks.

Spring: refers to that season in the northern hemisphere.

Streamers: elongated slender tail feathers that may be the innermost (e.g. in tropicbirds) or the outermost (e.g. in some terns).

Subadult: the latter stages of development of an **immature** bird which requires more than one year to reach maturity, prior to achieving full **adult** plumage.

Submoustachial-stripe: a stripe running back from the base of the bill below a **moustachial-stripe**.

Subsong: a quieter, less emphatic version of a bird's song.

Sulcus: a narrow line along the inside edge of the lower mandible.

Summer: refers to that season in the northern hemisphere.

Supercilium: a stripe immediately above the eye.

Superspecies: a grouping of species considered to be of common stock, which have probably at least achieved reproductive isolation.

Tarsus: the 'ankle' bone of the leg.

Tertials: a term applied to the innermost **secondaries,** or secondaries nearest the body in some groups (e.g. waders); true tertials attached to the upper arm bone are found only in some large, long-winged birds (e.g. albatrosses).

Tibia: the 'drumstick' bone of the leg.

Twinkling: flying lightly and swiftly with quick bursts of rapid wingbeats (as in some swifts).

Undertail-coverts: the feathers behind the **vent** covering the bases of the tail feathers.

Uppertail-coverts: the feathers behind the **rump** covering the upper surface of the bases of the tail feathers.

Wattle: a fleshy unfeathered (or mostly unfeathered) appendage on the head or neck.

Web: the flexible, broad side of the feather; also, the loose flap of skin between the toes (e.g. in seabirds and ducks).

Wingbar: a band across the wing, lighter in colour than the surrounding feathers.

Wing-coverts: the feathers of the upperwing covering the bases of the **flight feathers,** comprising the very short **lesser coverts** (nearest to the leading edge), the shortish **median coverts** and the longer **greater coverts** (furthest from the leading edge).

Wing formula: the relative length of the **primaries**.

Winter: refers to that season in the northern hemisphere.

Vent: the feathers at the base of the **underparts** forward of the **undertail-coverts**.

ORIGINS OF THE BIRDS OF SEYCHELLES

That there are seabirds in Seychelles is no surprise: these are birds capable of considerable feats of endurance and they can rest and refuel on the ocean. But how landbirds came to make their home here is a mystery, for these are tiny specks of rock thousands of kilometres from any of the major continents. The geological record indicates separation from other landmasses at least 65 million years ago; for some islands too, changes in sea level must have periodically drowned parts of their forest cover. But the careful records of birdwatchers have revealed some amazing survival stories for wind-blown landbirds, which, dragged off course by storm winds, have managed to reach the lost archipelago of Seychelles and live to tell the tale. Most of our resident landbirds must first have arrived as exhausted vagrants, unaware that they were to be the ancestors of a new race. Time and isolation then led to new variations and adaptations in these marooned colonists, until new species were born.

Those landbirds that are still identical to their continental cousins must be the most recent arrivals. The Yellow Bittern *Ixobrychus sinensis* is such an example: this species is widespread in South-East Asia, turning up as a vagrant in Australia from time to time. We know that Asian species can reach the islands: records of White-breasted Waterhen *Amaurornis phoenicurus*, Indian Pond-heron *Ardeola grayii* and Cinnamon Bittern *Ixobrychus cinnamomeus*, from Seychelles and nowhere else in the Afro-Malagasy region are dramatic proof of the capability of waterbirds to fly long distances. Similarly Common Moorhen *Gallinula chloropus* was once classed as an endemic race but is now considered to be race *orientalis*, which occurs in the Andaman islands, Malaysia, the Sundas and the Philippines. One of the most fascinating insights into the complex crossroads of the Seychelles is Green-backed Heron *Butorides striatus*, which occurs on virtually every island, no matter how small. In the granitic islands, race *degens* is most similar to the African races (nearest land 1600 km away in Somalia) whereas the distinct individuals in the Amirantes and Aldabra are race *crawfordi*, thought to be Asian in origin (3300 km to India). It is as likely that these southern atolls have been colonised from Madagascar (which makes the arrival of Asian birds in Madagascar even more remarkable). So this tough little heron has arrived twice, from two different directions!

A glance through the checklist of birds on Aldabra confirms this route for colonisation: Madagascar Kestrel *Falco newtoni*, Madagascar Turtle Dove *Streptopelia picturata*, Madagascar Coucal *Centropus toulou*, Comoro Blue Pigeon *Alectroenas sganzini*, Madagascar Nightjar *Caprimulgus madagascariensis*, Madagascar Bulbul *Hypsipetes madagascariensis*. The endemic Aldabra Drongo *Dicrurus aldabranus* is also believed to have originated from Madagascar. A mere 300 km of ocean separate the atoll of Aldabra from the northern tip of Madagascar. Some of these colonists have been isolated long enough to evolve their own unique features: Madagascar Sacred Ibis *Threskiornis (aethiopicus) bernieri* on Aldabra (race *abbotti*) is arguably distinct from its African relatives, whereas Aldabra Rail *Dryolimnas (cuvieri) aldabranus*, which looks very similar to its Madagascan cousins, is now quite flightless: it has a wing length (at 121 mm) 30 mm shorter than its closest relative, White-throated Rail *Dryolimnas cuvieri*. What better proof of isolation could there be?

Before man arrived, the Seychelles islands were completely lacking in mammalian ground predators, which made the reflex of flight unnecessary. All who visit comment on the tameness of the birds; the expensive investment in flight feathers and flight muscles can be reduced under these conditions. This trend is evident in the islands' moorhen: wing length of 161 mm compared to 180 mm in the nominate race, despite similar bodyweight. This species has obviously been here some time and is well on the way to flightlessness. Evolution confirms what we all know: you can afford to relax in this environment! The endemic Seychelles Warbler *Acrocephalus sechellensis* merely flits from branch to branch and has also evolved weaker wings. It is probably almost incapable of flying the narrow sound between Cousin and Praslin. It did once occur on other granitic islands, but on the larger islands, where humans settled first, its weak flight would have made it an easy target for rats or cats as well as suffering from the inevitable habitat destruction. DNA studies confirm its relationship to Swamp Warbler *A. gracilirostris* of Africa (perhaps via the stepping-stone of Madagascar), which has a similar relative, Madagascar Swamp Warbler *A. newtoni*. Seychelles Warbler came so close to extinction; others were less lucky. Aldabra Warbler, which probably colonised the atoll from Comoros or Madagascar (which has a similar relative, Tsikirity Warbler, almost certainly of African origin) was only discovered in 1967. It was last sighted in 1983 and was probably extinct shortly thereafter.

Another vulnerable species, which narrowly avoided extinction, is Seychelles Magpie-robin, *Copsychus sechellarum*. This is a member of an Asian family (e.g. three species *Copsychus* in India, a separate race on the Andamans, another species in Sumatra and Borneo, etc.) that managed to reach as far as Madagascar, where the endemic Madagascar Magpie-robin *C. albospecularis* also still survives. This family is renowned for its beautiful songs: is it just possible that early human seafarers carried individuals with them? The Magpie-robins are forest birds and are not known to make migrations. Another puzzle is the extinct Seychelles Parakeet *Psittacula (eupatria) wardi*, which was superficially very similar to Alexandrine Parakeet *P. eupatria* of India: why should this species make the long sea crossing to Seychelles?

The fodies, *Foudia*, are a genus confined to the western Indian Ocean, probably originally of African origin, and related to the weavers. These seem to have island-hopped to Seychelles: both Madagascar Fody

F. madagascariensis and Forest Fody *F. omissa* occur in Madagascar. Aldabra Fody *F. aldabrana* of Aldabra Atoll and Seychelles Fody of the granitics both appear closer in form to Forest Fody than to Madagascar Fody, which is a bird of more open areas. It is interesting how through isolation the Seychelles endemic has lost almost all bright colours; lack of competition may remove the need for the gaudy plumage of its ancestors. A similar trend is visible with Seychelles Sunbird *Nectarinia dussumieri* (probably the drabbest of its family) and Seychelles White-eye *Zosterops modesta* (plain ashy grey with little of the bright yellows and chestnuts of other relatives). Sunbirds occur throughout Africa and Asia, but the origin of the Seychelles species may be via the Comoros, where Humblot's Sunbird *N. humbloti* shows similar features (whitish tip to tail, orange-yellow pectoral-tuft and traces of maroon breast-band). By comparison, the sunbird forms on Aldabra (Souimanga Sunbird *N. sovimanga aldabrensis*), Assumption (Abbott's Sunbird *N. abbotti* race *abbotti*) and Astove and Cosmoledo (Abbott's Sunbird race *buchenorum*) are distinct but all similarly bright-coloured. They are all clearly related to Souimanga Sunbird of Madagascar and must have colonised from this route, but the Madagascan species are themselves thought to be related to Olive-backed Sunbird *N. jugularis* of South-East Asia. Similarly Madagascar White-eye *Z. maderaspatana* of the Aldabra group is undoubtedly of Madagascan origin but is considered to have earlier origins and be closely related to Oriental White-eye *Z. palpebrosa* of India. Seychelles White-eye, on the other hand, appears to be related to its cousins in the Mascarenes, Mauritius Olive White-eye *Z. cloronothus* and Réunion White-eye *Z. olivaceus*, though its earlier origin is unknown.

A route to colonisation from Madagascar can be seen in the granitics, where for example, Seychelles Kestrel *F. araea* is most closely related to Madagascar Kestrel, and Seychelles Paradise Flycatcher *Terpsiphone corvina* may be related to Madagascar Paradise Flycatcher *T. mutata*. However, despite the tendency for us to look to Africa for origins of Seychelles birds, an Asian ancestry continues to dominate the earlier origins of the endemic species. Seychelles Bulbul *Hypsipetes crassirostris* is probably derived from Comoros Bulbul *H. parvirostris*, itself probably of Asian origin, and likewise Madagascar Bulbul of Aldabra is almost indistinguishable in plumage from 12 other bulbul races scattered through India, Sri Lanka, Vietnam, Taiwan and China. Likewise, the blue pigeons *Alectroenas*, represented in the granitics by Seychelles Blue Pigeon *A. pulcherrima* and at Aldabra by Comoros Blue Pigeon are endemic to the western Indian Ocean but probably have earlier Asian or even Australasian origins. Seychelles Black Parrot is of another genus, *Coracopsis*, endemic to the Indian Ocean, and though earlier origins are not known with certainty, it is unlikely to be African.

Most convincing of all is the presence of Seychelles Swiftlet *Aerodramus elaphrus*. The *Collocalia* genus, to which it is closely related, is widespread from the Himalayas through to Australia, with three species in India, five in South-East Asia and three in Australia. That this considerable sea crossing is possible is confirmed by Seychelles records of Pacific Swift *Apus pacificus* and White-throated Needletail *Hirundapus caudacutus*, the former not recorded elsewhere in the Afro-Malagasy region.

The origin of Seychelles Scops-owl *Otus insularis* is almost as mysterious as the bird itself. An Australasian origin has been suggested on the basis of similarity of call to Moluccan Scops-owl *O. magicus*. If correct this would be the first proven avian link between the western Indian Ocean and Australasia. On the other hand, it was once considered a race of Madagascar Scops-owl *O. rutilus*. However, whatever its origins, morphological differences suggest it is certainly a full species.

DNA analysis will undoubtedly shed more light on the origins and kinship of the birds of this region. But what is clear is that evolution and speciation continue to develop even as we watch, and that bird records today will help explain the species of tomorrow. New birds arrive all the time: Black-crowned Night Heron *Nycticorax nycticorax* arrived here in 1992 and is now breeding; Long-tailed Cormorants *Phalacrocorax africanus* arrived in numbers on Aldabra in 1999 and may well breed one day. Darwin would have loved to witness this island evolution at work; but anyone with a pair of binoculars can help write a postscript to the *Origin of Species* here in Seychelles.

CHECKLIST OF THE BIRDS OF SEYCHELLES

Black-necked Grebe	*Podiceps nigricollis*	Vagrant
Wandering Albatross	*Diomedea exulans*	Uncertain
Black-browed Albatross	*Diomedea melanophris*	Uncertain
Southern Giant Petrel	*Macronectes giganteus*	Vagrant
Cape Petrel	*Daption capense*	Vagrant
Mascarene Petrel	*Pterodroma aterrima*	Uncertain
Barau's Petrel	*Pterodroma baraui*	Uncertain
Antarctic Prion	*Pachyptila desolata*	Uncertain
Slender-billed Prion	*Pachyptila belcheri*	Uncertain
Bulwer's Petrel	*Bulweria bulwerii*	Uncertain
Jouanin's Petrel	*Bulweria fallax*	Vagrant
Flesh-footed Shearwater	*Puffinus carneipes*	Vagrant
Wedge-tailed Shearwater	*Puffinus pacificus*	Breeds (non-endemic)
Audubon's Shearwater	*Puffinus lherminieri*	Breeds (non-endemic)
Wilson's Storm-petrel	*Oceanites oceanicus*	Vagrant
White-faced Storm-petrel	*Pelagodroma marina*	Vagrant
Black-bellied Storm-petrel	*Fregetta tropica*	Uncertain
Matsudaira's Storm-petrel	*Oceanodroma matsudairae*	Uncertain
Swinhoe's Storm-petrel	*Oceanodroma monorhis*	Vagrant
Red-billed Tropicbird	*Phaethon aethereus*	Vagrant
Red-tailed Tropicbird	*Phaethon rubricauda*	Breeds (non-endemic)
White-tailed Tropicbird	*Phaethon lepturus*	Breeds (non-endemic)
Abbott's Booby	*Sula abbotti*	Extinct
Masked Booby	*Sula dactylatra*	Breeds (non-endemic)
Red-footed Booby	*Sula sula*	Breeds (non-endemic)
Brown Booby	*Sula leucogaster*	Breeds (non-endemic)
African Darter	*Anhinga rufa*	Uncertain
Great Cormorant	*Phalacrocorax carbo*	Vagrant
Long-tailed Cormorant	*Phalacrocorax africanus*	Vagrant
Pink-backed Pelican	*Pelecanus rufescens*	Extinct
Great Frigatebird	*Fregata minor*	Breeds (non-endemic)
Lesser Frigatebird	*Fregata ariel*	Breeds (non-endemic)
Grey Heron	*Ardea cinerea*	Breeds (non-endemic)
Purple Heron	*Ardea purpurea*	Vagrant
Cattle Egret	*Bubulcus ibis ibis*	Breeds (non-endemic)
Cattle Egret	*Bubulcus ibis seychellarum*	Endemic race
Great White Egret	*Egretta alba*	Vagrant
Intermediate Egret	*Egretta intermedia*	Vagrant
Little Egret	*Egretta garzetta*	Vagrant
Dimorphic Egret	*Egretta dimorpha*	Breeds (non-endemic)
Squacco Heron	*Ardeola ralloides*	Uncertain
Indian Pond-heron	*Ardeola grayii*	Vagrant
Madagascar Pond-heron	*Ardeola idae*	Breeds (non-endemic)
Green-backed Heron	*Butorides striatus degens*	Endemic race
Green-backed Heron	*Butorides striatus crawfordi*	Endemic race
Black-crowned Night Heron	*Nycticorax nycticorax*	Breeds (non-endemic)
Yellow Bittern	*Ixobrychus sinensis*	Breeds (non-endemic)
Cinnamon Bittern	*Ixobrychus cinnamomeus*	Vagrant
Eurasian Bittern	*Botaurus stellaris*	Vagrant
Greater Flamingo	*Phoenicopterus ruber*	Breeds (non-endemic)
White Stork	*Ciconia ciconia*	Vagrant
Sacred Ibis	*Threskiornis aethiopicus*	Vagrant
Madagascar Sacred Ibis	*Threskiornis (aethiopicus/ bernieri)abbotti*	Endemic race
White-faced Whistling Duck	*Dendrocygna viduata*	Vagrant
Ruddy Shelduck	*Tadorna ferruginea*	Vagrant
Common Teal	*Anas crecca*	Uncertain
Mallard	*Anas platyrhynchos*	Vagrant
Northern Pintail	*Anas acuta*	Vagrant

Garganey	*Anas querquedula*	Annual migrant
Northern Shoveler	*Anas clypeata*	Vagrant
Ferruginous Duck	*Aythya nyroca*	Vagrant
Osprey	*Pandion haliaetus*	Vagrant
Western Honey Buzzard	*Pernis apivorus*	Vagrant
Black Kite	*Milvus migrans*	Vagrant
Yellow-billed Kite	*Milvus aegyptius*	Vagrant
Western Marsh Harrier	*Circus aeruginosus*	Vagrant
Booted Eagle	*Hieraaetus pennatus*	Vagrant
Lesser Kestrel	*Falco naumanni*	Vagrant
Madagascar Kestrel	*Falco newtoni*	Breeds (non-endemic)
Seychelles Kestrel	*Falco araea*	Endemic species
Western Red-footed Falcon	*Falco vespertinus*	Vagrant
Amur Falcon	*Falco amurensis*	Vagrant
Eurasian Hobby	*Falco subbuteo*	Vagrant
Eleonora's Falcon	*Falco eleonorae*	Vagrant
Sooty Falcon	*Falco concolor*	Vagrant
Peregrine Falcon	*Falco peregrinus*	Vagrant
Grey Francolin	*Francolinus pondicerianus*	Breeds (introduced)
Common Quail	*Coturnix coturnix*	Vagrant
Aldabra Rail	*Dryolimnas (cuvieri) aldabranus*	Endemic (species/race?)
Aldabra Rail	*Dryolimnas (cuvieri) abbotti*	Extinct race
Corncrake	*Crex crex*	Vagrant
Spotted Crake	*Porzana porzana*	Vagrant
Striped Crake	*Aenigmatolimnas marginalis*	Vagrant
White-breasted Waterhen	*Amaurornis phoenicurus*	Vagrant
Common Moorhen	*Gallinula chloropus*	Breeds (non-endemic)
Allen's Gallinule	*Porphyrula alleni*	Vagrant
Purple Swamphen	*Porphyrio porphyrio*	Extinct
Crab Plover	*Dromas ardeola*	Annual migrant
Eurasian Oystercatcher	*Haematopus ostralegus*	Vagrant
Black-winged Stilt	*Himantopus himantopus*	Vagrant
Stone Curlew	*Burhinus oedicnemus*	Uncertain
Common Pratincole	*Glareola pratincola*	Vagrant
Oriental Pratincole	*Glareola maldivarum*	Vagrant
Black-winged Pratincole	*Glareola nordmanni*	Vagrant
Pacific Golden Plover	*Pluvialis fulva*	Annual migrant
Grey Plover	*Pluvialis squatarola*	Annual migrant
Common Ringed Plover	*Charadrius hiaticula*	Annual migrant
Little Ringed Plover	*Charadrius dubius*	Vagrant
Lesser Sandplover	*Charadrius mongolus*	Annual migrant
Greater Sandplover	*Charadrius leschenaultii*	Annual migrant
Caspian Plover	*Charadrius asiaticus*	Vagrant
Oriental Plover	*Charadrius veredus*	Vagrant
Black-tailed Godwit	*Limosa limosa*	Vagrant
Bar-tailed Godwit	*Limosa lapponica*	Annual migrant
Little Curlew	*Numenius minutus*	Vagrant
Whimbrel	*Numenius phaeopus*	Annual migrant
Slender-billed Curlew	*Numenius tenuirostris*	Uncertain
Eurasian Curlew	*Numenius arquata*	Annual migrant
Common Redshank	*Tringa totanus*	Vagrant
Marsh Sandpiper	*Tringa stagnatilis*	Vagrant
Common Greenshank	*Tringa nebularia*	Annual migrant
Green Sandpiper	*Tringa ochropus*	Vagrant
Wood Sandpiper	*Tringa glareola*	Annual migrant
Common Sandpiper	*Actitis hypoleucos*	Annual migrant
Grey-tailed Tattler	*Heteroscelus brevipes*	Vagrant
Terek Sandpiper	*Xenus cinereus*	Annual migrant
Ruddy Turnstone	*Arenaria interpres*	Annual migrant
Red-necked Phalarope	*Phalaropus lobatus*	Vagrant
Pintail Snipe	*Gallinago stenura*	Vagrant
Swinhoe's Snipe	*Gallinago megala*	Uncertain

Great Snipe	*Gallinago media*	Vagrant
Common Snipe	*Gallinago gallinago*	Vagrant
Red Knot	*Calidris canutus*	Uncertain
Great Knot	*Calidris tenuirostris*	Vagrant
Sanderling	*Calidris alba*	Annual migrant
Red-necked Stint	*Calidris ruficollis*	Uncertain
Little Stint	*Calidris minuta*	Annual migrant
Temminck's Stint	*Calidris temminckii*	Vagrant
Long-toed Stint	*Calidris subminuta*	Vagrant
Pectoral Sandpiper	*Calidris melanotos*	Vagrant
Sharp-tailed Sandpiper	*Calidris acuminata*	Vagrant
Curlew Sandpiper	*Calidris ferruginea*	Annual migrant
Broad-billed Sandpiper	*Tryngites falcinellus*	Vagrant
Buff-breasted Sandpiper	*Tryngites subruficollis*	Vagrant
Ruff	*Philomachus pugnax*	Vagrant
South Polar Skua	*Catharacta maccormicki*	Vagrant
Antarctic Skua	*Catharacta antarctica*	Annual migrant
Pomarine Skua	*Stercorarius pomarinus*	Uncertain
Arctic Skua	*Stercorarius parasiticus*	Vagrant
Long-tailed Skua	*Stercorarius longicaudus*	Uncertain
Heuglin's Gull	*Larus heuglini*	Vagrant
Lesser Black-backed Gull	*Larus fuscus*	Vagrant
Great Black-headed Gull	*Larus ichthyaetus*	Uncertain
Brown-headed Gull	*Larus brunnicephalus*	Uncertain
Grey-headed Gull	*Larus cirrocephalus*	Uncertain
Black-headed Gull	*Larus ridibundus*	Vagrant
Whiskered Tern	*Chlidonias hybridus*	Vagrant
White-winged Tern	*Chlidonias leucopterus*	Annual migrant
Gull-billed Tern	*Gelochelidon nilotica*	Annual migrant
Caspian Tern	*Hydroprogne caspia*	Breeds (non-endemic)
Greater Crested Tern	*Thalasseus bergii*	Breeds (non-endemic)
Lesser Crested Tern	*Thalasseus bengalensis*	Annual migrant
Sandwich Tern	*Thalasseus sandvicensis*	Vagrant
Common Tern	*Sterna hirundo*	Annual migrant
Roseate Tern	*Sterna dougallii*	Breeds (non-endemic)
White-cheeked Tern	*Sterna repressa*	Uncertain
Black-naped Tern	*Sterna sumatrana*	Breeds (non-endemic)
Bridled Tern	*Sterna anaethetus*	Breeds (non-endemic)
Sooty Tern	*Sterna fuscata*	Breeds (non-endemic)
Little Tern	*Sterna albifrons*	Vagrant
Saunders' Tern	*Sterna saundersi*	Annual migrant
Brown Noddy	*Anous stolidus*	Breeds (non-endemic)
Lesser Noddy	*Anous tenuirostris*	Breeds (non-endemic)
Fairy Tern	*Gygis alba*	Breeds (non-endemic)
Feral Pigeon	*Columba livia*	Breeds (introduced)
European Turtle Dove	*Streptopelia turtur*	Vagrant
Oriental Turtle Dove	*Streptopelia orientalis*	Uncertain
Madagascar Turtle Dove	*Streptopelia picturata picturata*	Breeds (introduced)
Madagascar Turtle Dove	*Streptopelia picturata rostrata*	Endemic race
Madagascar Turtle Dove	*Streptopelia picturata coppingeri*	Endemic race
Madagascar Turtle Dove	*Streptopelia picturata saturata*	Extinct race
Barred Ground Dove	*Geopelia striata*	Breeds (introduced)
Comoro Blue Pigeon	*Alectroenas sganzini minor*	Endemic race
Seychelles Blue Pigeon	*Alectroenas pulcherrima*	Endemic species
Seychelles Black Parrot	*Coracopsis (nigra) barklyi*	Endemic (species/race?)
Grey-headed Lovebird	*Agapornis canus*	Extinct
Seychelles Parakeet	*Psittacula (eupatria) wardi*	Extinct
Ring-necked Parakeet	*Psittacula krameri*	Breeds (introduced)
Great Spotted Cuckoo	*Clamator glandarius*	Vagrant
Jacobin Cuckoo	*Oxylophus jacobinus*	Vagrant
Common Cuckoo	*Cuculus canorus*	Vagrant
Asian Lesser Cuckoo	*Cuculus poliocephalus*	Vagrant

Madagascar Coucal	*Centropus toulou insularis*	Endemic race
Barn Owl	*Tyto alba*	Breeds (introduced)
Eurasian Scops-owl	*Otus scops*	Vagrant
Seychelles Scops-owl	*Otus insularis*	Endemic species
Brown Fish Owl	*Ketupa zeylonensis*	Vagrant?
Eurasian Nightjar	*Caprimulgus europaeus*	Vagrant
Madagascar Nightjar	*Caprimulgus madagascariensis aldabrensis*	Endemic race
Seychelles Swiftlet	*Aerodramus elaphrus*	Endemic species
White-throated Needletail	*Hirundapus caudacutus*	Vagrant
Common Swift	*Apus apus*	Vagrant
Pacific Swift	*Apus pacificus*	Vagrant
Little Swift	*Apus affinis*	Vagrant
Common Kingfisher	*Alcedo atthis*	Uncertain
Blue-cheeked Bee-eater	*Merops persicus*	Vagrant
Madagascar Bee-eater	*Merops superciliosus*	Uncertain
European Bee-eater	*Merops apiaster*	Vagrant
European Roller	*Coracias garrulus*	Vagrant
Broad-billed Roller	*Eurystomus glaucurus*	Vagrant
Hoopoe	*Upupa epops*	Vagrant
Greater Short-toed Lark	*Calandrella brachydactyla*	Vagrant
Sand Martin	*Riparia riparia*	Vagrant
Mascarene Martin	*Phedina borbonica*	Vagrant
Barn Swallow	*Hirundo rustica*	Vagrant
Common House Martin	*Delichon urbica*	Vagrant
Yellow Wagtail	*Motacilla flava*	Vagrant
Citrine Wagtail	*Motacilla citreola*	Vagrant
Grey Wagtail	*Motacilla cinerea*	Vagrant
White Wagtail	*Motacilla alba*	Vagrant
Richard's Pipit	*Anthus richardi*	Uncertain
Tree Pipit	*Anthus trivialis*	Vagrant
Olive-backed Pipit	*Anthus hodgsoni*	Uncertain
Red-throated Pipit	*Anthus cervinus*	Vagrant
Red-whiskered Bulbul	*Pycnonotus jocosus*	Breeds (introduced)
Madagascar Bulbul	*Hypsipetes madagascariensis rostratus*	Endemic race
Seychelles Bulbul	*Hypsipetes crassirostris*	Endemic species
Red-backed Shrike	*Lanius collurio*	Vagrant
Lesser Grey Shrike	*Lanius minor*	Vagrant
Woodchat Shrike	*Lanius senator*	Vagrant
Rufous-tailed Rock Thrush	*Monticola saxatilis*	Vagrant
Seychelles Magpie-robin	*Copsychus sechellarum*	Endemic species
Common Redstart	*Phoenicurus phoenicurus*	Vagrant
Whinchat	*Saxicola rubetra*	Vagrant
Northern Wheatear	*Oenanthe oenanthe*	Vagrant
Isabelline Wheatear	*Oenanthe isabellina*	Vagrant
Madagascar Cisticola	*Cisticola cherina*	Breeds (non-endemic)
Aldabra Warbler	*Nesillas aldabranus*	Extinct
Sedge Warbler	*Acrocephalus schoenobaenus*	Vagrant
Seychelles Warbler	*Acrocephalus sechellensis*	Endemic species
Icterine Warbler	*Hippolais icterina*	Vagrant
Willow Warbler	*Phylloscopus trochilus*	Vagrant
Chiffchaff	*Phylloscopus collybita*	Uncertain
Wood Warbler	*Phylloscopus sibilatrix*	Vagrant
Blackcap	*Sylvia atricapilla*	Vagrant
Common Whitethroat	*Sylvia communis*	Vagrant
Spotted Flycatcher	*Muscicapa striata*	Vagrant
Seychelles Paradise Flycatcher	*Terpsiphone corvina*	Endemic species
Souimanga Sunbird	*Nectarinia sovimanga aldabrensis*	Endemic race
Abbott's Sunbird	*Nectarinia (sovimanga) abbotti*	Endemic (species/race?)
Abbott's Sunbird	*Nectarinia (sovimanga/abbotti) buchenorum*	Endemic (species/race?)

Seychelles Sunbird	*Nectarinia dussumieri*	Endemic species
Madagascar White-eye	*Zosterops maderaspatana aldabrensis*	Endemic race
Madagascar White-eye	*Zosterops maderaspatana menaiensis*	Endemic race
Madagascar White-eye	*Zosterops maderaspatana maderaspatana*	Breeds (non-endemic)
Seychelles Chestnut-flanked White-eye	*Zosterops semiflava*	Extinct
Seychelles White-eye	*Zosterops modesta*	Endemic species
Aldabra Drongo	*Dicrurus aldabranus*	Endemic species
House Crow	*Corvus splendens*	Breeds (introduced)
Pied Crow	*Corvus albus*	Breeds (non-endemic)
European Golden Oriole	*Oriolus oriolus*	Vagrant
Wattled Starling	*Creatophora cinerea*	Vagrant
Rose-coloured Starling	*Sturnus roseus*	Vagrant
Common Myna	*Acridotheres tristis*	Breeds (introduced)
Ortolan Bunting	*Emberiza hortulana*	Vagrant
Yellow-fronted Canary	*Serinus mozambicus*	Breeds (introduced)
Common Rosefinch	*Carpodacus erythrinus*	Vagrant
Common Waxbill	*Estrilda astrild*	Breeds (introduced)
Madagascar Fody	*Foudia madagascariensis*	Breeds (non-endemic/ introduced?)
Aldabra Fody	*Foudia (eminentissima) aldabrana*	Endemic (species/race?)
Seychelles Fody	*Foudia sechellarum*	Endemic species
House Sparrow	*Passer domesticus*	Breeds (introduced)

PLATE 1 EXTINCTIONS

More than 90% of avian extinctions during the last 300 years have been island endemics. Seychelles has lost fewer birds than many other islands such as Mauritius or Hawaii. This is partly due to a shorter period of human settlement and, in recent years, to more enlightened attitudes towards conservation. Nevertheless, it has had its share of tragic losses. Of the birds of Seychelles at the time of settlement in 1770, at least six species have gone forever, destroyed by a combination of habitat clearance, introduced predators and direct persecution, though one may have been a natural extinction. Another extinction, Grey-headed Lovebird *Agapornis canus*, survived for over 70 years as an introduction before disappearing for unknown reasons.

1 ALDABRA WARBLER *Nesillas aldabranus*   Text page 283
Only discovered in 1967. No young have ever been seen. After November 1975 only males were seen, and the last sighting was of a lone male in 1983. Described in 1985 as almost certainly the rarest, most restricted and most highly threatened species of bird in the world, by 1994 it was listed as extinct. Rats are prime suspects, being present on Malabar. Cats may also have played a part, and goats may have contributed in reducing the available habitat. On the other hand, this may be a natural extinction, the environment on Aldabra offering little suitable habitat.

2 SEYCHELLES PARAKEET *Psittacula (eupatria) wardi*   Text page 249
Endemic to Mahé, Praslin and Silhouette. It was probably common before the arrival of man. Prior found 'a considerable number of green parrots' in 1811. However, Newton noted in 1876 that due to the clearing away of the natural forests, the planting of coconuts and the ruthless killing of the parrots 'there cannot be much doubt that they are doomed to extinction'. Sadly, this prophecy proved correct. Marianne North painted two tame birds on Mahé in 1883, which were said to have come from Silhouette, the last record from the island. It was last recorded on Mahé in March 1893, a specimen shot by Abbott. It possibly survived to the turn of the 20th century.

3 PINK-BACKED PELICAN *Pelecanus rufescens*   Text page 160
Once bred on St Joseph Atoll in the Amirantes. Abbott collected a specimen from the area in August 1892 and described 'a small colony – perhaps 100 individuals – the only colony of pelicans in these seas'. By the 1930s the colony was extinct. Apart from records from Dahlak Archipelago in the Red Sea and Madagascar, there are no reports outside mainland Africa. The reason for this extinction is unknown but it was probably direct persecution or habitat destruction or both.

4 SEYCHELLES CHESTNUT-FLANKED WHITE-EYE *Zosterops semiflava*   Text page 295
Only ever recorded with certainty from Marianne, although there are also unsubstantiated reports from Mahé, South-east Island, Praslin, La Digue and Silhouette. It may well have been this species noted in 1768 by the Marion Dufresne expedition on South-east Island and recorded as 'several little yellow birds like canaries'. First described by Newton who collected specimens from Marianne in 1865. Abbott also took specimens in 1892. It was also probably a victim of habitat destruction when the original vegetation was replaced by a monoculture of coconuts. The exact date of extinction is unknown, but was probably very early in the 20th century.

5 PURPLE SWAMP-HEN *Porphyrio porphyrio*   Text page 195
The most likely identity of the legendary 'Poul Ble'. No-one has yet unearthed bones which may be checked by zoologists, nor are there surviving skins or accurate biological descriptions. The first mention of it was by the Marion-Dufresne expedition of 1768. An expedition party exploring the northwest of Mahé reported: 'There are in these rocks many giant tortoises, goats and also the "poules bleues", of which we killed one; they are like a large chicken, with blue feathers, the beak flat, wide and red, and the same colour feet...' Two days later, following the northwest and north coast of Mahé, they report: 'We see always many gamebirds and "poules bleues".'

6 ABBOTT'S BOOBY *Sula abbotti*   Text page 154
First described from Assumption by Ridgway from a specimen collected by Abbott. In 1895 Baty reported 'Boobies or Fous of different kinds are to be found in the trees all over the island'. In 1908 a settlement was built on Assumption to mine guano, the phosphate deposits that had accumulated over centuries under the tree-nesting seabird colonies. This destroyed the forest and the soil on which it depended; with the loss of its habitat the species was doomed to extinction. In 1916 Dupont wrote 'the boobies are all destroyed', and in a later visit wrote that all seabirds and landbirds had been destroyed by 1909. The last colony survives on Christmas Island, Indian Ocean.

PLATE 2 BLACK-NECKED GREBE AND CORMORANTS

Black-necked Grebe is a vagrant around December. African Darter has been reported from Aldabra but its status remains uncertain. Great Cormorant is a vagrant to the granitics, and Long-tailed Cormorant is a vagrant to both the granitics and Aldabra.

1 BLACK-NECKED GREBE *Podiceps nigricollis* 28-34 cm   **Text page 139**
A medium-sized grebe with a high forehead, sharp upturned bill and blood red eye.
1a Adult breeding: Black back, neck, breast and head with spray of yellow plumes behind eye. Chestnut flanks.
1b Adult non-breeding: Black mask over face contrasts with white chin and sides to nape. Grey neck, white breast, black back and grey-white flanks.
1c Immature: Similar to non-breeding but with some dark chestnut streaking on sides of neck and ear-coverts and a grey band across chest.
1d Adult non-breeding in flight: The only grebe with white secondaries and innermost primaries, but no white forewing on an otherwise all-dark wing.

2 AFRICAN DARTER *Anhinga rufa* 85-97 cm   **Text page 158**
A large dark cormorant-like bird with snake-like neck and long dagger-like bill.
2a Juvenile: Paler than adult with buff-brown upperparts, buff streaking on inner secondaries and scapulars. Chin, throat and foreneck whitish. White belly darkens to blackish in immature.
2b Female: Entirely brown neck but no or indistinct lateral stripe. Bill dark above, pinkish below.
2c Male in flight: Very long, thin sinuous neck, long thin bill and long rounded tail. Neck outstretched but slightly kinked. Strong flight with rapid wingbeats.
2d Male: Black sometimes with green iridescence, chestnut neck and white line below eye. Black-and-white plumes on scapulars and some white wing-coverts give striking scaly effect. Tail finely barred white. Bill pale grey (but variable). Non-breeding browner and duller.
2e Female swimming: Back submerged with sinuous head and neck visible above water.

3 LONG-TAILED CORMORANT *Phalacrocorax africanus* 51-60 cm   **Text page 159**
A small, long-tailed cormorant with a short, yellow bill.
3a Adult non-breeding: Fairly uniform dark brown upperparts, whitish below (sometimes dark). Eye brown.
3b Adult non-breeding in flight: Appears small-headed, short-necked and long-tailed. Rapid bursts of duck-like wingbeats.
3c Adult non-breeding swimming: Back often (not always) visible above surface. Much smaller than Great Cormorant, much shorter-billed and shorter-necked than African Darter.
3d Adult breeding: All-black with a greenish iridescence in bright light, white feather edging to back, rump and uppertail-coverts. Short crest. White facial plumes form an obvious eyebrow. Bill yellow with brown culmen and yellow facial skin. Eye red.

4 GREAT CORMORANT *Phalacrocorax carbo* 80-100 cm   **Text page 159**
A very large cormorant, all-black with brownish tints and a large yellow patch of bare skin at base of bill.
4a Adult non-breeding in flight: All-dark with large head; neck kinked. Heavy, goose-like flight.
4b Immature swimming: Back visible above surface. Much larger than Long-tailed Cormorant.
4c Adult breeding (race *lucidus*): Large white patch on thigh and cheek; white upper breast.
4d Adult non-breeding: Appears all-black but in bright light shows a bronze sheen to scapulars and wing-coverts.
4e Immature: Duller brown above than adult with dirty white from throat to undertail-coverts.

PLATE 3   PETRELS

Southern Giant Petrel and Cape Petrel are vagrants from the southern oceans, most likely around July to October. Northern Giant Petrel should be considered as a potential vagrant from which Southern Giant Petrel must be separated. Jouanin's Petrel is more frequently encountered in the northern Indian Ocean, reaching Seychelles waters October to March. Bulwer's Petrel is of uncertain status, more likely in the southern Indian Ocean December to March. Mascarene and Barau's Petrels breed on Réunion and are also of uncertain status in Seychelles waters.

1 SOUTHERN GIANT PETREL *Macronectes giganteus* Wingspan: 185-205 cm **Text page 141**
Like a giant all-dark fulmar, with an enormous, thickset green-tipped bill.
1a Head Northern Giant Petrel *M. hallii*: Bill pale with dark pink-orange tip. Pale eye. Rarely as pale-headed as Southern Giant and more 'capped'.
1b Head immature: Uniform blackish brown. Dark eye. Bill pinkish orange with greenish tip (as adult).
1c Northern Giant Petrel in flight below: WS:180-200 cm. Underparts uniform pale brown. Darker under-wing than Southern Giant Petrel with black leading edge to forewing.
1d Adult dark morph above: Mostly dark brown, whitish head and throat, pale breast.
1e Adult white morph above: All-white, flecked blackish.
1f Adult dark morph below: Head and breast whitish, contrasting with underparts (unlike Northern Giant), which darken towards the rear. Paler grey-brown underwing than Northern Giant Petrel with pale leading edge to forewing.

2 CAPE PETREL *Daption capense*   Wingspan:83-88 cm   **Text page 142**
Unmistakable black-and-white plumage, like a piebald fulmar.
2a Adult below: Pure white with black wing margins and black end to tail.
2b Adult above: Black head and tail. Wings black, flecked white, with white windows at base of primaries.

3 BULWER'S PETREL *Bulweria bulwerii* Wingspan: 67 cm   **Text page 145**
An all-dark, long-tailed petrel with pale diagonal wingbar and short, stubby black bill. Flight erratic and twisting, close to surface.
3a Adult above: Dark brown throughout except for broad, paler panel across secondary coverts, most distinct at carpal end. Long tail appears tapered (wedge-shaped when flared).
3b Adult below: Uniform dark brown, smaller than Jouanin's Petrel.

4 JOUANIN'S PETREL *Bulweria fallax* Wingspan: 76-83 cm   **Text page 145**
An all-dark, long-tailed bull-headed petrel with indistinct wingbar and thick black bill. Characteristic bounding flight, rising high above waves.
4a Adult below: Uniform dark brown, larger than Bulwer's Petrel.
4b Adult above: Dark brown throughout with pale diagonal bar in worn plumage. Long angular wings and long wedge-shaped tail, usually held closed in a tapering shape.

5 MASCARENE PETREL *Pterodroma aterrima* Wingspan: 88 cm   **Text page 142**
An all-dark petrel with white-flecked face and notched bill.
Adult above: Dark brown throughout. Slightly slimmer-winged and shorter-tailed than Jouanin's. Squarish tail. Short, heavier bill appears notched between tube and swollen tip.

6 BARAU'S PETREL *Pterodroma baraui* Wingspan: 96 cm   **Text page 143**
A sturdy, short-necked, pale grey gadfly petrel with white forehead, dark cap and white below. Short, thick bill.
6a Adult below: All-white underparts save for grey neck-collar, black trailing edge to wing and distinctive black chevrons at carpal joint.
6b Adult above: Distinctive white face and forehead contrast with dark cap. Pale grey wings with black 'M' formed by dark primaries and dark carpal bar (note, does not continue onto the rump as in other *Pterodroma*). Dark grey crown, nape and rump.

PLATE 4 SHEARWATERS

Both Audubon's and Wedge-tailed Shearwaters breed in the granitics and Amirantes, with the former also in the Aldabra group. Flesh-footed Shearwater is a vagrant likely May to October.

1 AUDUBON'S SHEARWATER *Puffinus lherminieri* Wingspan: 64-74 cm   **Text page 148**
A small, rather dumpy black-and-white shearwater with rounded wings.
1a Adult above: Entirely dark brownish black.
1b Adult below: All-white, with brownish undertail-coverts and broad dark margins to underwing.
1c Adult swimming: Small size; black-and-white plumage.
1d Head (race *nicolae*): Granitics. Slim, black bill with indistinct tube.
1e Chick: brown-grey to pearl grey above, white below.
1f Head (race 'colstoni'): Aldabra. Thicker, longer bill, much broader at base.

2 WEDGE-TAILED SHEARWATER *Puffinus pacificus* Wingspan: 97-104 cm   **Text page 147**
A large, long-winged all-dark shearwater with flesh-pink legs. Wedge-shaped tail usually held closed.
2a Adult swimming: Large; all-dark.
2b Head: Bill dark grey-brown, paler purplish in centre. Throat ashy grey.
2c Pale morph: (Very rare.) White below, with broad dark trailing edge to wing, mottled on flanks and dark undertail.
2d Chick: Entirely covered in soft, grey down.
2e Adult above: Uniform dark brown, often paler on wing-coverts.
2f Adult below: Uniform dark brown including entire underwing.

3 FLESH-FOOTED SHEARWATER *Puffinus carneipes* Wingspan: 99-107 cm   **Text page 146**
A large, long-winged all dark shearwater with a pale dark-tipped bill and pink legs. Broad, pointed tail.
3a Adult above: Uniform dark brown.
3b Adult below: Uniform dark brown, save for silvery roots to underwing primaries.
3c Head: Bill pale horn with black tip.

PLATE 5 STORM-PETRELS

Wilson's Storm-petrel and Swinhoe's Storm-petrel are vagrants throughout Seychelles, most likely October to November. White-faced Storm-petrel is a vagrant around May. Matsudaira's Storm-petrel and Black-bellied Storm-petrel are of uncertain status due to confusion with other similar species.

1 MATSUDAIRA'S STORM-PETREL *Oceanodroma matsudairae* Wingspan: 56 cm **Text page 151**
A large, all-dark storm-petrel with a forked tail. Slow, nightjar-like flight.
1a Adult above: Uniform dark brown-black including rump. Broad, coffee-coloured bar on upperwing-coverts. Flight feathers black-brown with prominent white primary shaft streaks forming a visible pale wing-patch. Tail deeply forked, but often held closed in flight.
1b Adult below: Uniform dark brown.

2 WHITE-FACED STORM-PETREL *Pelagodroma marina* Wingspan: 41-43 cm **Text page 151**
A large storm-petrel with distinctive head pattern and broad-centred wings. Unique pendulum-like flight.
2a Adult above: Brownish back and wing-coverts; broad whitish wingbar contrasts with black-brown flight feathers. Pale grey rump-patch and black tail. Distinctive head pattern: a white muzzle and broad white supercilium contrast with dark mask, cap and pale nape.
2b Adult below: Pure white including all underwing-coverts with broad dark trailing edge and tips to underwing.

3 SWINHOE'S STORM-PETREL *Oceanodroma monorhis* Wingspan: 44-46 cm **Text page 152**
A medium-sized, all-dark storm-petrel with forked tail, and no white rump. Lively, petrel-like flight.
3a Adult above: Uniform dark brown-black. Head and back with smoky grey cast. Wingbar broad, pale grey at carpal end, fading to narrow, brown at trailing edge. Primary shafts with narrow white streaks, visible at close range. Tail shallowly forked, but often held closed.
3b Adult below: Uniform dark brown but with pale grey bar on underwing-coverts.

4 BLACK-BELLIED STORM-PETREL *Fregetta tropica* Wingspan: 46 cm   **Text page 151**
Dark brown above with white rump; white below, with black ventral-stripe. Erratic low flight, legs dangling.
4a Adult below: White underwing-coverts and white flanks contrast with black upper chest and broad black stripe down centre of belly, black flight feathers and tail.
4b Adult above: Dark brown with a striking, broad white rump. Narrow, pale wingbar contrasts with blackish flight feathers. Sooty black head with white chin.

5 WILSON'S STORM-PETREL *Oceanites oceanicus* Wingspan: 38-42 cm   **Text page 150**
A small, sooty-brown storm-petrel with pale wing-panel and projecting legs. Low, swallow-like flight.
5a Adult below: All-dark with faint pale wingbar.
5b Adult above: Black-brown throughout except for broad white rump. Broad, rounded wings with distinct short, pale grey band on upperwing secondary coverts. Feet project beyond square tail, yellow webs visible at close range.

PLATE 6 TROPICBIRDS

White-tailed Tropicbird is most common in the granitics but rare between Mahé and Aldabra where it again breeds in good numbers. Red-tailed Tropicbird only breeds on Aride in the granitics, but in good numbers on Aldabra and Cosmoledo. Red-billed Tropicbird is a vagrant.

1 WHITE-TAILED TROPICBIRD *Phaethon lepturus* Wingspan: 90-95 cm   **Text Page 153**
White with black wingbars, long white tail-streamers and yellow bill.
1a Adult in flight: Clean white plumage with black carpal bar and black wedge on the primaries. Long white tail-streamers and diagnostic yellow bill.
1b Adult head: Yellow bill, black eye-stripe beyond eye.
1c Immature in flight: Coarse barring on the upperwing, paler bill, no tail-streamers.
1d Courtship flight: Male flies close above female, lowering tail.
1e Race *fulvus* (Christmas Island Tropicbird): Plumage suffused with golden yellow throughout.
1f Chick: Fluffy grey-white; black chevrons appear on first wing-coverts.
1g Nest site: Nests mainly on the ground, often with overhanging rock or vegetation.

2 RED-TAILED TROPICBIRD *Phaethon rubricauda* Wingspan: 104-119 cm   **Text Page 153**
A large, pure white tropicbird with dark red tail bristles and heavy red bill.
2a Immature in flight: Heavy barring on upperwing, pale yellowish bill, no tail-streamers. Less black in outerwing than other tropicbirds.
2b Adult in flight: Gull-like, with body and wings pure white and narrow dark red tail-streamers (not obvious at a distance).
2c Adult head: Heavy red bill, eye-stripe short or absent behind eye.
2d Courtship flight: Male flies 'backwards' behind female, honking loudly.
2e Adult breeding: Nests on ground, acquiring pink flush to plumage at height of breeding.

3 RED-BILLED TROPICBIRD *Phaethon aethereus* Wingspan: 99-106 cm   **Text Page 152**
Very large with fine black barring on back, white tail and red bill.
3a Immature: Barred above like adult, but with black collar across nape, yellow bill and no tail-streamers.
3b Adult: White, with heavy black wedge on primaries, dense black barring on back and wing-coverts, giving grey effect at distance. Long white tail-streamers.
3c Head (race *indicus*): Heavy orange-red bill with black cutting edges. Eye-stripe long and flared behind eye (but often reduced in this race).

PLATE 7 BOOBIES

Red-footed Booby breeds on Aldabra, Cosmoledo and Farquhar and is the most likely species to encounter in the southern atolls. Masked Booby breeds on Boudeuse and Desnoeufs in the Amirantes group, and in good numbers on Cosmoledo. Brown Booby still breeds in small numbers on Cosmoledo, but is the scarcest of the three breeding species.

1 RED-FOOTED BOOBY *Sula sula* Wingspan: 91-101 cm   **Text page 156**
All-white with black flight feathers, pink-lilac bill and red feet.
1a Adult at nest: Black flight feathers contrast with white head, body and tail. Lilac bill, pinkish at base and around eye. Nests in trees.
1b Adult above: Black flight feathers contrast with white head, body and tail.
1c Adult below: White underparts save for black flight feathers. Pinkish feet, turning red in breeding season.
1d Immature above: Wholly grey-brown with a darker pinkish purple bill.
1e White-tailed brown morph: Dark brown wings, grey-brown head, white tail.

2 MASKED BOOBY *Sula dactylatra* Wingspan: 142-154 cm   **Text page 155**
Large, with black-and-white wings, black tail, yellow bill and black feet.
2a Juvenile: Dark brown head with narrow white collar, pale brown upperparts, white upper breast and belly.
2b Immature above: Dark brown head with broad white collar. White gradually becomes more extensive in wing-coverts and rump.
2c Adult above: Black secondary coverts form heavy black wing margin, black tail; remainder of wing-coverts and body white.
2d Adult below: Pure white with contrasting black flight feathers and tail. Feet dark brown or black.
2e Adult at nest: Pure white body and wing-coverts contrast with black flight feathers. Diagnostic black tail. Yellow bill, black facial mask. Dark feet. Nests on ground.
2f Immature below: White upper breast, belly and underwing-coverts. Legs and feet dark.

3 BROWN BOOBY *Sula leucogaster* Wingspan: 132-150 cm   **Text page 157**
Chocolate brown above, with white belly, yellow bill and yellow feet.
3a Adult below: Dark chocolate brown upper breast contrasts with pure white belly and underwing-coverts.
3b Adult above: Dark chocolate brown throughout.
3c Adult at nest: Dark chocolate brown head, back, wings and tail. White lower breast and belly. Nests on ground.
3d Immature above: Dull brown, paler than adult. No white collar (unlike immature Masked).
3e Immature below: Underparts mottled brown. Pale centre to underwing.

2f

PLATE 8 FRIGATEBIRDS

Both Great and Lesser Frigatebirds breed on Aldabra Atoll, with seasonal roosts at Farquhar, Coëtivy, Alphonse and St Joseph Atoll in the coralline islands and on Cousin, Bird and Aride Island in the granitics. Great Frigatebird also breeds in small numbers at Cosmoledo. Although Lesser outnumbers Great on Aldabra, it is much the rarer of the two in the granitic islands.

1 GREAT FRIGATEBIRD *Fregata minor* Wingspan: 205-230 cm   **Text page 160**
Mainly black, without white axillaries extending to underwing but with pale bar in upperwing.
1a Adult female below: Black with white throat and white breast-patch, pale grey bill and pink eye-ring visible at close range.
1b Adult male below: All-black with red gular pouch (may not be visible). Black eye-ring visible at close range.
1c Immature chasing: In the granitics, most chases are by immatures victimising terns and other species.
1d Immature below: Head white or buff, belly white and bill and eye-ring grey.
1e Head female: White throat and upper breast, pink eye-ring, blue-grey bill.
1f White-headed immature: Aldabra birds show white head. All morphs with grey eye-ring and greyish bill.
1g Buff-headed immature: Immatures with buff heads from populations outside Seychelles often seen in the granitics.
1h Male breeding display: Inflates red gular pouch and displays to passing females, pointing bill skyward, turning wings over, shaking bill and making an extraordinary reeling or drumming sound.
1i Adult above: Broad pale upper wingbar in both sexes. Frequently chases boobies in Aldabra group.

2 LESSER FRIGATEBIRD *Fregata ariel* Wingspan: 175-195 cm   **Text page 162**
White axillaries extend to underwing. Upperwing all-black.
2a Adult female below: Black throat, and white breast-patch, also with white axillary 'spurs'. Pink eye-ring.
2b Adult male below: All-black with white 'spurs' in armpits.
2c Immature below: Head white or orange-buff, white breast-patch, black belly. Also has axillary 'spurs'.
2d Male above: All-black with no distinct upperwingbar.

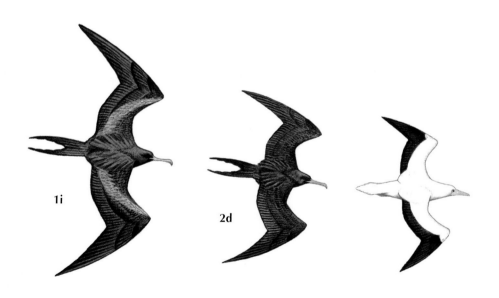

PLATE 9 HERONS AND ALLIES

Grey Heron breeds throughout Seychelles. Purple Heron and White Stork are vagrants, mainly to the granitics. Madagascar Sacred Ibis and Greater Flamingo both breed on Aldabra Atoll, where Sacred Ibis has occurred as a vagrant.

1 GREY HERON *Ardea cinerea* 90-98 cm   **Text page 163**
A large mainly ash grey heron with black-and-white head.
1a Adult in flight: Huge wings, slow flight. Whitish head and neck contrast with ashy grey wing-coverts and black flight feathers.
1b Adult breeding: Pinkish red flush to bill and legs. Large, untidy twig nest built in trees.
1c Adult: White crown and forehead. Yellow bill. Dull orange-brown or yellowish legs.
1d Juvenile: No black plumes on head, whitish face and neck, sides grey. Grey crown and forehead. Bill horn and legs greenish grey.

2 PURPLE HERON *Ardea purpurea* 78-90 cm   **Text page 164**
A fairly large, dark brown heron with a buff-brown neck. Bill longer and more slender than in Grey Heron.
2a Immature: Duller than adult with sandy buff wing-coverts and underparts.
2b Adult: Back and wing-coverts slate grey, chest and belly dark chestnut, neck buff-chestnut, crown black.
2c Adult in flight: Dark crown and neck contrast with red-brown hue on wing-coverts and dark flight feathers. Chestnut band on underwing.

3 WHITE STORK *Ciconia ciconia*   100-102 cm   **Text page 174**
A huge, white bird with red legs, long, straight red bill and black wings.
3a Adult in flight: Black flight feathers contrast with white in rest of plumage. Long red legs project beyond tail.
3b Adult: Head, neck, back and belly white. Wing-coverts white contrasting with black flight feathers.

4 SACRED IBIS *Threskiornis aethiopicus* 65-89 cm   **Text page 174**
A large, white ibis with dark eye and black trailing edge to wing. Bill thicker than in Madagascar Sacred Ibis.
4a Adult: White with black head and neck. Very stout curved, black bill. Brown eye.
4b Adult in flight: White wing with black trailing edge to wing.
4c Immature: Similar to Immature Madagascar Sacred Ibis but with darker head and neck.

5 MADAGASCAR SACRED IBIS *Threskiornis (aethiopicus/bernieri) abbotti*   70-85 cm
                                                                                                                          Text page 175
A large, white ibis with pale eye and all-white wings. Bill more slender than in Sacred Ibis
5a Adult in flight: Entirely white wings (sometimes shows black tips to outer primaries).
5b Adult: White back, underparts and wings. Long, lacy black plumes over tail. Striking pale blue or white eye.
5c Immature: Dirty white feathers are flecked black on head and neck. Milky coffee brown plumes. Eye brown.

6 GREATER FLAMINGO *Phoenicopterus ruber* 120-145 cm   **Text page 172**
A huge pinkish waterbird with long neck and legs.
6a Head Lesser Flamingo *Phoenicopterus minor*: 80-105 cm. Bill dark red, tipped black. Shorter base and more abrupt bend to bill gives more fierce expression. Eye and surrounding skin dark red.
6b Adult head: More extensive black tip to pink bill than in Lesser. Yellow eye.
6c Adult: Head and body white flushed with pink.
6d Nest: Builds a mound of mud. Chick covered in grey-brown down.
6e Adults in flight: Wing-coverts bright crimson and flight feathers black. Long pink legs extend well beyond tail.
6f Juvenile: Head and neck grey-brown, back streaky brown, underparts and tail dirty white. Legs dirty brown, bill grey with black tip.

1a

1b

2a

2b

1d

2c

1c

3a

4a

4c

4b

5a

4c

3b

5b

5c

6e

6b

6c

6d

6f

PLATE 10 EGRETS

Cattle Egret, race *seychellarum*, breeds in the granitics, and race *ibis* breeds in the outer islands. Dimorphic Egret occurs throughout the Aldabra group, breeding in good numbers on Aldabra Atoll. Little Egret is an irregular migrant. Intermediate and Great White Egrets are vagrants.

1 CATTLE EGRET *Bubulcus ibis* 40-56 cm   **Text page 164**
A pure white, short-legged egret tending to feed in open, grassy areas or scavenge at dumps, etc.
1a Adult breeding (race *seychellarum*): Buff-gold plumes on crown, breast and lower back. Displaying birds have crimson legs, incubating birds pinkish legs.
1b Head breeding (race *seychellarum*): Yellow bill with crimson base and scarlet eye-ring in displaying birds. Incubating birds have yellow bill, mauve-grey round eye.
1c Head non-breeding: Yellow bill with grey-blue round eye.
1d Adult non-breeding: Pure white plumage, dull yellowish black legs.
1e Nest site: Breeds colonially in trees, often mangroves.
1f Immature: Pure white plumage, and blackish legs.
1g Head immature: Dull grey-yellow bill, yellow around eye.

2 LITTLE EGRET *Egretta garzetta* 55-65 cm   **Text page 166**
A slender, white egret with long black bill, black legs and yellow feet.
2a Adult non-breeding head: Bill black with pink base to lower mandible, grey lore.
2b Adult non-breeding: White plumage, brown-green legs and yellow feet.
2c Adult breeding: Pure white plumage with plumes on nape, upper breast and scapulars, blacker bill and brighter yellow feet.
2d Adult white morph head: Two long white plumes on nape. Lore grey-blue. Black bill with pale horn base on lower mandible.

3 DIMORPHIC EGRET *Egretta dimorpha*   55-65 cm   **Text page 166**
Has a larger, heavier bill than in Little Egret, longer more slender neck and is more exclusively marine.
3a Immature: Grey to slate grey plumage. Yellow feet.
3b Immature head: Dark grey bill, pinkish at base of lower mandible. Lore dark grey with yellow line from base of mandible to below eye.
3c Dark morph in flight: Dark wings with white 'thumb' at carpal joint.
3d Dark morph head: Lore yellow, bill black, throat white.
3e Adult dark morph: Slate grey with white throat. Pied individuals also occur. Black legs with yellow feet.
3f Adult white morph head: Lore yellow (sometimes red at onset of breeding), black bill usually with yellow base to lower mandible.
3g Adult white morph: Pure white plumage, black legs with yellow feet.

4 INTERMEDIATE EGRET *Egretta intermedia* 65-72 cm   **Text page 166**
Longer-necked than Cattle or Little Egret, S-shaped neck lacking sharp kink of Great White Egret, and has black feet.
4a Adult non-breeding head: Bill yellow, often black at tip. Lore yellow. Gape line extends only to below eye. Head lacks plumes.
4b Adult non-breeding: Brownish black legs.
4c Adult breeding head: Lore green. Bill yellow.
4d Adult breeding: Pure white plumage, long scapular plumes and shorter plumes on breast. Legs yellow above tarsal joint, black below, with black feet.

5 GREAT WHITE EGRET *Egretta alba* 85-102 cm   **Text page 165**
Grey Heron size: taller than all other egrets, with dagger-like yellow bill, with sharp kink to snake-like neck, black feet.
5a Adult breeding: Black legs. Long white plumes on back and breast.
5b Adult non-breeding: Pure white, with yellow bill and all-black legs.
5c Adult breeding head: Bill blackish with yellow base. Green lore.
5d Adult non-breeding head: Bill yellow with black tip. Lore yellow-green. Gape line extends behind eye.

PLATE 11 POND-HERONS

Green-backed Heron breeds throughout Seychelles, present on almost every island large and small. Madagascar Pond-Heron breeds on Aldabra, and is a vagrant to the granitics. Indian Pond-Heron is a vagrant to Bird and the granitics. The status of Squacco Heron is uncertain.

1 GREEN-BACKED HERON *Butorides striatus* 40-48 cm   **Text page 169**
A small, dark short-legged heron that typically feeds along the shoreline.
1a Adult in flight: Dark, slatey green-black wings and tail; bright orange soles to feet obvious, projecting beyond tip of tail.
1b Immature: Brownish, with white and buff spotting on the wing-coverts.
1c Adult (race *crawfordi*): Outer islands. Back lighter than *degens* with smoky blue-purple cast; wing-coverts with broad creamy white edging; ashy grey hindneck and sides to upper breast.
1d Adult (race *degens*): Granitics. Back green-black, wing-coverts edged rufous-buff, grey-brown sides to upper breast.

2 MADAGASCAR POND-HERON *Ardeola idae* 45-48 cm   **Text Page 168**
Pure white in breeding plumage; non-breeding with dark mantle, streaked paler and very heavy streaking below.
2a Adult non-breeding in flight: Dark sepia brown back with buff streaks, contrasting with white wings and tail.
2b Immature: Brown streaks on wing-coverts and cream streaking on scapulars. White wings with brown-grey smudging on outer primary webs and tips of inner primaries; primary shafts dark. Bill orange-yellow with black tip.
2c Adult breeding: Pure white, with long back plumes. Bill, greenish blue with black tip and legs pink.
2d Adult non-breeding: More heavily marked and back darker compared to other *Ardeola*. Face, nape and upper breast yellow-ochre with heavy, dark brown streaking, becoming thrush-like spots on the throat. Bill dirty yellow with black tip, legs yellowish.

3 INDIAN POND-HERON *Ardeola grayii* 42-45 cm   **Text Page 168**
Uniform grey-brown back in non-breeders with heavy streaking below.
3a Adult non-breeding in flight: Unstreaked grey-brown back, contrasting with white wings and tail (with brownish tips to primaries).
3b Immature: Head and neck streaked ochre. Brown and buff spotting on wing-coverts, breast more spotted than streaked. Outermost primaries smudged grey, next three with grey tips, dark primary shafts.
3c Adult breeding: Head and neck yellowish buff with long white nape plumes. Back bright purple-brown, underparts white, tinged buff on breast.
3d Adult non-breeding: Drab grey-brown back. Head, neck and chest heavily streaked dark grey-brown.

4 SQUACCO HERON *Ardeola ralloides* 42-47 cm   **Text Page 167**
Pale uniform buff-brown back in adult with less heavy streaking below compared to other *Ardeola*.
4a Adult non-breeding in flight: Buff-brown back (paler than other *Ardeola*) contrasts with pure white wings and tail.
4b Immature: Dull yellow-brown back; head and neck buff, more heavily streaked brown. Wings smudged grey on primaries.
4c Adult breeding: Bright golden buff on head (long black-and-white nape plumes) buff sides of breast, back and long scapular plumes fawn-brown with a pinkish blush. Bill greenish blue with black tip.
4d Adult non-breeding: Dull buff-brown on mantle and scapulars. Crown and sides of neck streaked. Retains vestiges of breeding mane. Bill yellow with black tip.

PLATE 12 BITTERNS AND NIGHT HERON

Yellow Bittern is resident, breeding in small numbers in scattered freshwater marshes of Mahé, Praslin and La Digue. Cinnamon Bittern and Eurasian Bittern are vagrants from the north. Black-crowned Night Heron is a recent colonist occurring on Mahé and Silhouette.

1 YELLOW BITTERN *Ixobrychus sinensis* 30-40 cm   **Text page 170**
A small skulking heron, pale buff with black crown and flight feathers. Yellow bill (reddish in breeding season) and legs.
1a Adult female: Crown streaked rufous. Chestnut streaks on foreneck. Darker brown back than male.
1b Adult male Foxy orange mane on neck, paler buff back and wing-coverts.
1c Immature: Back and crown heavily mottled with dark brown; neck and breast with bold, rusty streaking.
1d Adult in flight: Black crown, tail and flight feathers contrast with paler wings and back.

2 CINNAMON BITTERN *Ixobrychus cinnamomeus* 40-41 cm   **Text page 172**
A small uniformly chestnut bittern with no contrast on wing-coverts or tail.
2a Adult in flight: Uniform dark rufous throughout, back and tail tan-rufous, wing-coverts yellower, and flight feathers with a smoky pink cast.
2b Adult female: Similar to male, generally browner with darker crown, bolder dark marks on side of chest and darker streaks on neck.
2c Adult male: Uniform dark rufous plumage on back, wings and tail. Breast warm rufous-buff.
2d Immature: Dark mottling on upperparts, heavily streaked below.

3 EURASIAN BITTERN *Botaurus stellaris* 70-80 cm   **Text page 172**
A large, heron-like bird, buff-brown with complex darker streaking.
3a Immature: Paler cap and moustache than in adult.
3b Adult: Tawny-buff stippled and barred with black. Black crown and moustache contrast with rufous cheek. Heavily black-barred back and wings.
3c Adult in flight: Dark-barred, rufous flight feathers contrast with pale buff, speckled wing-coverts and rump. Mantle and scapulars blotched black.

4 BLACK-CROWNED NIGHT HERON *Nycticorax nycticorax* 55-65 cm   **Text page 170**
A chunky heron midway in size between Grey and Green-backed Herons.
4a Second-year Crown greyish black, plumes absent. Grey-brown back and pale grey-brown underparts. Eye red-brown. Bill black, yellow at base of lower mandible. Legs yellow.
4b Adult in flight: Compact silhouette, wings broad and rounded. Light grey wings contrast with black back.
4c First-year Mantle and scapulars uniform grey-brown, breast streaked. Eye orange. Bill black with yellow streaking. Legs yellow.
4d Juvenile: Dark brown with large white spots on wing-coverts. Back with finer buff spots and streaks. Eye orange-brown. Bill dirty yellow with black tip. Legs dirty yellow-green.
4e Adult breeding: Pale grey wings and underparts contrast with black crown (with white head plumes) and back. Lore blue. Eye red. Blue-black bill contrasts with white forehead and cheeks. Legs pinkish red.
4f Adult non-breeding: No head plumes, duller cap and back. Eye red. Bill black. Legs yellowish.

PLATE 13 DUCKS

Garganey is the only annual migrant duck, occurring throughout the islands. Other ducks are vagrants mostly in the granitics, except for White-faced Whistling Duck which is likely mainly in the Aldabra group.

1 GARGANEY *Anas querquedula* 37-41 cm   **Text page 177**
A small duck with rapid flight. Pale grey forewing and green speculum bordered white, prominent on inner bar.
1a Immature: Mottled dark brown, brown spots and streaks on lower breast. Pale lines above and below dark eye-stripe. **1b Female:** Pale streaks above and below eye and white spot at base of grey bill (also shown in eclipse male). **1c Male in flight:** Light grey wings with green speculum; white trailing secondary edge broader than white wingbar. **1d Adult male breeding:** Chestnut cheeks and neck and bold white stripe from eye to nape. Speckled chestnut breast and finely patterned silver flanks.

2 COMMON TEAL *Anas crecca* 34-43 cm   **Text page 176**
A small duck with rapid flight. Dark wings, green speculum bordered white, prominent on trailing edge.
2a Male in flight: Dark brown forewing and flight feathers; green speculum bordered by white wingbar, broader than white trailing edge. **2b Adult female:** Mostly mottled dark brown; white stripe at base of tail; pale face lacks distinctive features. **2c Adult male:** Bright chestnut head with green mask from eye down neck. Grey back and flanks, white belly with black chest spots and obvious creamy yellow patch on side of rump, bordered black.

3 NORTHERN SHOVELER *Anas clypeata* 44-52 cm   **Text page 178**
A medium-sized duck with large flared bill and blue-grey wing-panels.
3a Eclipse male: Similar to female, more uniform above, more streaked below; lacks green speculum.
3b Adult male: Striking chestnut flanks contrast with white breast and shoulder, and green-black head.
3c Adult female: Uniform buff-brown with darker mottling. Bill black with orange stripe along lower edge.
3d Male in flight: Appears front-heavy. Blue-grey forewing contrasts with dark flight feathers, green speculum framed by broad white wingbar. Dark belly and all-white underwing-coverts.

4 MALLARD *Anas platyrhynchos* 50-65 cm   **Text page 176**
A medium-sized duck with yellow-orange bill and purple-green speculum.
4a Male in flight: Pale grey forewing, with broad purple-blue speculum bordered black and white, the trailing edge border thickest. **4b Adult female:** Uniform mottled buff-brown with pale supercilium and dark eye-stripe. Orange-brown bill. **4c Adult male:** Metallic green head, chocolate brown breast, grey back and flanks, and black rump and tail-curls contrast with white tail feathers.

5 NORTHERN PINTAIL *Anas acuta* 50-66 cm   **Text page 177**
A medium-sized, elegant duck with long thin neck and pointed body shape.
5a Adult male: Dark chocolate head contrasts with white breast and neck-stripe. Grey flanks, cream side patch and long black tail. **5b Male in flight:** Distinctive silhouette with long narrow neck. Banded underwing pattern. Speculum green-bronze, bordered by broad white trailing edge. **5c Immature:** Like female but plainer above, more spotted beneath cinnamon-coloured head. **5d Adult female:** Brown with darker mottling and warm, buff-coloured head.

6 RUDDY SHELDUCK *Tadorna ferruginea* 61-67 cm   **Text page 176**
A large, rusty brown duck with white head, conspicuous white wing-patches and black tail.
6a Male in flight: Rusty brown body contrasts with black-and-white wings and black tail. **6b Adult male:** Back, breast and belly uniform bright rusty brown. Pale face and black neck-collar (absent in winter).
6c Adult female: Buff head and neck contrast with whiter face. No neck-collar.

7 WHITE-FACED WHISTLING DUCK *Dendrocygna viduata* 46-48 cm   **Text page 175**
Unmistakable upright, long-necked duck with black-and-white head pattern.
7a Adult: Dark chestnut back and breast, black-and-white flank barring and black belly. Black head with a white face. **7b Adult in flight:** All-black wings with chestnut leading edge to inner wing and black rump, tail and belly. **7c Immature:** Buff-grey head and neck and paler than adult on the breast.

8 FERRUGINOUS DUCK *Aythya nyroca* 38-42 cm   **Text page 178**
An all-dark brown diving duck with broad white wingbar and white undertail.
8a Adult male: Dark chestnut head, breast and flanks, black back and diagnostic white undertail. Eye white. **8b Male in flight:** Broad white wingbar from secondaries to wingtip. Bill slate grey with paler edges.
8c Adult female: Browner and duller than male. Eye brown.

PLATE 14 ACCIPITERS

All species are vagrants during the northwest monsoon. All have occurred in the granitics, except Yellow-billed Kite, which is a vagrant to Aldabra. Western Honey-Buzzard has also occurred in the Amirantes.

1 WESTERN HONEY-BUZZARD *Pernis apivorus* Wingspan: 130-150 cm   **Text page 179**
A slim-tailed, small-headed hawk with distinct bands on tail. Tail long with rounded corners.
1a Immature above: Pale patches at base of primaries and base of uppertail feathers.
1b Immature below: Darker tail, 3-4 tail-bands and heavier secondary barring.
1c Adult below: Chestnut blotches on breast and underwing (forming bars in the female). Individuals may vary from dark brown below to all-white. Diagnostic black carpal-patches. Broad bar at tip and two narrower bars at base.
1d Adult male above: Ashy grey with broad black trailing edge to upperwing (female darker brown, lacking trailing edge).

2 BLACK KITE *Milvus migrans* Wingspan: 135-155 cm   **Text page 180**
An all-dark raptor with dark bill, grey crown, long, angled wings and long, shallowly forked tail.
2a Adult below: Rufous wash to underparts.
2b Adult above: Dark brown throughout with paler greyish diagonal bands on upperwing and paler crown.
2c Immature below: Paler on breast with noticeable pale bases to primaries on underwing.
2d Adult head: Pale greyish crown, black bill, yellow cere.
2e Immature above: Paler, light brown to buff on upperwing.

3 YELLOW-BILLED KITE *Milvus aegyptius* Wingspan: 132-150 cm   **Text page 181**
All-dark kite with all-yellow bill, rusty wash on head, long angled wings and shallowly forked tail.
3a Adult head: Dark rusty crown, white face, all-yellow bill and cere.
3b Immature below: Bill darker and duller. Paler, buff spotting on throat and chest, creamy streaks on belly.
3c Adult below: Rufous wash on head, more rufous throat and chest. Whitish primary bases form 'windows'.
3d Adult above: Back and wings dark brown, rufous head, lightly streaked, rufous brown tail with faint dark bars.

4 WESTERN MARSH HARRIER *Circus aeruginosus* Wingspan: 110-130 cm   **Text page 181**
A large, broad-winged harrier, female dark brown, male grey wings and tail.
4a Immature below: Similar to female, browner overall and darker underwing-coverts. Never a breast-band.
4b Female below: Chocolate brown with creamy throat-patch and often pale breast-band.
4c Immature above: Resembles female, but seldom with obvious cream edge to wing. Rufous fringes to crown and paler tips to coverts.
4d Female above: Dark brown with diagnostic creamy buff crown and leading wing edge.
4e Male below: Underparts rufous, streaked dark brown.
4f Worn male: Browner on crown and face. Grey wings with black primaries. Grey tail with dark brown back rump and wing-coverts.
4g Female head on: Distinct creamy cap and wing edges.
4h Female at rest: Chocolate brown with creamy crown, throat and carpal-patch.

5 OSPREY *Pandion haliaetus* Wingspan: 145-170 cm   **Text page 179**
An unmistakable, large-winged raptor, dark brown above and white below.
5a Adult below: All-white except for black carpal-patches, primaries and secondary coverts. Tail smudged grey and banded. Dark breast-band, broader in female.
5b Adult at rest: Dark brown-black above. Head all-white except for dark eye-band.
5c Immature above: Similar to adult but lighter brown, paler tips to back and wing-coverts, creating a distinct speckled effect.

6 BOOTED EAGLE *Hieraaetus pennatus* Wingspan:100-121 cm   **Text page 182**
A small, buzzard-like eagle, occurring in either dark brown or pale forms.
6a Dark adult above: Dark brown with pale patch across upperwing-coverts and scapulars.
6b Dark adult below: Dark brown with paler undertail-coverts and tail base.
6c Pale adult above: All-dark wings contrast with pale crown and distinct whitish scapulars and wing-panels.
6d Pale adult below: Striking white underparts (with spotting on underwing-coverts) contrast with black flight feathers and tail.
6e Intermediate morph below: Rufous wash to body and wing-coverts.

PLATE 15 FALCONS I

Seychelles Kestrel is endemic to the granitic islands, occurring on Mahé, Praslin, Silhouette, North and some satellite islands. Madagascar Kestrel breeds only on Aldabra Atoll. Other species are vagrants, mainly to the granitics (except Eleonora's Falcon which is more frequent at Aldabra) during the northwest monsoon.

1 SEYCHELLES KESTREL *Falco araea* Wingspan: 40-45 cm **Text page 184**
A very small kestrel; the only resident bird of prey in the granitics.
1a Adult at rest: Small and compact with blue-grey head, unspotted breast and rounded wingtips.
1b Adult above: Chestnut back with dark crescents on the wing-coverts, dark slate grey head and banded tail. **1c Adult below:** Diagnostic unspotted pinkish underparts. **1d Immature above:** Pale chestnut back with dark barring, streaked chestnut head. Tail tipped buff. **1e Immature below:** Underparts with dark streaks and spots.

2 MADAGASCAR KESTREL *Falco newtoni* Wingspan: 45-50 cm **Text page 184**
A small, pale kestrel, whitish below with heavy spots, particularly on flanks.
2a Female below: Heavily spotted underparts (but variable, sometimes unspotted). **2b Female above:** Similar to male, but head and tail browner and more heavily spotted. **2c Male below:** Underparts pale with thrush-like black spots, heavier on flanks. Lower belly and undertail-coverts white. **2d Male at rest:** Rufous chestnut above, pale below with heavy spots. **2e Male above:** Rufous brown back with dark crescents, streaked rufous head and banded tail.

3 LESSER KESTREL *Falco naumanni* Wingspan: 58-72 cm **Text page 183**
A typical kestrel with unspotted back and tail, white claws and no barring.
3a Immature at rest: Pale face and faint moustachial-stripe. Pale barring on flight feathers. At close range, white claws diagnostic. **3b Male above:** Plain tan back contrasts with diagnostic blue-grey greater coverts; blue-grey hood and tail. Tail plain save for heavy black terminal band, with slightly projecting central tail feathers. **3c Male below:** Underwing silvery white, breast buff with lines of fine thrush-like spots.
3d Female above: Crown pale rufous. Chestnut back with black spotting. Tail with faint barring. **3e Female below:** Some spotting on underwing-coverts.

4 RED-FOOTED FALCON *Falco vespertinus* Wingspan: 65-75 cm **Text page 185**
A fast, long-winged dark falcon with dark or buffish underwing-coverts and red-orange legs, eye-ring and cere.
4a Female below: Head, underparts and underwing-coverts ginger-orange. **4b Female above:** Orange crown and nape. Barred tail, upperparts and upperwing. **4c Immature below:** Pale underparts, streaked reddish brown. Entire underwing heavily barred, coverts having a distinct buff wash. **4d Male in flight:** Blue-grey above with contrasting silvery grey primaries and blackish tail. Dark grey underwing.
4e Immature head: Sandy brown crown lacking grey tones. Mantle not so dark as in Amur Falcon. Short untapered moustache.

5 AMUR FALCON *Falco amurensis* Wingspan: 65-70 cm **Text page 186**
A fast, long-winged dark falcon; with white underwing-coverts and red-orange legs, eye-ring and cere.
5a Female below: White underparts, heavily spotted with black, underwing-coverts mainly white with some fine black spotting. Yellowish buff only on thighs and undertail-coverts. **5b Male in flight:** Dark sooty grey above lacking contrast with grey tail, but primaries silvery grey, as Red-footed Falcon. From below, unmistakable white underwing-coverts contrast with black flight feathers and grey body. **5c Female above:** Grey crown and nape, black moustachial-stripe. Barred tail, upperparts and upperwing. **5d Immature below:** Similar to female but whiter underwing and flight feathers, unspotted, buff trousers and heavy, sepia/black streaking or blotching on breast and flanks. **5e Immature head:** Grey-blue crown. Dark mantle. Long tapered moustache.

4e

5e

PLATE 16 FALCONS 2

All the following occur as vagrants during the northwest monsoon and are possible throughout the islands.

1 EURASIAN HOBBY *Falco subbuteo* Wingspan: 69-84 cm   **Text page 186**
A long-winged falcon with red thighs and clear black-and-white face.
1a Adult above: Dark slate grey with black flight feathers. Black hood contrasts with white cheeks and throat.
1b Adult below: Clean white breast streaked with black. Underwing and tail finely barred and streaked black. Chestnut red 'trousers'.
1c Immature above: Pale buff tips to feathers give a greyer hue.
1d Immature below: Breast with buff ground colour, coarser brown streaking and buffy thighs. Underwing with uniform warm buff wash throughout.

2 ELEONORA'S FALCON *Falco eleonorae* Wingspan: 90-105 cm   **Text page 187**
A dark, long-winged, long-tailed falcon with paler flight feathers of underwing contrasting with coverts.
2a Adult pale morph below: Rufous brown, heavily streaked underparts. Diagnostic underwing pattern: darker underwing-coverts contrast with paler flight feathers.
2b Adult dark morph above: Uniform dark slate grey on back with coffee hue.
2c Adult dark morph below: Sooty black with paler rear wing.
2d Immature above: Greyish to brownish, with pale flecking.
2e Immature below: Breast paler, more buff than adult, buff barring on tail and flight feathers, but underwing-coverts still contrast darker. Narrow subterminal tail-band.

3 SOOTY FALCON *Falco concolor* Wingspan: 78-90 cm   **Text page 188**
A slim, long-winged falcon; longer central feathers forming a wedge-shaped tail.
3a Adult below: Uniform pale grey throughout, including all-grey head and tail. Primaries slightly darker.
3b Adult above: Ashy grey, only contrast being darker head, distinct black primaries and black tail tip.
3c Immature above: Dark, dirty brown with pale half-collar and tail tip. Dark hood, cheeks and throat dirty brown.
3d Immature below: Underparts with buff ground colour and dark, heavy blotching. Chin and throat white. Underwing uniformly pale buff-grey, finely barred. Tail finely barred with broad subterminal band.

4 PEREGRINE FALCON *Falco peregrinus* Wingspan: 80-120 cm   **Text page 188**
A large, deep-chested, short-tailed falcon, slate grey above, barred below.
4a Adult male above: Slate grey with paler rump and darker flight feathers.
4b Adult male below: Paler, finely barred on breast and underwing. Tail with dark bars, broader at tip. Black hood and moustache contrast with white cheeks, throat and upper breast.
4c Immature female above: Dark brown-black with paler tips to coverts. Clear buff tip to tail. Facial pattern less distinct.
4d Immature female below: Heavier, dark, vertical streaking on the breast.

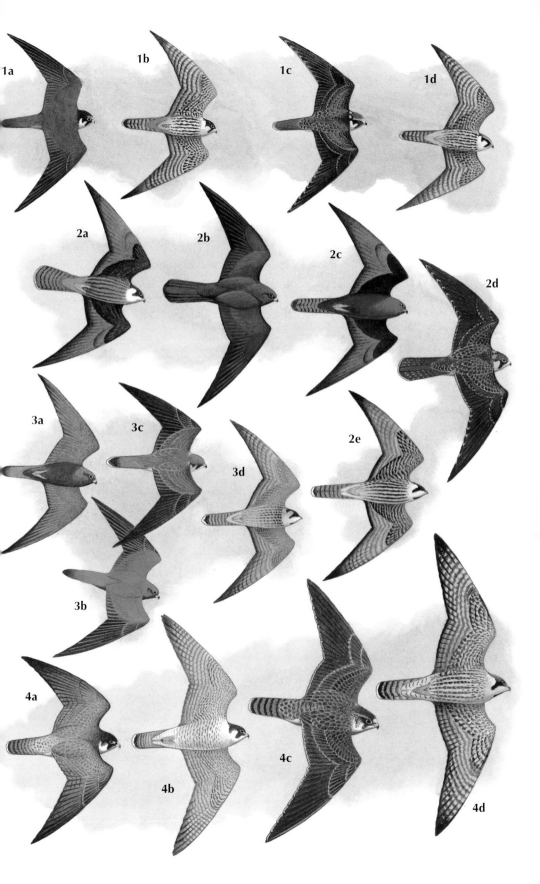

PLATE 17 RAILS AND CRAKES

Grey Francolin is introduced on Desroches and Coëtivy. Common Moorhen is resident throughout the granitics. Aldabra Rail is endemic to Aldabra. Other species are vagrants. Allen's Gallinule has occurred both on Aldabra and in the granitics, Striped Crake only at Aldabra; Corncrake and White-breasted Waterhen only at Bird and the granitics.

1 COMMON MOORHEN *Gallinula chloropus* 32-35 cm **Text page 193**
Black and brown, white undertail-patches, yellow-green legs and red shield.
1a Adult: Body blackish with white arc on flank. Red shield, bill with yellow tip. **1b Chick:** Fluffy black down, blue forecrown, red-and-yellow bill, black legs. **1c Immature:** Dull brown with a white throat and dirty green bill and legs.

2 GREY FRANCOLIN *Francolinus pondicerianus* 33-35 cm **Text page 189**
A plump grey, barred partridge with rufous undertail.
2a Adult male: Head rufous brown with creamy buff supercilium and buff gorget. Dull red legs. **2b Adult female:** Orange legs lack spurs. **2c Immature:** Similar to adult but less clearly marked on head and throat.

3 ALLEN'S GALLINULE *Porphyrula alleni* 22-24 cm **Text page 194**
A small gallinule; blue-purple and olive green above in adult, brownish in immature.
3a Immature: Ginger brown face, neck and breast and dark mottled back. Dull bill, olive brown shield and brownish legs. **3b Adult:** Blue-black head, olive green back and iridescent blue-purple underparts. Red eye and bill, turquoise shield, legs dark red to crimson.

4 ALDABRA RAIL *Dryolimnas (cuvieri) aldabranus* 30-32 cm **Text page 190**
The only resident rail of Seychelles.
4a Adult: Dark brown, tinged olive green above and chestnut below. Warmer rufous brown head and neck with white chin and throat. Bill dark, base red. **4b Juvenile:** Lighter brown than the adult and lacks white throat. Black legs. **4c Chick:** Black down, becoming dull olive brown.

5 STRIPED CRAKE *Aenigmatolimnas marginalis* 18-21 cm **Text page 192**
A tiny, dark brown rail with white back stripes and rufous-buff undertail.
5a Adult male: Sides of head and neck buff-brown, throat white, breast orange-buff. Back and wing-coverts dark brown, striped with white. Diagnostic deep rufous buff undertail-coverts. Bill dirty green, legs and feet dirty olive-grey. **5b Adult female:** Face and neck light grey, breast and flanks light grey scalloped with white. Rufous buff undertail.
5c Immature: Similar to male, less olive above, lacking stripes, rufous tinge to face/breast and legs blue-grey.

6 SPOTTED CRAKE *Porzana porzana* 22-24 cm **Text page 192**
A small, brown rail with white spots, buff undertail, red-and-yellow bill and olive green legs.
6a Adult in flight: White spots on wings and white leading edge to first primary. **6b Immature:** Browner below; buff, unspotted face and white throat. Lacks grey on supercilium, throat and breast. **6c Adult:** Green-brown above with black centres to feathers of crown, back and rump and white flecks throughout. Face and underparts slate-grey spotted white. Flanks olive green with white barring. Undertail pale buff.

7 WHITE-BREASTED WATERHEN *Amaurornis phoenicurus* 32 cm **Text page 193**
A black gallinule with white face and breast, and long, yellow-green legs.
7a Immature: Dirty white face and breast flecked with black or brown, upperparts dark olive brown and undertail-coverts duller. Eye yellow-green. **7b Adult:** White face, throat, breast and belly contrast with black crown, hindneck, back, wings and tail. Blackish on flanks extends down to legs in irregular border. Undertail-coverts bright cinnamon brown. Eye red. Bill yellow-green, frontal shield red.

8 CORNCRAKE *Crex crex* 27-30 cm **Text page 192**
A mottled brown and black rail with bright rust-red wings.
8a Adult in flight: Striking chestnut red wings only visible in flight. **8b Adult:** Tawny buff ground colour, streaked with rows of black spots. Flanks grey-white with chestnut barring. Bill pink-brown, legs pinkish grey.

9 COMMON QUAIL *Coturnix coturnix* 16-18 cm **Text page 189**
A tiny, tailless gamebird with streaked brown back and paler buff underparts.
9a Adult: Mottled black and buff back with broad, pale buff streaks. Rufous chest with paler buff to cream belly, bold black and cream darts on flanks. Streaked crown, thin buff supercilium. Male with black 'bridle' from throat to neck, pale or rufous cheek-patches and bolder flank marking. **9b In flight:** Short wings, uniform pale buff and brown, contrast with darker, streaked back.

9a

9b

PLATE 18 WADERS 1

Crab Plover occurs annually in small numbers in the granitic islands, and in large numbers in the lagoons of the coralline atolls mainly October to April. Eurasian Oystercatcher is a vagrant to the granitics and Aldabra. Black-winged Stilt is a vagrant to the granitics. Stone Curlew is of uncertain status.

1 CRAB PLOVER *Dromas ardeola* 33-36 cm   **Text page 195**
A black-and-white wader with a massive black bill and long, bluish grey legs.
1a Adult: Pure white head and body, white wings with black flight feathers, black mantle, white rump and tail.
1b Immature: Dark streaking on hindneck, grey mantle and dirty smudging on wings and tail.
1c Adult in flight: Pure white below, white wings with black flight feathers, white rump. Legs trail well beyond tail.
1d Group: Highly gregarious, often in flocks or small parties of mixed age.

2 EURASIAN OYSTERCATCHER *Haematopus ostralegus* 40-46 cm   **Text page 196**
A pied, chunky wader, black above, white below with red bill and pinkish legs.
2a Adult non-breeding: Black head, back and tail and white underparts. White neck-collar (absent in adult breeding). Bright orange-red bill, long heavy and straight. Red eye and eye-ring. Relatively short, pink legs.
2b Adult in flight: Wings and body white below save for black hood. Upperwing black with broad white wingbar, white rump and black tail-band.
2c Immature: Blackish brown with pale fringes to back and wing-coverts, white neck-collar. Duller bill, browner eye, less distinct, duller eye-ring and paler, greyer legs.

3 BLACK-WINGED STILT *Himantopus himantopus* 35-40 cm   **Text page 196**
A slender black-and-white wader with long pink legs and thin, needle-like bill.
3a Adult: Pure white head and body, all-black wings cover white rump and tail. Pink legs. Red eye.
3b Immature: Brownish upperparts with buff fringes to coverts. Dusky crown and hindneck, white trailing edges to wings. Dull pink legs. Yellow-brown eye.
3c Adult in flight: Body white below. Wings all-black above and below, save for white axillaries. Long legs trail way behind the white rump and tail.

4 STONE CURLEW *Burhinus oedicnemus* 40-44 cm   **Text page 197**
A curlew-like wader with heavy plover-like bill, pale yellow legs and huge yellow eye in square head.
4a Immature: Rufous buff fringes to scapulars and tertials. Black-and-white wingbars and white supercilium less distinct than in adult.
4b Adult: Pale, mottled buff-brown plumage with fine streaking on head and breast, heavier streaking on back and wing-coverts. Horizontal white bar across wing-coverts, paler in female, heavily bordered black in male. Black flight feathers with white panels.
4c Adult in flight: Upperwing with broad white carpal bar, edged black. Flight feathers black with two white flashes in innermost and outermost primaries. Underwing white with black trailing edge, lower breast and belly white.

PLATE 19 WADERS 2 (PRATINCOLES)

All pratincoles are vagrants, mainly to the granitics between October and December.

1 COMMON PRATINCOLE *Glareola pratincola* 23-26 cm   **Text page 197**
A long-tailed pratincole with chestnut underwing and white trailing edge.
1a Adult in flight: Shows strong contrast between coffee brown mantle/coverts and black flight feathers. Diagnostic combination of dark chestnut underwing-coverts and white trailing edge to secondaries. Long outer tail feathers.
1b Adult non-breeding: At rest, tail is longer than wings. Pale cream throat with indistinct black border.
1c Head adult breeding: Distinct cream throat bordered black. Compared to Black-winged, shows more extensive red base to bill and less distinct blackish lore.
1d Immature: Buff tips and dark chevrons on back and wing-coverts, and tips to secondary feathers buff rather than white.

2 BLACK-WINGED PRATINCOLE *Glareola nordmanni* 23-26 cm   **Text page 198**
A dark-winged pratincole with completely dark brown underwing.
2a Adult in flight: Almost uniform dark brown above with white rump. Diagnostic combination of dark brown underwing-coverts and all-dark secondaries above. Shorter tail feathers than Common Pratincole.
2b Adult non-breeding: At rest, tail feathers as long as wingtips. Pale cream throat edged black.
2c Head adult breeding: Distinct cream throat bordered black. Little red at base of bill. Distinct blackish lore.
2d Immature: Buff tips and dark scaling on back and wing-coverts; may show some chestnut flecking on underwing and narrow buff tips to secondaries, but wing mostly dark brown above and below.
2e Head Madagascar Pratincole *Glareola ocularis*: 23-25 cm. Black crown and narrow white eye-stripe.
2f Madagascar Pratincole in flight: Very dark black-brown above with very short tail, narrow white rump and thick black terminal band, reminiscent of Northern Wheatear. Bright chestnut underwing-coverts and diagnostic chestnut patch on belly.

3 ORIENTAL PRATINCOLE *Glareola maldivarum* 23-24 cm   **Text page 198**
A short-tailed pratincole with chestnut underwing and all-dark upperwing.
3a Adult in flight: Dark brown above with white rump. Diagnostic combination of chestnut underwing-coverts and dark trailing edge to secondaries. Tail feathers short.
3b Adult non-breeding: At rest, wings are longer than tail. Pale cream throat-patch edged black, lower breast suffused orange-buff.
3c Immature: Buff tips to back and wing-coverts; may show a very narrow buff trailing edge to the secondaries.

PLATE 20 WADERS 3

Greater and Lesser Sandplovers are both annual migrants, occurring throughout the islands between September and April. Caspian Plover and Oriental Plover are vagrants during the northwest monsoon.

1 LESSER SANDPLOVER *Charadrius mongolus* 19-21 cm   **Text page 199**
A small, drab plover with more upright posture than Greater Sandplover, legs set well back on body.
1a Adult male breeding: Black face-mask, white throat and orange breast-patch with a fine black border on the upper margin. Sexes similar, face-mask and breast-patch more extensive in male and somewhat duller in female.
1b Adult non-breeding: Grey-brown above, white below with grey-brown sides to breast. Shorter legs than both Greater Sandplover and Caspian Plover, showing little tibia.
1c Adult in flight: White wingbar, white underwing and white tail sides. Toes do not project beyond tail.

2 GREATER SANDPLOVER *Charadrius leschenaultii* 22-25 cm   **Text page 199**
A small drab plover with more horizontal posture than Lesser Sandplover, legs set centrally.
2a Adult non-breeding: Grey-brown above, white below with grey-brown sides to breast. Longish legs, showing more tibia and with more horizontal posture than Lesser Sandplover.
2b Adult in flight: White wingbar, white underwing and white tail sides. Toes project beyond tail.
2c Adult male breeding: Black face-mask extending beyond ear-coverts, white forehead, narrower orange breast-patch with a white wedge against the wing. Female duller and greyer.

3 ORIENTAL PLOVER *Charadrius veredus* 22-25 cm   **Text page 200**
A small, elegant, very upright long-necked plover with no wingbar and dark underwing.
3a Adult breeding: White face, forehead and sides of neck. Chestnut breast-band heavily bordered black on lower margin.
3b Adult in flight: No wingbar (or only a trace) and diagnostic dark grey-brown underwing. Toes project beyond tail.
3c Adult non-breeding: Upright stance, long neck. Grey-brown above, white below. Pale supercilium and white face. Wings longer than tail at rest. Legs pale yellow or orange.

4 CASPIAN PLOVER *Charadrius asiaticus* 18-20 cm   **Text page 200**
A small, elegant, long-necked plover with faint wingbar and pale underwing.
4a Adult male breeding: White face and prominent white supercilium, broadest behind the eye. Chestnut breast-patch with black edge on lower margin. Female breeding has grey-brown breast with little chestnut and no black rim.
4b Adult in flight: Narrow but obvious wingbar and inner primary flash. Greyer underwing than sandplovers but not so dark or brown as Oriental Plover. No white obvious on tail sides. Toes project beyond tail.
4c Adult non-breeding: Dark grey-brown above and white below. Unbroken grey-buff band across upper breast. Clear white supercilium, much bolder at rear than in sandplovers. Long, yellowish brown legs showing as much tibia as Greater Sandplover.

PLATE 21 WADERS 4

Ringed Plover, Pacific Golden Plover and Grey Plover are all annual migrants recorded throughout the islands during the northwest monsoon (though a few may remain year round). Little Ringed Plover and Great Knot are vagrants to the granitics and Bird, likely around the same period.

1 PACIFIC GOLDEN PLOVER *Pluvialis fulva* 23-26 cm   **Text page 202**
An elegant plover with golden yellow upperparts and rump, grey underwing and indistinct wingbar. Fine black bill.
1a Adult non-breeding: Upperparts suffused golden yellow, breast and belly smudged dark grey and lower belly white.
1b Adult male breeding: All-black underparts from face to lower belly, clearly edged in white from forehead to flanks. Female shows less black below.
1c Adult in flight: Upperwing-coverts mottled grey-golden, no obvious wingbar and underwing uniform grey throughout. Toes project slightly.

2 GREY PLOVER *Pluvialis squatarola* 27-30 cm   **Text page 203**
A large, grey, plump plover, with white rump, black armpits and white wingbar. Bill heavier than in Pacific Golden Plover.
2a Adult non-breeding: Mottled grey and white above, contrasting with white belly. Face pale,
2b Adult non-breeding in flight: Grey upperwing with bold white wingbar, white rump and grey tail. Diagnostic black armpits show up on white underwing. Toes do not project.
2c Adult male breeding: Bold black underparts from face to flanks, with broad white border from forehead to sides of breast. White belly. Female shows some white in underparts.

3 LITTLE RINGED PLOVER *Charadrius dubius* 14-17 cm   **Text page 202**
A very small plover with distinct eye-ring, no wingbar and pinkish legs.
3a Adult non-breeding: Blackish brown face-mask pointed at base. Yellow eye-ring and buffish supercilium.
3b Immature: Indistinct buff supercilium. Whitish forehead.
3c Adult in flight: Grey-brown above, black flight feathers, no obvious wingbar.

4 COMMON RINGED PLOVER *Charadrius hiaticula* 18-20 cm   **Text page 201**
A small plover with no eye-ring, white wingbar and short orangey legs.
4a Adult non-breeding: Blackish brown face-mask, rounded at base. No eye-ring and a white supercilium.
4b Immature: White supercilium and forehead. Incomplete breast-band. Black appears on head December onward.
4c Adult in flight: Grey-brown above, blackish brown flight feathers and bold white wingbar.

5 GREAT KNOT *Calidris tenuirostris* 26-28 cm   **Text page 204**
A deep-chested wader, with heavy straight bill, grey above, spotted grey below.
5a Adult non-breeding: Plain grey above with dark streak to feather centres. Lower breast and belly white with black spotting on upper breast sides and grey chevrons on flanks. Legs dark slate or greenish grey.
5b Adult non-breeding in flight: Grey upperwing with black primary coverts, and white rump (finely speckled black) contrasting with black tail.
5c Adult breeding: Heavy black spotting, dense on upper breast, sparse along white flanks; bright rufous scapulars on dark, mottled back. Legs olive green.

PLATE 22 WADERS 5 (SNIPE)

Common Snipe, Pintail Snipe and Great Snipe are vagrants to the granitics likely during northwest monsoon. Pintail Snipe is also vagrant to Aldabra. Swinhoe's Snipe is of uncertain status.

1 COMMON SNIPE *Gallinago gallinago* 25-27 cm   **Text page 205**
A slim, long-billed snipe with white trailing edge to the wing.
1a In flight above: Broad white trailing edge to secondaries. Fast and sharp-winged, rising rapidly with erratic flight when disturbed.
1b Adult: Long tail extends beyond primary tips. Supercilium narrower than eye-stripe at base of bill.
1c In flight below: Broad white barring on underwing-coverts.
1d Tail: Terminal band rufous, with a dark wedge either side.

2 PINTAIL SNIPE *Gallinago stenura* 25-27 cm   **Text page 205**
A short-billed, short-tailed snipe with no white trailing edge to wing and dark underwing.
2a Adult: Pale bars of tertials wider than dark bars, and primaries are covered by tertials. Primary tips reach almost to tip of short tail. Supercilium broader than eye-stripe at base of bill.
2b In flight above: Wings appear more rounded, and lack white trailing edge to the secondaries of Common Snipe with paler midwing-panel. Toes project beyond short tail more conspicuously than in other snipe. Climbs less high with less erratic escape route than Common.
2c In flight below: Narrow grey barring on the underwing gives all-dark appearance.
2d Tail: Terminal band rufous, with black upper rim, double-barred at edge. Pin-like outer tail feathers, sometimes visible at close range (diagnostic). Little white visible.

3 SWINHOE'S SNIPE *Gallinago megala* 27-29 cm   **Text page 206**
A plain, dark snipe with longish bill and tail, no white trailing edge and dark underwing.
3a Adult: Pale bars of tertials generally of similar width to dark bars. Primaries project beyond tertials. Supercilium broader than eye-stripe at base of bill.
3b In flight above: No white trailing edge to wing. Broad, pale rufous median-covert panel on the upper-wing like Pintail Snipe. Tips of toes project slightly beyond tail. Slow when flushed, with heavy direct flight.
3c In flight below: Uniform dark barring on the underwing.
3d Tail: Terminal band rufous centrally, dark at the edges. More white at corners compared to Pintail.

4 GREAT SNIPE *Gallinago media* 27-29 cm   **Text page 206**
A tubby snipe with heavy barring on belly and prominent white tail margins.
4a In flight above: Two distinct white wingbars either side of a dark midwing-panel. Indistinct, narrow white trailing edge to wing. Slow, heavy flight when flushed, preferring to land again soon.
4b Adult: Larger, shorter-billed and more pot-bellied than Common Snipe. Two distinct rows of white formed by tips to wing-coverts. Flanks and most of belly heavily barred.
4c In flight below: Underparts heavily barred across belly with dark chevrons.
4d Tail: Diagnostic white outer tail feathers, most obvious on landing and take off.

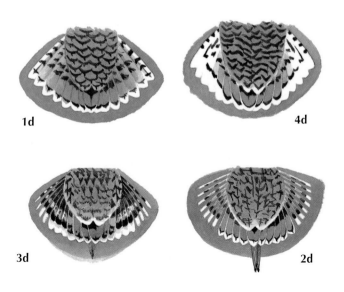

1d   4d

3d   2d

PLATE 23 WADERS 6

Bar-tailed Godwit, Whimbrel and Eurasian Curlew are annual migrants throughout the islands mainly during the northwest monsoon, a few possible year round (particularly Whimbrel). Black-tailed Godwit and Little Whimbrel are vagrants during the northwest monsoon.

1 BAR-TAILED GODWIT *Limosa lapponica* 37-41 cm   **Text page 207**
A large, grey-brown wader with white rump. Long, straight bill, slightly upturned at the tip. Shorter legs than Black-tailed Godwit.
1a Adult in flight (race *baueri*): Spotted rump and barred underwing. Toes project only slightly.
1b Adult male breeding: Back black speckled with rufous. Supercilium, face, neck, breast and belly all a deep chestnut-orange, darker and more extensive than in Black-tailed Godwit. Bill all-dark.
1c Adult non-breeding: Dull grey-brown and streaked above, pale below with barred tail. Pale supercilium extends behind eye.
1d Adult in flight (race *lapponica*): White wedge up lower back and white underwing. Toes project only slightly.

2 LITTLE CURLEW *Numenius minutus* 29-32 cm   **Text page 210**
A tiny curlew with dark crown-stripes and warm sandy buff body colour. Thin bill, fairly straight on basal half, decurved towards tip.
2a Adult in flight: Rump mottled brown like rest of upperparts. Flight feathers black and unspotted, unlike Whimbrel or Curlew, contrasting with pale upperwing-coverts on inner wing. Underwing-coverts entirely grey-brown.
2b Adult: Dark crown with a buff mid-stripe. Distinctive buff eye-stripe and pale lore give pale-faced appearance. Warmer sandy brown than Whimbrel.

3 BLACK-TAILED GODWIT *Limosa limosa* 40-44 cm   **Text page 207**
A large, wader with distinctive black-and-white flight pattern. Very long straight bill. Longer legs than Bar-tailed Godwit.
3a Adult male breeding: Back black speckled with rufous. Brick red neck and upper breast, breast and belly white heavily streaked with black. Orange base to bill.
3b Adult non-breeding: Uniform grey above, off-white below with black-and-white tail. Short eye-stripe in front of eye only.
3c Adult in flight: Broad white wingbars contrast with black flight feathers, white square rump above black tail. White underwing. Toes project conspicuously.

4 EURASIAN CURLEW *Numenius arquata* 50-60 cm   **Text page 209**
A very large, uniformly mottled brown wader with exceptionally long, downcurved bill. Long blue-grey legs.
4a Adult in flight (race *orientalis*): White underwing, axillaries contrasting with flanks. Paler overall and averages larger than nominate.
4b Adult: Uniform patterned dull brown plumage. Pale crown visible at close range.
4c Adult in flight (race *arquata*): Barred axillaries.

5 WHIMBREL *Numenius phaeopus* 40-46 cm   **Text page 208**
A large, mottled brown wader with long, downcurved bill and striped crown. Bill and legs shorter than in Eurasian Curlew.
5a Adult in flight: White wedge up lower back contrasts with barred brown back and wings. Dark crown visible at close range. Underwing often darker than in Eurasian Curlew, wingbeats more rapid.
5b Adult: Shape like Curlew with same patterned brown back and wings. Two dark stripes over crown, with narrow pale line between, giving a diagnostic head pattern.

4c

PLATE 24 WADERS 7

Common Greenshank and Wood Sandpiper are annual migrants throughout Seychelles during the north-west monsoon. Marsh Sandpiper, Common Redshank and Green Sandpiper are vagrants to the granitics around the same time. Spotted Redshank is a potential vagrant.

1 COMMON GREENSHANK *Tringa nebularia* 30-34 cm   **Text page 211**
A medium-sized grey-and-white wader; with slightly upturned bill and long, greenish legs.
1a Adult breeding: Scapulars show mixed black and grey mottling, heavier black streaking on head, neck and upper breast.
1b Adult in flight: Broad white gusset in lower back contrasts with grey back and wings and barred grey tail. Toes project slightly.
1c Adult non-breeding: Uniform grey above (finely scalloped and spotted black close to). Finely streaked grey crown and neck contrast with pure white breast and belly.

2 COMMON REDSHANK *Tringa totanus* 27-29 cm   **Text page 210**
A grey-brown wader with straight, red-based bill and orange-red legs.
2a Adult breeding: Head, face and back darker brown, streaked. Neck, breast and flanks more heavily marked throughout with fine black streaking.
2b Adult in flight: Striking pattern of white on lower back and trailing edge of wings contrasts with black primaries and barred tail.
2c Adult non-breeding: Uniform grey-brown above and paler beneath. Bill straight, dark with a red base.
2d Head Spotted Redshank non-breeding *Tringa erythropus*: 29-32 cm. Longer bill than Common Redshank, slightly 'hooked' at the tip, entirely black on upper mandible. White neck. Obvious white supercilium outlined by black lore.

3 MARSH SANDPIPER *Tringa stagnatilis* 22-25 cm   **Text page 210**
A slender grey-and-white wader, with long, thin straight bill and very long grey-green legs.
3a Adult in flight: Narrow white gusset up back contrasts with barred tail and dark wings. Toes project conspicuously.
3b Adult breeding: Back warm grey uniformly speckled with black; neck and upper breast with small black spots, extending onto flanks. Legs yellowish.
3c Adult non-breeding: Pale grey above; fine white scaling and faint dark streaking on back and wing-coverts. Prominent long, white supercilium gives capped appearance.

4 GREEN SANDPIPER *Tringa ochropus* 21-24 cm   **Text page 211**
A small, dark sandpiper with bold white rump and blackish underwing. Legs dark, shorter than in Wood Sandpiper.
4a Adult in flight above: Square white rump contrasts with very dark wings and broad black bars on tail. Toes project very slightly.
4b Adult: Dark green-brown back and wings with fine pale spotting. Dark upper breast contrasts sharply with clean white underparts. Dark head with wide pale supercilium in front of eye, absent or indistinct behind eye.
4c Adult in flight below: Blackish underwing contrasts with white belly.

5 WOOD SANDPIPER *Tringa glareola* 19-21 cm   **Text page 212**
An elegant sandpiper with speckled brown back and pale underwing. Legs dirty yellowish green, longer than in Green Sandpiper.
5a Adult: Grey-brown back, finely speckled with light spots (larger than in Green Sandpiper). Streaked and barred breast and flanks are not sharply demarcated from white belly. Streaked dark crown with pale supercilium conspicuous behind eye.
5b Adult in flight above: Uniform grey-brown upperparts contrast with white rump and narrow black bars on tail. Toes project conspicuously.
5c Adult in flight below: Grey flight feathers contrast with white belly and underwing-coverts.

PLATE 25 WADERS 8

Terek Sandpiper, Common Sandpiper and Ruddy Turnstone are annual migrants throughout Seychelles during the northwest monsoon with a few present year round (especially Ruddy Turnstone). Grey-tailed Tattler and Red-necked Phalarope are vagrants to Bird, Denis and the granitics, mainly during the northwest monsoon.

1 TEREK SANDPIPER *Xenus cinereus* 22-24 cm   **Text page 212**
A small mud-grey wader with long, upturned bill and yellowish legs. Distinctive dashing feeding action.
1a Adult non-breeding: Uniform dirty grey above with two darker bars across the mantle and shoulder. Grey on upper breast contrasts with clean white lower breast and belly. Bill long for body size, yellow-orange at base and noticeably upcurved. Legs short and yellowish or yellowish orange.
1b In flight: Dark primaries and white secondaries recall Common Redshank. Rump and tail grey-brown. Broad, white trailing edge to wing.

2 COMMON SANDPIPER *Actitis hypoleucos* 19-21 cm   **Text page 213**
A small short-necked, long-tailed wader with distinctive exaggerated tail-bobbing action.
2a Adult in flight: Wings held curved down, dark with a prominent white wingbar.
2b Adult non-breeding: Olive-brown back, head and upper breast, contrasting with white underparts. The white forms a sharp-edged 'wedge' between dark upper breast and wing. Black eye-stripe and white supercilium. Tail extends well beyond wings.
2c Immature: Buff tips and fringes to wing-coverts give a 'scaly' plumage.

3 GREY-TAILED TATTLER *Heteroscelus brevipes* 24-27 cm   **Text page 214**
A uniform ashy grey wader with dark, straight bill and yellowish legs.
3a Adult non-breeding: Uniform ashy grey back, wings and head; paler grey on face, breast and belly. Bill long and straight, dark with yellowish base and long nasal groove. Legs dull yellow, occasionally greyish.
3b In flight: Upperwing, tail, rump and underwing uniform grey, contrasting with white belly.

4 RUDDY TURNSTONE *Arenaria interpres* 21-25 cm   **Text page 214**
A stocky wader with black chest-collar, rufous back and short orange legs. Bill short, black and slightly upturned.
4a Adult non-breeding in flight: Blackish brown flight feathers and tail contrast with white back-stripes and white wingbars.
4b Adult male breeding: Vivid chestnut-orange and black pattern to back, and head strongly marked with contrasting white-and-black banding.
4c Adult non-breeding: Mottled rufous brown upperparts and clean white underparts. Head and face brown. Diagnostic black chest-band with small white lapels.

5 RED-NECKED PHALAROPE *Phalaropus lobatus* 18-19 cm   **Text page 215**
A tiny, slender wader, grey and white with black mask and needle-like bill. Swims in search of food. Bill black, straight, and finely tapered.
5a Female breeding: Scapulars black with rufous buff fringes. Dark grey head with bright chestnut neck (male duller with pale rufous neck and flanks).
5b Adult non-breeding: Ashy grey above with white 'scaling' to back feathers. Neck, head and underparts white save for black strap through eye and cheek.
5c In flight: Wings dark with distinct white wingbar, tail grey with white edges to rump.

PLATE 26 WADERS 9 (STINTS)

Little Stint and Sanderling are annual migrants October to April. Temminck's Stint and Long-toed Stint are vagrants, mainly to the granitics September onward. Red-necked Stint is of uncertain status.

1 RED-NECKED STINT *Calidris ruficollis* 13-16 cm **Text page 216**
A small, tubby black-legged stint; rufous scapulars and grey lesser coverts. Bill short and black, slightly thickened at tip. Legs very short and black.
1a Head adult breeding: Rufous face and upper breast. Rust-red scapular edges contrast with pale grey wing-coverts. Clean white below.
1b Adult non-breeding: Light grey above, with narrow black feather centres, clean white below.
1c Immature: Faint rufous wash on crown and mantle, rufous fringes to black-centred scapulars contrast with pale grey wing-coverts and tertials. May show faint V on mantle, but this tends to 'fade out' down the back.

2 LITTLE STINT *Calidris minuta* 12-14 cm **Text page 216**
A tiny, short-tailed, black-legged stint; all-rufous back with clear white 'braces' (in immature). Bill short and black, fine at tip. Legs short and black.
2a Head adult breeding: Rufous cheeks, nape sides of breast. Rust-red margins to scapulars continue onto tertials and rufous edgings to wing-coverts.
2b Adult non-breeding: Light grey-brown above, with broader dark feather centres, clean white below.
2c Immature: Appears 'red-necked'! Pale forehead and supercilium. Clear white 'braces' form a distinct V on the chestnut-orange back. Rust-coloured fringes to lesser coverts and tertials. Clean white lower breast and belly.
2d Tail: White sides to rump and grey sides to tail.

3 TEMMINCK'S STINT *Calidris temminckii* 13-15 cm **Text page 217**
A small stint with dull brown back, short downcurved bill and yellowish legs that appear too short for body, giving a 'sawn-off' silhouette.
3a Head adult breeding: Patterned back: black-centred scapulars mixed with grey feathers give a mottled effect. Finely streaked head and upper breast.
3b Adult non-breeding: Brown-grey above, white below. Folded wings shorter than long tail.
3c Immature: Grey-brown forehead and no obvious supercilium. Buff-grey upperparts appear 'scaly' due to fine black crescents on scapulars. Warm buff scalloping on lower wing-coverts. White lower breast and belly.
3d Tail: Diagnostic white outer tail feathers.

4 LONG-TOED STINT *Calidris subminuta* 12-13 cm **Text page 217**
A small stint like a tiny Ruff: 'braces', downcurved bill and long, yellowish legs. Often adopts an erect stance, appearing long-necked.
4a Head adult breeding: Rufous crown, pale rufous wash on cheek and upper breast, strong black and rust-red patterning on scapulars and tertials.
4b Adult non-breeding: Dark grey above with dark feather centres, pale grey streaked upper breast, white underparts. Dark forehead and lore. Dark streaked crown and cheeks and pale supercilium.
4c Immature: Bright rufous mottled back contrasts with clean white below. Dark-centred back feathers with bright rufous edging, and clean white V on edge of mantle. Streaked rufous crown, buff sides to breast with fine dark streaking.

5 SANDERLING *Calidris alba* 20-21 cm **Text page 215**
A small, tubby wader, silvery grey above, white below with short black legs.
5a Head adult breeding: Head, neck and upper breast rufous, heavily spotted black and sharply demarcated from white underparts.
5b Adult: Pale grey above with bright white face and underparts. Black 'wrist' shows as black smudge on folded wing.
5c Immature: Black centres to scapulars give a chequered pattern to the back. Fine black streaking on crown and face give head a darker look. Pale buff wash to sides of breast and mantle.

PLATE 27 WADERS 10 (SANDPIPERS)

Curlew Sandpiper is an annual migrant throughout Seychelles, mainly October to April, a few remaining year round. Other species are vagrants likely mainly in the granitics and Bird around the same period.

1 PECTORAL SANDPIPER *Calidris melanotos* 19-23 cm **Text page 217**
A drab sandpiper with diagnostic dark chest sharply contrasting with white belly.
1a Adult non-breeding: Drab grey-brown above. Streaking on face, neck and upper breast gives distinct 'corduroy' effect. Patterned upper breast changes abruptly to white belly.
1b Immature: Brighter, warmer brown due to rufous buff fringes to black-centred back feathers and tertials. Prominent white V-shapes on mantle and scapulars.
1c Adult non-breeding in flight: Mottled brown back, dark flight feathers, indistinct wingbar. Tail and centre of rump black, sides of rump white.

2 SHARP-TAILED SANDPIPER *Calidris acuminata* 17-20 cm **Text page 218**
A brownish sandpiper with distinct brown cap and no abrupt demarcation in plumage on lower breast.
2a Adult non-breeding in flight: Pale white wingbar; sides of rump white streaked black.
2b Immature: Brighter than adult, with chestnut fringes to back, warm ochre-buff flush to breast, which is streaked at sides, unstreaked in centre.
2c Adult non-breeding: Breast markings extend as streaks or chevrons on flanks and undertail-coverts. Dark chestnut cap and broad white supercilium. Obvious eye-ring. Legs dirty grey-green.

3 BROAD-BILLED SANDPIPER *Limicola falcinellus* 16-17 cm **Text page 219**
A small sandpiper with diagnostic bill shape, mostly broad and straight but with distinct downward kink at the tip and a split supercilium.
3a Juvenile: Snipe-like plumage, with chestnut fringes to mantle feathers, and 'tramlines' over back. Spotting on upper breast. Black crown with paler stripes, forming diagnostic 'forked' supercilium.
3b Head adult breeding: Black crown-stripes. Back and scapulars black, fringed rufous and white, neck and breast streaked with black, white below.
3c Adult non-breeding: Pale crown-stripe, broad white supercilium and darker eye-stripe form diagnostic, double, 'forked' supercilium.
3d Adult non-breeding in flight: Thin white wingbar; narrow white sides to rump.

4 CURLEW SANDPIPER *Calidris ferruginea* 18-23 cm **Text page 218**
A dumpy but elegant, sandpiper with white rump, black legs and long downcurved black bill.
4a Adult non-breeding in flight: Bold white wingbar and white rump contrast with black tail, diagnostic.
4b Adult non-breeding: Pale grey above, white below. Prominent white supercilium.
4c Immature: Darker than adult, with 'scaly' back and wing-coverts, and a buff flush to the crown, back and particularly the upper breast.
4d Adult male breeding: Back and scapulars mottled black, white and rufous, wing-coverts grey. Face, neck and underparts rich brick red.

5 RUFF *Philomachus pugnax* male: 26-30 cm; female: 20-25 cm **Text page 220**
Distinctive silhouette: fairly long-neck, small head and pot-belly. Female much smaller than male.
5a Adult non-breeding in flight: Scaly black and buff upperparts contrast with dark flight feathers. Broad white edges to rump.
5b Adult male non-breeding: Grey-brown above with dark feather centres giving a 'scaly' look. Dark bill with pale base, straight with a noticeably downcurved tip. Leg dirty yellowish green to dull orange.
5c Immature female: Brighter than adult: dark feather centres and rufous fringes give a 'tortoiseshell' effect to back and wing-coverts. Crown rufous, breast with warm sandy buff flush.

6 BUFF-BREASTED SANDPIPER *Tryngites subruficollis* 18-20 cm **Text page 219**
A buff-coloured wader with short, straight all-black bill and yellowish legs.
6a Adult in flight: Mottled black and buff above, sides of rump narrowly white. No obvious wingbar. Underwing white (save for dark crescent on primary coverts) contrasting with uniform warm buff breast.
6b Adult: Face, neck, breast and belly warm buff, unstreaked. Upperparts scaly, with dark feather centres fringed buff. Streaked cap. Black eye stands out in plain face. Neat spots on side of breast. Feathering on lower mandible extends further along bill than on lore.
6c Immature: Even more scaly than adult, with darker centres to wing-coverts, fringed whitish buff.

PLATE 28 SKUAS

All skuas are non-breeding visitors to Seychelles. Only Antarctic Skua is annual, while South Polar and Arctic Skuas are vagrants. The status of Long-tailed Skua is uncertain. There have been unconfirmed reports of other species, and a scarcity of confirmed records may in large part be due to a scarcity of pelagic observers. Time of year may be an indicator, southern species (Antarctic and South Polar) being more likely in the austral winter (May to September) and the other species at the opposite time of year.

1 POMARINE SKUA *Stercorarius pomarinus* Wingspan: 125-135 cm   **Text page 222**
A large, powerful skua with broad-based wings, barrel chest, thick neck, two-tone bill, and spoon-shaped central tail feathers (when present).
1a Immature: Grey-brown or black-brown. Plain unstreaked head. Pale hindcollar. Double white wing-patches on underwing. Evenly barred underparts.
1b Pale adult non-breeding: Generally darker than Arctic, breast-band more obvious and larger white wing-flashes, most extensive on underwing. Diffuse brown cap. Underwing-coverts paler than body (unlike South Polar).
1c Dark adult non-breeding: Dark brown with brownish or blackish cap. Paler cheeks and hindcollar. Smaller pale area at base of primaries than pale morph.
1d Pale adult at rest: Can appear small, but is large-headed and heavy-chested compared to Arctic and Long-tailed Skuas. Bill heavier than in Arctic. Pale base to bill, but no pale patch above base.

2 ARCTIC SKUA *Stercorarius parasiticus* Wingspan: 110-120 cm   **Text page 222**
A relatively small skua with small triangular head, rounded underbody, slender wings and slender bill. Pointed tail projections (when present).
2a Immature: Plumage variable black to brown, less grey than Long-tailed Skua. Pale hindneck. Barred underwing, wing-flashes visible above and below. Bill darker at tip.
2b Pale adult non-breeding: Underparts mainly whitish (but variable), sharply demarcated from dark underwing. Diffuse brown cap. Brown breast-band (variable in extent).
2c Dark adult non-breeding: Uniform black-brown to grey-brown including hindneck and sides of neck. Darker flight feathers. Prominent white wing-flashes.
2d Pale adult at rest: Appears gull-like. Fairly narrow bill often with no obvious pale base. Small pale patch above base of bill.

3 LONG-TAILED SKUA *Stercorarius longicaudus* Wingspan: 105-115 cm   **Text page 223**
A graceful skua with long narrow wings and tail and often long tail-streamers.
3a Immature: Mantle and upperwing-coverts grey-brown. Paler hindneck. Whitish belly and darker breast-band. Tail pointed with rather rounded tail feathers.
3b Adult non-breeding: Variable dark cap, uniformly dark underwing. Pale underparts with darker barring on breast and flanks. Shorter tail-streamers than breeding (sometimes absent).
3c Adult breeding: Cold grey above, black flight feathers and trailing edge to wing. Neat black cap. Little or no white in wing. Pale breast merges to grey belly.
3d Adult non-breeding at rest: Slender tern-like build. Rounded head. Fairly small, two-toned bill.

4 ANTARCTIC SKUA *Catharacta antarctica* Wingspan: 150-155 cm   **Text page 221**
A large, powerful skua with little contrast between warm, dark brown upperparts and underparts.
4a Adult at rest: Wingtips and tail length about equal. Large, powerful black bill.
4b Adult above: Blackish brown, heavily streaked and mottled, lacking contrast.
4c Adult below: Underwing dark brown.

5 SOUTH POLAR SKUA *Catharacta maccormicki* Wingspan: 125-135 cm   **Text page 220**
A medium sized, heavily built skua, usually showing some contrast above and below and a diagnostic pale nape.
5a Pale adult at rest: Wingtips project beyond short tail. Black bill less hooked and less powerful than in Antarctic.
5b Dark adult: Greyer and colder than Antarctic Skua. Pale 'nose-band'. More white in upperwing than Pomarine Skua.
5c Immature: Greyer than adult with smaller white wing-flashes. Pale head and nape.
5d Pale adult below: Pale underparts contrast with dark underwing.
5e Pale adult above: Pale head and nape contrast with dark upperparts. Little pale mottling.

PLATE 29 GULLS

All gulls are vagrants to Seychelles, most likely to occur from December to January. Black-headed Gull is the only small gull known to occur, but many records fail to eliminate other possibilities, particularly Grey-headed Gull and Brown-headed Gull.

1 BLACK-HEADED GULL *Larus ridibundus* 34-37 cm   **Text page 227**
A small gull with slender reddish bill and legs.
1a Adult winter: White head with blackish ear-spot. Bill dull red, tipped black. Legs dull red. Brown eye.
1b Head adult spring: Chocolate brown hood. Bright red bill.
1c First-winter: White head with blackish ear-spot. Bill and legs yellowish with black tip to bill. Dark eye.

2 GREAT BLACK-HEADED GULL *Larus ichthyaetus* 57-61 cm   **Text page 225**
A large, deep-chested gull with long, sloping forehead, powerful, banded bill and white eye-crescents.
2a Head adult spring: Combination of black hood (sometimes acquired by January), banded bill and large size diagnostic.
2b Adult winter: Whitish head with dark mask outlining eye-crescents. Crown to hindneck streaked.
2c Second-winter: More extensive grey wing than first-winter. Bill similar to adult winter.
2d First-winter: Grey mantle and scapulars. Dark ear-coverts. Bill yellow-grey with broad black band at or near tip.

3 HEUGLIN'S GULL *Larus heuglini* 60-70 cm   **Text page 224**
A large, bulky gull with big head and powerful bill.
3a Adult winter (race *taimyrensis*): Larger than nominate race. Dull medium to dark grey upperparts. White head with little streaking. Legs may be yellow or pinkish.
3b Head adult spring (race *taimyrensis*): Dark grey upperparts. White head.
3c Head adult spring (race *heuglini*): Blackish upperparts. White head.
3d First-winter (race *heuglini*): Whitish head and underparts. Dark diamond shapes on scapulars. Solidly dark-centred tertials with white fringes. Black bill. Pinkish legs.
3e Third-winter (race *heuglini*): Much grey in upperparts. Banded bill. Pink (or yellow) legs.
3f Adult winter (race *heuglini*): Slate grey to black upperparts. Greyish streaks on crown and neck, especially lower hindneck. Yellow legs.

4 LESSER BLACK-BACKED GULL *Larus fuscus* 52-60 cm   **Text page 224**
A fairly large gull with relatively slim build and long wings.
4a First-winter: Mottled blackish brown and scaly above. Scapulars poorly marked. Black bill. Pinkish legs.
4b Second-winter: Head and underparts whiter, streaked around eye, crown and hindneck. Pale bill tipped black.
4c Adult winter (race *graellsii*): Dark grey upperparts. Yellow legs. Yellow bill with red gonydeal spot.
4d Adult winter (race *fuscus*; Baltic Gull): Black upperparts.

PLATE 30 GULLS IN FLIGHT

1 LESSER BLACK-BACKED GULL *Larus fuscus* Wingspan: 135-150 cm **Text page 224**
A large but elegant dark-backed gull.
1a Adult: Dark upperwing shows little or no contrast with black wingtips.
1b First-winter: Mottled blackish brown and scaly above, paler on head and neck. Dark mantle contrasts with white rump and tail. Solid black tail-band.
1c Second-winter: Head and underparts whiter, mantle darker, streaked around eye, crown and hindneck.

2 HEUGLIN'S GULL *Larus heuglini* Wingspan: 142-160 cm **Text page 224**
A large, fairly heavy dark-backed gull.
2a First-winter (race *heuglini*): Dark diamond shapes on scapulars. Dark-centred tertials with white fringes. Broad tail-band.
2b Third-year (race *heuglini*): Greyer above. Retains at least traces of wide black tail-band.
2c Adult winter (race *heuglini*): Slate grey to black upperparts.
2d Adult winter (race *taimyrensis*): Duller medium to dark grey upperparts.

3 BLACK-HEADED GULL *Larus ridibundus* Wingspan: 100-110 cm **Text page 227**
A fairly small gull with white leading edge to upperwing and pale underwing.
3a First-winter: Blackish brown trailing edge to wing; white leading edge. White rump and tail with narrow, black subterminal band.
3b Adult: Prominent, extensive white leading edge. Tail white.

4 GREAT BLACK-HEADED GULL *Larus ichthyaetus* Wingspan: 150-170 cm **Text page 225**
A large, deep-chested gull with long, but relatively narrows wings.
4a Adult: Upperwing grey with white leading edge and black subterminal arc across outer primaries. Underwing mainly white. Tail white.
4b First-winter: Grey mantle and scapulars. Pale midwing-panel on upperwing. White tail with sharply demarcated black band.
4c Second-winter: Wing pattern intermediate between first-winter and adult, more extensively dark on wing including primaries and primary coverts than adult. Generally retains at least a partial tail-band.
4d Third-winter: As adult with incomplete remnants of tail-band and less white in tips of primaries.

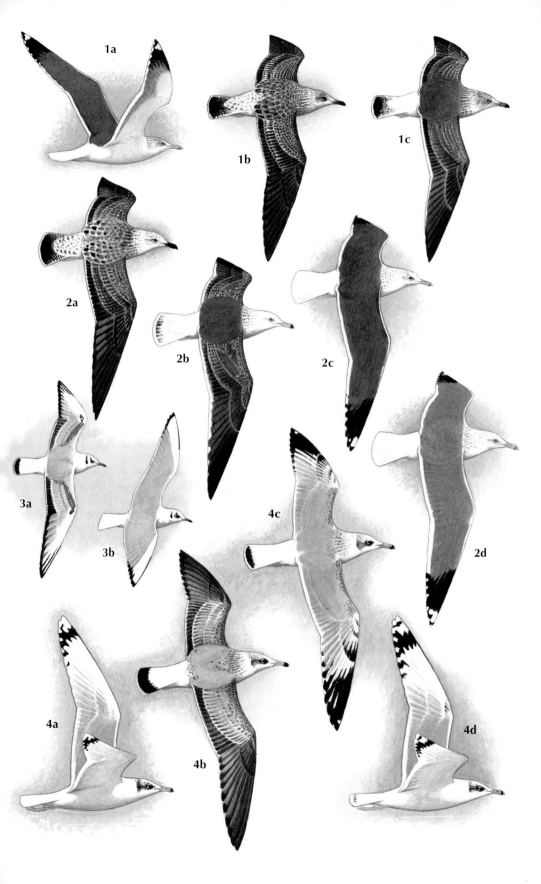

PLATE 31 TERNS 1

Greater Crested Tern occurs throughout Seychelles, breeding on outer islands, notably Aldabra and Cosmoledo. Lesser Crested Tern is a non-breeding visitor throughout the islands, mainly December to April. Sandwich Tern is a Palearctic vagrant, while Gull-billed Tern is annual, both likely around October to December. Caspian Tern breeds at Aldabra and is a vagrant to the granitic islands.

1 GREATER CRESTED TERN *Thalasseus bergii* 46-49 cm   **Text page 230**
A large tern with large yellow bill, big angular head, short shaggy crest and long legs.
1a Adult breeding: Shaggy black crest; yellow bill; long black legs.
1b Juvenile: Scapulars and upperwing feathers tipped dark brown.
1c Colony: Breeds in dense colonies on small rat-free islets.
1d Adult non-breeding: Large size distinguishes from all except even larger Caspian Tern.
1e Head adult non-breeding: Forehead, lore and forecrown white (but less extensive than in Lesser Crested Tern). Bill paler and duller than in breeding adult.
1f Adult in flight: Long, narrow, pointed wings. White-tipped secondaries form pale trailing edge.

2 LESSER CRESTED TERN *Thalasseus bengalensis* 36-41 cm   **Text page 231**
A medium-sized, elegant tern with slender orange bill, rounded head, neat crest and short legs.
2a Juvenile: Brown scalloping on scapulars. Leading edge of wing mottled brown. Yellowish bill. Legs dull orange to dusky.
2b Head adult non-breeding: Forehead and forecrown white, mottled grey. Bill dull orange.
2c Adult non-breeding in flight: Grey rump, uppertail-coverts and tail.
2d Adult breeding: Orange bill, not so heavy as in Greater Crested, and has shorter black legs.

3 SANDWICH TERN *Thalasseus sandvicensis* 36-41 cm   **Text page 231**
A large, graceful whitish tern with long, slender wings, long bill and fairly short legs.
3a First -winter: No pale tip to bill. Narrow black scaling on mantle and scapulars, soon lost with age.
3b Adult breeding: Black cap, prominent crest. Bill black with yellow tip.
3c Head adult non-breeding: White forehead, black crown with white streaks.
3d Adult non-breeding in flight: Pale grey upperwing, white underwing with dark webs and tips to outermost primaries. Upperparts very pale grey but some contrast with white rump and tail.

4 GULL-BILLED TERN *Gelochelidon nilotica* 35-38 cm   **Text page 229**
A large, heavy, gull-like, whitish tern with short, very slightly forked tail, short bill and fairly long legs.
4a Head adult non-breeding: Dark eye-patch. All-black, short, stout bill.
4b Adult breeding: Black cap extends to hindneck. No crest.
4c Adult non-breeding in flight: No contrast between silver-grey upperparts and rump. Deep grey outer primaries. Dark trailing edge to upper and under primaries.

5 CASPIAN TERN *Hydroprogne caspia* 47-54 cm   **Text page 229**
An enormous tern with powerful build, large head, broad wings and large, black-tipped red bill.
5a Colony: Nests in small groups on sand or coral shingle just above high tide line.
5b Head adult non-breeding: Forehead and crown streaked white. Bill duller with broader black ring at or near tip.
5c Adult in flight: Entirely black primaries of underwing contrast with pale grey upperwing. Short, shallowly forked tail.
5d Juvenile in flight: White tail barred grey and brown at tip. Dusky bar on secondaries. Blackish primaries on underwing.
5e Juvenile: Brownish black cap, streaked buff. Bill dull orange-red, legs yellow-brown. Brown scaly marks on mantle, scapulars, tertials and tail.
5f Adult breeding: Black cap; small crest at nape.

PLATE 32 TERNS 2

Roseate Tern breeds on Aride, African Banks, Etoile, and possibly elsewhere, April to August. Common Tern and White-winged Tern are annual in small numbers during the northwest monsoon. Whiskered Tern is a vagrant, and White-cheeked Tern of uncertain status.

1 ROSEATE TERN *Sterna dougallii* 33-38 cm   **Text page 232**
The only medium-sized breeding tern with a black cap and red bill and legs. Elegant with long tail-streamers, and narrow wings.
1a Juvenile: Dark brown nape and ear-coverts. Scaly golden brown saddle. Dark carpal bar. Whitish secondaries. Rosy tinge to upper breast. Black bill and legs.
1b Chick: Fawn above, whitish below often with rosy tinge.
1c Head adult non-breeding: White forehead. Long black bill. Dark carpal bar.
1d Adult in flight: Pale grey upperwing with blackish primaries. Pure white underwing.
1e Adult breeding: Pale grey upperparts and white underparts, often tinged rosy. Long tail-streamers project well beyond tail.
1f Head adult breeding: All-red bill at height of season, half red early and late season.

2 COMMON TERN *Sterna hirundo* 31-35 cm   **Text page 232**
A medium-sized tern, all plumages (except juvenile) having a blackish wedge on primaries of grey upperwing, and whitish underwing with broad, diffuse blackish trailing edge.
2a Adult non-breeding: Dark carpal bar. Tips of tail and wings about equal.
2b Head adult breeding: White wedge between black cap and red bill, tipped black. Underparts greyish white.
2c First-winter: Dark carpal bar. Grey upperparts sometimes with brownish cast. Pale orange base to bill.
2d Adult in flight: Dark wedge in upperwing primaries. Underwing whitish with broad diffuse black trailing edge to primary tips. Grey upperparts contrast with white tail.

3 WHITE-CHEEKED TERN *Sterna repressa* 32-34 cm   **Text page 234**
A medium-sized tern, smaller, shorter and narrower-winged than Common Tern, jizz more like a marsh tern.
3a Adult non-breeding: Long, slender, black bill (longer than in Whiskered Tern).
3b Head adult breeding: Prominent white facial streak contrasts with dark grey upperparts/underparts. Black cap. Red bill.
3c First-winter: Dark carpal bar. Brown scaling on upperparts. Broad, blackish subterminal crescents to scapulars and tertials. Grey rump and tail.
3d Adult in flight: Darker grey upperwing than Common. Upperparts dark grey, concolorous with rump/tail. Dark trailing edge to underwing, paler and more diffuse in midwing-panel.

4 WHITE-WINGED TERN *Chlidonias leucopterus* 20-23 cm   **Text page 228**
A small, compact marsh tern without a prominent tail fork.
4a Adult non-breeding: Head white with blackish ear-coverts and dark-streaked crown.
4b First-winter in flight: Grey back with dark brown marks on upper mantle and scapulars contrasting with white (or very pale grey) rump and pale grey inner wing.
4c Adult non-breeding in flight: Pale grey wings with contrasting blackish outer primaries and secondaries. Pale grey upperparts contrast with whitish rump.
4d Head adult breeding: Black head and underparts contrast with white wing.

5 WHISKERED TERN *Chlidonias hybridus* 23-25 cm   **Text page 227**
Larger and stockier than other marsh terns. Longer wings than White-winged Black, with heavier, stubby bill. Shorter tail than Common Tern with more shallow fork, and broader, shorter, less angled wings.
5a Adult non-breeding: Solid black mark through the eye and across nape, fairly straight and does not extend below eye. Hindcrown streaked black, forecrown white.
5b First-winter in flight: Upperparts and rump grey. Darker more extensive black on head and crown compared to White-winged Tern. Narrow dusky tail-band.
5c Adult non-breeding in flight: Upperparts very pale and clear for a marsh tern, lacking contrast with rump.
5d Head adult breeding: Dark grey above and below with black cap, contrasting with white cheeks; deep red bill.

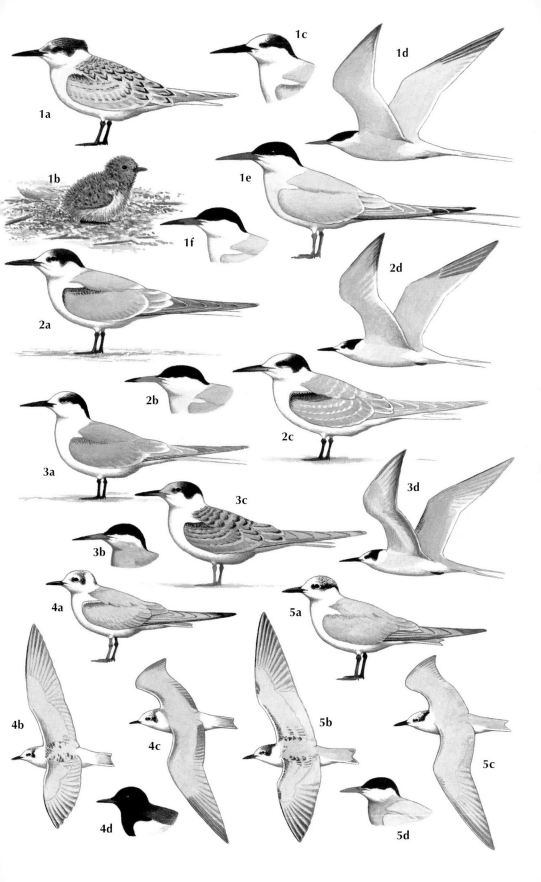

PLATE 33 TERNS 3

Sooty Tern, the most common seabird of Seychelles, breeds at scattered locations throughout Seychelles, present March to October. Bridled Tern breeds in much smaller numbers, mainly in the granitic islands, in a unique eight-month cycle. Saunders' Tern is a common non-breeding visitor August to April. Little Tern also occurs as a non-breeding visitor around the same time.

1 SOOTY TERN *Sterna fuscata* 36-39 cm **Text page 236**
A medium to large, noisy tern, blackish above, white below.
1a Juvenile: Plumage mainly blackish brown, feathers tipped whitish, most heavily on scapulars. Whiter on undertail-coverts.
1b Group in flight: Forms noisy groups over nest sites in breeding season.
1c Reflection from sea surface: Sometimes gives white underside a startling turquoise-blue, parakeet-like appearance.
1d Adult in flight: Long-winged with very long tail-streamers, white at tip. Underwing shows strong contrast between blackish flight feathers and white underwing-coverts.
1e Chick: Blackish brown, paler below.
1f Adult: Rounded white forehead-patch is restricted to area in front of eye. Black loral-stripe narrows from eye to base of bill.

2 BRIDLED TERN *Sterna anaethetus* 34-36 cm **Text page 235**
A small, slim-winged tern; smaller, paler, browner and less uniform above than Sooty Tern.
2a Adult: Pale sepia brown above, greyer on mantle and tail, white below. White forehead extends behind the eye as a narrow, white supercilium. Black loral-stripe does not narrow towards the bill.
2b First-year: Similar pattern to adult, less clear cut, with upperparts darker and slightly scaly and white in outer tail feathers absent or indistinct.
2c Adult in flight: Grey mantle above; underwing less sharply contrasting compared to Sooty Tern.
2d Juvenile: Brownish crown. Feathers of upperparts dark brown tipped whitish. Sides of breast and flanks washed greyish brown.

3 SAUNDERS' TERN *Sterna saundersi* 21-23 cm **Text page 238**
A very small, white, active tern, usually only safely separable from Little Tern in adult breeding plumage, rarely seen in Seychelles.
3a First-winter: Similar to adult non-breeding, head dark with crown distinctly spotted dark, upperparts scaly and with dusky carpal bar. Bill black sometimes with yellowish base.
3b In flight: More extensive black on primaries than Little Tern, outermost primaries having black shafts. Upperparts paler grey than in Little Tern, lacking contrast with rump and tail (but less distinct in non-breeding).
3c Adult non-breeding: White forehead-patch extends behind eye.
3d Adult breeding: Short square forehead-patch restricted to in front of eye.
3e Face: Oval-shaped white forehead-patch.

4 LITTLE TERN *Sterna albifrons* 22-24 cm **Text page 238**
A very small, white, highly active tern very similar to Saunders' Tern except in breeding plumage.
4a Adult non-breeding: White forehead-patch extends behind eye. White rump.
4b In flight: Black on outermost primaries less extensive than in Saunders', with whitish shafts. Upperparts grey, contrasting with paler rump and white tail.
4c Adult breeding: Triangular white forehead-patch extends to or behind eye.
4d Face: V-shaped white forehead-patch.

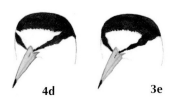

4d   **3e**

PLATE 34 TERNS 4

Brown Noddy breeds throughout Seychelles, mainly during the southeast monsoon, but during the north-west monsoon in the Aldabra group. Lesser Noddy breeds in the granitic islands during the southeast monsoon with smaller numbers on Bird and Desnoeufs. Fairy Tern breeds throughout the islands year round, except in the Aldabra group where it breeds during the northwest monsoon. Black-naped Tern breeds on coral atolls in the outer islands during the northwest monsoon.

1 BROWN NODDY *Anous stolidus* 38-40 cm   **Text page 239**
Larger and stockier than Lesser Noddy with heavier bill.
1a Adult: Greyish white forehead and crown sharply demarcated from black lore.
1b Chick: Generally dark brown.
1c Adult in flight: Underwing dark brown with paler grey centre.
1d Juvenile: Crown less distinct. Pale edges to feathers of upperparts and wing-coverts.
1e Display: Pairs nod to each other.
1f Pale chick: A significant minority are white or intermediate fawn.

2 LESSER NODDY *Anous tenuirostris* 30-34 cm   **Text page 240**
Smaller and more slender than Brown Noddy with grey crown less sharply defined and long, fine bill.
2a Adult: Mainly dark grey-brown with whitish grey crown merging gradually at lore.
2b Nest: Untidy nest, always in trees.
2c Feeding flock: Feeds in dense rafts flying about 2 m above surface.
2d Juvenile: Browner than adult with more contrasting cap.
2e Adult in flight: Underwing uniformly dark, unlike Brown Noddy.

3 FAIRY TERN *Gygis alba* 28-33 cm   **Text page 241**
The only pure white tern in the world.
3a Courtship: Pairs soar and swoop in tandem.
3b Juvenile: Broad brownish tips to feathers of saddle, rump and undertail-coverts; underparts white.
3c Chick: Mainly brown; well-developed claws to cling to bare branch of nest site.
3d Adult: Entirely white except for black eye-ring. Bill black, blue at base.

4 BLACK-NAPED TERN *Sterna sumatrana* 30-32 cm   **Text page 234**
A pale grey and white tern confusable with Fairy Tern in bright light.
4a Adult in flight: Pale upperwing with black outer web to outermost primary.
4b Chick: Mainly whitish below and heavily mottled grey-brown above.
4c Adult: Very pale grey above, white below with black band from eye to eye, broadest at nape.
4d Juvenile: Crown streaked grey-brown. Poorly defined blackish brown eye-stripe. Scapulars mottled grey-brown. Dusky carpal bar.

PLATE 35 TURTLE DOVES

Madagascar Turtle Dove is common throughout the granitics, Bird and Denis, where the nominate race (presumed to be introduced) has interbred with race *rostrata*. The nominate race also occurs in the Amirantes. Race *coppingeri* is common at Aldabra and also occurs on Cosmoledo. European Turtle Dove is a vagrant. The status of Oriental Turtle Dove is uncertain.

1 MADAGASCAR TURTLE DOVE *Streptopelia picturata* 25-32 cm   **Text page 243**
A stocky, ground-feeding dove.
1a Adult in flight: Brownish wings, purple-brown back, grey-brown white-tipped tail.
1b Adult (race *picturata*): Grey head. Mauve-grey above, maroon below, paler on belly.
1c Adult (race *rostrata*): Purple-red head. Smaller than nominate race.
1d Adult (race *coppingeri*): Pale purple head and breast, darker on back, greyish on underparts.
1e Immature (race *picturata*): Upperparts dull sepia brown-grey, wing-coverts brown with rufous tips. Breast dull brown-grey, buff wash on lower breast, belly whitish.

2 EUROPEAN TURTLE DOVE *Streptopelia turtur* 26-28 cm   **Text page 242**
A slender, graceful dove with black-and-white patch on side of neck.
2a Adult: Grey nape and crown. 3-4 black bars on neck usually on whitish background. Scapulars and wing-coverts with broad chestnut borders and small well-defined black centres. Obvious red eye-ring.
2b In flight above: Relatively narrow wings. Blue-grey panel on inner wing. Some brown in blue-grey rump.
2c Immature: Less distinctive and browner than adult. Brownish breast does not extend to belly, feathers with diffuse pale ochre fringes. Brown primaries with diffuse pale edges.
2d In flight below: Breast tinged lilac, belly whitish.
2e Tail: Edged white.

3 ORIENTAL TURTLE DOVE *Streptopelia orientalis* 33-35 cm   **Text page 243**
Longer, darker, broader-winged and heavier than European Turtle Dove with shorter tail.
3a Adult: Brownish nape contrasts with grey crown. 4-6 black bars on neck, usually against bluish background. Scapulars and wing-coverts with narrow rufous borders and large diffusely-defined, grey-black centres. Breast dark brownish pink, lower breast greyish. Indistinct eye-ring.
3b In flight above (race *meena*): Smaller, darker blue-grey patch and two pale wingbars. Rump blue-grey.
3c Immature: Black primaries with more defined rufous edges. Brownish breast may extend to belly, feathers with orange-pink fringes. Browner nape, darker mantle and bluer rump than European.
3d Tail (race *orientalis*): Edged grey.
3e Tail (race *meena*): Edged white.

PLATE 36 GROUND DOVE, BLUE PIGEONS AND PARROT

Seychelles Blue Pigeon is endemic to the larger granitic islands. Comoro Blue Pigeon, the Aldabra equivalent, is an endemic race found only on Aldabra. Seychelles Black Parrot is endemic to the Praslin group, occurring on both Praslin and Curieuse. Barred Ground Dove is a common introduced resident in the granitics and some outer islands.

1 BARRED GROUND DOVE *Geopelia striata* 20-23 cm **Text page 245**
A small, ground-feeding, long-tailed dove.
1a Adult in flight: Small size and long tail. Tangerine webs to flight feathers.
1b Adult: Greyish above, pink below with heavy black-and-white barring on sides of head, neck and flanks.
1c Juvenile: Paler and less tidy than adult with shorter tail.
1d Display: Bows, tail feathers raised and spread.

2 SEYCHELLES BLUE PIGEON *Alectroenas pulcherrima* 23-25 cm **Text page 246**
A medium to large, sturdy, broad-winged arboreal pigeon of the granitics.
2a Display: Plume feathers of head and neck fluffed out.
2b Display flight: Rises high above canopy then plummets steeply, wings held rigid, forward and downward.
2c Adult: Dark blue with whitish bib. Bare red skin around eye and top of head.
2d Juvenile: Dark grey head and bib; dark pink wattle around eye. Greenish brown with pale yellow or buff fringes to many feathers.

3 SEYCHELLES BLACK PARROT *Coracopsis (nigra) barklyi* 26-31 cm **Text page 248**
The only parrot occurring naturally in Seychelles, confined to Praslin and Curieuse. Often betrays its presence with piercing whistle.
3a Juvenile: Duller and paler than adult. Pale tips to wing feathers. Yellow tinge to bill.
3b Adult: Entirely dark grey-brown. Dark bill slightly paler in breeding season.
3c Adult in flight: Strong flight with intermittent gliding.

4 COMORO BLUE PIGEON *Alectroenas sganzini minor* 24-25 cm **Text page 245**
A medium to large arboreal pigeon of Aldabra.
4a Adult: Deep blue with silvery white head, neck and breast.
4b Display: Plume feathers of head and neck fluffed out.
4c Juvenile: Lacks blue and red of adult, mainly olive green.

5 RING-NECKED PARAKEET *Psittacula krameri* 38-42 cm **Text page 249**
Unmistakable. A large, mainly green parrot with long, graduated tail.
5a Adult male: Rose pink collar and black chin-stripe. Pale blue cheeks and nape. Red bill.
5b Male in flight: Long pointed tail. Swift, direct flight.
5c Head adult female: Collar absent or indistinct pale green. Lacks black chin-stripe and blue on head.

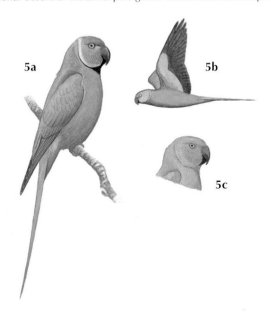

5a

5b

5c

PLATE 37 CUCKOOS AND COUCAL

Cuckoos are vagrants to Seychelles at times of passage (mainly October to December, but also March to April). Common Cuckoo is most frequently recorded, followed by Asian Lesser Cuckoo. Most reports fail to rule out other potential confusion species which should be considered (see SIMILAR SPECIES in main text). Great Spotted Cuckoo and Jacobin Cuckoo are more distinctive, and are easily separated from each other by size and plumage. Madagascar Coucal is an endemic race confined to Aldabra Atoll.

1 COMMON CUCKOO *Cuculus canorus* 32-34 cm   **Text page 251**
A falcon-like cuckoo with long, wedge-shaped tail, and many thin wavy bars on underparts.
1a Adult male: Grey upperparts and neck. Heavily barred belly.
1b Immature: Dark brown head, upperparts and wings; heavily barred on wings and tail, feathers tipped white. Blackish brown tail barred tawny. White spot on nape.
1c Adult female grey morph: Similar to male, tinged brown on breast. (Note, in hepatic phase has rich, chestnut brown upperparts, barred blackish brown, underparts washed rufous.)
1d Male in flight: Tail and rump concolorous with back.

2 ASIAN LESSER CUCKOO *Cuculus poliocephalus* 28 cm   **Text page 252**
Similar to Common Cuckoo but smaller with shorter darker tail and more boldly barred below.
2a Adult male: Fewer, bolder more distinct bars on underparts compared to Common Cuckoo.
2b Immature: Crown feathers tipped brown, nape brown. Flight feathers and scapulars spotted brown; tail generally barred brown.
2c Male in flight: Slate grey back contrasts with darker tail and rump.
2d Adult female: Similar pattern of barring to male. Brownish wash to breast and sides of neck.

3 GREAT SPOTTED CUCKOO *Clamator glandarius* 38-40 cm   **Text page 250**
A large, crested, long-tailed cuckoo.
3a Immature in flight: Crest black and less obvious than in adult. Upperparts darker, spotting duller than in adult, wing darker with prominent rusty patch in primaries. Throat and breast warm buff.
3b Adult: Dark above with bold white spots. Yellowish on throat and breast, creamy white below.

4 JACOBIN CUCKOO *Oxylophus jacobinus* 31-33 cm   **Text page 250**
A medium-sized, crested, pied, long-tailed cuckoo.
4a Adult in flight: White windows in black wings. White tips to long, black tail feathers.
4b Adult: Black above, white below with a prominent white patch at base of primaries. Long, graduated tail is black with white tips.
4c Juvenile: (Only vestiges of this plumage likely in Seychelles.) Upperparts dark brown, underparts washed buff.

5 MADAGASCAR COUCAL *Centropus toulou insularis* 38-41 cm   **Text page 253**
A large, long-tailed bird confined to Aldabra. A reluctant flier.
5a Adult in flight: Broad chestnut wings and long black tail.
5b Adult non-breeding: Dark greyish brown, browner on wings, streaked and spotted whitish. Tail with narrow chestnut bars.
5c Adult breeding: Glossy black above and below with chestnut wings. Long black, fan-shaped tail.
5d Immature: Blackish upperparts, speckled creamy on mantle and crown. Wings cinnamon-rufous barred dusky brown. Dusky underparts.

PLATE 38 OWLS AND NIGHTJARS

Seychelles Scops-owl, confined to Mahé, is the only endemic owl, considerably smaller than the introduced Barn Owl. Brown Fish Owl is a vagrant of unknown origin (possibly ship-assisted). Madagascar Nightjar is the only resident nightjar, confined to Aldabra. Eurasian Nightjar and Eurasian Scops-owl are vagrants, most probable at Bird and the granitics October to December.

1 BARN OWL *Tyto alba* 33-35 cm   **Text page 254**
A medium-sized owl with pale, ghostly appearance.
1a Adult in flight: Appears very pale overall. White underwing.
1b Adult: Golden buff and grey above with fine white vermiculations and dark grey flecks. Pale golden brown below with dark brown speckling.

2 EURASIAN SCOPS-OWL *Otus scops* 19-20 cm   **Text page 255**
A small grey-brown owl with large head showing ear-tufts, long wings, whitish braces and shoulder-stripe.
2a Adult in flight: Wings appear proportionately long and narrow.
2b Grey adult (race *turanicus*): Silver-grey with some rufous tones above, white braces still obvious. Often sits bolt upright, ears raised.
2c Brown adult: Mainly brown, with some rufous areas and white speckles. Slightly paler below with bold blackish streaks and cross-barring. Tufts sometimes lowered, top of head appearing squarish.

3 SEYCHELLES SCOPS-OWL *Otus insularis* 21-23 cm   **Text page 256**
A small, chestnut brown forest owl with strange, rasping call.
3a Display: Pairs duet to each other on a tree branch.
3b Adult: Dark chestnut brown above, paler below mottled black and pale brown.
3c In flight: Wings appear fairly broad and short.
3d Juvenile: Heavily barred above and below.
3e Nest site: May nest in holes in endemic trees such as bwa rouz.

4 BROWN FISH OWL *Ketupa zeylonensis* 54-57 cm   **Text page 257**
A large, brownish owl with short ear-tufts and short tail.
Adult: Dull brown upperparts. Fine dark streaking on crown, mantle and scapulars. Fine, brown streaks on dull buff underparts.

5 MADAGASCAR NIGHTJAR *Caprimulgus madagascariensis* 21 cm   **Text page 258**
A medium-sized, cryptically coloured nightjar.
5a Male in flight: White wingbar. Top of two pairs of outer tail feathers white.
5b Female in flight: Darker wingbar may be indistinct, spots on outer tail feathers pale buff.
5c Female at rest: Upperparts grey-brown mottled black, upperwing brown. Grey head and neck, pale brown cap, cheeks, chin and throat. Orange-brown line extends from under throat to behind ear.

6 EURASIAN NIGHTJAR *Caprimulgus europaeus* 26-28 cm   **Text page 257**
A large, long-winged, long-tailed nightjar.
6a Male in flight: Conspicuous white patches on wings and tail.
6b Female in flight: Lacks white wing-patches. Tail-patches absent or buffish.
6c Female at rest: Blackish carpal bar. Broad whitish bar in upperwing-coverts.

3e

PLATE 39 SWIFTS

Seychelles Swiflet, the only resident swift, is endemic, breeding Mahé, Praslin and La Digue. Common Swift is a vagrant throughout Seychelles, though many reports fail to rule out other potential vagrants. Other swifts are recorded mainly from the granitics to Amirantes. All migrant records are spread between September to May, with the majority at times of passage, October to December or March to April.

1 SEYCHELLES SWIFTLET *Aerodramus elaphrus* Wingspan: 28 cm   **Text page 258**

A small dark swift, typically seen in small parties hawking for insects early morning or late afternoon.

1a Adult above: Dark grey-brown, slightly paler on crown and rump; blackish, slightly forked tail. Rather broad outer wing.

1b Adult below: Body paler brown than wings.

1c Nest colony: Bracket-shaped nests closely packed together on sloping cave roof. Single white egg.

2 COMMON SWIFT *Apus apus* Wingspan: 42-48 cm   **Text page 260**

A medium-sized mainly dark swift with short, forked tail.

2a Adult below (race *apus*): Blackish brown, paler on the throat and slightly paler on flight feathers.

2b Adult above (race *apus*): Uniform blackish brown. Long, scythe-shaped wings.

2c Adult above (race *pekinensis*): Browner than nominate race with larger white throat-patch and paler forehead and greater contrast between dark outer primaries and inner wing.

2d Immature above: Blacker than worn adult with more prominent throat-patch and white forehead. Contrasting dark eye-patch. Narrow white feather fringes on flight and undertail feathers.

3 PACIFIC SWIFT *Apus pacificus* Wingspan: 48-54 cm   **Text page 261**

A large, elegant, blackish swift with fast, powerful flight, wings swept back and with no obvious carpal joint.

3a Adult below: Underparts grey-brown, appearing scaly at close range. (Note, tail appears pointed when closed.)

3b Adult above: Black to dark brown, sometimes contrastingly blacker on tail and uppertail-coverts. Prominent white rump wrapped around deeply forked tail.

4 WHITE-THROATED NEEDLETAIL *Hirundapus caudacutus* Wingspan: 50-53 cm
          Text page 259

A large, powerful swift; appears cigar-shaped with pointed tail when held closed. Open tail is square with tiny spine-like protrusions visible at very close range (rarely possible in field).

4a Adult above: White forehead and lore. Pale patch on back.

4b Adult below: Dark brown with strongly contrasting white throat and white U-shape on rear underparts.

5 LITTLE SWIFT *Apus affinis* Wingspan: 34-35 cm   **Text page 261**

A small, chunky, sooty black swift. Wings long, but blunter than in other migrant swifts.

5a Adult above: Black with a broad, white rump-patch wrapped around tail. Tail broad and square (more rounded when open).

5b Adult below: White patch on throat. Tail paler than body (sometimes appearing translucent).

PLATE 40 BEE-EATERS, ROLLERS AND HOOPOE

Blue-cheeked Bee-eater is recorded throughout Seychelles at times of both spring and autumn passage. European Bee-eater is a vagrant to the Aldabra group around the same times. Broad-billed Roller is near-annual in the Aldabra group and vagrant elsewhere, usually October to December. European Roller is a vagrant throughout Seychelles October to March. Hoopoe is also a vagrant, unlikely to be confused, though size, colour tone and wing pattern should be noted as an indicator of race.

1 BLUE-CHEEKED BEE-EATER *Merops persicus* 27-31 cm   **Text page 262, 263**
A large, slender, graceful bee-eater with mainly grass-green plumage.
1a Adult: Green crown, concolorous with upperparts. Black face-mask, pale blue supercilium and cheek and whitish forehead. Red eye.
1b Madagascar Bee-eater *M. superciliosus*: 30-33 cm. Brown crown, upperparts less vivid green. Black mask bordered white, underwing buff (not rufous).
1c Immature: Duller than adult, with shorter tail-streamers and dark brown eye; lacks pale blue on head, forehead greenish, yellowish at bill. May retain a scaly appearance until midwinter.
1d Adult in flight above: Bright grass-green throughout. Elongated central tail feathers.
1e Adult in flight below: Underwing rufous, with narrow, dark trailing edge.

2 BROAD-BILLED ROLLER *Eurystomus glaucurus* 29-30 cm   **Text page 265**
A chunky, large-headed roller with large, hooked, bright yellow bill.
2a Adult in flight: Long wings mainly dark blue except for cinnamon forewing.
2b Immature: Duller and more washed out than adult, upperparts browner. Culmen spotted brown.
2c Adult (race *glaucurus*): Mainly cinnamon above, deep lilac below. Dark grey-blue undertail-coverts and vent.

3 EUROPEAN ROLLER *Coracias garrulus* 31-32 cm   **Text page 264**
A mainly pale azure roller with striking wing pattern.
3a Immature: Duller than adult, whitish on face, brown on back. Cheeks streaked brown. Throat and breast washed brownish, streaked white.
3b Adult: Cinnamon upperparts pale blue head, neck and underparts.
3c Adult in flight: Wings flash jet black and vivid blue.

4 HOOPOE *Upupa epops* 26-28 cm   **Text page 265**
A distinctive pinkish or orangey bird with black-and-white barred wings and tail, prominent crest and long, curved bill.
4a Adult (race *epops*): Pinkish brown body; crest tipped black with whitish subterminal band.
4b Adult (race *epops*) in flight: Black-and-white barred wings. White-banded primaries.
4c Male in flight (race *africana*): Extensive white patch in secondaries (female secondaries similar to nominate). Entirely black primaries (in both sexes).
4d Adult (race *africana*): More richly coloured dark orange than nominate. Crest lacks white subterminal band.

5 EUROPEAN BEE-EATER *Merops apiaster* 27-29 cm   **Text page 263**
A bright multi-coloured, elegant bee-eater with long wings and pointed central tail feathers.
5a Adult: Combination of chestnut cap and black-bordered yellow throat diagnostic.
5b Immature: Duller than adult. Greenish tone to crown, nape and back. Short tail-streamers (if present).
5c Adult in flight: Graceful flight, gliding on outstretched wings. Translucent pale rufous underwing with thick, black trailing edge.

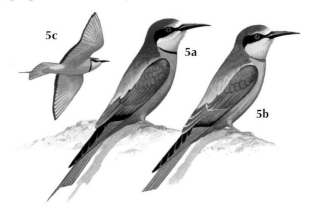

PLATE 41 MARTINS AND SWALLOW

All species are vagrants to Seychelles. Barn Swallow is near-annual, Sand Martin less frequent, and others vagrant. Despite the relative frequency of Barn Swallow, most records received by SBRC have been accepted with a caveat regarding other, albeit less likely, vagrants. Likewise, some incomplete descriptions of Sand Martin and Common House Martin have been accepted with a similar caveat (see SIMILAR SPECIES in main text). Mascarene Martin has usually been seen October to November (the end of the breeding season on Madagascar), and other species throughout northwest monsoon, but most frequently at times of passage.

1 SAND MARTIN *Riparia riparia* Wingspan: 27-29 cm   **Text page 266**
A small brown-and-white martin, often seen over water or in open grassy areas.
1a Adult: Combination of brown breast-band on white underparts, brown upperparts and squarish tail diagnostic. White throat.
1b Immature: Pale edges to feathers of upperparts, most obvious on inner wing-coverts and rump, give paler appearance compared to adult. Buffish grey throat.
1c Adult in flight above: Uniform greyish brown. Slightly forked tail.
1d Adult in flight below: All-dark underwing. White breast with brown chest-band.

2 MASCARENE MARTIN *Phedina borbonica* Wingspan: 30 cm   **Text page 267**
The only martin with boldly striped underparts likely in Seychelles.
2a Adult: Dark brown above, white below streaked dark brown with white vent.
2b Adult in flight: Shallow fork to dark brown tail.

3 BARN SWALLOW *Hirundo rustica* Wingspan: 32-34 cm   **Text page 267**
A blue, white and rufous swallow with long tail-streamers.
3a Adult: Combination of rufous forehead/throat and steel blue upperparts/breast-band diagnostic. Underparts clear white.
3b Immature: Duller than adult with shorter tail-streamers. Throat more buff, underparts may be whiter.
3c Adult in flight below: White breast and belly. Black chest-band, red throat.
3d Adult in flight above: Uniform steely blue-black. White 'windows' in tail-streamers. Deeply forked tail.

4 COMMON HOUSE MARTIN *Delichon urbica* Wingspan: 26-29 cm   **Text page 268**
A dumpy blue-and-white martin. Combination of unstreaked, white rump, bluish upperparts and white underparts diagnostic.
4a Adult: Glossy blue upperparts, blackish tail.
4b Immature: Lacks gloss to upperparts. Tertials tipped whitish. Greyish wash to throat and sides of breast.
4c Adult in flight below: All white breast and throat. Pale underwing.
4d Adult in flight above: Dark blue-black with white rump, which may have greyish wash in winter. Fairly deeply forked tail.

PLATE 42 PIPITS AND SHORT-TOED LARK

Tree Pipit is near-annual during the northwest monsoon. Short-toed Lark and Red-throated Pipit have been recorded as vagrants around the same time. There have been reports of Olive-backed Pipit and Richard's Pipit but their status remains uncertain.

1 GREATER SHORT-TOED LARK *Calandrella brachydactyla* 13-14 cm   **Text page 266**
A small, stocky lark; feeding on the ground and walking with a jerky gait.
1a Adult non-breeding (race *longipennis*): Sandy grey above with fine, narrow shaft streaks to mantle and scapulars, clear white below. Slender black moustache. Smudge at side of neck, sometimes faintly streaked. Shorter bill than in nominate race.
1b Adult non-breeding (race *brachydactyla*): Pale or warm buff above with heavier streaks on mantle and scapulars. Slightly longer bill than race *longipennis*. May appear to have chestnut cap.
1c Adult breeding: More distinct dark patch on side of breast.

2 RED-THROATED PIPIT *Anthus cervinus* 15 cm   **Text page 270**
A robust, medium-sized heavily streaked pipit, often showing a reddish throat. Yellowish or brownish pink legs with fairly long hindclaw. Yellowish base to fine bill.
2a First-winter: Heavily streaked upperparts and rump on brown ground colour. White mantle braces. Creamy white below, with bold streaking on breast and flanks. Black wedge-shape on throat. No contrast between white of breast and belly.
2b Head breeding male: Pale pink to brick red on throat, upper breast, sides of neck and supercilium. Pale brown upperparts.
2c Female winter: Breast, chin and supercilium buffish. Little or no pink on throat. Prominent white mantle-stripes.

3 TREE PIPIT *Anthus trivialis* 15 cm   **Text page 269**
A slender, medium-sized pipit with plain rump. Pinkish legs with short hindclaw. Bill heavier than in Red-throated with pink base.
3a Adult winter: Olive brown upperparts streaked blackish brown. Unstreaked rump. Buff wash to breast, contrasting with paler belly. Pale supercilium and dark eye-stripe.
3b Fresh autumn adult: Warmer buff on throat, breast and flanks contrasting with white belly. Heavily streaked on breast, finer on flanks. Narrow, pale edges to tertials.

4 OLIVE-BACKED PIPIT *Anthus hodgsoni* 14.5 cm   **Text page 270**
A compact, elegant, medium-sized pipit. Compared to Tree Pipit, more striking head pattern, more greenish and less streaked above, more heavily streaked below.
4a Adult winter: Greenish olive wash to crown, mantle and scapulars (olive grey in worn plumage), upperparts only faintly streaked. Greenish olive fringes to greater coverts, tertials and secondaries. Supercilium more prominent than in Tree Pipit, creamy white behind and buff in front of eye with black edge to crown. Distinct white spot and black patch on rear ear-coverts. Thickly striped underparts.
4b Fresh autumn adult: Distinctly brighter than Tree Pipit; very prominent supercilium and breast streaking.

5 RICHARD'S PIPIT *Anthus richardi* 18 cm   **Text page 268**
A large, stocky, long-legged, long-tailed pipit with very long hindclaw and stout bill.
5a Adult: Warm buff breast and flanks contrast with white centre of belly, upper breast streaked. Broad rusty tips to tertials.
5b First-winter: Pale supercilium and lore. Dark eye-stripe, moustachial-stripe and broad, blackish malar-stripe. Wing-coverts with triangular black centres, edged white.
5c Erect stance: Pauses between short dashes to stand erect, neck stretched, occasionally dipping tail.

PLATE 43 WAGTAILS

All wagtails are vagrants, possible throughout Seychelles during the northwest monsoon; several different races of Yellow Wagtail are known to occur.

1 YELLOW WAGTAIL *Motacilla flava* 17 cm   **Text page 271**
A small wagtail with relatively short tail and upright stance. Olive tone to upperparts.
Heads of Spring males:
1a *M. f. simillima* **(Eastern Blue-headed Wagtail):** Grey crown, black ear-coverts and lore, white supercilium.
1b *M. f. beema* **(Sykes's Wagtail):** Pale grey head, broad white supercilium and cheek-stripe.
1c *M. f. leucocephala* **(White-headed Wagtail):** White head.
1d. *M. f. thunbergi* **(Grey-headed Wagtail):** Slate grey crown, blackish ear-coverts, usually no supercilium.
1e. *M. f. lutea* **(Yellow-headed Wagtail):** Yellow head and throat; pale yellow-green ear-coverts.
1f. *M. f. flava* **(Blue-headed Wagtail):** Blue-grey head, yellow throat, white supercilium.
1g. *M. f. feldegg* **(Black-headed Wagtail):** All-black head.
1h Female: Olive green or brownish above; yellow below, whiter on chin/upper breast.
1i Male winter: Olive green above, pale yellow below (brighter in spring).
1j First-winter: Olive brown above, more distinct white wingbar than adult, whiter below; pale yellowish undertail-coverts (may be absent in autumn). May show faint necklace of spots on breast. Pale base to lower mandible. Dark forehead and lore.

2 GREY WAGTAIL *Motacilla cinerea* 18-19 cm   **Text page 272**
The wagtail with the shortest legs and longest tail. Horizontal stance; constantly wags tail. In all plumages, pinkish brown legs (black in all other wagtails) and broad, white wingbars.
2a Female winter: Grey above, contrasting with blackish wings. White on throat, supercilium and sub-moustachial-stripe. Breast yellow, sides to underparts white. Yellow vent.
2b Male winter: Grey above, yellow below (more intense on undertail-coverts) with black throat of breeding male reduced to small speckled area.
2c First-winter: Similar to female but duller. Yellow restricted to undertail-coverts and rump, rest of underparts pale, tinged buffish or very pale yellowish buff on breast. Bill black (note, pink base in juvenile).

3 WHITE WAGTAIL *Motacilla alba* 18 cm   **Text page 273**
A grey, black-and-white, fairly long-tailed wagtail.
3a Female winter: Entire crown and nape grey; lore and cheek mottled dusky.
3b Male winter: Grey above, rear crown black, forehead white mottled blackish. White face and underparts with black bib. White wingbars and white tertial edges.
3c First-winter: Similar to female, dirty yellow-grey on face and sides of breast. Greyish black bib.

4 CITRINE WAGTAIL *Motacilla citreola* 17 cm   **Text page 272**
A small wagtail with grey back lacking any olive tones, broad double white wingbars and usually a yellow head. Never shows any yellow on undertail-coverts.
4a First-winter: Dull ash grey above, with wide wingbars. Whitish below, sides of breast and flanks washed grey. Pale surround to grey cheek; pale lore. Yellow appears first on head in late winter.
4b Female winter: Broad yellow supercilium curves around ear-coverts to join yellow throat and underparts. Pale forehead and lore.
4c Head Spring male: Yellow head with black half-collar. Grey mantle, yellow underparts.

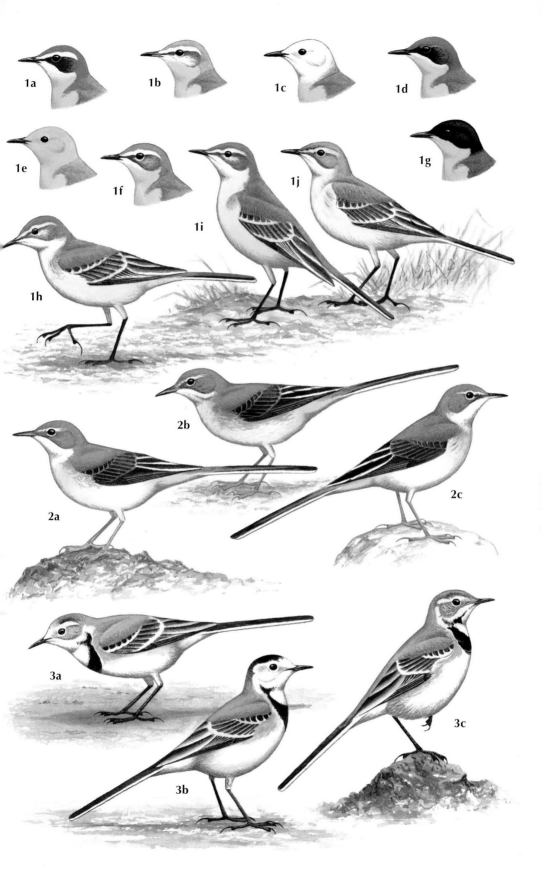

PLATE 44 BULBULS AND ROCK THRUSH

Seychelles Bulbul, endemic to the granitic islands, is a common resident in suitable woodland at all levels, though less frequent at sea level. Madagascar Bulbul is confined to Aldabra where it is fairly common. Red-whiskered Bulbul is introduced to Assumption where it is the most common species. Rufous-tailed Rock Thrush has been recorded as a vagrant around October.

1 SEYCHELLES BULBUL *Hypsipetes crassirostris* 24-25 cm   **Text page 274**
A large, stocky, noisy, mainly grey-brown bulbul.
1a Adult: Grey-brown upperparts with an olive green hue. Paler below. Shaggy, black crest. Bright orange bill. Eye reddish brown.
1b Adult in flight: Dark above, paler below. Pale edge to tail.
1c Juvenile: Similar to adult but with chestnut flight feathers and dark brown bill.
1d Group: Noisy family groups move through trees.

2 MADAGASCAR BULBUL *Hypsipetes madagascariensis* 21-22 cm   **Text page 274**
A slender, medium-sized, grey-brown bulbul.
2a Adult in flight: Dark above, paler below.
2b Juvenile: Duller than adult, browner below with duller browner bill. Chestnut margins to tail/flight feathers.
2c Adult: Dark grey upperparts, browner on wings. Paler below. Shaggy black crest and black lore. Orange bill.

3 RED-WHISKERED BULBUL *Pycnonotus jocosus* 17-22 cm   **Text page 273**
A slender, medium-sized, noisy bulbul with a prominent crest.
3a Adult: Mainly brown above, white below with glossy back crown and erect crest. Red patch behind eye, and red vent.
3b Juvenile: Lighter brown on crest and nape compared to adult, crest smaller, with paler orangey vent and no red patch behind eye.

4 RUFOUS-TAILED ROCK THRUSH *Monticola saxatalis* 18.5-20 cm   **Text page 277**
A small, long-bodied, short-tailed thrush with long wings and longish bill.
4a Female winter: Mottled grey-brown with crescent-shaped barring. Creamy centre to throat.
4b Immature: Similar to adult female, greyer above, less warm buff below. More intensely mottled.
4c Male summer: Blue neck and head, pale bluish white rump, rusty underparts and tail.
4d Male winter: Intensely mottled grey-brown with blue-grey tones on head and upperparts except for centre of back which is whiter. Short orange-red tail.

PLATE 45 ORIOLE AND SHRIKES

Eurasian Golden Oriole, the only vagrant oriole, might occur almost anywhere in Seychelles around October/November or March. Shrikes are also vagrants, which could occur particularly on spring migration with records of Red-backed and Lesser Grey Shrike in the Aldabra group and Woodchat Shrike in the granitic islands. Immature Red-backed may be confused with Isabelline Shrike or Brown Shrike, both potential vagrants. Likewise, Masked Shrike should be considered as well as Woodchat Shrike.

1 EURASIAN GOLDEN ORIOLE *Oriolus oriolus* 24 cm **Text page 299**
A fairly large, long-winged, long-tailed passerine.
1a Head male (race *kundoo*): Blacker around and behind eye than nominate.
1b Male in flight: Contrasting yellow-and-black plumage. Flight rapid, powerful and undulating, wings set forward and long tail obvious.
1c Male (race *oriolus*): Bright yellow with black wings, tail and lore. Crimson eye.
1d Female (race *oriolus*): Greenish above, tinged yellowish on rump, uppertail-coverts and flanks. Underparts very pale grey streaked dull brown, most heavily on breast.
1e Immature: Similar to adult female; more heavily streaked below, greener above. Median coverts tipped yellowish. Bill brown, becoming pinkish later.

2 RED-BACKED SHRIKE *Lanius collurio* 17 cm **Text page 275**
A small, compact, short-billed shrike.
2a Head male (race *kobylini*): Grey nape extends to back; mantle duller and darker. No supercilium.
2b Male (race *collurio*): Blue-grey head, black mask, thin white supercilium. Chestnut back.
2c Male in flight: Black tail with white sides at base, grey rump. Sometimes a small white patch at base of primaries. Flight rapid and agile undulating over long distances.
2d Adult female: Dull rufous brown above, buff below with brown crescent-shaped markings. Warm brown ear-coverts; creamy white supercilium. Dark brown tail tipped and edged white.
2e First-winter: Rusty brown and heavily vermiculated above, pale whitish below with some scalloping. Rufous tail (tipped and edged white) lacks contrast with upperparts. Darker and more barred, with less patterned face than Isabelline Shrike or Brown Shrike (potential vagrants).

3 LESSER GREY SHRIKE *Lanius minor* 20 cm **Text page 276**
A medium-sized, long-winged shrike with erect posture and a thick stubby bill.
3a Head Spring Adult: Pink tinge on breast and flanks. Black mask extends to forehead. Bill all-black.
3b Adult: Black mask; black barring on forehead. Grey back. Black wings with prominent white wing-patch.
3c First-winter: Upperparts may retain some scaling of juvenile. No black on forehead. Bill pale at base.

4 WOODCHAT SHRIKE *Lanius senator* 18 cm **Text page 277**
Small to medium-sized shrike with diagnostic chestnut rear crown and nape.
4a Head breeding male: Smarter than non-breeding, thick solid black mask.
4b Adult non-breeding (race *niloticus*): Chestnut rear crown and nape. Black mask admixed with some chestnut feathers. Black upperparts. White underparts, rump, uppertail and scapulars.
4c First-winter: May retain some juvenile scalloping, especially on the upper breast. Paler rufous on crown and nape than adult, forehead flecked brown-black, mask less distinct. Pale rump, wing-patch and shoulder similar to adult pattern.

PLATE 46 MAGPIE-ROBIN AND ALLIES

Seychelles Magpie-robin was once common throughout the granitics until its extinction on all islands except Frégate. It is now making a comeback under a BirdLife programme involving re-introduction to several islands. All other allied species illustrated here are vagrants to Seychelles, most likely to be encountered in the granitic islands, except for Northern Wheatear, which is more frequent at Aldabra (and also recorded in the granitics), mainly October to November.

1 SEYCHELLES MAGPIE-ROBIN *Copsychus sechellarum* 22-24 cm   **Text page 278**
A black-and-white thrush-like bird, endemic to the granitic islands.
1a Juvenile: Lacks sheen and has ginger edges to some white feathers of wing-patch.
1b Adult: All-black with a deep blue sheen, except for large white patch in wing.
1c Adult in flight: Prominent white wing-patches in otherwise all-black plumage.
1d Adult male display: Puffs out feathers in defence of territory.
1e Association with tortoise: May follow giant tortoises (or humans) feeding on invertebrates disturbed in leaf litter.

2 COMMON REDSTART *Phoenicurus phoenicurus* 14 cm   **Text page 280**
A small, elegant chat with reddish rump and tail, blackish in centre of tail.
2a Male breeding (race *samamiscus*): Brighter and cleaner than non-breeding. Compared to nominate, darker above, deeper orange below with prominent white wing-panel.
2b Male non-breeding (race *phoenicurus*): Ash grey back; black face to upper breast, pale fringes to feathers on throat and forehead; orangey red breast.
2c Immature male: Grey-brown back, grey-black on throat. Pale orange, mottled breast. Faint white supercilium.
2d Female non-breeding: Browner above, paler orange below than male; whitish on throat and belly.

3 WHINCHAT *Saxicola rubetra* 12.5 cm   **Text page 281**
A small, stocky chat with broad supercilium, dark cheek panel with pale lower border and white tail triangles and white wing-panel in adult.
3a Adult non-breeding: Bold white supercilium. Brownish streaked crown and upperparts with buff tips to feathers, paler on rump. Throat and breast orange-buff.
3b Immature: Similar to adult non-breeding, but warmer buff with prominent buffish supercilium, but no white wing-panel.
3c Spring adult male: Sides of head brownish black bordered white above and below. Throat and breast bright orange-buff.

4 ISABELLINE WHEATEAR *Oenanthe isabellina* 16.5 cm   **Text page 282**
A large pale, plain, sandy brown wheatear with short tail and long wings. Upright stance.
4a Adult non-breeding: Uniform plumage, lacking contrast between wings and back. Black alula stands out in pale wing. Sexes similar, male with blacker lore.
4b Tail: Broad black terminal band.
4c Immature: May retain duller brown flight feathers of juvenile, faint mottling on upperparts and dark markings on breast. White in front of eye.

5 NORTHERN WHEATEAR *Oenanthe oenanthe* 14.5-15.5 cm   **Text page 281**
A stocky wheatear, a little darker than Isabelline with shorter legs, slighter bill, longer tail and less upright posture.
5a Tail: Black T-shaped tip.
5b Male breeding (race *libanotica*): Paler grey crown and back compared to nominate; tail-band narrower; buff wash of underparts confined to throat and upper breast. Black wings and mask.
5c Male non-breeding (race *oenanthe*): Back and crown buffish, sometimes with grey tones; pale buff throat and breast. Brownish mask. Long, white supercilium.
5d Female non-breeding: Paler than male with no black on head, shorter supercilium, buff in front of eye. Wing dark brown.

PLATE 47 WARBLERS (1)

Seychelles Warbler is endemic, breeding on Aride, Cousin and Cousine. Madagascar Cisticola is resident on Cosmoledo and Astove. Sedge Warbler and Blackcap are both vagrants to the granitic islands around November.

1 SEYCHELLES WARBLER *Acrocephalus sechellensis* 13-14 cm **Text page 285**
A small brownish warbler with long legs and long bill. Often draws attention to itself with melodious call from within vegetation.
1a, 1b Adult: Dull greenish brown above, dingy white below.
1c Head adult: Eye red-brown.
1d Head subadult: Eye grey-brown.
1e Head juvenile: Eye blue-grey.
1f Nest: In fork of branch or between plant stems. One, sometimes two chicks.
1g Aggressive display: Loud scolding where skink or other predator threatens nest.

2 MADAGASCAR CISTICOLA *Cisticola cherina* 12 cm **Text page 282**
A small, highly active, brownish warbler of low vegetation.
2a Female: More boldly streaked on head and upperparts, browner on underparts compared to male.
2b Male: Tawny brown above streaked blackish brown. Pale below. White supercilium.
2c Male in flight: Pale brown, unstreaked rump. White tip to tail.
2d Juvenile: Rusty above, with heavy brown streaking. Rusty wash on flanks. Yellow wash to chin, throat and upper breast in female.

3 SEDGE WARBLER *Acrocephalus schoenobaenus* 13 cm **Text page 284**
A small, brown streaked warbler of damp vegetation.
3a Adult: Boldly streaked upperparts, unmarked below. Long, creamy supercilium. Crown heavily streaked blackish, paler brown in centre. Long primary projection.
3b Adult in flight: Rump tawny and unstreaked, contrasting with brown tail.
3c First-winter: Flight feathers fresher than adult in autumn. Usually some fine, dark streaking in upper breast. Inseparable from adult after post-juvenile moult in winter quarters.

4 BLACKCAP *Sylvia atricapilla* 13 cm **Text page 289**
A robust, capped woodland warbler.
4a Head female: Rusty brown crown. Upperparts browner than in male.
4b Male: Black crown to eye. Grey-brown upperparts; sides of head and underparts ash grey.
4c Head first-winter male: Brown feather tips in the crown (especially on the forehead) and dull nape.

PLATE 48 WARBLERS (2)

Hippolais, Phylloscopus and *Sylvia* warblers have been recorded as vagrants during October to December and February to March in the granitics and at Aldabra.

1 ICTERINE WARBLER *Hippolais icterina* 13.5 cm   **Text page 286**
A medium-sized warbler with long wings, longish broad bill, blue-grey legs, sloping forehead and narrow, square tail.
1a Adult spring: Short yellow supercilium, pale lore. Greenish olive above, darker on wings with pale yellow wing-panel. Long primary projection. Yellow underparts.
1b Adult winter: More grey-green above, pale yellowish white below. Indistinct wing-panel.
1c First-winter: Greyer than adult with little yellow and indistinct yellow-buff wing-panel.

2 WILLOW WARBLER *Phylloscopus trochilus* 10.5-11.5 cm   **Text page 287**
A small, slender, long-winged warbler (primary projection equals tertial length) with pale supercilium, usually pale legs and pale base to bill.
2a First-winter: Bright olive brown above, more extensive yellow below than adult.
2b Adult winter (race *acredula*): Pale olive above, whitish below. Long pale yellow supercilium. Some yellow and dusky streaks on throat/breast.
2c Adult spring (race *acredula*): Olive brown above, dull yellow on throat/breast merging to dull white on belly.
2d Adult winter (race *yakutensis*): Grey-brown above, dull white below. Grey breast, white supercilium.
2e Chiffchaff *P. collybita*: 10-11 cm. Darker, more compact, more round-headed, shorter-winged (primary projection about half tertial length). Supercilium less distinct. Legs usually darker.

3 WOOD WARBLER *Phylloscopus sibilatrix* 12 cm   **Text page 289**
Larger and brighter than Willow Warbler with shorter tail. Long primary projection.
3a Adult: Bright yellowish green above. Yellow face, throat and breast sharply demarcated from white belly. Broad, bright yellow supercilium and dark eye-stripe.
3b First-winter: Greyer above than adult. Paler yellow on throat/breast.

4 COMMON WHITETHROAT *Sylvia communis* 14 cm   **Text page 290**
A medium-sized, slender, long-tailed warbler with white throat and eye-ring, pale rufous edges to wing feathers, white edges to tail and pale base to bill.
4a Head Lesser Whitethroat. *S. curruca*: 13 cm. Dark ear-coverts give masked appearance. Dark bill.
4b Adult male (race *icterops*): Grey crown and ear-coverts. Pinkish wash to breast.
4c Adult female (race *volgensis*): Grey-brown crown and ear-coverts. Whitish underparts, darker on flanks.
4d First-winter: Browner above. Eye darker.

PLATE 49 FLYCATCHERS AND DRONGO

Seychelles Paradise Flycatcher is endemic to La Digue and some neighbouring islands (where sometimes breeds). Spotted Flycatcher is near-annual in the Aldabra group and vagrant elsewhere during the north-west monsoon. It is the only migrant flycatcher recorded with certainty, though reports sometimes fail to rule out other possibilities. Aldabra Drongo is endemic to Aldabra.

1 SEYCHELLES PARADISE FLYCATCHER *Terpsiphone corvina*
Male 20-37 cm (tail 7-22 cm); female 17-20 cm   **Text page 291**
An unmistakable endemic, displaying extreme sexual dimorphism.
1a Adult female: Black head, chestnut brown upperparts, wings and tail. Creamy white underparts.
1b Adult male: Entirely black, with a deep purple sheen.
1c Male in flight: Very long central tail feathers.
1d Immature: Similar in pattern to female but dingier and scruffier in appearance.
1e Nest: Exposed at end of hanging branch.

2 SPOTTED FLYCATCHER *Muscicapa striata* 14.5 cm   **Text page 290**
A small indistinctive flycatcher with an upright stance.
2a Head Brown Flycatcher *M. dauurica*: 11.5 cm Large dark eye, prominent white eye-ring and whitish loral-stripe. Unstreaked forecrown. Pale yellow base to lower mandible.
2b Adult (race *neumanni*): Grey above, slightly darker on primaries and tail; streaked on crown, upper flanks, breast and lower throat. Pale edges to tertials.
2c Head adult (race *sarudnyi*): Paler more sandy brown upperparts. Streaking less distinct.
2d Immature: Broad pale margins to wing feathers.

3 ALDABRA DRONGO *Dicrurus aldabranus* 24-26 cm   **Text page 297**
An all-dark bird with long forked tail, confined to Aldabra.
3a Adult: Entirely black with slight greenish blue sheen, duller on flight feathers and tail. Red eye.
3b Immature: Grey-brown above, paler below. Brown eye.
3c Nest: Neat cup-shaped construction, often over or near water.

PLATE 50 WHITE-EYES AND SUNBIRDS

Seychelles White-eye is endemic to Mahé and Conception. Madagascar White-eye occurs in the Aldabra group with separate races on Aldabra, Cosmoledo and Astove. Seychelles Sunbird is a very common endemic found throughout the granitic islands. Souimanga Sunbird occurs on Aldabra Atoll. Abbott's Sunbird, often considered a race of Souimanga, occurs on Assumption (the nominate race) and Cosmoledo/Astove.

1 SEYCHELLES WHITE-EYE *Zosterops modestus* 10-11 cm   **Text page 296**
A small grey-brown bird, most active early morning and late afternoon, often drawing attention by pleasant, soft contact call.
1a, 1b Adult: Upperparts olive grey to brown. Broad white eye-ring. Bill blackish
1c Flight: Weak undulating flight, birds calling to each other.

2 MADAGASCAR WHITE-EYE *Zosterops maderaspatana* 10 cm   **Text page 295**
A small, yellowish green bird that moves within and between trees constantly calling softly.
2a Adult (race *aldabrensis*): Aldabra. Yellow forehead, chin and throat. Upperparts and wings dark olive green. White eye-ring. Underparts pale grey, yellow vent and undertail-coverts. Eye reddish brown.
2b Adult (race *maderaspatana*): Astove. Greener above than Aldabra birds.
2c Head Juvenile: Eye greyish brown. Eye-ring tinged yellowish.

3 SEYCHELLES SUNBIRD *Nectarinia dussumieri* 11-12 cm   **Text page 293**
A small, noisy, active, dark, dumpy bird of the granitics, with long, curved bill.
3a Nest: Hanging domed nest with narrow side entrance.
3b Male: Dull slate grey with dark blue iridescence on throat. Yellow pectoral-tufts.
3c Female: Duller than male, lacking pectoral-tufts and iridescence.
3d Male feeding: Note, sometimes shows orange pectoral-tufts.

4 SOUIMANGA SUNBIRD *Nectarinia sovimanga* 10 cm   **Text page 292**
The only sunbird of Aldabra atoll. Active, noisy and common throughout the atoll.
4a Female: Dark grey-brown above. Pale grey below.
4b Male: Head and neck metallic green. Maroon chest-band. Yellow pectoral-tufts. Upper abdomen black. Lower abdomen dull yellowish.
4c Juvenile: Black chin and throat. No maroon band in male.

5 ABBOTT'S SUNBIRD *Nectarinia (sovimanga) abbotti* 11 cm   **Text page 293**
The only sunbird of Assumption, Cosmoledo and Astove. An active, noisy, common species.
5a Male (race *abbotti*): Broad maroon chest-band. Dark brown abdomen.
5b Female (race *abbotti*): Dark brown above, a little paler below.
5c Male (race *buchenorum*): Broad maroon chest-band. Black abdomen.
5d Female (race *buchenorum*): Dark brown above and below, slightly paler on abdomen.

PLATE 51 CROWS AND STARLINGS

Common Myna is introduced and very common throughout the larger granitic islands (except Aride), Bird and Denis. Rose-coloured Starling and Wattled Starling are vagrants recorded from scattered locations. Pied Crow is resident in the Aldabra group. House Crow has been accidentally introduced to the granitics, where control measures have to date successfully limited its spread.

1 COMMON MYNA *Acridotheres tristis* 23 cm **Text page 301**
A common, noisy, distinctive resident of the granitics.
1a Head 'King Myna': Bare yellow skin on head and neck.
1b Adult: Glossy black head and upper breast merges with dark brown body. Yellow skin around eye; yellow bill and legs.
1c Juvenile: Duller and paler than adult, lacking gloss.
1d Adult in flight: Large white wing-patches and white tip to tail.

2 ROSE-COLOURED STARLING *Sturnus roseus* 21.5 cm **Text page 300**
Typical starling in structure and behaviour. Unmistakable in adult plumage.
2a Adult non-breeding in flight: Pale rump and back contrast with black wings and tail.
2b Adult non-breeding: Grey-brown pinkish-tinged body. Shaggy crest.
2c Immature: Sandy brown upperparts, paler on rump and underparts, becoming whitish on throat. Dark brown wings with buffish fringes. Bill pale yellow, darker at tip.
2d Adult breeding: Pink body, black head, wings and tail. Long, drooping crest (shorter in female).

3 WATTLED STARLING *Creatophora cinerea* 19-20 cm **Text page 300**
A greyish starling with pointed dark wings and pale rump.
3a Head male breeding: Bright yellow head with black wattles on forehead, chin and centre of crown.
3b Male non-breeding in flight: Prominent white rump contrasts with black flight feathers and tail.
3c Male non-breeding: Pale grey body contrasts with black wings and tail. Greater coverts and primary coverts whitish. Pale pink bill, black at base.
3d Female non-breeding: Little or no white in blackish brown wing.
3e Immature: Brownish grey body, dark brown wings. Yellowish behind eye and on side of throat. Brownish bill.

4 PIED CROW *Corvus albus* 45 cm **Text page 299**
A medium-sized, black-and-white crow. The only crow in the Aldabra group.
4a Adult in flight: Black and white contrast obvious.
4b Adult: Glossy black head, neck and throat. White hindneck, upper back and belly. Lower back, tail and wings black.
4c Juvenile: Lacks gloss on black head. Some white feathers tipped black.

5 HOUSE CROW *Corvus splendens* 41-43 cm **Text page 298**
A medium-sized, mainly black crow. The only crow in the granitics.
5a Adult: All-black with deep purple gloss, except for grey patch on neck and upper back.
5b Juvenile: Grey patch darker above and more restricted below than in adult. Lacks gloss, more dark brown-black.

PLATE 52 FINCHES AND ALLIES

Common Waxbill is resident on Mahé, La Digue and Alphonse. House Sparrow is resident in the Amirantes with occasional records from the granitics. Yellow-fronted Canary is resident on Assumption. Common Rosefinch and Ortolan Bunting are vagrants to the granitics, likely around October to November. The latter species needs to be separated from Cretzchmar's Bunting, a potential, albeit less likely, vagrant.

1 COMMON WAXBILL *Estrilda astrild* 11-13 cm **Text page 304**
A small finch of open, grassy areas, usually seen in flocks.
1a Adult: Bright red bill and eye-stripe. Grey-brown above with dark brown bars. Pale buff below, whitish on chin with fine black bars on flanks and breast and pinkish wash. May have red patch in centre of belly.
1b Juvenile: Blackish bill. Fainter barring than adult with very pale pink wash.
1c Feeding flock: Feeds in tall grasses, constantly twittering.

2 HOUSE SPARROW *Passer domesticus* 14-15 cm **Text page 309**
A familiar bird, commensal with man, represented in Seychelles by race *indicus*.
2a Female Madagascar Fody: Smaller with more slender legs, lacks sandy brown in upperparts and white patch in wing.
2b Male breeding: Distinctive grey-and-chestnut head pattern with black lore and bib. Warm brown upperparts with black streaking. Grey-brown rump. Greyish underparts. Black bill.
2c Female: Grey-brown above, streaked blackish on upperparts. Underparts pale brown, whitish in centre. Brown bill.

3 YELLOW-FRONTED CANARY *Serinus mozambicus* 11.5 cm **Text page 303**
A small arboreal finch usually in small parties.
3a Adult male: Bright yellow forehead and supercilium, underparts and rump. Greenish wash to sides of breast. Upperparts greenish brown with dark brown streaks. Tail black edged yellow. Black eye-stripe and moustachial-stripe.
3b Head adult female: Duller and greyer than male with less striking face pattern.
3c Juvenile: Similar to adult female, with dull greenish yellow rump and only a trace of yellow on underparts. Dark flecking on breast.

4 COMMON ROSEFINCH *Carpodacus erythrinus* 14.5-15 cm **Text page 303**
A medium-sized stocky finch with round head and heavy bill, showing little contrast above and below.
4a Head male breeding: Bright red head and breast.
4b Male non-breeding: Uniform grey-brown, slightly paler on underparts; faintly streaked. Reddish crown tinged brown. Buffish brown wing-coverts.
4c Adult female: Uniform grey-brown above, slightly paler below. Streaked brown on forehead and forecrown, fainter on chin and throat, heavier on breast and upper flanks. Grey-white wingbars. Pale grey-brown bill.
4d Immature: Similar to female, tinged olive, more heavily spotted. Brown bill tinged pinkish at base. Pale buff double wingbars.

5 ORTOLAN BUNTING *Emberiza hortulana* 16-17 cm **Text page 302**
A streaked, brown, dumpy bunting with a yellowish throat and pale yellowish eye-ring.
5a Head first-winter Cretzchmar's Bunting *E. caesia*: 16 cm. Whiter throat and eye-ring than Ortolan. (Note also, often rufous tinge to rump and vent.)
5b First-winter: Head grey-brown with pale yellow throat and yellowish buff eye-ring. Very bold streaking on mantle, breast, upper flanks and malar-stripe. Broad reddish brown edges to outer tertials, widening towards centre.
5c Female non-breeding: Duller, paler, less olive, more browner and streakier than male. Pale yellow-buff throat (never rusty orange as in Cretzchmar's). Brownish rump. Pale buff vent.
5d Head male breeding: Olive grey head and breast-band surrounds yellow throat and moustache.
5e Male non-breeding: Pale yellow throat, olive brown malar-stripe, yellow eye-ring, pinkish bill and pinkish buff underparts. Boldly streaked mantle. Head and upper breast greyish olive, belly and flanks orange-brown.

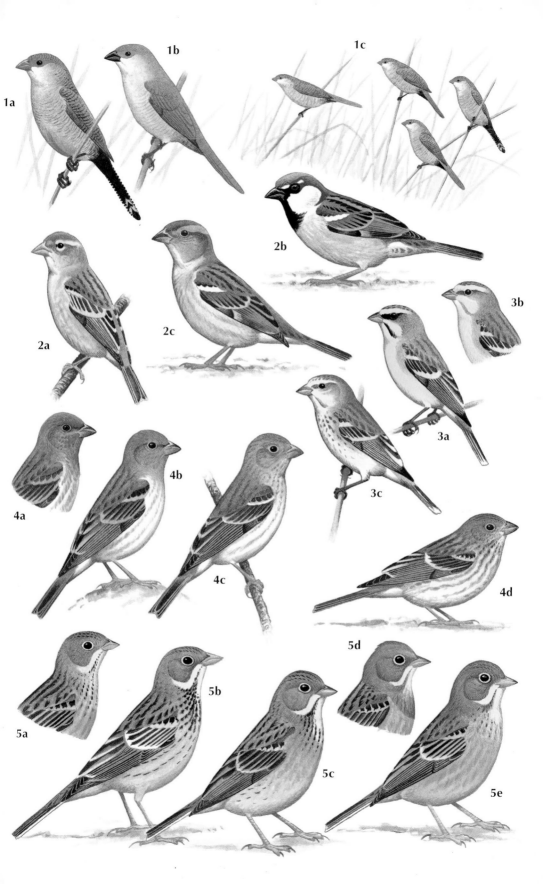

PLATE 53   FODIES

Madagascar Fody occurs throughout the granitic islands and on some of the outer islands. Aldabra Fody is endemic to Aldabra Atoll. Seychelles Fody is endemic to the granitic islands occurring on Cousin, Cousine and Frégate and is introduced on D'Arros.

1 MADAGASCAR FODY *Foudia madagascariensis* ·12-13 cm   **Text page 304**
The 'sparrow' of Seychelles, being by far the most common small passerine in the granitics.
1a Male breeding: Bright red with brown wings and tail. Black mask. (Note, often greyish on belly and vent even at height of breeding season.) Black bill.
1b Feeding flock: Attracted to seeding grasses or spilt food.
1c Male moulting: Retains variable amounts of dull red in grey-brown plumage. Bill becoming pinkish brown.
1d Male display: Fluffs out feathers, most exaggerated on back and rump and on lower breast and belly.
1e Female: Darker olive brown and more streaked above. Uniform paler brown below. Bill pinkish brown.
1f Flavistic male: Scarlet replaced by golden yellow.
1g Male in flight: Red body and rump contrast with brown wings and tail.

2 SEYCHELLES FODY *Foudia sechellarum*   13 cm   **Text page 308**
A small, dark brown dumpy endemic fody.
2a Male in flight: More compact than Madagascar Fody. May show small, white patch in wing.
2b Male breeding: Dark brown, darker on back and wings. Yellow crown, face and chin.
2c Male display: Fluffs out breast and back feathers, beating wings high above back, head outstretched.
2d Female: Similar to male, lacking yellow markings.
2e Predation: Predates Fairy Tern (in particular) by rolling unattended eggs from branch.

3 ALDABRA FODY *Foudia (eminentissima) aldabrana*   15-16 cm   **Text page 306**
The only fody of Aldabra.
3a Male breeding: Scarlet head and breast sharply demarcated from yellow belly. Black mask. dark olive brown back and wings with heavy, black streaking.
3b Male display: Fluffs out feathers of breast and rump.
3c Female: Yellow supercilium. Yellowish brown head, upperparts and rump, heavily streaked on back. Underparts paler yellowish grey. Undertail-coverts yellowish.
3d Male breeding in flight: Red head and rump, brown wings.
3e Male moulting: Scruffier, retaining some red feathers.

BLACK-NECKED GREBE *Podiceps nigricollis*   Plate 2

Other name: Eared Grebe   **French:** Grèbe à cou noir   **Creole:** Greb Likou Nwanr

DESCRIPTION Length 28-34 cm; wingspan 56-60 cm. A medium-sized grebe with a distinctive head shape reminiscent of a peaked cap, a high forehead and a sharp, upturned bill. **Adult non-breeding** has grey-white flanks, contrasting with black back, white breast and grey neck. Head pattern is distinctive, with arc of white up either side of nape and white chin, leaving a black mask over the face from forehead to lower ear-coverts. In flight, it is the only grebe with white secondaries but no white patch on the forewing. Legs grey-green on inner edge brown on outer. Bill dull grey, sometimes with white tip, eye blood red. **Adult breeding** has chestnut flanks contrasting with black back, breast, neck and head, a spray of yellow plumes forming an arc from the eye to the lower ear-coverts; black bill. **Immature** is similar to non-breeding adult, with some dark chestnut streaking on sides of neck and ear-coverts, plus grey band across breast.
VOICE Generally silent outside breeding season. Alarm call 'whit-whit-whit'.
BEHAVIOUR Frequents inland waters and in winter, sheltered bays or estuaries. Gregarious outside breeding season, but vagrants are likely to be solitary. Flight weak and fluttering over short distance but strong over long distance, long neck drooped below body, large feet projecting. Dives for food (immersing directly with no initial jump) but will also catch insects from the surface. Relies less on fish than other grebe species.
RANGE Nominate race *nigricollis* has a scattered breeding distribution in Europe from Spain to Denmark, but the main breeding areas are Ukraine to Kazakhstan, wintering southward to East Africa and east Asia. Race *gurneyi* breeds southern Africa and is mainly sedentary; it is smaller and greyer above with variably rufous lesser wing-coverts. Most Asian birds winter Turkey, Egypt and Persian Gulf. Another race breeds southwest USA.
STATUS Vagrant to granitic islands and Platte around December. Only the nominate race, by far the most likely, has been recorded with certainty.
SIMILAR SPECIES No other grebe recorded but Little Grebe *Tachybaptus ruficollis* and Great Crested Grebe *P. cristatus* are potential vagrants. **Little Grebe** (L25-29 cm, WS40-45 cm) nominate race breeds Europe to Caspian Sea, wintering Turkey to Iraq. Small and stubby, with ochre coloration to cheeks, neck and breast, and all-dark wings. Race *capensis* breeds sub-Saharan Africa, Madagascar, India, Sri Lanka, and is more rufous with extensive white in the secondaries to inner primaries. **Great Crested Grebe** (L46-51 cm, WS85-90 cm) nominate race breeds Europe and Asia. Russian birds winter along coasts of Iraq, Iran and Pakistan and across northern India. Large, with long neck, long dagger-like bill and prominent white on the inner wing and secondaries. Breeding adult has black crest and chestnut and black frills on sides of head. Race *infuscatus* breeds discontinuously from Ethiopia to South Africa. Mainly sedentary, but moves in response to rains. Lacks white lore of Palearctic nominate race and maintains breeding plumage year round.

WANDERING ALBATROSS *Diomedea exulans*   Illus. p. 140

Other names: White-winged (or Snowy) Albatross   **French:** Albatros hurleur   **Creole:** Albatros Vagabon

DESCRIPTION Length 110-135 cm; wingspan 275-305 cm. A large, white albatross with narrow black margins to underwing, massive pink bill and feet projecting well beyond tail in flight. **Adult** has white head and body. Long wings are also white with black flight feathers and a variable amount of black on outer wing-coverts, decreasing with age. Underwing white with narrow black margins to rear wing and forewing and black wingtip. Often shows black edges to white tail. Pink legs and a massive pink bill with yellow tip. Sexes similar, female showing less white in upperwing and retaining some dark marks on body, tail and head. **Immature** has chocolate brown body and a white face-mask. Underwing as adult, a constant identification feature. Underparts progressively whiten with age, becoming speckled. Above whitens on mantle and back then cap then tail, some white present at all ages, upperwing developing a white wedge, gradually extending to the shoulder. May take 20 years or more to achieve full adult plumage.
VOICE Normally silent; may utter a harsh croak when competing for food.
BEHAVIOUR Pelagic and generally solitary at sea. Flight effortless, gliding, soaring and banking on stiff, outstretched wings, legs projecting well beyond tail. Often follows ships. Feeds mainly on squid, taken from surface.
RANGE Circumpolar, breeding throughout islands of Antarctica, including south Indian Ocean islands of Marion, Prince Edward, Crozet and Kerguelen. Normal range outside breeding season to Tropic of Capricorn. Breeders disperse December, but only breeds every other year, so non-breeders might occur anywhere in any month. Taxonomy complex, but two races generally recognised, *dabbenena* on Gough and Tristan da Cunha and the nominate elsewhere.
STATUS Uncertain; reported from Aldabra, but the possibility this may have been some other species of albatross could not be ruled out[1].
SIMILAR SPECIES Size and underwing pattern distinguish from **Southern Giant Petrel** and **Northern Giant Petrel**. **Amsterdam Albatross** *D. amsterdamensis*, (L107-122 cm, WS300 cm) is a potential vagrant (possi-

bly conspecific); breeding in very small numbers on Amsterdam Island in southern Indian Ocean, record-ed as a vagrant to Rodrigues. As large as Wandering, but wholly dark brown above with white face, diag-nostic dark-tipped pink bill and pure white below save for dark breast-band and tail.
Reference: 1. Travis (1959).

Wandering Albatross (left), and Black-browed Albatross

BLACK-BROWED ALBATROSS *Diomedea melanophris* Illus. above

Other name: Black-browed Mollymawk **French:** Albatros à sourcils noirs **Creole:** Albatros Soursi Nwanr

DESCRIPTION Length 80-95 cm; wingspan 213-246 cm. In mollymawks (smaller dark-backed albatross-es), feet do not extend beyond tail. Underwing pattern important to distinguish species at long range, and bill colour at closer range. **Adult** has wings and saddle dark brown-black. Head, rump and body white with dark grey tail. Underwing mostly black, with white axillaries and greater coverts, forming a white inner panel. Bill pale tangerine-yellow with red tip. Legs bluish white. Dark eye-shadow and eye-stripe (hence name), eye brown. Sexes similar, female smaller and shorter-billed. **Immature** has grey flecking on the white underwing-panel, dark grey neck-collar and uniform dirty grey bill. Bill greyish with black tip, gradually acquiring colour of adult. Bill colour and mainly dark underwing distinguish from other grey-headed albatrosses.

VOICE Normally silent at sea except for occasional cries when competing for food, sometimes calling 'waak' 3-4 times.

BEHAVIOUR Pelagic and mainly solitary, but may join flocks of other species of seabird. Highly aerial, gliding effortlessly on stiff, straight wings using opened webbed feet as rudders. Follows ships and is the least shy albatross, sometimes occurring in coastal waters. Dives for food, sometimes submerging, taking mainly squid; often takes offal or discarded food from fishermen.

RANGE Circumpolar, breeding at Kerguelen and Heard islands in Indian Ocean. Rarely recorded within tropics, but vagrant northward to temperate latitudes. Most common albatross off South Africa; vagrant to Kenya. Two races recognised, *impavida* south of New Zealand, and the nominate elsewhere.

STATUS Uncertain; as the most numerous and widespread albatross, it must be considered a likely vagrant.

SIMILAR SPECIES Shy, Grey-headed, Yellow-nosed and Sooty Albatrosses are all potential vagrants, breeding on islands in the southern Indian Ocean. **Shy Albatross** *D. cauta* (L90-99 cm, WS220-256 cm) race *salvini* (Salvin's Albatross; breeds Crozet) has grey head and white cap, all-black primaries and an ivory strip along top and base of grey bill, which is tipped yellow on upper and black on lower mandible. Race *cauta* (Chatham Island Albatross; breeds off Tasmania, wintering off South Africa, vagrant to Kenya and Tanzania) is a more likely vagrant; white cap but head grey only to eye, the bill pale greenish grey tipped black. Both races have diagnostic black wedge where leading edge of underwing meets body and white underwing with narrow black margins in all plumages. **Grey-headed Albatross** *D. chrysostoma* (L81 cm, WS180-220 cm) has a circumpolar breeding distribution including Kerguelen, Crozet and Prince Edward Islands; normally remains below 31°S but sometimes reaches South Africa. A medium-sized albatross with very heavy black margins on leading edge of forewing forming a distinct W, all-grey 'hooded' appearance and distinct black bill. **Yellow-nosed Albatross** *D. chlororhynchos* (L71-81 cm, WS200-256 cm) breeds south Atlantic and southern Indian Ocean (St Paul, Amsterdam, Prince Edward and Crozet islands), dispersing March to August. A slim, black-and-white albatross with long, black bill and tail, narrow trailing edge to underwing and distinct bill pattern. Southern Indian Ocean race *bassi* has grey cheek and throat. **Sooty Albatross** *Phoebetria fusca* (L84-89 cm, WS203 cm) breeds south Atlantic and Indian Ocean (St Paul, Marion, Prince Edward, Crozet and Kerguelen), dispersing April to August, but rarely further north than 30°S. A small, elegant, completely dark brown albatross with long, narrow wings, long, tapering tail and graceful, buoyant flight.

SOUTHERN GIANT PETREL *Macronectes giganteus*   **Plate 3**

Other names: Antarctic Giant Petrel, Glutton, Giant Fulmar, Sea Goose, Nelly, Stinker
French: Fulmar géant or Pétrel géant   **Creole:** Fouke Zean Disid

DESCRIPTION Length 85-100 cm; wingspan 185-205 cm. An all-dark, round-tailed seabird, like a small albatross; flight and proportions like a huge fulmar, but with an enormous, pale, thickset bill. **Dark morph** is mostly dark grey-brown above, typically with clean whitish head and throat and paler breast and underparts darkening towards the rear (though these features are variable). Underwing grey-brown with pale leading edge to forewing usually visible only on underwing, but also on upperwing in worn individuals. Bill pinkish orange with a greenish tip, sometimes appearing all-pale from long range (distinct from the Northern Giant Petrel, in which the darker bill tip is visible at distance). Eye pale grey. Strong black legs; can stand and walk without difficulty, unlike petrels and albatrosses. Sexes similar, female slightly smaller. **White morph** constitutes *c*15% of the population and is all-white with black spotting throughout plumage, pink legs and uniform horn-coloured bill. **Immature** is uniform blackish brown at first with a dark brown eye. Dark eye gradually becomes paler and adult plumage is progressively acquired until first breeding at 6-10 years of age, with some immature features retained even then. Underparts and upperparts generally uniform (Northern Giant underparts paler than upperparts). Bill similar to adult and may be the only means to distinguish many from Northern Giant.
VOICE Silent at sea; sometimes quarrels over food accompanied by loud croaking.
BEHAVIOUR Pelagic and solitary outside breeding season. Gliding, stiff-winged flight, but not as effortless as albatrosses, interspersed with more flapping. A major scavenger of the southern oceans, but unlike other petrels, unique in that it feeds along coastlines like the larger gulls. The huge bill, used to tear off the skin of dead seals, whales and penguins, is also capable of killing smaller seabirds including petrels and prions, which are battered into the sea and then drowned. Smaller female tends to scavenge less, preferring live prey including squid, fish and occasionally krill. Will follow ships when offal is flung into sea.
RANGE Circumpolar, breeding Antarctica, south Chile, south Argentina and east to Heard Island. Adults rarely move from coasts where they feed, but juveniles move eastwards around the southern oceans, regular to 25°S, accidental further north. Immatures disperse further afield and may reach the tropics by following cold water currents north up the coast of South Africa or South America. In the Indian Ocean most frequently recorded off South Africa. Usually considered monotypic.
STATUS Vagrant to granitic islands; recorded July to October.
SIMILAR SPECIES The closely related **Northern Giant Petrel** *M. hallii* (L81-94 cm, WS180-200 cm) is a potential vagrant. It has a circumpolar breeding distribution, including sub-Antarctic islands Marion, Crozet, Kerguelen and Heard, immature birds often straying north. Adult is all-dark brown, with darker underwing than Southern Giant Petrel and dark (not pale) leading edge to inner wing. Underparts uniform dark brown, without any contrast from breast to tail and more contrastingly paler than upperparts. Head pale grey or white, but with some mottling at all ages. Face pale with freckled cheek and darker crown, giving a 'capped' effect. Bill horn with reddish tip, which at distance may give impression of an all-dark drooping bill. No white morph. Immature is all-dark, underparts often somewhat contrastingly paler (unlike Southern Giant immature), sooty black, but still with red-tipped bill. Paler eye than Southern Giant Petrels of the same age. Immature **Wandering Albatross** is larger and is mainly white on the underwing.

CAPE PETREL *Daption capense*   Plate 3

Other names: Cape Pigeon, Pintado, Cape Fulmar, Pied (or Spotted, or Black and White) Petrel
French: Damier du Cap   **Creole:** Domino

DESCRIPTION Length 37-40 cm; wingspan 83-88 cm. Striking black-and-white coloration renders virtually unmistakable. Flight similar to a fulmar but with more rapid wingbeats. **Adult** has black head, pure white underbelly, black tail and black leading edge to wing, back mostly white, the bulk of the wing black flecked with white, with a large white 'window' at the base of the primaries. Bill and legs black. Sexes similar, female slightly smaller than male. **Immature** is similar to adult but black parts of plumage may look greyish.
VOICE Generally silent at sea. An accelerating fulmar-like cackle: 'kak-kak-kak-kk-kk-kk-kkr'.
BEHAVIOUR Pelagic, occurring in southern oceans in flocks of thousands where food is abundant but vagrants are likely to be solitary. Flight fulmar-like, with bursts of wingbeats, interspersed by gliding, wings held stiffly. Follows ships, particularly fishing vessels, feeding behind trawlers as they haul up their nets. Often alights on the sea at these times, snatching prey from the surface; the jerky head movements of feeding birds earned them the name 'pigeon'. Also associates with feeding whales. Feeds mainly on crustaceans (especially krill) though it will also take jellyfish larvae, squid, fish, offal and carrion.
RANGE Breeds Antarctica, in particular off Antarctic peninsula (nominate race *capense*) and islands of New Zealand (race *australe*; smaller with less white in upperparts in fresh plumage). Nearest breeding colonies to Seychelles are Crozet, Kerguelen and Heard. Starting in March there is an extensive northward dispersal, reaching its furthest extent in August to September. Common visitor to South Africa; the commonest petrel off Cape of Good Hope. Rarely recorded further north than 20°S (Beira, Mozambique) but vagrant to Kenya.
STATUS Vagrant to Aldabra group; recorded September.
SIMILAR SPECIES Unmistakable with good views, but Southern Fulmar *Fulmarus glacialoides* and Antarctic Petrel *Thalassoica antarctica* should be considered as potential vagrants. **Southern Fulmar** (L46-50 cm, WS114-120 cm), a circumpolar breeder, is fairly regular off South Africa between April and January; considerably larger, it resembles a slim-billed fulmar with typical stiff-winged fulmar flight, silver grey upperparts contrasting with dark flight feathers and black-tipped pink bill with a yellowish subterminal band. **Antarctic Petrel** (L40-46 cm, WS101-104 cm) is another circumpolar breeder, dispersing widely from March. A striking dark brown and white petrel, silhouette and flight like a fulmar and distinguished by broad white subterminal trailing edge to wing.

MASCARENE PETREL *Pterodroma aterrima*   Plate 3

Other names: Réunion (or Mascarene Black) Petrel   **French:** Pétrel noir de Bourbon
Creole: Fouke Bourbon

DESCRIPTION Length 36 cm; wingspan 88 cm. An all-dark, slim-winged petrel with a shearwater-like silhouette and short, rounded tail. **Adult** is entirely blackish brown with long, slim wings and a rather short rounded tail. Leg is bi-coloured with flesh-coloured tarsus extending on to the innermost web, the remainder of foot black. Gap between tubenose and thickened bill tip gives a notched appearance to heavy, black bill. Sexes similar. **Immature** description unknown. Probably similar to adult.
VOICE Recorded as making a subdued 'ti-ti' call.
BEHAVIOUR There is almost nothing known about this species.
RANGE Breeds Réunion, perhaps in mountain forest. Until 1970, known only from four specimens collected and preserved at the end of the 19th Century. Then in 1970 and 1974 single birds were captured on Réunion. There have since been a few more sightings, but the breeding haunts have yet to be discovered. Sub-fossil remains suggest it once occurred also on Rodrigues. Considered by some authorities to be a race of Tahiti Petrel *P. rostrata*. Sometimes placed in a separate genus, *Pseudobulweria*. Monotypic.
STATUS Uncertain; there are unconfirmed reports from waters around the southern islands of Seychelles.
SIMILAR SPECIES This and **Jouanin's Petrel** are so rare, any birdwatcher lucky enough to find an all-dark petrel should take careful measurements using the following table of measurements (mm):

	Wing	Tail	Tarsus	Middle Toe
Mascarene	232–253	94–100	36–39	45–47
Jouanin's	232–246	111–124	32–35	36–40

Bulwer's Petrel is smaller than both Jouanin's and Mascarene. Both Bulwer's and Jouanin's have long, tapered tails. Confusion is also possible with **Wedge-tailed Shearwater**, which is larger, has a longer, greyish (not all-black) bill and a wedge-shaped tail.

BARAU'S PETREL *Pterodroma baraui*   **Plate 3**

French: Pétrel de Barau   **Creole:** Fouke Barau

DESCRIPTION Length 38 cm; wingspan 96 cm. A striking petrel with a black 'M' on grey background of upperwing and pure white beneath with thin black bar on white underwing. **Adult** is pale grey above with a broad black 'M' formed by thick carpal bar and black primaries, rump dark grey and tail black. Blackish grey hood and nape contrast with white forehead and chin. Narrow, black strap from eye to cheek contrasts with white face. Underparts pure white, save for grey side collar to the neck, narrow black line from the carpal joint and black margins to the trailing edge of the wing. Eye brown. Short, thick black bill. Legs pink, upper part of foot pink, lower two thirds black. Sexes similar. **Immature** similar to adult.

VOICE Silent at sea.

BEHAVIOUR Pelagic outside breeding season, May to September. Gregarious when gathering in loose flocks offshore at Réunion but in Seychelles waters likely to be solitary. In flight, rapid wingbeats alternate with long glides close to the surface of the water. Sometimes associates with other seabirds around fishing boats but rarely follows ships. Feeds by dipping to or seizing from surface.

RANGE Breeds Réunion, November to April, and a nest also recorded Rodrigues. At this time, breeding birds are restricted to waters around Réunion, non-breeders dispersing southeastward towards Amsterdam Island and Australia. Recorded off Oman, Cocos Keeling, Sumatra, Madagascar and Sri Lanka. Monotypic.

STATUS Uncertain; there are unconfirmed reports from waters around the outer islands.

SIMILAR SPECIES None with this striking underwing pattern, but Soft-plumaged Petrel *P. mollis* and White-headed Petrel *P. lessonii* should be considered potential vagrants. **Soft-plumaged Petrel** (L32-37 cm, WS83-95 cm) breeds Prince Edward, Amsterdam, Kerguelen and Crozet Islands (race *dubia*) and Antipodes, Gough, Tristan da Cunha (race *mollis*), dispersing in June. A dark-winged, grey-backed petrel with distinctive 'M' on back; white underparts, grey neck-collar (thicker in race *dubia*) and dark eye-patch. Typical 'gadfly' flight, with fast wingbeats interspersed with glides. **White-headed Petrel** (L40-46 cm, WS109 cm) breeds Crozet, Kerguelen, possibly Prince Edward and southern islands of New Zealand, dispersing July to November. White head, dark eye-patch and white rump and tail contrast with darker wings. White breast and undertail contrast with dark underwing. Heavy, black bill. In flight wings held bowed, angled forward, swinging side to side in high arcs.

ANTARCTIC PRION *Pachyptila desolata*   **Illus p. 144**

Other names: Dove Prion, Bank's (or Blue) Dove-Petrel, Snowbird, Whalebird
French: Prion de la Désolation   **Creole:** Prion Antarktik

DESCRIPTION Length 25-27 cm; wingspan 58-66 cm.
A small petrel with bold, open M-shape across wings and back, heavy blue bill and blue-grey nape extending to form a dark partial neck-collar. **Adult** is blue-grey above, with a distinctive black 'M' across the wings and a broad black tip to wedge-shaped tail. Longer-winged than other prions. Underwing white with dusky tips and trailing edge. Underparts white with broad, dark grey pectoral band where the nape colour extends onto sides of neck and breast. Black central wedge on the undertail. Dark cap, white supercilium and thick black eye-stripe. Noticeably heavy, pale blue duck-like bill. Feet pale blue with cream webs. Sexes similar. **Immature** similar to adult.

VOICE Silent at sea.

BEHAVIOUR Pelagic outside breeding season. Vagrants are likely to be solitary. Flight rapid, close to surface of sea with much erratic twisting and turning interspersed with occasional short glides in which wings held slightly bowed. Most typical form of feeding is 'hydroplaning' in which birds face into wind with wings fully open, effectively hovering on the sea surface with their feet in the water, dropping to seize food items (small fish and crustaceans) from just below the surface.

RANGE Nominate race breeds on Crozet and Kerguelen Islands in the southern Indian Ocean; other races, much less likely in Seychelles, breed west to the South Sandwich Islands and east to New Zealand. Vagrants recorded off Somalia, Kenya, Mozambique, Madagascar and Mauritius.

STATUS Uncertain; possibly the most likely prion in East African waters, and by extension, probably Seychelles. Prions of unknown species have been reported in Seychelles, most frequently from the Aldabra group.

SIMILAR SPECIES Slender-billed Prion is very similar and is probably the next most likely prion in Seychelles. Dark patches of neck/breast are much less conspicuous, generally much paler, M-shape less distinct, black tail tip narrower, white lore, supercilium more conspicuous and outer tail feathers pale above. Other possibilities include Broad-billed Prion *P. vittata*, Salvin's Prion *P. salvini* and Blue Petrel *Halobaena caerulea*, and these should also be considered. **Broad-billed Prion** (L25-30 cm, WS57-66 cm) breeds St Paul, Tristan da Cunha, Gough, New Zealand and Chatham, and has been recorded off Madagascar and Mozambique. There is a published record of a prion seen off Tanzania in July, thought to be this species[1]. Has a massive black bill, dark cap, black smudge through the eye and a black wedge below on the terminal tail-band. **Salvin's Prion** (L25-28 cm, WS57-58 cm) breeds on sub-Antarctic islands of Marion, Prince Edward, St Paul, Amsterdam and Crozet in the southern Indian Ocean. Heavy, blue bill,

less dark cap, grey nape extends down onto the sides of the breast to form a partial collar. Appears smaller-headed and shorter-winged. **Blue Petrel** (L26-32 cm, WS58-71 cm) breeds sub-Antarctic islands from Cape Horn to south of New Zealand including Marion, Prince Edward, Crozet and Kerguelen Islands and occasionally recorded off South Africa between May and October. Black bill, black hood forming partial neck-collar, blue-grey upperparts with heavy black 'M' across wings and diagnostic white terminal tail-band. Note, taxonomy of prions is uncertain; some authors consider Antarctic, Broad-billed and Salvin's Prion as a single species *P. vittata*.
Reference: 1. Voous (1966).

SLENDER-BILLED PRION *Pachyptila belcheri*   **Illus. below**

Other names: Thin-billed Prion, Narrow-billed Prion   **French:** Prion de Belcher   **Creole:** Prion Labek Fen

DESCRIPTION Length 25-26 cm; wingspan 56 cm. The palest prion with white lore and eyebrow, slim bill (diagnostic, but difficult to observe), thick grey eye-stripe, weak 'M' and white outer tail feathers. **Adult** appears pale grey at sea, with a weakly defined 'M' across the back and wings. Tail has a narrow blackish smudge at the tip, interrupted by white outer tail feathers. Undertail has a blackish central wedge, rest of underparts pure white. Face pattern distinctive and may be discernible at long range with white lore, broad, long white supercilium and grey stripe behind the eye. Bill narrow, rather long and blue-grey. Feet pale blue with cream webs. Sexes similar. **Immature** similar to adult.
VOICE Silent at sea.
BEHAVIOUR Pelagic outside breeding season. Vagrants are likely to be solitary. Flight fast and erratic, rather wader-like. Food mostly small crustaceans; some small fish and squid also taken.
RANGE Breeds on sub-Antarctic islands of Crozet and Kerguelen in the southern Indian Ocean (between September and March). Also breeds south Atlantic and south Chile. Regularly recorded off the coasts of Australia and New Zealand, suggesting birds disperse eastward after breeding. Monotypic.
STATUS Uncertain; recorded East Africa and Madagascar, suggesting this may be the possible identity of some prion sightings in Seychelles waters.
SIMILAR SPECIES Antarctic Prion is very similar. Dark patches of neck/breast are much larger and bolder, plumage generally darker, M-shape more distinct, black tail tip broader and lacks pale on outer tail feathers. Potential vagrants that should be considered include Fairy Prion *P. turtur* and Fulmar Prion *P. crassirostris*, though generally neither disperses very far northward. **Fairy Prion** (L23-28 cm, WS56-60 cm) breeds Falklands to New Zealand including Marion, Prince Edward and Crozet Islands in the southern Indian Ocean. It is distinguished by slim blue bill, 'white-faced' appearance with pale head, back and rump and thick black terminal band above and below tail. **Fulmar Prion** (L24-28 cm, WS60 cm), sometimes considered a race of *P. turtur*, breeds islands off New Zealand and Heard Island in southern Indian Ocean (race *eatoni*, slightly smaller than nominate). Has paler head with heavier bill, similar thick black terminal tail-band, but more distinct 'M' across pale wings.

Adult Slender-billed Prion (two figures on left), and Antarctic Prion

BULWER'S PETREL *Bulweria bulwerii*   Plate 3

French: Pétrel de Bulwer   **Creole:** Fouke Bulwer

DESCRIPTION Length 26 cm; wingspan 67 cm. An all-dark petrel, smaller than Jouanin's Petrel, with smaller head, long wings, distinct, diagonal wingbar and long, tapering tail. **Adult** is all-dark brown above and below save for pale diagonal wingbar, visible at long range and most distinct towards wingtip. Tail wedge-shaped when opened, but appears long, narrow and tapered when closed, similar to Jouanin's Petrel but lacking 'step'. Bill black, short and stubby. Leg colour variable from grey to bright pink. Sexes similar. **Immature** similar to adult.

VOICE Silent at sea.

BEHAVIOUR Pelagic and normally solitary at sea. In flight, weaves and twists close to surface of sea, wings held forward and slightly bowed, bill pointed downward. Rarely follows ships. Feeds at night mainly on plankton.

RANGE Breeds April to September in Pacific, dispersing to Indian Ocean, mainly south of Mascarenes December to early March. There are April records between 0 and 10°S, 50-70°E. A few remain in the Indian Ocean during the breeding season. Monotypic.

STATUS Uncertain; there have been unconfirmed reports during southeast monsoon from the outer islands.

SIMILAR SPECIES Mascarene Petrel is larger, all-dark, with squarer tail and very heavy bill, swollen at the tip. **Jouanin's Petrel** is larger with heavier head, less slender tail with slight 'step', thicker bill and a more sweeping flight in high arcs above the sea. **Matsudaira's Storm-petrel** is wholly dark brown with pale upperwingbar, deeply forked tail, white patch at the primary bases (visible at close range) and slow, lazy flight. **Swinhoe's Storm-petrel** is smaller, with shorter forked tail, shorter wings, greyish head, mantle and underwing-coverts and a faster, more erratic flight.

Top to bottom: (1) Mascarene Petrel, (2) Jouanin's Petrel, (3) Bulwer's Petrel, (4), Matsudaira's Storm-petrel, (5) Swinhoe's Storm-petrel

JOUANIN'S PETREL *Bulweria fallax*   Plate 3

French: Pétrel de Jouanin   **Creole:** Fouke Jouanin

DESCRIPTION Length 30-32 cm; wingspan 76-83 cm. The most frequently reported all-dark petrel in Seychelles waters; larger, more bull-headed and more broad-winged than Bulwer's Petrel, with similar long, tapering tail. **Adult** is uniformly dark brown throughout, showing a pale upperwingbar in worn plumage. Structure is important to differentiate from other all-dark petrels, with long, angular wings and long wedge-shaped tail that is usually held closed to give a long tapering shape with a slight step due to the projection of central tail feathers. Also, heavy, rounded head and thick black bill give a heavier appearance than Bulwer's Petrel. Legs pink, feet flesh-coloured with blackish edges to webs. Sexes similar. **Immature** similar to adult.

VOICE Silent at sea.

BEHAVIOUR Pelagic and solitary. Powerful flight, making wide, sweeping curves high (15-20 m) above the sea, wings held slightly forward. In calm conditions or on take off from sea, flight more direct and skua-like, with continuous steady wingbeats. May rest on sea in middle of day and most sightings in Seychelles waters are from 5 pm to dusk.

RANGE Breeding grounds unknown, but Oman or the offshore Kuria Muria islands are possible sites; birds with down have been found in the desert 70 km north of Salalah, Oman. Thought to fledge November/December. Noted in abundance off island of Masirah, northern Oman March to August. Once considered a race of Bulwer's Petrel, now accepted as a separate species, named by Christian Jouanin only in 1955. Confirmation for this came from parisitology; each petrel has a different species of *Halipeurus* feather louse. Very restricted range, seen most commonly from the Gulf of Aden to the mouth of the Persian Gulf and south to the Seychelles; vagrant Kenya. Monotypic.

STATUS Vagrant; mainly recorded between Aldabra, Farquhar and Amirantes, October to March. Possibly under-recorded and may even be annual.

SIMILAR SPECIES Bulwer's Petrel is smaller, with smaller head, more slender wings and tail (lacking any 'step'), shorter bill and a lower flight over the sea. **Mascarene Petrel** appears shorter-tailed and slimmer-winged with slightly wedge-shaped tail and 'notched' bill. **Swinhoe's Storm-petrel** is smaller, shorter-winged with shorter forked tail (tail fork may not be obvious, appearing square-ended but lacking tapered appearance of Jouanin's) and lacks bounding flight of Jouanin's. **Wedge-tailed Shearwater** is also all-dark brown, but much larger, with longer wings and the heavier flight of a large shearwater. In addition, the following are potential vagrants, all of which have shorter, less tapering tails compared to Jouanin's Petrel: **Great-winged Petrel** *Pterodroma macroptera* (L41 cm, WS97 cm) breeds from south Atlantic to Australia including southern Indian Ocean islands dispersing October to November. Recorded as far north as Zambezi River, Mozambique and vagrant to Madagascar and Somalia. Uniform, dark brown petrel with long, slim wings, short square-cut tail, distinctive pale 'muzzle' at base of short, stout bill very obvious head-on. Erratic, zigzag flight with high wheeling arcs above sea. **Herald Petrel** *Pterodroma arminjoniana* (L35-39 cm, WS88-102 cm) breeds Round Island, Mauritius and possibly also Réunion (as well as elsewhere in the tropical and subtropical Pacific and Atlantic). A long-winged, small-headed petrel occurring in three colour morphs from ashy grey to all-dark brown, all with pale skua-like patches on underwing. Strong, fast flight with, high sweeping arcs over sea. **Kerguelen Petrel** *Aphrodroma brevirostris* (L80-82 cm, WS80-82 cm) breeds Marion, Prince Edward, Crozet and Kerguelen and islands off New Zealand; vagrant Natal and Somalia. A small, dumpy, all-dark petrel with blunt, rounded tail and shiny plumage, especially reflective on underwing. Fast, scudding flight; soaring high above sea, racing glides interspersed with very rapid wingbeats. **Grey Petrel** *Procellaria cinerea* (L48-50 cm, WS115-130 cm) has circumpolar breeding distribution including southern Indian Ocean islands of Marion, Prince Edward, Crozet and Kerguelen; vagrant to Natal and Mozambique. A bulky shearwater-like petrel with large, round-headed, drab grey upperparts and underwing contrasting with white throat, chest and belly and dirty yellow bill with brighter yellow tip. **White-chinned Petrel** *Procellaria aequinoctialis* (L51-58 cm, WS134-147 cm) is also circumpolar in distribution breeding sites including Prince Edward, Marion, Crozet and Kerguelen Islands; recorded as vagrant off Kenya coast. A large, bulky, entirely dark brown shearwater-like petrel, looking black in most lights with heavy, rotund body. White chin and pale bill visible at close range.

FLESH-FOOTED SHEARWATER *Puffinus carneipes* Plate 4

Other name: Pale-footed Shearwater **French:** Puffin à pieds pâles **Creole:** Fouke Labek Pal

DESCRIPTION Length 40-45 cm; wingspan 99-107 cm. An all-dark shearwater with silvery roots to the underwing primaries and a pale horn bill with black tip (diagnostic). **Adult** has body and wings dark brown above and below; primaries appearing the darkest, but with silvery bases to the underwing primaries. Rounded tail is darker than body. Bill colour diagnostic: pale pink-horn with black culmen and tip. Legs flesh-pink, similar to Wedge-tailed Shearwater. In flight, wings are held straighter than Wedge-tailed Shearwater and overall impression is of a heavier, shorter-tailed bird with less angled wings. Sexes similar. **Immature** similar to adult.

VOICE High-pitched calls when squabbling over food.

BEHAVIOUR Pelagic and usually solitary. Flight is more ponderous than Wedge-tailed Shearwater, wing-beats slower and general progress is lower over the water. Makes shallow clumsy dives or snatches food at surface, feeding mainly on squid and small fish. Often follows ships.

RANGE Breeds St Paul, southern Indian Ocean to New Zealand September to April, dispersing May and moving north into Pacific and north and west across Indian Ocean towards Madagascar and Seychelles. Some pass further north into the Arabian Sea area, others may reach as far west as South Africa; vagrant Somalia. Also breeds islands off south coast of Australia and off North Island, New Zealand. Monotypic.

STATUS Vagrant throughout Seychelles, May to October. Probably under-recorded.
SIMILAR SPECIES Wedge-tailed Shearwater (very similar but far and away the more abundant), has dark roots to underwing primaries and dark grey bill, more angled wings, more slender head and body and narrower, more pointed tail. See under Wedge-tailed for other possible confusion species.

WEDGE-TAILED SHEARWATER *Puffinus pacificus* Plate 4

Other names: Moaning Bird **French:** Puffin fouquet **Creole:** Fouke Dezil

DESCRIPTION Length 41-46 cm; wingspan 97-104 cm. The commonest large shearwater in Seychelles waters, dark brown above and below, with dark grey bill and wedge-shaped tail. **Adult dark morph** appears uniformly all-dark brown throughout. Underwing is uniformly dark but from above, upperwing-coverts are slightly paler than flight feathers (noticeable in worn birds). Tail long and wedge-shaped, (though usually held closed in flight) giving a long tapered shape to the body. Bill dark grey and slender, legs pale flesh (pale pink-mauve in the hand). Sexes similar. **Pale morph** has been very rarely recorded in Seychelles. It is paler above, white below, mottled brown on flanks and undertail and with a broad dark trailing edge to underwing. **Chick** is covered in soft, grey down but at fledging, **immature** resembles adult and is indistinguishable at sea.
VOICE Silent at sea; birds flying in to their breeding burrows make eerie moanings and wailings ('oo-woooo-urr') that rise and fall in pitch.
BEHAVIOUR Adults leave breeding colonies an hour before dawn and at first light can be seen streaming out to sea towards favoured feeding areas. May be seen again in the evenings close to breeding islands, settled on the sea in loose 'rafts' waiting for cover of darkness before flying in to burrows. Long wings and large size give an easy, leisurely flight; moves across the sea in long sweeping arcs with wings well stretched. Feeding takes place largely on the wing, but may also seize food from the surface or plunge into fish shoals. They are highly proficient divers and depth gauges attached to feathers of birds on Cousin have shown they can dive to depths of over 60 m, though most remain within 20 m of the surface[1]. Follows fishing boats and also associates with dolphins and terns where fish are bunched near the surface. Diet consists of smaller fish or squid.
BREEDING May breed in any month, but there is a general pattern of prospecting for burrows in August, laying in September, with chicks from December to February. The single egg is 60 mm by 40 mm, dirty white and only seen above ground if washed out by rain. Sexes share incubation, each shift lasting several days. Incubation usually takes 50-54 days. Chicks are fed on regurgitated smelly, fat-rich fish oil. Chicks fatten rapidly until they exceed the weight of adults by up to 20% and fledge after three months.
RANGE Breeds throughout the Indo-Pacific, including islands off southwest Madagascar, through Seychelles to Maldives, east to Cocos-Keeling and Christmas Island and throughout the Pacific. Populations near Equator probably sedentary, while those at outer edge of the tropics migrate north or south depending on the season. Monotypic.
STATUS Breeds only on rat-free islands and therefore, in the granitic islands, virtually confined to Cousine (tens of thousands), Aride (c.19,500 pairs[2]) and Cousin (c.14,500 pairs[3]), with smaller colonies on Récif, Mamelles and some of the smaller granitic islets. Also breeds Bird. In the Amirantes there are colonies at St Joseph Atoll, Desnoeufs, Marie-Louise and possibly elsewhere. Absent south of Amirantes though 'fouquets' recorded on Cosmoledo in 1878 may have referred to this species[4].
THREATS AND CONSERVATION The collection of birds for food has resulted in extinction on some islands (e.g. St François) and poaching remains a threat at all remaining breeding sites. Rats have also contributed to a decline and continued vigilance against the further accidental spread of rats is a major conservation concern. Wright's Skink *Mabuya wrightii* is a natural predator of eggs (accounting for the loss of 10% of all eggs laid on Cousin[5]).
SIMILAR SPECIES Flesh-footed Shearwater is darker and has diagnostic pale horn bill with black tip, silvery roots to the underwing primaries, broader less pointed tail, heavier jizz and slower flight, low over the sea. **Jouanin's Petrel** is much smaller, with slimmer wings and more agile flight. The following potential vagrants should also be considered: **Sooty Shearwater** *P. griseus* (L40-51 cm, WS94-109 cm) breeds off the coasts of southeast Australia, New Zealand and tip of South America, dispersing north into Atlantic, Pacific and Indian Oceans as far as Eilat, Israel. A smaller, stocky shearwater, uniformly sooty-brown save for silvery grey underwing-panels unlike uniformly dark underwing of Wedge-tailed Shearwater; narrow dark grey bill and blackish feet. **Great Shearwater** *P. gravis* (L43-51 cm, WS100-118 cm) breeds southern Atlantic from Falklands to Tristan da Cunha, dispersing north into Atlantic in May but some move into Indian Ocean up east coast of South Africa. Large, dark grey-brown above, white below with dark belly-patch, distinct white collar contrasts with dark cap, white rump crescent contrasts with dark tail. **Cory's Shearwater** *Calonectris diomedea* (L45-48 cm, WS100-125 cm) breeds Mediterranean, Azores, Canary Islands and Cape Verde Islands, dispersing round the tip of South Africa into southern Indian Ocean; recorded as a vagrant as far north as Madagascar. Large, flat-headed, and shorter-tailed than Wedge-tailed Shearwater, mottled grey-brown above and white below, narrow white band across the rump and diagnostic pale yellow bill visible at close range. **Streaked Shearwater** *Calonectris leucomelas* (L48 cm, WS122 cm) breeds islands off Japan and northern China, dispersing south to Australia, but vagrant to Sri

Lanka and Maldives. Large, brown above, white below, with a mottled white head, white undertail-coverts, bill long and thin, pale yellow with a grey tip. Also possible may be **Great-winged Petrel** *Pterodroma macroptera*, **White-chinned Petrel** *Procellaria aequinoctialis* and **Grey Petrel** *Procellaria cinerea* (see under Jouanin's Petrel).

References: 1. Burger (2000c), 2. Betts (1998), 3. Burger and Lawrence (2000), 4. Rivers (1878), 5. Penny (1974).

AUDUBON'S SHEARWATER *Puffinus lherminieri*   Plate 4

Other name: Dusky-backed Shearwater   **French:** Puffin d'Audubon   **Creole:** Riga

DESCRIPTION Length 27-33 cm; wingspan 64-74 cm. The most common small shearwater in Seychelles, blackish above and white below, with rounded wings. **Adult** is brown-black above including cap to below eye, white beneath, with brownish undertail-coverts. Underwing is white in centre with brownish black edges, (much broader margins than Manx or Little Shearwater, potential vagrants, where the border is well-defined and narrow). Bill slim, relatively short, with no obvious 'tube'; legs black-pink and feet pink. Sexes similar. **Chick** is covered in soft, grey down but at fledging, **immature** resembles adult and is indistinguishable at sea.

VOICE Uncanny cackling, screaming calls at breeding colonies ('ke-whirrr-irrr'), like a baby being throttled.

BEHAVIOUR Normally only seen well out to sea, though in the breeding season will feed closer to breeding colonies. Agile over the sea, banking and jinking close to the waves with rapid wingbeats followed by short glides. Normally only seen close to land at dusk, when birds collect in rafts of thousands off their breeding islands. These flocks wait for cover of darkness before flying in to their burrows, departing again before dawn. On the granitic islands they clamber to vantage points on rocks or (unusually for a shearwater) climb trees. Birds flutter up sloping tree trunks, wings outstretched, gaining purchase with claws and the hooked tip of the beak, to a fork at 5-10 m. Here, they pause, preen and call, before taking off through the canopy[1]. Where shoals of fish are abundant may feed by 'pursuit plunging', dropping into the water from flight, or may settle on the sea and lower its head to seek prey beneath the surface. May catch fish or squid by 'pursuit diving' using wings to swim. Depth gauges attached to feathers of birds on Cousin have shown they can dive to depths of 35 m, most remaining within 10 m of the surface[2].

BREEDING Nests on rat- and cat-free islands and islets of coral atolls. There is no clear breeding season in the granitic islands, incubating birds being found in any month. The single white egg, 47 mm by 35 mm, is laid in a rock crevice or, where there is deeper soil, at the end of a long burrow. Incubation lasts seven weeks and chicks fledge after 8-12 weeks, by which time they may have grown to a weight distinctly greater than adults' (on Cousin an average peak of 250 g for chicks has been recorded, compared to 160-200 g for adults[2]). Birds are thought to be mature at eight years and may survive to over 20 years of age.

RANGE Breeds throughout tropical oceans with many recognised races including *nicolae* in Seychelles, Maldives and probably Chagos, *bailloni* (Baillon's Shearwater) in the Mascarenes and *persicus* (Persian Shearwater) in Arabian Sea and Red Sea. Bill lengths of birds collected at Aldabra are intermediate between race *nicolae*, which breeds in the granitic islands, and *bailloni* of the Mascarenes. These were described as a separate race, *colstoni*, in 1995[4]. However, later DNA work found *colstoni* indistinguishable from *nicolae* (Bretagnolle and Austin, unpublished result)[5]. Fairly sedentary (hence large numbers of races) though pelagic dispersal not well known.

STATUS Resident in Seychelles waters, but in the granitic islands breeds only on Aride (c57,000 pairs[6]), Cousin (5,000-10,000 pairs[7]), Cousine (a few hundred pairs) and Récif. In the Amirantes it breeds at St Joseph Atoll (Fouquet and Benjamin), Desnoeufs and possibly elsewhere. It formerly bred at D'Arros and Desroches where it may have been wiped out by rats or humans, and Marie-Louise where there are no rats, but a small human population. At Aldabra it was first found to breed in 1967-68 on many of the lagoon islets (50-100 pairs[8]). It may breed Cosmoledo where 'fouquets'(possibly this species) were recorded in 1878[9] and it was also heard at night but no nests found in 1980s at Menai and Grand Ile[10].

THREATS AND CONSERVATION In the past the most serious threat would have been persecution by humans on the islands, killing and taking shearwaters for food. Once the islands were settled and rats came ashore, cats were brought in to deal with the rats. This combination of rats and cats must have wiped out the shearwaters from many of their former breeding islands. Both surviving breeding strongholds, Aride and Cousin, are now nature reserves, but the wardens must remain vigilant against the accidental arrival of rats on these precious islands.

SIMILAR SPECIES Both Little Shearwater *P. assimilis* and Manx Shearwater *P. puffinus* are potential vagrants. **Little Shearwater** (L25-30 cm, WS58-67 cm), sometimes considered conspecific, breeds Atlantic, Pacific and at St Paul in Indian Ocean. Race *tunneyi* (of southwest Australia and possibly this race at St Paul) occurs regularly off South Africa, as does *elegans* of Tristan da Cunha and Gough Island. A very small shearwater, black above, white below, white undertail-coverts, black eye visible in white face, bluish feet. **Manx Shearwater** (L30-38 cm, WS76-89 cm) breeds in the north Atlantic, normally wintering off South America, but a few occur off South Africa. Overall colour scheme very similar, but larger, with longer, sharper wings with broad black margins on the trailing edge of the underwing and white undertail-coverts. **Mascarene Shearwater** is blacker above, lacking any brownish hue to the upper and under-tail-coverts, lacks large nasal tubes and has bluish grey not pink legs (but note, there is considerable doubt over validity of this species).

References: 1. Bullock (1995), 2. Burger (2000c), 3. Penny (1974), 4. Shirihai and Christie (1996), 5. Bretagnolle *et al.* (in press), 6. Betts (1998), 7. Burger and Lawrence (2000), 8. Diamond (1971), 9. Rivers (1878), 10. Mortimer and Constance (2000).

Top to bottom: (1) Swinhoe's Storm-petrel, (2) Matsudaira's Storm-petrel, (3) Bulwer's Petrel, (4)Jouanin's Petrel, (5) Mascarene Petrel, (6) Wedge-tailed Shearwater, (7) Southern Giant Petrel

MASCARENE SHEARWATER *Puffinus atrodorsalis*

Described as a valid species from a specimen collected in Durban in 1987 and one at Eilat in 1992. Sightings at sea have been claimed between Aldabra, Comoros, Mombasa and Mozambique, off the west side of Madagascar and as far east as Réunion [1]. It is doubtfully separated from Audubon's Shearwater from which it is said to differ by having upperparts wholly black lacking any brown hue. However, this plumage resembles the juvenile of Audubon's Shearwater race *bailloni* from Réunion and dates claimed for sightings of Mascarene Shearwater correspond to the period when juvenile-like plumage of Audubon's Shearwaters from Réunion might be expected[2].
References: 1. Shirihai *et al.* (1995), 2. Bretagnolle and Attié (1996).

WILSON'S STORM-PETREL *Oceanites oceanicus*   Plate 5
Other name: Yellow-webbed Storm-petrel   **French:** Océanite de Wilson   **Creole:** Satanik Wilson

DESCRIPTION Length 15-19 cm; wingspan 38-42 cm. A small storm-petrel, barely larger than a wheatear, with rather short but broad, rounded wings and legs that extend beyond the tail. Mainly sooty brown with a pale grey upperwing-panel and white rump. **Adult** is sooty brown throughout except for a broad white band on the rump, which extends down onto flanks and undertail-coverts. A pale grey band on the upperwing secondary coverts contrasts with darker primaries. Between May and October plumage may look paler than in breeding birds; generally browner and wing-panels may appear even paler as whitish grey. Tail appears square-ended or slightly rounded. Legs and feet dark with yellow webs between toes (only visible on feeding birds at close range) projecting well beyond tail, an important feature of this species. Bill black. Sexes similar. **Immature** similar to adult, but more worn around October (the time of most sightings) when adults are fresher.
VOICE Silent at sea, but almost inaudible rapid squeaking when feeding.
BEHAVIOUR Pelagic outside breeding season. Where food is abundant may gather in flocks of several thousand, but Seychelles vagrants are generally solitary. Low, direct swallow-like flight over the sea. Feeds on tiny shrimps, small squid, very small fish and planktonic crustaceans. Yellow webs may help attract or disturb prey, for the most common form of feeding involves a 'pattering' technique in which the bird appears to walk on water. Often bounces on surface, legs dangling and wings held in shallow V. May also 'stand' on the surface, facing into wind with wings spread, feet suspended in the water like drag anchors. The idea that this attracts fish is borne out by frequency of leg wounds in this species, presumably caused by predatory fish. Will also dive into water in pursuit of prey. Most petrels have a strong sense of smell and may locate prey this way, particularly huge concentrations of krill. Favours fatty or oily foods, gathering over offal or even feeding on whale faeces. May follow dolphins, whales and fish to feed on scraps or upon small fish driven to the surface.
RANGE Breeds on Antarctic mainland and sub-Antarctic islands including Crozet and Kerguelen. Nominate breeds Cape Horn to Kerguelen, race *exasperatus* (slightly longer wings and tail) elsewhere. With a possible world population of several million individuals, may well be the most numerous seabird on earth. Breeds November to April, dispersing north including clockwise round the Indian Ocean, reaching the Gulf of Aden in May, off India in July/August and Australia by October and November. This might constitute a round trip of 16,000 km in a season.
STATUS Vagrant throughout Seychelles October to November, though with more observations it may be found to occur annually throughout southern winter.
SIMILAR SPECIES European Storm-petrel *Hydrobates pelagicus* and Leach's Storm-petrel *Oceanodroma leucorhoa* both migrate as far as South Africa and should be considered as potential vagrants to Seychelles. **European Storm-petrel** (L14-18 cm, WS36-39 cm) breeds from Iceland to Canaries, dispersing south in September to winter mostly off western coast of South Africa, some reaching Natal. Small, black above and below with clean white rump (white restricted to upperside) and square-ended tail. Wings appear short and more angled at carpal joint with diagnostic narrow white bar on underwing-coverts. Flight more irregular and jinking, more closely hugging surface. Feet do not project beyond tail. **Leach's Storm-petrel** (L19-22 cm, WS45-48 cm) breeds in north Atlantic from Nova Scotia to Britain, and in north Pacific from Japan to California. Atlantic birds (of nominate race) disperse south in September, some wandering round the Cape as far as Kenya, Red Sea and Persian Gulf. Medium-sized, long-winged storm-petrel, wings more angled than Wilson's, with a forked tail. Black flight feathers contrast with smoky grey-brown upperparts, small V-shaped white rump-patch, which does not extend onto undertail-coverts and is split by a greyish centre (difficult to observe) and pale bar on upperwing-coverts, which shows up well at a distance (more obvious than Wilson's). More erratic flight. Feet do not project beyond tail.

WHITE-FACED STORM-PETREL *Pelagodroma marina*   Plate 5

Other names: Frigate Petrel **French:** Océanite Frégate **Creole:** Satanik Figir Blan

DESCRIPTION Length 20-21 cm; wingspan 41-43 cm. A large storm-petrel with broad-centred wings and a distinctive face pattern; brownish above with a grey rump and white below. **Adult** is very distinctive with brown-grey back and wing-coverts and ashy grey rump, contrasting with dark flight feathers and square-cut sooty brown tail. The only storm-petrel with entirely white underparts and a distinct grey shoulder-patch. Underwing is white with broad blackish trailing edge. Striking face pattern with dark crown, white supercilium, forehead and lore and a dark sooty brown band from behind the eye to the ear-coverts. In flight, long legs project beyond tail and are black with yellow webs (the latter not normally visible at sea). Bill black. Sexes similar. **Immature** similar to adult with slightly paler head, lighter rump and more obvious upperwing-panel.
VOICE Silent at sea.
BEHAVIOUR Pelagic and solitary away from breeding colonies. Has a unique, pendulum-fashion flight, legs projecting well beyond tail, drooping only when landing on surface. Does not patter at the surface in the same manner as other storm-petrels, but weaves erratically from side to side sometimes banking and belly-flopping onto the surface. Dips and makes shallow plunge for food, mostly krill, crustaceans, cephalopods and even small fish. Attracted to offal and floating oil, presumably located by sense of smell.
RANGE Race *hypoleuca* breeds in north and south Atlantic (e.g. Cape Verde and Tristan da Cunha), but Indian Ocean birds originate from breeding sites off west and south coasts of Australia and New Zealand. Individuals of race *dulciae* have been collected off the Horn of Africa in June and July. Birds breed in southern spring and summer (September to December) and then disperse to western Indian Ocean with the onset of southeast monsoon in May. Most Indian Ocean records in May, June and July as far as Gulf of Aden.
STATUS Vagrant throughout Seychelles around May. Probably under-recorded and could prove to be annual in small numbers.
SIMILAR SPECIES Black-bellied Storm-petrel is blackish above with a white rump, white below with black upper breast and thick black central breast-stripe.

BLACK-BELLIED STORM-PETREL *Fregetta tropica*   Plate 5

Other names: Gould's (or Striped) Storm-petrel
French: Océanite à ventre noir **Creole:** Satanik Vant Nwanr

DESCRIPTION Length 20 cm; wingspan 46 cm. A large, stocky storm-petrel, blackish above, white below with a black central breast-stripe, white rump and underwing white in centre, edged black. **Adult** is sooty brown above appearing black at a distance with an inconspicuous pale wingbar (formed by the grey-brown secondary coverts) and a white rump. Underwing is white in the centre with broad black edges. Chin and sides of belly are white, with black upper breast and a thick black central stripe (variable in width and occasionally absent) running from the breast down the belly to the black undertail-coverts. Small, slender black bill; legs and feet black, toes extending beyond square tail in flight. Sexes similar.
Immature similar to adult.
VOICE Silent at sea.
BEHAVIOUR Pelagic and solitary away from breeding colonies. Distinctive erratic flight close to the surface with legs dangling. Feeds mainly by pattering at the surface, apparently finding food by smell, and belly-flopping similar to White-faced Storm-petrel. Feeds on squid and small fish.
RANGE Circumpolar, nominate race breeding on southern islands including Indian Ocean islands of Crozet and Kerguelen, dispersing north between May and October.
STATUS Uncertain; there have been a number of unconfirmed reports which almost certainly relate to this species.
SIMILAR SPECIES White-bellied Storm-petrel *F. grallaria* (L19-20 cm, WS44 cm) is a potential vagrant. Several races breed on islands south of Tropic of Capricorn, including *leucogaster* on St Paul and possibly Amsterdam (sometimes included with Black-bellied Storm-petrel). Upperparts paler than Black-bellied, black head, throat, upper breast and undertail-coverts contrasting with white lower breast, belly and underwing. Legs black, extending beyond the slightly forked tail.

MATSUDAIRA'S STORM-PETREL *Oceanodroma matsudairae*   Plate 5

Other name: Sooty Storm-petrel **French:** Océanite de Matsudaira **Creole:** Satanik Matsudaira

DESCRIPTION Length 24-25 cm; wingspan 56 cm. A large, uniform dark brown storm-petrel with broad, pale sepia wing-panel, diagnostic white patch at the primary bases and no white rump. **Adult** is a wholly dark brown storm-petrel, appearing black at distance. At closer range, a distinct pale upperwingbar is visible, broad and uniformly diffuse sepia brown from tertials to the leading edge of the wing. The black outer primaries have white shaft bases, which form a distinct white patch, visible at 100-150 m range. The long tail is deeply forked, but in flight is more usually folded so that this is not an obvious feature. Bill and feet

151

black. Sexes similar. **Immature** similar to adult.

VOICE Silent at sea.

BEHAVIOUR Pelagic and solitary away from breeding colonies. Slow, nightjar-like flight, with a series of flaps and glides and occasional bursts of twisting flight. Feeds mainly on the wing, dipping to snatch food from the surface of the sea, or hanging above the water with the wings held in a V above the body. Will rest on the sea in calm weather. Feeds on plankton.

RANGE Only known breeding sites are two islets of Bonin Island and North Volcano Island southeast of Japan, though little known of numbers. Breeds January to early June, then probably moves south of Indonesia through the Timor Sea and thence west into the Indian Ocean. Most records have been within 5° of the Equator, a zone of light winds and high zooplankton concentration where there is considerable upwelling of cooler water. Monotypic.

STATUS Uncertain; there are a number of unconfirmed records.

SIMILAR SPECIES Swinhoe's Storm-petrel has very similar brownish plumage, but is smaller with a faster, more petrel-like flight, smoky-grey cast to head and mantle and a paler wing-panel, obvious only at carpal end. Also has white primary shafts, though finer, forming a pale patch in wing visible only at very close range.

SWINHOE'S STORM-PETREL *Oceanodroma monorhis*   **Plate 5**

Other name: Ashy Storm-petrel   **French:** Océanite de Swinhoe   **Creole:** Satanik Swinhoe

DESCRIPTION Length 19-20 cm; wingspan 44-46 cm. A medium-sized, all-dark petrel with grey upper-wing-panels and a forked tail. **Adult** is dark brown-black with no white rump and a longish forked tail (the latter not obvious at sea). In good light the smoky greyish cast to head and mantle can be seen. Shows a pale diagonal bar on the upperwing-coverts, prominently whitish grey at the carpal joint, fading to brown at the trailing edge and contrasting with black flight feathers. Note, also has whitish grey edgings to the tertials. Thin white shafts at the root of the primaries form a pale wing-patch, less obvious than in Matsudaira's Storm-petrel. Pale grey on the underwing-coverts. Bill and legs black. Sexes similar. **Immature** similar to adult.

VOICE Silent at sea.

BEHAVIOUR Pelagic and solitary away from breeding colonies. Fast, buoyant tern-like flight, bounding and swooping over the water. Feeds mostly in flight, dipping to the surface of the sea to snatch small items of food. Not known to 'patter' like other storm-petrels.

RANGE Breeds on islands off southern Japan and in the Yellow Sea (South Korea and China), migrating south and west into the northern Indian Ocean and Arabian Sea. Recorded off Somalia and as far up the Red Sea as Eilat. Monotypic. Sometimes treated as conspecific with Leach's Storm-petrel.

STATUS Vagrant throughout Seychelles; recorded around October to December.

SIMILAR SPECIES Matsudaira's Storm-petrel is similarly all-dark black-brown, but is larger, with a slow, nightjar-like flight. It lacks the smoky-grey upperparts, has broader, uniformly pale sepia-brown wing-panel and more prominent pale patch in the wing formed by the white shafts at the base of the primaries.

RED-BILLED TROPICBIRD *Phaethon aethereus*   **Plate 6**

Other name: Short-tailed Tropicbird   **French:** Phaéton à bec rouge   **Creole:** Payanke Labek Rouz

DESCRIPTION Length 90-105 cm (tail 45-56 cm); wingspan 99-106 cm. A large tropicbird with long white tail-streamers, large red bill and fine black barring on mantle and back. **Adult** is the largest trop-icbird, with a white tail and large red bill obvious even at distance. Black mask is slightly larger than in other tropicbirds, occasionally joining at the nape in worn plumage (note race *indicus* has black cutting edges to orange-red bill and a smaller mask than other races). Mantle, back, scapulars and rump all have fine black barring giving a grey effect at distance. A thick black wedge on the outer edge of the primaries extends onto the primary coverts. Legs yellowish. Sexes similar; male tail-streamers averaging longer. **Juvenile** is more densely barred above than other tropicbirds with broader black eye-stripes meeting over the nape. Yellow bill, sometimes tipped black. No tail-streamers. Post-juvenile moult may be completed 12-18 months after fledging.

VOICE Generally silent at sea. May give loud, shrill screams.

BEHAVIOUR Feeds far out at sea, generally solitary, sometimes in pairs. Though they may rest on the sur-face of the sea, tropicbirds are capable of considerable feats of endurance, flying far from their breeding islands. Powerful, pigeon-like flight action with slower wingbeats on broader wings compared to White-tailed Tropicbird. Favourite prey is squid and small fish, especially flying fish. Prey captured by plunging into the sea like boobies, but flying fish are sometimes taken in flight.

RANGE The most restricted of all the tropicbirds, occurring along the west coast of America from Baja California to the Galápagos, the Caribbean, tropical Atlantic and Red and Arabian Seas. Three races recog-nised, of which only race *indicus* (sometimes considered a separate species) of Red Sea to Persian Gulf is likely to reach Seychelles. Range does not overlap with Red-tailed Tropicbird, suggesting they are too sim-ilar ecologically to co-exist.

STATUS Vagrant to Bird, granitic islands and possibly Amirantes. One remained at Bird for almost a year, displaying to White-tailed Tropicbirds during this time.
SIMILAR SPECIES White-tailed Tropicbird also has a white tail, but a yellow bill, no barring on the back, narrower wings and bold black bars across the secondary coverts, giving a cleaner black-and-white plumage. **Red-tailed Tropicbird** also has a red bill, but is pure white with thin, bristle-like red tail-streamers quite unlike the other tropicbirds.

RED-TAILED TROPICBIRD *Phaethon rubricauda*   Plate 6

Other name: Silver Bosunbird   **French:** Phaéton à brins rouges   **Creole:** Payanke Lake Rouz

DESCRIPTION Length 78-81 cm (tail 30-35 cm); wingspan 104-119 cm. A pure white, heavy-bodied tropicbird, with broad gull-like wings, large red bill and an indistinct thin red tail. **Adult** is a large, heavy, all-white tropicbird with broad wings. The thin, dark red tail-plumes are almost bristle-like and only visible at close quarters, leaving the impression at a distance of a large pure white broad-winged gull. Breeding birds may have a pink flush to the plumage. Heavy red bill is obvious even at distance. Legs blue-grey. Sexes similar. **Juvenile** has upperparts and upperwing-coverts barred dark grey, heavier than in White-tailed Tropicbird; pale bill turns red within six months of fledging. Black shafts in outer primaries less extensive than in other immature tropicbirds. No tail-streamers.
VOICE Utters a harsh disyllabic 'kelek'. During courtship flight, makes a loud, goose-like 'aahnk aahnk' call. Normally silent away from breeding grounds.
BEHAVIOUR The most pelagic of the three tropicbirds, only returning to land to breed. May be solitary or in pairs or small groups. Powerful pigeon-like flight. Feeds further out to sea than White-tailed Tropicbird, specialising more on larger (8-10 cm) flying fish.
BREEDING Several weeks before laying, birds return to breeding islands and make elaborate courtship flights over prospective nest sites. Male appears to fly in a backward circle around the female by stalling as he flies above her, hovering with feet spread wide to brake his flight. The pair then descend to a nest site for copulation. On Aride, selects sheltered sites high up on the open cliffs where take off may be easier. On Aldabra, tends to nest under vouloutye and bwa tabak bushes[1]. The single large egg (dark red-purple with many fine blotches and speckles) is 10% of the female's weight. Some shelter from the sun is crucial, nest studies show that overheating is one of the commonest causes of breeding failure. On Aride, a granite rock overhang is usually selected while on Aldabran islets, most nest-sites are under bushes or other vegetation. Incubation lasts 40-46 days; individuals may spend up to 16 days on the egg without a changeover. Incubating birds make a loud horse-like whinny if disturbed. Chick is fed in turns by the adults and makes a loud rattling call whenever birds land with food. Fledging takes place 90 days after hatching, and studies on Aldabra suggest that roughly half of the eggs laid result in fledged young[2]. Breeding appears to be on an 11-month cycle. Breeding at Aldabra occurs throughout the year, but is synchronised with rainfall, especially during northwest monsoon (January to April) when vegetation growth provides shady nesting sites, with virtually no nesting activity in the dry months, August to September[2].
RANGE Breeds across the Indo-Pacific with race *rubricauda* breeding throughout Seychelles and in Mauritius and Comoros. Non-breeding distribution is not well known, but more pelagic and wide-ranging than other species. Three other races are recognised.
STATUS Breeds only on Aride in the granitic Seychelles (up to six pairs[3]). In the Aldabra group, breeds on lagoon islets of Aldabra (c2,000 pairs[4]) and Cosmoledo (50-200 pairs[5]). Formerly bred Astove (recorded 1836 but never noted subsequently[6]) and on Assumption.
THREATS AND CONSERVATION Breeding only on rat- and cat-free islands and islets; on some other islands they have probably been exterminated by these introduced predators. Breeding is regularly attempted on Wizard (Grand Ile), Cosmoledo, where young are taken by cats[7]. Their restricted breeding range in the granitic islands may also be due to this factor, likewise no doubt their disappearance from Astove and Assumption.
SIMILAR SPECIES Red-billed Tropicbird is distinguished by fine black barring on the back and white, not red tail-streamers. **White-tailed Tropicbird** is distinguished by lighter, yellow bill and white tail-streamers.
References: 1. Diamond (1975c), 2. Prys-Jones and Peet (1980), 3. Bullock (1989), 4. Betts (2000b), 5. Rocamora and Skerrett (1999), 6. Stirling (1843), 7. Mortimer and Constance (2000).

WHITE-TAILED TROPICBIRD *Phaethon lepturus*   Plate 6

Other names: Yellow-billed Bosunbird, Yellow-billed Tropicbird, Long-tailed Tropicbird
French: Phaéton à bec jaune   **Creole:** Payanke Lake Blan

DESCRIPTION Length 70-82 cm (tail 33-45 cm); wingspan 90-95 cm. The most elegant tropicbird, white with black bands on the upperwing, prominent long white tail-streamers and a yellow bill. **Adult** is an unmistakable, large, clean white seabird with long white tail-plumes. Black secondary coverts form a black diagonal bar on each wing and the outer webs of the primaries form a black wedge on the wingtips. Shows a thick black bar through the eye, common to all tropicbirds. The large bill is yellow, unlike other adult tropicbirds. Legs yellowish. Sexes similar. **Juvenile** is coarsely barred grey-black on upperparts and upperwing-coverts with a very pale dull yellow bill and no tail-streamers.

VOICE A shrill, grating whinny rarely heard except at breeding sites; also a loud 'kek-kek-kek' (higher-pitched more rapid and longer than Red-tailed call) and a strident scream.

BEHAVIOUR May be solitary or in pairs or small groups. Sometimes rests on the sea in calm weather, but more often in steady, purposeful but graceful flight over the sea. When food is sighted below the surface, birds plunge like boobies from 25 m or more. Tropicbirds have internal airsacs round the head, throat and neck to absorb some of the impact of these spectacular dives. Though flying fish are also taken, studies on Aldabra show more squid are captured (4-6 cm long) than is the case for Red-tailed Tropicbird[1]. Prey held crossways in the bill.

BREEDING Birds soar high over breeding islands on clear, windy days, making courtship flights in groups of twos and threes. Male flies close above the female, occasionally bending his tail-plumes down to touch hers. The white stiletto-like shapes are unforgettable against a clean blue sky. Pairs mate for life and return to traditional nest sites, some sites used by more than one pair[2]. May breed at any time of year, studies on Cousin[2] and Aride[3] suggesting a mean interval between successful breeding attempts of nine months. On Aride, the interval between nesting failure or successful fledging and the next breeding attempt were both measured as c5 months[3]. The ideal site is both in the shade and sheltered by rock or tree roots from flash flooding after heavy rain. Such sites may be fiercely contested by rivals, involving duels on the ground with beaks locked and wings spread, in which their immaculate plumage may end up spattered with blood. Pairs occasionally nest in the hollow trunks of trees up to 5 m above the ground, in the leaf bases of palms or even high on inland cliffs as on Mahé. The single coffee brown egg is incubated alternately by both parents and takes 40-42 days to hatch. Freshly hatched chicks are tiny white powder-puffs and weigh only 30g; the female leaves at once to fetch a first meal and at this stage chicks are vulnerable to predation by crabs. Adults return at intervals of 1-2 days with food and swoop in to the nest sites through holes in the canopy following habitual flight paths. The chick when full-grown is often 20% heavier than the adult and is abandoned at this stage (70-85 days after hatching). It then loses weight until it is hungry enough to make its first flight, unaccompanied, out to sea. The only serious threat to chicks at this stage is seeds of Pisonia on some granitic islands, which may entangle a bird that fails to clear the forest floor on its maiden flight. One study at Aldabra found roughly half of eggs laid result in fledged young[4], but this is probably a long-lived bird.

RANGE Globally this is the most widespread tropicbird, occurring in the Caribbean, the tropical Atlantic and throughout the Indian and Pacific Oceans. Nominate *lepturus* breeds throughout the Indian Ocean except for Christmas Island (race *fulvus*) and Europa (race *europae*).

STATUS The main breeding islands in the granitics are Aride (c300 pairs during each monsoon season[5]), Cousin (c. 500 pairs during southeast monsoon[6] and c1,500 pairs during northwest[7]) and Cousine (450-850 pairs[8]) which are cat- and rat-free. Population estimates represent active breeders and annual totals may be much higher. A few pairs do nest high on the larger islands such as Mahé, La Digue, Silhouette and Frégate where they fly well inland and use mature trees or ledges on the highest cliffs. There are usually one or two pairs on the smaller, uninhabited islands such as Mamelles, Ile aux Cocos, etc. Elsewhere they breed on Aldabra (c2,000-2,500 pairs, mainly on small lagoon islets in solution holes in the raised coral[9,10,11]), probably on Cosmoledo (where there are sight records of adults) and possibly Assumption and Astove. A few breed Bird and Denis. There are sight records from the Amirantes where a few pairs may breed. There are sight records of tropicbirds on Cousin of birds with plumage tinged golden yellow, which were assumed to be race *fulvus* (Christmas Island Tropicbird). However, another possibility might be *europae* of Europa in the Mozambique Channel, which has a high frequency of golden morphs[12]. Alternatively, it may be a rare local variation within the Seychelles population.

THREATS AND CONSERVATION Tropicbirds are true seabirds with small webbed feet at the rear of the body that make them clumsy on land. At nest sites, they have little fear of man and were easily killed by early mariners and settlers. Their feathers were once taken for hats. Some persecution by humans still goes on, but as the most important breeding sites of Aride, Cousin and Aldabra are now nature reserves, the worst threat they still face is predation by cats or rats on the islands where small populations still survive.

GENERAL In use from about the 17th Century, 'payanke' is probably the first real Creole name for any bird.

SIMILAR SPECIES Red-tailed Tropicbird has heavier red bill, all-white plumage and red tail-streamers. **Red-billed Tropicbird** has diagnostic combination of red bill, white tail-streamers and barred upperparts.

References: 1. Cheke (1982), 2. Phillips (1987), 3. Davis (2000), 4. Betts (1998), 5. Bowler and Hunter (2000), 6. Burger and Lawrence (1999a), 7. Burger (2000b), 8. Wright and Passmore (1999), 9. Prys-Jones and Peet (1980), 10. Diamond (1975c), 11. Betts (2000b), 12. Le Corre and Jouventin. (1999).

ABBOTT'S BOOBY *Sula abbotti*   Plate 1

French: Fou d'Abbott   **Creole:** Fou Bef

DESCRIPTION Length 79 cm. A long-winged, white-bodied booby with all-black wings and tail and diagnostic spotting on the rump and flanks. **Adult** has head, neck, mantle, breast, vent and underwing pure white, upperwing-coverts black-brown, mottled with white, flight feathers black with white edges to the tips of the secondaries and black tail. Diagnostic black flecks on the white rump and likewise in black feathers on the flanks form a scaly patch. Bill black-tipped, pink in female, blue-grey in male. Black facial skin, legs and feet dark grey. **Immature** has coffee brown wing-coverts and grey bill.

VOICE The deepest and loudest voice of any sulid, resonant and commanding[1]. A cry like a bull (from which it gets its Creole name).

BEHAVIOUR Little recorded on feeding ecology, though this must be distinct as its bill has a hooked tip unlike other booby species. Early accounts mention the large serrated bill, the inside like a saw.

BREEDING A tree-nesting booby. On Christmas Island birds form loose colonies scattered through the forest, with nests about 30 m apart. There is a long incubation, c57 days and an unusually long period of dependence in the chicks, up to nine months (compared to five months in other boobies). The extended period of chick growth runs into the wet, windy monsoon months when a large proportion of young birds die of starvation. Breeds once every two years and may only successfully raise one chick every five years[1].

RANGE Now known only from Christmas Island in the eastern Indian Ocean. Occurred in historical times as a breeding species on Assumption, Rodrigues and possibly Iles Glorieuses and Chagos[2]. Fossil records from the Solomons and Marquesas in the Pacific and from Mauritius in the Indian Ocean suggest a wide distribution at one time. It was probably this species mentioned from Rodrigues in 1726 by Tafforet ('nesting in trees, with a cry like a bull'[3]) and by Pingré in 1763 ('its eyes handsome large and black'[4]), the last record there. Though there has been some suggestion it bred at Iles Glorieuses[1,5], some authors consider tree-nesting boobies seen by Abbott to have been Red-footed Booby[2,6]. Subsequent to extinction in Seychelles, there has been a report claimed from c100 km due west of Farquhar in 1957 but this is questionable. More convincing are reports from Chagos, which may indicate some possibility the species may one day be recorded as a vagrant to Seychelles. Last remaining breeding site on Christmas Island is now protected as part of the Christmas Island National Park (administered by Australia) but it remains endangered. Some authorities treat this species as distinct from other boobies, in a monospecific genus *Papasula*. Monotypic.

STATUS Extinct; formerly bred Assumption. First described from Assumption by Ridgway from a specimen collected by Abbott[7]. In 1895 Baty reported 'Boobies or Fous of different kinds are to be found in the trees all over the island'[8]. Fryer collected two specimens in 1908 and wrote 'a fresh kind of Fou was found: it was larger than the ordinary with a grey back tinged with pink: dark eyes and wings and long coverts black: its cry was very like a cow'[9]. Fryer collected two specimens from Assumption in 1908. He found the species 'inhabits the large dune, never descending to low parts of the island'[10] (the largest trees grew here[8]). In 1908 a settlement was built on Assumption to begin harvesting guano. The guano industry mined the surface of the coral atolls for the phosphate deposits that had accumulated over centuries under the tree-nesting seabird colonies. This destroyed the forest and the soil on which it depended; with the loss of habitat it was doomed to extinction. In 1916, Dupont wrote 'the boobies are all destroyed'[11] and in a later visit wrote that all seabirds and landbirds had been destroyed by 1909[12].

References: 1. Nelson (1971), 2. Bourne (1976), 3. Tafforet (1726), 4. Pingré (1763), 5. Gibson-Hill (1950), 6. Benson *et al.* (1975), 7. Ridgway (1893), 8. Baty (1895), 9. Fryer (1908), 10. Fryer (1911), 11. Dupont (1916), 12. Dupont (1929).

MASKED BOOBY *Sula dactylatra*   Plate 7

Other names Blue-faced (or White) Booby   **French:** Fou masqué   **Creole:** Fou Zenero

DESCRIPTION: Length 80-90 cm; wingspan 142-154 cm. The largest booby, with black tail and flight feathers, white body and wings; yellow bill, dark feet. **Adult** has a pure white body, black flight feathers and black greater primary and secondary coverts and a black tail. Bill is yellow in male, greenish yellow in female, with a contrasting black 'mask' around yellow eye, bill base and gular pouch. Legs and feet black. Sexes similar, except for the slight difference in bill colour only detectable at close range. **Juvenile** is pale brown with white upper breast and white hind-collar. Bill is whitish horn, black at the base with face and gular skin bluish or blackish. With age, the white collar becomes more extensive, upperparts and wing-coverts become speckled white gradually acquiring dazzling white plumage of adult by c33 months.

VOICE A high-pitched double honk, sometimes given at sea. Noisy at breeding colonies, sexes having different calls: in early and late breeding season, male may whistle from a vantage point; female bellows loudly, becoming more frenzied during interactions.

BEHAVIOUR Pelagic outside breeding season, when generally solitary or in small groups. One ringed on Boudeuse in 1976 was recovered at Porbander in northwest India seven months later. Flight powerful with rapid wingbeats. Feeds on shoaling fish, especially larger flying fish, making steep gannet-like dives for its prey. Holds the diving record, having been seen to dive from 100 m above the sea; heights of 10-30 m are more usual. Unlike other boobies, rarely follows ships, but frequently associated with tuna shoals and whales. Catches larger fish (up to 40 cm) than other boobies. One ringed on Boudeuse in 1976 was found in the stomach of a large grouper near D'Arros in 1988, almost certainly caught as the bird was diving for food[1].

BREEDING Breeds during northwest monsoon October to May. Pair bonds are weak, even after successful breeding. Courtship includes brief flights in which wings are held high; also at rest there is much head shaking and bill thrusting. Nests are made on flat ground or occasionally gentle slopes or even cliffs. The nest itself is a simple mound of excreta, with a slight depression at the top to hold the eggs. Two eggs are laid and incubated for 44 days, hatching about five days apart. Eggs blue with a chalky white coating that rubs off, staining the nest. It is normal for only one chick to survive, the larger chick murdering or expelling its sibling in first few days. Chick is continuously guarded and sheltered from direct sun by a parent for first 3-4 weeks, each parent taking a turn for an average of 18 hours. Fledging takes around 17 weeks. Post-fledging care may last a further 4-9 months. Birds breed at 2-3 years of age, not always at the island of their birth and may live up to 23 years.

RANGE Pantropical breeding across the Indian Ocean and Caribbean and Atlantic. Race *melanops* breeds in western Indian Ocean.

STATUS Breeds on Cosmoledo (5,000-6,000[2,3] pairs) and Boudeuse (3,000 pairs[2,3]) with a few pairs on Desnoeufs. On Cosmoledo, breeds mainly on South Island, Pagode, East North Island, West North Island, Goëlettes and Ile du Trou[4]. Formerly bred on Bird, African Banks, Etoile, Providence, Farquhar and Assumption[3]. May forage up to 1000 km from the nearest land and frequently reported in the granitics, particularly Aride (usually immatures, which may travel further than adults from their natal colony).

THREATS AND CONSERVATION The western Indian Ocean race is threatened and declining. Where causes of extinction are known, man has been responsible through habitat destruction (coconut plantations on previously open land) and direct exploitation, with one possible exception (Etoile, where erosion has occurred)[3]. With the decline of the coconut industry, further habitat loss does not appear a problem at present, but adults and young are still killed and eggs taken by local fishermen and sometimes by poachers entering Seychelles illegally from Comoros (particularly at Cosmoledo). Fishermen sometimes deliberately harass birds to force them to disgorge their catches, which are used for bait (accounting for the Creole name meaning 'generous booby'). There is no active protection at the few remaining colonies where poaching may occur and birds may be taken for food[4].

SIMILAR SPECIES Adults are most likely to be confused with **Red-footed Booby**, but this is smaller, has white tail, less black on wings (but has a diagnostic black carpal-patch) and red feet. Immature resembles **Brown Booby**, which lacks white collar and pale patch on upperparts. One potential vagrant, **Cape Gannet** *S. capensis* (L85-90 cm, WS160-170 cm), should be considered. Breeds South Africa and Namibia and may wander up East African coast as far as Zanzibar. Adult very similar to Masked Booby, but with pure white greater coverts, head and hindneck tinted buff-yellow, bill blue-grey, eye sky blue. Immature dark brown mottled or speckled white (not uniform brown).

References: 1. Feare (1989), 2. Feare (1978a), 3. Rocamora and Skerrett (1999), 4. Feare (1978b).

RED-FOOTED BOOBY *Sula sula*   Plate 7

Other name: Tree Booby   **French:** Fou à pieds rouges   **Creole:** Fou Bet, Fou Rozali

DESCRIPTION Length 66-77 cm; wingspan 91-101 cm. The most common booby of the Aldabra and Farquhar groups with white tail, body and wings, white tail feathers, red legs and feet and a lilac bill with a pinkish base. Seychelles population is almost entirely white morph. **Adult white morph** has white head, body and tail and white wings with black primaries and secondaries. Bill is pale blue; pink facial skin above and below bill base are separated by bright blue skin in front and around the eye; black gular skin extends as a black line outlining rim of lower pink patch. Brown morph, common in parts of the Pacific, has not been recorded in Seychelles, but the **white-tailed brown morph**, with white scapulars, vent, rump and tail, the rest of the plumage brown, is very occasionally encountered at Aldabra. Such birds are distinguished from immature by white rear, and colour of facial skin and bare parts, which are the same for all morphs. Sexes similar. **Juvenile** is entirely dark grey-brown, with dark purple or blackish bill, purplish black facial skin and yellowish pink or pinkish legs and feet. By **second-year**, birds have uniform dusky brown plumage flecked with white. Adult plumage is usually acquired by third year.

VOICE Male has a thin whistle; female gives a hoarse honking sound at the breeding colonies. When competing for food or roost sites, birds make a hoarse, guttural rasping croak.

BEHAVIOUR Heat regulation techniques involve exposing the webs of the feet, or excreting on them to cool by evaporation. Birds also hang the wings loosely while gular-fluttering. Highly social and gregarious both at breeding colonies and at sea. Skims low over the surface to overhaul gliding flying fish, but will also dive in the manner of gannets to seize them. May feed at night (hence large eye), unlike other boobies, when squid are close to the surface, especially during moonlight. Frequently loses catch to frigatebirds, which will attack boobies. Regularly roosts on ship's masts, using as a vantage point for fishing, birds diving and chasing flying fish panicked into flight by the vessel's bow wave. Flying fish are their favourite prey (100% of their diet in the dry southeast season), but will also take squid (20% of their diet in the rainy northwest season).

BREEDING The only booby species in Seychelles to nest in trees. At Aldabra, frequently nests alongside frigatebirds. Birds arrive at their breeding sites more than a month before egg laying. In courtship, male points bill skyward; female joins in with a similar display, strengthening the pair bond. The season is extensive with peak laying periods at Aldabra in southeast monsoon August and September and in northwest monsoon November to February[1]. The August/September peak coincides with the nest building period of frigatebirds which rob boobies of nesting material and is only possible away from the frigatebird colonies such as Bras Takamaka, Aldabra. The majority of boobies nest with frigatebirds and lay in northwest monsoon when frigatebirds have ceased robbing for nesting material[1]. Nest is a loose platform of sticks cemented together by guano in the top of mangroves with a lining of freshly picked leaves with twigs. The single white egg is incubated for c45 days. As boobies have no brood patch, they stand on the eggs using the webs of their feet for incubation. Chicks are covered in white down and grow slowly, taking anything from 14 to 20 weeks to fledge. Even after fledging, they continue to be dependent on adults for food for up to a further six months. With adults committed for up to 15 months to incubation, fledging and post-fledging parental care, it is not possible to breed successfully every year. However as food supplies are relatively constant, birds may recommence the cycle again as soon as they are able. Tends to lay fewer eggs than other booby species; if food is scarce, it may not lay at all, or may desert. Normally breeds at 2-3 years old and the maximum age recorded for a wild bird is 23 years. Brown morph is confined to the Pacific and reports from Seychelles refer to juvenile white morphs, distinguishable from adult brown morph on colour of bill and facial skin.

RANGE Breeds in all tropical seas with race *rubripres* (slightly larger than the nominate) in the Indian Ocean to the central Pacific Ocean. In the Indian Ocean, breeds Seychelles, Chagos, Christmas Island, North Keeling Atoll and Tromelin. Most Indian Ocean and all Seychelles colonies are entirely or mainly white morph[1]. However, over 95% of the population of Europa and one third of that of Tromelin are white-tailed brown morph[2] as were 90% of the now extinct colony of Iles Glorieuses[3].

STATUS Breeds Aldabra (c10,000 pairs[4,5]) and Cosmoledo (c15,000 pairs[6]), with a smaller colony at South Island, Farquhar (50-70 pairs[6]). The largest colonies at Aldabra are at the eastern end of Malabar and at Bras Takamaka, Grande Terre. Population estimates represent a point in time ad year-round figures may be much higher. The Cosmoledo population is the largest in the world outside of Galápagos. The Aldabra population was estimated at 6,000-7,000 pairs in 1967/68[1] but a census in 2000 estimated this had increased by about 50% with nesting more widespread, attributing this as probably due to the cessation of exploitation[4,5]. At Cosmoledo, most of the population breeds on South Island with other colonies on Pagode, Ile aux Chauve-Souris, West North Island, East North Island, Menai, Goëlettes, Ile du Trou and elsewhere[7,8]. Formerly bred on Assumption, Astove, North Island (Farquhar), St Pierre, Platte and in the granitic islands at Petite Soeur and probably elsewhere (including possibly Booby Island between Aride and Praslin). May forage up to 150 km from breeding colonies. A few occur annually as non-breeding visitors in the granitic islands.

THREATS AND CONSERVATION The extinctions of many colonies has been due to man, through habitat destruction (removal of nesting trees for timber, for coconut plantations or, in the case of Assumption and St Pierre, for guano mining) and direct exploitation[9]. Seychelles second largest colony, at Aldabra Atoll, has some degree of protection due to the status of this site as a nature reserve and World Heritage Site, though patrolling breeding sites is extremely difficult for the small number of staff based on the atoll. Elsewhere, there is no effective protection and both birds and eggs are still taken for food. The substantial increase in population since Aldabra received protection in 1967 shows the impact that might be achieved should protection of the outer islands ever be widened. Birds at Farquhar nest remarkably high in trees, in contrast to elsewhere, possibly due to human predation at lower sites[10]. The existence of colonies of white-tailed brown morph birds close to Seychelles at Europa, Tromelin and formerly at Iles Glorieuses may suggest there are strong isolating mechanisms between nearby islands[2]. This has strong conservation implications for each colony.

SIMILAR SPECIES Adult **Masked Booby** is larger with a yellow bill, blackish feet, no black carpal-patch and a black tail. Immature **Brown Booby** resembles immature Red-footed, but has yellowish legs and usually a more evident breast line.

References: 1. Diamond (1974), 2. Le Corre (1999), 3. Nicoll (1906), 4. Burger (2000a), 5. Burger and Betts (2000), 6. Rocamora and Skerrett (1999), 7. Skerrett (1996b), 8. Mortimer and Constance (2000), 9. Feare (1978a), 10. Skerrett (1999b).

BROWN BOOBY *Sula leucogaster*   Plate 7

French: Fou brun **Creole:** Fou Kapisen

DESCRIPTION Length 64-74 cm; wingspan 132-150 cm. The rarest booby of Seychelles with yellow bill and feet, and mainly chocolate brown plumage sharply demarcated from white belly. **Adult** has chocolate brown back, wings, tail and head, extending down the neck to a clear line on the upper breast, the lower breast and belly pure white. This plumage pattern explains the Creole name which means Capuchin Booby, a reference to the black-and-white robes of Capuchin monks. Eye dark brown, feet and bill yellow, with a pale blue face in the breeding season. Sexes similar, but female significantly larger with paler facial skin and a yellow eye-ring (blue in male). **Immature** is dull brown, paler than adult, with dirty white underparts, generally similar in pattern to adults, unlike other boobies. Eye pale grey; bill grey with yellow-green base and legs dull yellow.

VOICE Male gives a loud, high-pitched wheezing whistle, bill held closed. Female utters a harsh, loud honk.

BEHAVIOUR Less pelagic than other boobies, adults generally remaining close to breeding colonies year round, though immatures wander widely. Sometimes roosts in trees. Tends to feed more communally than other boobies when food is abundant, often flying in loose V-formations. Also often solitary or in pairs.

Strong, direct flight, interspersed with gliding. Prey includes flying fish, squid, halfbeaks and mullet. Fishes mostly by plunge diving often from low angles. Will also snatch prey from surface or steal from other booby species (particularly Red-footed), or even harass frigatebirds on occasion.

BREEDING Nests on the ground, March to December and typically in smaller colonies than other boobies. In courtship, both sexes point and shake bills, the male sometimes thrusting its bill inside that of the female simulating feeding. Also may point bill skyward, both at rest and in flight. Pairs often circle nest site, with male to rear of female uttering wheezy whistles and pointing bill skyward. Nest site is a shallow depression lined with grass and twigs collected mainly by the male. Usually lays two eggs (sometimes one and very occasionally three), pale bluish or greenish with a chalky white coating. Incubation lasts c43 days. Unless food is unusually abundant, normally only one chick survives, fledging after 12 to 15 weeks. Young are brooded on webs of feet for up to three weeks, each parent taking a turn and then guarded up to five weeks. Both parents feed young by regurgitation. Adults continue to feed juveniles for up to nine months after they have left the nest. Once independent, young birds may wander widely in their second year, ultimately even moving between colonies. Adults are more sedentary and tend to remain close to the breeding colony throughout the year. Birds normally breed at 2-3 years old.

RANGE Pantropical throughout Indian, Pacific and Atlantic Oceans. Race *plotus* (larger and more uniformly dark above than the nominate) breeds Red Sea, Indian Ocean to central Pacific Ocean.

STATUS Breeds Cosmoledo, mainly on South Island with a few pairs on West North Island (Ile du Nord) and a total population of c60 pairs[1,2]. Probably extinct on Farquhar where formerly bred Ile Lapin, Ile Déposé and Ile du Milieu until at least 1963 (G. Gendron pers. comm.). Last recorded at Lapin in 1976 (Feare pers. comm.). Also formerly bred Desnoeufs where declined from 20 pairs in 1955 to only two pairs in 1974[3] and one pair in 1988 (I. Peterson pers. comm.), the last breeding record. At Cosmoledo there were no reliable published accounts of breeding prior to 1996; formerly bred Pagode, Goëlettes and Grand Polyte[4]. Also used to breed on Bird, African Banks, St Joseph Atoll and Boudeuse[3]. Individuals appear in the granitic islands occasionally; single birds have been recorded roosting on Aride between August and March.

THREATS AND CONSERVATION Despite being the most numerous and widespread of booby species worldwide, it is the rarest breeding species of Seychelles and most vulnerable. Cosmoledo has no protection and is frequently visited by poachers. Introduced predators such as cats have severely affected some colonies and disturbance as a result of tourism may become a problem.

SIMILAR SPECIES Juvenile **Masked Booby** has paler brown upperparts, a white collar extending across nape and white upper breast. The underwing pattern also differs (see plate 7). Immature **Red-footed Booby** has similar brown plumage, but has red or pinkish feet and lacks dividing line on breast.

References: 1. Rocamora and Skerrett (1999), 2. Skerrett (1999b), 3. Feare (1978a), 4. Mortimer and Constance (2000).

AFRICAN DARTER *Anhinga rufa*   Plate 2

Other name: Snakebird   **French:** Anhinga d'Afrique   **Creole:** Zwazo Koulev

DESCRIPTION Length 85-97 cm; wingspan 116-128 cm. A large dark cormorant-like bird with a snake-like neck and dagger-like bill. Unmistakable silhouette, with long body, long, broad tail, sinuous neck and pointed bill. **Male** is mainly black with an oily green iridescence, the neck chestnut with a white lateral line below the eye, black flight feathers, tail and mantle. Long, wide tail is strongly graduated and finely barred white. Black-and-white plumes on scapulars and white feathering on wing-coverts give a striking scaly effect. Non-breeding birds lack iridescence, are browner and duller, particularly on head and neck, the white neck line less distinct. Legs brownish grey, eye orange-red. Bill dull grey-yellow, bare gular patch creamy. **Female** has neck entirely brown with white lateral stripe indistinct or absent. Legs and eye as male, but bill dark grey on upper mandible, pinkish on lower. **Immature** is much paler than adult with buff-brown upperparts, long buff streaking on inner secondaries and scapulars. Chin, throat and foreneck whitish. Belly white to buff-brown in juvenile, soon becoming blackish in immature.

VOICE Normally silent away from breeding grounds.

BEHAVIOUR Frequents inland lakes and rivers at breeding grounds, but also uses estuaries and mangrove swamps where perches and roosts in trees. Generally solitary, sometimes in pairs. Perches erect, often with wings outstretched and tail fanned. Strong flight with rapid wingbeats followed by a short glide, head slightly kinked, sometimes soaring with tail spread. When swimming, body is often completely submerged, leaving only the head and neck above water giving the impression of a swimming snake. Takes off with long, splashing run over surface, often preferring to scramble out of water onto rocks or branches, and walks with clumsy gait. Rarely pursues prey underwater, but waits until a fish swims past, spearing side of belly with a 'darting' strike. Prey is mainly quite large fish, but will also take frogs, snakes, crustaceans, molluscs and terrapins.

RANGE Breeds sub-Saharan Africa and Madagascar. The African race *rufa* and Madagascar race *vulsini* are both likely vagrants to Seychelles. The latter differs from the nominate race in being duller, browner and slightly larger with broader white lines on lower scapulars and tertials. Sometimes considered conspecific with Indian Darter *A. melanogaster*.

STATUS Uncertain; there is a published record from Aldabra, almost certainly this species, but unfortunately notes and photographs did not survive to the time SBRC was formed.

SIMILAR SPECIES Great Cormorant is larger and lacks sinuous appearance. **Long-tailed Cormorant** is smaller and lacks sinuous appearance. Sometimes swims darter-style, with only head and neck above surface, but is readily distinguished by smaller size and shorter less-pointed bill.

GREAT CORMORANT *Phalacrocorax carbo*   **Plate 2**

Other names: White-necked Cormorant (East Africa), Cormorant, White-breasted Cormorant, Black Shag
French: Grand cormoran   **Creole:** Gran Kormoran

DESCRIPTION Length 80-100 cm; wingspan 130-160 cm. A very large cormorant, plumage all-black with brownish tints, white patches in breeding adults. **Adult** gives a general impression of all-black plumage, but good light shows a metallic bronzy cast to scapulars and wing-coverts. Large yellow patch of bare skin around the eye and bill base, clearly visible at distance. Bill otherwise grey with black culmen. Eye bottle-green, legs black. **Adult breeding** shows a large white patch on the cheek and on the thigh and white throat and upper breast in the African race; non-breeding lacks the white cheek and thigh patches. **Immature** is duller and browner above than adult with dirty white from throat to undertail-coverts.
VOICE Normally silent away from breeding grounds.
BEHAVIOUR Rarely wanders far from the coast, preferring shallow water in which to feed. Fairly gregarious but vagrants are likely to be solitary. Heavy flight, with goose-like silhouette but neck kinked; powerful wingbeats interspersed with short glides. Has a typical forward leap out of the water before diving for prey. Regularly adopts a characteristic 'wing-drying' posture when not actively feeding. Prey mostly bottom-dwelling fish, but on rocky coasts will take wrasse and similar species.
RANGE Breeds eastern Canada, across Eurasia and parts of Africa to Australia and New Zealand. Sedentary in most of range but a migrant or partial migrant in northern populations. Race *lucidus* of East Africa (smaller and greener than the nominate race of Canada to northern Europe and with white extending from the neck right down onto the breast and belly) occurs on large inland lakes and makes local dispersals according to cycles of drought and rainfall. Race *sinensis* of central Europe to China (has extensive white feathering down the sides of the neck) might also occur, being fairly migratory, moving south into South-East Asia.
STATUS Vagrant to granitic islands; recorded January.
SIMILAR SPECIES African Darter is smaller and more slender. **Long-tailed Cormorant** is much smaller (almost half the size of Great Cormorant), long-tailed and short-necked, with a short, yellow bill and red eye. **Cape Cormorant** *P. capensis* (L61-64 cm, WS109 cm) is a potential vagrant. It breeds round the South African coast, dispersing northwards from January to August, as far as Mozambique. Smaller and short-tailed, almost entirely black, iridescent green in bright light, with bright yellow facial skin and throat pouch and turquoise eye.

LONG-TAILED CORMORANT *Phalacrocorax africanus*   **Plate 2**

Other names: Reed Cormorant, Reed Duiker   **French:** Cormoran africain   **Creole:** Kormoran Lake Long

DESCRIPTION Length 51-56 cm (race *pictilis* up to 60 cm); wingspan 80-90 cm
A small, long-tailed cormorant with a short, yellow bill. **Adult breeding** is all-black (plumage with a greenish cast close to), feathers of back, rump and uppertail-coverts with distinct white edges and moderately long all-black tail. Short crest on forehead, black with white bases to bristly feathers. White facial plumes form an obvious eyebrow. Eye red. Legs black. Bill yellow with brown culmen and some barring on lower mandible and yellow facial skin. Sexes similar, female averages larger. **Adult non-breeding** lacks crest and white eyebrow, plumage browner and underparts may be white though in southern Africa (and sometimes East Africa) may have dark breast feathers. Distinctive pattern of rounded black spots on back. Eye dull red. **Immature** similar to non-breeding adult, white or buff-white below, upperparts more uniform dark brown, eye brown.
VOICE Generally silent outside breeding season. At roosts may give a bleating 'ka-ka-ka-ko'.
BEHAVIOUR Occurs in native habitat mainly on small rivers, swamps and ponds, sometimes on coastal wetlands and mangrove swamps. Seychelles vagrants have favoured mangrove sites. Roosts in reeds, mangroves or dead trees, perching upright, often with wings outstretched. Flies with bursts of rapid duck-like wingbeats followed by a short glide. Fairly gregarious, but vagrants may be solitary or associate with egrets and herons at roosts. Swims low in water, sometimes with body submerged similar to African Darter. Feeds mainly on small fish and frogs caught by pursuit-diving.
RANGE Nominate race resident throughout sub-Saharan Africa. Race *pictilis* (larger with larger, less rounded white spots on upperparts) resident throughout Madagascar. Mainly sedentary, but may move in response to changes in local conditions such as drought or flooding.
STATUS Vagrant to Aldabra and the granitics recorded in most months.
SIMILAR SPECIES Great Cormorant is considerably larger, almost double the size of Long-tailed Cormorant. **African Darter** is also larger, with longer bill and more sinuous appearance.

PINK-BACKED PELICAN *Pelecanus rufescens*   **Plate 1**

French: Pélican gris   **Creole:** Pelikan Roz

DESCRIPTION Length 125-132 cm; wingspan 216-290 cm. A very large, heavy bird, uniform greyish white with dark flight feathers and an enormous, pouched bill. **Adult** is greyish white throughout with a tufted head, browner wing-coverts and white rump. Shows a patch of stiff, bristly feathers on the chest with a yellowish hue. Huge, straight, pale pinkish grey bill with loose gular pouch, black lore and grey or pinkish legs and feet. In flight, back, wing-coverts and rump appear off-white, secondaries brownish grey and primaries black. Eye-ring pink above eye, yellow below, pouch deep yellow with fine dusky lines, dark shaggy crest and vivid pink-red feet. Sexes similar, upper eyelid of female deep orange at onset of breeding season (yellow in male). **Adult breeding** plumage is suffused with pink, especially on back, wing-coverts, rump and belly. **Juvenile** is mainly brown with white rump, grey bill and off-white legs.
VOICE Generally silent away from breeding colonies. At breeding sites, utters guttural notes.
BEHAVIOUR Prefers freshwater, but occurs in saltwater lagoons. Roosts in trees when not feeding, then fishes most actively in morning and evening. Gregarious but usually feeds singly; where suitable shoals occur, small groups will co-operate by swimming in a ring and 'corralling' fish, thus driving them into the shallows. They then plunge head and neck underwater to scoop huge mouthfuls of fish and water into the flexible pouch.
BREEDING Breeds colonially in tall trees. Reported in 1905 with young in October nesting in large numbers in coconut and other large trees on the eastern island of St Joseph Atoll (but misidentified at the time as Dalmatian Pelican *P. crispus*, a ground nester)[1]. In Africa, a small nest (50-60 cm across) of twigs is built in which usually two (sometimes one or three) pale blue eggs are laid and incubated for c30 days, young fledging in c12 weeks.
RANGE Breeds throughout sub-Saharan Africa. Makes short migrations outside the breeding season, some moving inland from the coast, others, e.g. from Ethiopia, moving north into Egypt or crossing the Red Sea into Arabia. A former colony in Madagascar was wiped out in the early 1960s, presumably by local villagers. Monotypic.
STATUS Extinct. In 1882 H.M.S. Alert charted 'Pelican Island' on St Joseph Atoll in the Amirantes, 250 km southwest of Mahé, presumably so named as birds were noted there. Abbott collected a specimen from the same area in August 1892, which proved to be Pink-backed Pelican: 'a small colony – perhaps 100 individuals – the only colony of pelicans in these seas'[2]. By the 1930s the colony had gone and the pelican was extinct in Seychelles. Apart from records from Dahlak Archipelago in the Red Sea and Madagascar, there are no reports outside mainland Africa. A published report of two Dalmatian Pelican on Bijoutier[3] refers to birds introduced by the owner and resident for several years but later killed by labourers from neighbouring Alphonse for food (G. Gendron pers. comm.), a possible clue to the demise of the Pink-backed Pelican on St Joseph Atoll.
References: 1. Gardiner and Cooper (1907), 2. Ridgway (1895), 3. Loustau-Lalanne (1963).

GREAT FRIGATEBIRD *Fregata minor*   **Plate 8**

Other names: Man O'War Bird, Seahawk, Pacific Frigatebird
French: Frégate du pacifique   **Creole:** Gran Fregat

DESCRIPTION Length 85-105 cm; wingspan 205-230 cm. Frigatebirds have an unmistakable silhouette: tiny head and body compared to the huge wings, which form a unique 'M' shape, bent again just before they meet the body. At all ages, lacks the white 'armpits' of Lesser Frigatebird. **Male** is all-black, with a metallic green gloss on the head and back and a broad pale bar on the upperwing. Red gular pouch rarely shows up in flight unless partially inflated. The tail is long and forked like a swallow. Black eye-ring, pale to dark grey bill, long and thin with a pronounced hook at the tip. Reddish brown legs are tiny and pigeon-like. **Female** is mainly blackish brown with a white throat and breast-patch. Also shows a faint greenish gloss on the back and a pale bar on the upperwing. Pink eye-ring and blue-grey bill often with pink tinge. Legs reddish pink. Female generally larger than male. **Immature** upperparts are blackish brown with black breast-band and pale wingbar. White belly, greyish bill and eye-ring. Aldabra juveniles have a white or creamy head. Elsewhere in their range the head may be orangey buff and birds in this plumage are frequently seen in the granitics, notably at Aride. Several stages of plumage development occur in the five years it takes individuals to reach full adult plumage. Distinguishable from Lesser at all ages by absence of white spurs extending to underwing.
VOICE A variety of noises including a warble, bill-snapping and rattling and reeling sounds. Displaying male makes a 'horse-whinny', 'whoo-hoo, hooo-ooo' accompanied by vigorous head-shaking. On arrival at the nest, male makes a loud 'teeu-teeu-tutu' and female a 'wiick-wickwick-wick' call. Generally silent at sea and roost sites.
BEHAVIOUR Highly aerial, soaring effortlessly and dispersing huge distances. May hang in the air, wings angled, with only the occasional deep wingbeat but in active flight capable of great speed and dramatic twists and turns, using tail as a rudder. Gregarious at all times. Will often perch for hours, sometimes with wings held open and inverted. The main prey are flying fish, particularly *Exocoetus volitans* and

Cypselurus furcatus. These are snatched from just above the surface of the sea as they flee underwater predators. Though well able to feed directly on surface fish in this way, frigatebirds are most famous for their spectacular aerial piracy in which one (or several) pursue a hapless seabird relentlessly until it disgorges the fish in its crop. This is then caught in mid-air before it hits the sea. At the main breeding colonies at Aldabra, Red-footed Booby is most often mugged in this way. At out-of-season roosts such as Aride, shearwaters, terns and tropicbirds are all victims[1]. However, the importance of piracy as a food source is often exaggerated; clearly 10,000 pairs of frigatebirds (the joint total of both species) at Aldabra could not rely on a similar number of Red-footed Boobies to feed both colonies! In one study at Aldabra, only 18% of chases by frigatebirds resulted in food being obtained[2]. Females chase more than males. When males do chase, it is more often for nesting material than food. By contrast, an Aride study found, males chased more than females and most attacks were by juveniles, (suggesting Aride might be a training ground) with a higher success rate than at Aldabra of 38.5%[1]. Frigatebirds have also been recorded as taking and eating the eggs and chicks of other birds though at Aldabra this is a rare event in marked contrast to studies at Christmas Island, Pacific Ocean[3] or the Galápagos[4]. Will also take carrion, dead fish and hatchling turtles if they chance upon them. Birds frequently dive to drink on the wing from brackish pools at the eastern end of Aldabra, a habit recorded at very few locations worldwide.

BREEDING Male displays to passing females by inflating its incredible scarlet throat-pouch, pointing the bill skyward, turning wings over and vibrating them and shaking bill to make an extraordinary reeling or drumming sound. Female shows interest merely by hovering overhead, in contrast to Lesser. Sometimes, displaying male responds to appearance of another male overhead with a subdued display. If arriving male lands too close, it is chased away by first male, aggressively snapping and stabbing with its bill. Builds loose stick nests exclusively on top of mangroves inside the lagoons[5]. Nests appear to be grouped in clusters of about 20, possibly due to female choice. Males outnumber females so that male rivalry is intensified and females select a male in a cluster where there is less likelihood of the site being stolen (a major cause of nesting failure)[6]. Male collects nesting material, mainly twigs and leaves of mangroves and Pemphis which female incorporates into the nest. Occasionally material is robbed from other frigatebirds in flight, from unguarded nests or from Red-footed Boobies. Nests generally face northwest, maximising protection from the strong southeast monsoon winds of May to October. The timing of breeding depends on sufficient food to lay down the fat reserves to see birds through the long ritual of courtship and nest building, during which there is little time to feed. The main courtship activity lasts a fortnight with a marked peak in activity in August and September, and the single white egg is laid August to January and incubated for up to 120 days. Both sexes share incubation approximately equally. Young take between four and six months to fledge and are fed by parents for up to 18 months. The length of the breeding cycle combined with the delay in the onset of moult at least until late in the fledging period means successful breeders cannot breed annually. They may not be fully mature until 8-10 years old, but can be long-lived birds, the record being 34 years.

RANGE Breeds in the south Atlantic Ocean, Indian Ocean and Pacific Ocean. Race *aldabrensis* breeds in a few breeding colonies in the western Indian Ocean including Aldabra, Europa, Cargados Carajos and Tromelin.

STATUS Breeds at Aldabra (4,000 pairs[5]) on Malabar with 10-20 pairs at Cosmoledo on Menai and South Island. Formerly bred Astove[7], Assumption[8] and St Pierre[9] and probably elsewhere. Present at breeding colonies throughout the year, but also wanders widely and regularly seen throughout Seychelles. Records kept on Aride show that roosting birds are present in all months (lowest numbers *c*200 birds in May and June) but that numbers build up steadily in August and September to reach a peak in October and November (maximum more than 4,500[10]). As the start of the main breeding season is late August to October, these birds must represent late fledging birds from the previous year, which will presumably miss out breeding for that season. Roosting birds on Aride include many immature individuals with reddish heads, unlike the Aldabra form, indicating birds may be from even further afield than this, the nearest breeding site. Roosts also occur on Cousin, Bird, Coëtivy (about 2,500 birds) and St Joseph Atoll (mainly on Ressource; over 5,000 birds). The combined total population of Great and Lesser Frigatebirds at Aldabra (perhaps 30,000 birds including non-breeders) could make this the world's second largest breeding site after McKean Island which has 40,000 birds[11]. However, a number of other Pacific islands such as Enderbury, Phoenix and Howland have totals similar to Aldabra. The Aldabra population, first censused in 1976-77 was probably roughly stable between that date and 2000[12,13].

THREATS AND CONSERVATION Colonies have been wiped out by collection of adults and young for food. Some poaching continues, including at major roosts. Aldabra colony is reasonably assured of survival by its remoteness and protected status. The population on Cosmoledo is endangered and poaching could wipe it out completely. Restriction of nest sites to mangroves at Aldabra may be due to the presence of rats. At Cosmoledo and elsewhere in the world, Great Frigatebirds breed in Pemphis, common at Aldabra, but mangroves growing in water may give more protection from predation than terrestrial vegetation.[2] Breeding sites are prone to disturbance including from tourists. This has caused whole colonies at Aldabra to desert in the past, though strict rules on visiting colonies have now been implemented by SIF.

GENERAL Frigatebirds are extraordinary for a number of reasons, but their overall design breaks many records in the bird world. They have a wingspan well over 2 m, yet weigh no more than 1.6 kg (about the same as a Mallard). Their bones are incredibly light, no more than 5% of their total weight and this, combined with their enormous wings, gives them the lowest wing-loading (weight per wing area) in the bird world. They are seabirds (recent DNA analysis suggests that they are related to shearwaters, divers and penguins) but are quite unable to swim, having so little waterproofing that they would drown in minutes. They do not even have webs on their feet, which are so small that they cannot walk on land, merely perch. All this for a bird that spends the bulk of its life over the sea, feeding at times up to 2000 km from land.

SIMILAR SPECIES Lesser Frigatebird is significantly smaller, but this is almost impossible to judge even among wheeling birds close to the observer. Most reliable character is the white mark in the axillaries, present in both sexes. This shows as white 'armpits' extending onto the underwing, whereas in Great Frigatebirds, white patches of females and immatures are confined to the belly. In the female the black comes further up the belly than on the female Great and the throat is black rather than grey. Viewed from above, wings are all-black (Great shows a pale bar).

References: 1. Birch (2000), 2. Diamond (1975a), 3. Schreiber and Ashmole (1970), 4. Nelson (1967), 5. Reville (1983), 6. Reville (1988), 7. Stirling (1843), 8. Baty (1895), 9. Bergne (1901), 10. Bowler and Hunter (1999), 11. Sibley and Clapp (1967), 12. Burger (2000a), 13. Burger and Betts (2000).

LESSER FRIGATEBIRD *Fregata ariel*   Plate 8

Other name: Least Frigatebird   **French:** Frégate ariel   **Creole:** Pti Fregat

DESCRIPTION Length 70-80 cm; wingspan 175-195 cm. Similar outline to Great Frigatebird, significantly smaller, but size difficult to judge even among many wheeling birds. From below, best distinguished by white spurs on 'armpits' and above, by absence of pale wingbar. **Male** is all-black with white spurs on axillaries. Legs reddish brown to blackish. Bill grey to black. **Female** also has white spurs as extensions to underwing of white chest-patch, plumage otherwise black including throat (which is white in female Great Frigatebird) and black gusset on lower margin of white breast-patch. Legs pink or reddish. Bill blue-grey. Female averages larger. **Immature** is less white on belly than immature Great Frigatebird, white of breast extending to underwing as adult. White or buff head, many having a strong reddish suffusion (unlike Great Frigatebird at Aldabra, though other populations have a reddish tone).

VOICE As Great Frigatebird, only vocal at roosts and breeding sites. In display, male makes a rapid whistle. When returning to nest, male gives a short whistle and female calls 'chik-chuk-chuk'.

BEHAVIOUR Similar to Great Frigatebird in behaviour and diet (though Lesser has a marked preference for squid). Indeed, the overlap between these two species, particularly in feeding techniques, is greater than any two other seabird species (e.g. Red-tailed and White-tailed Tropicbirds). Differences in size, display and a slight difference in nest site may explain absence of interbreeding[1].

BREEDING Breeds in the dry season, June to October. The stronger southeast winds that blow at this time create a more reliable food supply and allow birds to wander further from the breeding colony. On Aldabra, the display period is much shorter at only 26 days compared to 91 in the Great Frigatebird. Male gives a similar display, but accompanied by a whistling call and unlike Great Frigatebird, female registers interest in a group with a distinctive 'goose-neck' action. In this, the female stretches the head forward and downward, sometimes with wings vibrated rapidly and tail wagged from side to side for a few seconds before flying on or descending to join a male[1]. Unlike Great Frigatebird, nests are evenly spaced in groups of up to 250[2]. Also, males advertising for mates do not significantly outnumber females[3]. Nest is built in mangroves and also terrestrial shrubs including Pemphis[3]. May nest on lower branches than Great Frigatebird. Studies suggest that the denser they nest, the lower is breeding success. Materials used are similar, mainly mangrove and Pemphis leaves and twigs, but Lesser also uses vines which grow over the mangroves. Breeding colonies can be quite chaotic and occasionally unpaired males arriving late can be quite disruptive, even taking over occupied nests and killing nestlings. A single white egg is laid between April and October and incubated for 45 days; fledging takes five to six months and there is a period of up to six months afterwards in which young birds continue to be dependent on the female for food. Birds may use different nests each year and almost certainly form a new pair each season. This is unusual among seabirds, which usually mate for life. Failure rates are high; probably only 25% of all nests fledge young and only about 5% of chicks reach maturity.

RANGE Breeds in a similar global distribution to the Great Frigatebird, but does not breed around the Tropic of Cancer as does Great. Race *iredalei* breeds in the western Indian Ocean at Aldabra and Cargados Carajos. Most breeding sites are near or below the Equator, although birds still wander well north, reaching the coast of India.

STATUS Breeds Aldabra (6,000 pairs[3]). Formerly bred Cosmoledo and possibly elsewhere. Some turn up at roosts in the granitic islands but are vastly outnumbered by Great Frigatebirds. Possibly Lesser Frigatebirds are more sedentary or perhaps have a different path of migration outside the breeding season. At Farquhar, in contrast to elsewhere in Seychelles, there is a major roost which comprises around 90% Lesser Frigatebird[4]. One ringed as a chick at Aldabra in August 1969 was recovered exhausted near Bombay in June 1970, 4400 km to the northeast. The Aldabra population, first censused in 1976-77 was probably roughly stable between that date and 2000[5,6].

THREATS AND CONSERVATION Exploitation has led to extinction on some islands including Cosmoledo. The sole surviving colony at Aldabra is reasonably secure given the atoll's protected status.

However, elsewhere roosting birds are still sometimes taken for food. As with Great Frigatebird, colonies may desert due to disturbance, and SIF have enforced strict rules for visitors to colonies at Aldabra.

SIMILAR SPECIES Great Frigatebird female and immature have white of underparts variable in extent, but always confined to the belly. Male is all-black below, lacking the white 'armpits' of Lesser Frigatebird. From above, shows a pale wingbar.

References: 1. Diamond (1975a), 2. Reville (1991), 3. Reville (1983), 4. Skerrett (1996b), 5. Burger (2000a), 6. Burger and Betts (2000).

GREY HERON *Ardea cinerea* — Plate 9

French: Heron cendré   **Creole:** Floranten Sann

DESCRIPTION Length 90-98 cm; wingspan 175-195 cm. A large, mainly grey heron with black-and-white head, pale neck and yellow bill and legs. The only large heron resident in Seychelles. **Adult** is a very tall long-legged bird with long, dagger-like bill, S-shaped neck and predominantly ash grey plumage. Very broad wings; the back and wing-coverts pale grey contrasting with black flight feathers. Prominent white flashes on leading edge of wing show up clearly when seen in flight head-on. White cheeks, throat and sides of neck contrast with black line from eye to crown. Black streaks on neck continue onto breast. Bill yellow (with a reddish flush in breeding season), tinged green around eye. Legs and feet dull orange-brown becoming pinkish in breeding season. **Juvenile** appears more uniform grey than adult (including grey forehead and crown), whitish on face and upper neck only, lacking black plumes. Bill horn coloured, becoming yellow in second year; legs greenish grey.

VOICE Flight call a loud, guttural 'graank'. At nest site, a deep grunting 'urr-urr-urr-urr'.

BEHAVIOUR Feeds in a variety of habitats, including grassland, inland marshes, streams, mangroves, open seashore estuaries and reef flats. Solitary away from breeding colonies. Ponderous, heavy flight with slow wingbeats, legs trailing, neck coiled and rounded at the keel. Territorial when feeding, individuals spaced out 10 m or more along the shoreline. Preferred fish include eels, common in lagoons and reef flats. Can catch fish up to 11 kg and up to two-thirds its own length. Will also catch and eat frogs, crabs, shellfish, snakes, mice, hatching turtles and even seabird chicks such as Sooty Terns from the edge of breeding colonies. At Aldabra, Diamond found a pellet thought to be from a Grey Heron, containing Brown Noddy remains and three rings from chicks he had ringed[1].

BREEDING Breeding is usually colonial, occasionally solitary, in large untidy stick nests in trees, but will make its nest on the ground on occasion. Colonies may include other species (e.g. breeds with Cattle Egret on Hodoul Island; at Cosmoledo breeds with Red-footed Booby[2]). Complex displays involve a wing-flap greeting and mutual preening. On Mahé birds come into breeding condition in April and are actively nest building from May into June. Young may be in the nest from as early as May to as late as December if there is protracted rainfall. At Aldabra, young birds have been noted from August to April, indicating breeding is also extensive. Likewise at Farquhar and in the Amirantes, though information is sparse, the breeding season appears to be extensive. Male establishes nest site, defending against intruders with a lunge of the bill, plumes erect, and gives a yelping advertising call. Clutch size as usually two, but sometimes up to four[3]. Eggs are pale blue in colour. On Mahé two chicks per nest seems typical. Incubation is only 21 days, but chicks may take up to seven weeks to fledge, remaining for a further two to three weeks in the nest while their parents continue to feed them.

RANGE Nominate *cinerea* breeds across most of Africa and across Europe and Asia merging with race *jouyi*, which occurs from Japan to Indonesia. Most northern populations are migratory. Other races include notably *firasa* in Madagascar and Comoros (which has longer bill and tarsus than nominate).

STATUS Breeds throughout Seychelles. In the granitic islands breeds only on Mahé and its satellite island of Hodoul, records from all other islands appearing to be mostly of passage birds. Measurements of birds at Aldabra show bill length exceeding maximum for the nominate *cinerea* (as in Madagascar race *firasa*) but tarsus shorter than *firasa* (as in *cinerea*). Birds are probably best regarded as intermediate between the two races (though one bird measured from Amirantes stands as nominate)[1]. Grey Herons were once taken for food and killed for sport. Guns were confiscated following the coup d'etat of 1977. Due at least in part to the tourism boom following the opening of the international airport, which led to increased affluence and sophistication, the habit of routinely killing wild birds disappeared. This came too late to save the Grey Heron, which was wiped out as a breeding species in the granitics by 1970. During the early to mid-1980s, the occasional vagrant Grey Heron would be seen around Victoria, Mahé and along the east coast of Mahé. Sightings of these huge birds drew remarks from many Seychellois who had forgotten the existence of the 'Floranten'. Breeding commenced towards the end of the 1980s at the same site as the main Cattle Egret colony, just south of Le Chantier roundabout. Numbers built up and were estimated at 24 birds by 1993 and 40 by 1995, by which time herons were once more a common sight around Victoria and a second heronry had been founded on Hodoul Island in Victoria Yacht Basin. The Le Chantier colony was abandoned shortly thereafter, possibly due to disturbance. Grey Herons were observed nest building on Cerf in 1995, but abandoned possibly due to harassment from domestic guard dogs that patrol the be[ ] chasing birds (W. Travis pers. comm.). Also breeds Amirantes; sites used to include African Banks, b[ ] longer do so, probably due to loss of suitable nesting trees. Elsewhere they still breed at the southern[ ] especially Aldabra, with a population of a few hundred pairs, and on Cosmoledo.

THREATS AND CONSERVATION The main heronry of the granitics on Hodoul Island is threatened by development. Land reclamation in 2000 increased the size of the island with a view to further development and the same programme in-filled most of the feeding area used by herons on mudflats north of Victoria. In the outer islands, the species is now fairly secure and birds are no longer killed for food.

SIMILAR SPECIES Purple Heron is smaller, with buff-chestnut neck and buff, not grey wing-coverts. The following potential vagrants should also be considered: **Madagascar Heron**, *A. humbloti*, (L100 cm), breeds in Madagascar and possibly the Comoro Islands. Entirely black with dark grey neck and blackish dirty yellow bill and legs. **Black-headed Heron**, *A. melanocephala*, (L96 cm), resident throughout Africa. All-black with a white chin and throat and blackish legs. **Goliath Heron**, *A. goliath*, (L135-140 cm, WS210-230 cm) breeds throughout mainland Africa, vagrant to India and Madagascar. Huge, with a purple-brown head, neck, lower breast and belly, white throat, dark grey back and wing-coverts and blackish legs.

References: 1. Benson and Penny (1971), 2. Skerrett (1996b), 3. Penny (1974).

PURPLE HERON *Ardea purpurea*   Plate 9

French: Héron pourpré   **Creole:** Floranten Mov

DESCRIPTION Length 78-90 cm; wingspan 120-150 cm. A fairly large heron, dark grey-brown with sinuous buff-chestnut neck and slate grey back. **Adult** is smaller than Grey Heron, but overall impression is of a much thinner, darker, browner bird. Black crown and neck plumes, black line through cheek and black line down side of neck. Neck chestnut brown, chest and belly chestnut to black. Wing-coverts dark slate grey with brownish hue when breeding. In flight, wing-coverts show buff-red to purple-brown, with bright buff-coloured wing-flashes on the leading edge. Bill, lore and legs yellowish, brighter in adult breeding. **Juvenile** is a browner and duller version of the adult, lacking grey on upperparts, with sandy buff upperwing-coverts and underparts and without prominent neck-stripes and neck plumes. **Subadult** resembles adult but is browner above, retaining broad sandy buff margins to upperwing-coverts.

VOICE Generally silent outside breeding season. Call similar to Grey Heron, but higher-pitched and less throaty.

BEHAVIOUR Much more skulking than Grey Heron, preferring the cover of reedbeds or dense vegetation. Mainly solitary outside breeding season. In flight, neck is coiled, appearing snake-like and more angular at the keel than Grey Heron, long legs trailing; flight not as heavy, lighter body sometimes rising on downward wingbeat. Extremely long toes are adapted to walking on floating vegetation and in reeds: these show as more splayed toes in flight. Food consists of fish (smaller species than Grey Heron) and large insects (e.g. dragonflies and crickets). Will also take small mammals and frogs.

RANGE Nominate race breeds throughout southern Europe and southern Russia, wintering in sub-Saharan Africa. The Indian/South-East Asian race *manilensis* is mainly resident; it is paler with less black on the neck. There is also a breeding population in East Africa and a Madagascan race *madagascariensis*, darker with less distinct gular streaking.

STATUS Vagrant to the granitic islands and Bird, usually between October and February, but also as late as May and as early as August. Race unknown, but the highly migratory nominate race appears most likely.

SIMILAR SPECIES Grey Heron is slightly larger, greyer and paler, with white crown and cheeks and whitish neck, with no buff tints in the plumage. See Grey Heron for other potential vagrants herons.

CATTLE EGRET *Bubulcus ibis*   Plate 10

Other name: Buff-backed Heron   **French:** Héron garde-boeufs   **Creole:** Madanm Paton

DESCRIPTION Length 40-56 cm; wingspan 88-96 cm. A small, short-legged, white heron tinged buff on lower back, common away from water. **Adult breeding** is mainly white with plumes on crown, breast and lower back buff-gold (race *seychellarum*) or rufous buff (race *ibis*). Lore and legs acquire a pinkish purple hue with scarlet around eye. Feathering under bill gives appearance of a prominent jowl. In flight, appears stockier than other egrets, with bill projecting less far beyond tail. Bill yellow and relatively short compared to relatives. Leg colour is highly variable, crimson in displaying birds, pinkish in incubating birds. Male brighter, with darker rusty-buff crown. **Adult non-breeding** is entirely white; the bill and lore pale yellow, grey-blue around the eye, legs yellowish black. **Juvenile** is also white, sometimes with a grey tinge, bill dusky to dull yellow and legs blackish, soon changing to blue-grey or greenish grey.

VOICE In flight, a short, harsh 'aark' or disyllabic 'aark-ak' and at breeding colonies a duck-like 'ga-ga-ga-ga'.

BEHAVIOUR Unlike other herons, often found away from water and in granitics appears to show a preference for areas of human activity and cultivation. Highly gregarious at all times. Flight direct, with shallow, rapid wingbeats. On Mahé most obvious at Victoria market, waiting for scraps around fish stalls, fishing along culverts or streams through the town and at low tide, and on east coast mudflats near the inter-island quay. Very common at Mahé's Providence rubbish tip where feeds on flies. Also feeds in open pastures and its name enshrines its association with livestock; feeding cattle disturb grasshoppers and other insects which are a source of food. It has been estimated that by following cattle or tractors, Cattle Egret can increase the number of prey they catch by 50%. At Aldabra associates to some extent with giant tortoises. On seabird islands such as Desnoeufs and Goëlettes (Farquhar Atoll) preys on eggs and chicks, particularly of Sooty Terns. May also chase Sooty Tern chicks forcing them to disgorge fish[1].

BREEDING In the granitics, the main colony was at Le Chantier but moved to nearby Hodoul Island in the mid-1990s, breeding colonially with Grey Herons, mainly April to October. At Farquhar, also breeds April to October on Goëlettes, in bushes next to a large Sooty Tern colony, the timing possibly to coincide with the terns' breeding season. In contrast, at Aldabra breeds September to February on Ile aux Aigrettes in a mixed colony with Little Egret and Madagascar Pond-heron. Display takes place on or around the untidy nest of twigs, the male prancing around the female with wings open and bowed, raising plumes on neck to reveal bright blue bare skin along the nape. Clutch size 1-3 (lower than in continental regions), eggs very pale blue. Both sexes share incubation, which takes 22-26 days, fledging a further 30 days. Chicks have white down. Young are independent about 15 days after fledging.

RANGE Race *ibis* breeds over most of Africa other than desert areas, from southwest Europe to the Volga Delta, South America (invaded 1930s), North America (first bred Florida 1953, Canada 1962) and in the western Indian Ocean including Madagascar, Comoros, Mauritius, Aldabra group and Farquhar group. Asiatic race *coromandus* breeds from Pakistan and India south to Maldives and spread in 20th century east into China and Japan and via New Guinea to Australia, reaching New Zealand in 1963. Shorter-winged race *seychellarum* breeds in Seychelles' granitic islands/Amirantes and is intermediate between the other two races, possibly due to continuous genetic infusion from both.

STATUS Race *seychellarum* is resident in the granitic islands, Coëtivy and Amirantes with race *ibis* resident Aldabra, Cosmoledo, Assumption, Astove and Farquhar (mainly on North and Goëlettes). Localised on Mahé (mainly in the Providence-Victoria-Beau Vallon area), Cousin and Praslin (often seen at Amitie airstrip and Côte d' Or), resident on Frégate and Bird and occasional on other granitics. In the Amirantes, occurs notably at Desnoeufs (c.200 birds) where numbers may increase during the Sooty Tern season (May to September). Numbers have probably increased since human occupation. At Aldabra, numbers may also have increased (Abbott saw only one during a three-month stay in 1895) though birds are most common in the east and south of the atoll whereas the only settlement is on Picard, western Aldabra.

THREATS AND CONSERVATION May have benefited from the arrival of man and the increase in opportunities for scavenging such as at dumps and markets (and even hotel restaurants). This benefit is probably more marked than in other countries because of the lack of competitors such as gulls. However, development can pose a threat as evidenced by the abandonment of the Le Chantier colony near Victoria. The future of the main colony at Hodoul Island is in doubt following reclamation work in 2000, which increased the size of the island with a view to developing the area. The Ministry of Environment has authorised culling of Cattle Egrets on Praslin and Mahé, where it has been alleged that the increase in populations feeding at dumps close to airports might pose a threat to aircraft. Culling at Amitie, Praslin in 1999 led to a noticeable decrease in the size of the colony breeding on nearby Cousin.

SIMILAR SPECIES Little Egret and the white morph of **Dimorphic Egret** are distinguished by black bill, longer black legs and yellow feet. **Great White Egret** and **Intermediate Egret** are considerably larger with black bill, legs and feet.

Reference: 1. Feare (1975c).

GREAT WHITE EGRET *Egretta alba*   Plate 10

Other names: (Great) White Heron, Great (or White or Large) Egret.
French: Grande aigrette   **Creole:** Gran Zegret

DESCRIPTION Length 85-102 cm; wingspan 140-170 cm. A large, pure white heron, similar in size to Grey Heron, with long snake-like neck and long, black legs. **Adult** is all-white, huge size rendering confusion with other species unlikely, except possibly Intermediate Egret. Bare facial skin dull yellow or yellowish green with black gape line extending to behind yellow eye (unlike Intermediate). Sharp kink in long neck. Lacks feathering under bill (unlike Cattle Egret). In flight, shows large bulge in breast, legs project considerably beyond tail tip. Very long black legs and feet. Long, dagger-like yellow bill. **Adult breeding** develops long white plumes on nape, bill becoming partially or entirely black, lore bright green, eye sometimes red. **Immature** similar to non-breeding adult, but bill yellow with a blackish tip.

VOICE A rapid, rasping 'ar-ar-ar-ar'; also a deep guttural croak.

BEHAVIOUR Feeds in wetlands, meadows, by streams and pools; vagrant birds in Seychelles also recorded on open mudflats or coral beaches or inside the reef at low water. Mainly solitary (particularly vagrants). Flight much slower than in other egrets, with heavy wingbeats, slender body sometimes raised by power of downstroke. Diet includes small fish, insects, lizards, small mammals and young birds.

RANGE Nominate race of Europe to Asia is a partial migrant south to North Africa, other races breeding North and South America and from India to New Zealand. However, the race most likely in Seychelles is *melanorhychos* of sub-Saharan Africa and Madagascar. Frequently placed in genus *Casmerodius*, and

DNA studies suggest a closer link with *Ardea* than *Egretta*.

STATUS Vagrant to granitic islands and Aldabra mainly October to February, but one bird remained at Mahé for a full year.

SIMILAR SPECIES Intermediate Egret is somewhat smaller but size can be difficult to judge in a lone individual; neck thicker and less kinked, bill stubbier and gape line does not extend behind eye. **Dimorphic Egret** (white morph) and **Little Egret** are also all-white but much smaller and have yellow feet.

INTERMEDIATE EGRET *Egretta intermedia*   Plate 10

Other names: Yellow-billed (or Plumed) Egret   **French:** Aigrette intermédiare or Héron intermédiaire
Creole: Zegret Labek Zonn

DESCRIPTION Length 65-72 cm; wingspan 105-115 cm. A fairly large all-white heron with yellow bill, yellow-brown tibia and black tarsus and feet. Smaller, shorter-billed and shorter-necked than Great White Egret but larger than other egrets. **Adult** has bare facial skin at the base of the gape, which does not extend in a line to beyond the eye, as in Great White Egret, and a smooth, S-shape to the neck. Eye yellow; bill yellow, often tipped blackish. Legs yellowish brown above the tarsal joint (sometimes difficult to observe) and blackish below, with black feet. **Adult breeding** has long white plumes on back and breast (latter absent in Great White) and, briefly, bill orange-red tipped yellow, bright green lore, red eye and pinkish tibia. **Immature** similar to non-breeding adult.

VOICE Usually silent away from breeding grounds. Sometimes gives a low 'quawrk'.

BEHAVIOUR Frequents inland waters with abundant vegetation, rarely coastal. Vagrants are likely to be solitary, but forms loose flocks elsewhere. Buoyant flight with slow wingbeats. Usually feeds in shallow water, walking slowly. Diet includes small fish, frogs, insects and sometimes lizards.

RANGE Race *brachyrhyncha* breeds throughout sub-Saharan Africa. Race *intermedia* breeds India, South-East Asia, and race *plumifea* east Indonesia to Australia. Mainly sedentary, but with some dispersal in the dry season. Vagrant to Marion Island, southern Indian Ocean. DNA studies suggest more closely related to *Ardea* than *Egretta* and sometimes placed in monospecific genus *Mesophoyx*, or in *Casmerodius*.

STATUS Vagrant to granitic islands; recorded March to April (race unknown).

SIMILAR SPECIES Great White Egret is larger has all-black legs, longer more kinked neck, longer pointed bill, gape line and facial mask extending to behind eye. **Little Egret** is smaller and has black legs with yellow toes.

LITTLE EGRET *Egretta garzetta*   Plate 10

Other names: Lesser (or Spotless) Egret   **French:** Aigrette garzette   **Creole:** Pti Zegret

DESCRIPTION Length 55-65 cm; wingspan 86-104 cm. A medium-sized, slender, white heron with long, black bill, grey lore, black legs and yellow feet. **Adult** has all-white plumage, black bill with pale flesh base to lower mandible, lore grey or bluish grey, yellow eye and black legs with yellow feet obvious in flight. There is a very rare grey morph, which varies in tone from, blue-grey to black, chin and throat usually white. **Adult breeding** has two long, white plumes on the nape, and plumes on upper breast and scapulars, a blacker lower mandible, and brighter pale yellow feet. **Immature** is all-white, developing short dorsal plumes during first year, bill, legs and feet duller.

VOICE A harsh croak 'haaarch' when disturbed.

BEHAVIOUR May frequent inland marshes or coasts and tidal mudflats. Infrequent at drier sites, unlike Cattle Egret. Gregarious, but in Seychelles generally solitary or in twos or threes. Rather leisurely flight, sometimes gliding, head retracted and feet projecting well beyond tail. Diet includes fish, amphibians, insects and occasionally lizards and vegetation.

RANGE Nominate *garzetta* breeds southern Europe to South-East Asia, northern populations migrating southward to tropical Africa and Indian subcontinent. Other races breed South-East Asia to Australia. Often considered conspecific with Dimorphic Egret though in East Africa latter appears to be entirely coastal, unlike Little Egret.

STATUS Near-annual in the granitics; race unknown. Recorded in all months, but mainly October to May.

SIMILAR SPECIES White morph **Dimorphic Egret** is very similar, but lore is yellow (not grey). **Western Reef Egret** (see under Dimorphic Egret) is a potential vagrant, which should also be considered.

DIMORPHIC EGRET *Egretta dimorpha*   Plate 10

French: Aigrette dimorphique   **Creole:** Zegret Blan e Nwanr

DESCRIPTION Length 55-65 cm; wingspan 86-104 cm

A medium-sized black or white slender heron, with black bill and yellow lore; resident in Aldabra group. **Adult** occurs in two strikingly different colour phases: white morph is pure white with yellow lore and a black bill, black legs and yellow feet. Dark morph is blackish with an all-dark bill, white throat and similar leg colours and has a white flash on the 'thumb' of each dark wing, obvious in flight. 'Pied' individuals occur, with a mixed garb of white and grey, often with more extensive white in the wings or a white head, neck or both. Breeding pairs commonly include one white, one dark. Slightly longer-necked than Little Egret, the bill larger and heavier. At onset of breeding, lore may turn coral red and legs become suffused red. **Juvenile** is grey to slate grey, becoming mottled when moulting into adult plumage. Lore dark grey with yellow line from base of mandible to below eye. Dark grey bill with pinkish base to lower mandible. May show a variable amount of yellow on tarsus.

VOICE A deep, low grunt uttered on take off. Loose groups appear to communicate with a low rasping call also heard at nest site[1].

BEHAVIOUR Solitary, in pairs or in small groups along coasts. Also feeds inside the lagoon on sand or mud bars as tide rises when may be seen in large numbers spaced out round the shore and among mangroves but rarely found at inland pools. Leisurely flight, sometimes gliding, head retracted and feet projecting well beyond tail. Curiously, white morph birds are more wide-ranging in Madagascar where white morph is common inland but black morph is almost entirely coastal[2]. In East Africa, both morphs are coastal[3]. At Aldabra, there is some evidence of a higher population of black morphs within the lagoon and channels compared to the coast[1]. The reasons for these differences are unknown. Feeds mainly on crabs, small fish and eels up to 30 cm. Generally stalks prey, frequently moving feet in jerks, possibly to cause fish to move or alternatively to remain still until bird is within range[1]. Prey are harpooned or seized with a rapid strike of the head; the sinuous neck, like a compressed spring is straightened with lightning speed. If the bird misses, however, the sedate stalk degenerates into a flailing scramble through the shallows in pursuit of prey. Also sometimes stands or runs with wings spread while hunting.

BREEDING Season is extensive, but with a peak of egg-laying around November. In courtship, plumes on the head and neck are raised and a dance is performed in which one bird dives below the neck of another. Nest is a loosely woven platform of twigs 30 to 40 cm in diameter. Sometimes builds a new nest in an old one. Breeds colonially, nests often packed close together in low bushes. Lays 2-3 pale blue eggs. Incubation lasts 21-25 days, fledging 40-45 days.

RANGE Breeds and resident Madagascar, Comoros, Aldabra and East Africa. Seychelles birds have been claimed as a separate race, *assumptionis*, on the basis of a longer bill but this has not been generally recognised, there being considerable overlap. However, bill does average longer and wing shorter[4].

STATUS Breeds Aldabra (1,000-3,000 pairs[5]), with smaller numbers on Astove and Cosmoledo. Extinct Assumption (last noted as a breeding species in 1906[6]). White morph seems to outnumber the dark by about three to one (in contrast with East Africa where black outnumbers white at least two to one[2]).

THREATS AND CONSERVATION Extinction at Assumption was undoubtedly due to human factors including destruction of habitat. Never seriously exploited in Seychelles for the plume trade, as have other egret species, and the stronghold at Aldabra appears secure given the protected status of the atoll. May be susceptible at least to disturbance at Astove and Cosmoledo, which are regularly visited by poachers.

SIMILAR SPECIES Little Egret is very similar (sometimes considered conspecific), but lore is grey (not yellow) in adult non-breeding. **Western Reef Egret** *E. gularis* (L55-68 cm, WS88-112 cm) is a potential vagrant. Breeds Red Sea, Persian Gulf to India, dispersing south to Kenya. It also occurs in both black and white morphs, the former a mainly coastal bird, the latter sometimes inland. Less graceful than Dimorphic or Little with longer, more powerful yellowish, brownish or brownish olive (never black) bill with darker culmen, almost always slightly decurved at tip.

References: 1. Benson and Penny (1971), 2. Morris and Hawkins (1998), 3. Zimmerman *et al.* (1996), 4. Benson (1967), 5. Rocamora and Skerrett (1999), 6. Nicoll (1909).

SQUACCO HERON *Ardeola ralloides*   Plate 11

French: Crabier chevelu   **Creole:** Gas Loulou

DESCRIPTION Length 42-47 cm; wingspan 80-92 cm. A small, skulking heron with buff-brown upperparts, paler than other *Ardeola*, but with similar white wings and tail. **Adult non-breeding** has buff crown and nape with black streaking, sides of neck buff with finer dark brown streaking. Mantle, scapulars and inner wing-coverts dull buff-brown and unstreaked, contrasting dramatically in flight with white wings and tail. Also has much buff in underparts. Head and neck fawn-coloured with heavy darker striping and back and inner wing-coverts dull brown-buff. Retains vestiges of breeding mane. Bill pale green-yellow with black tip. Legs dull yellow-green. **Adult breeding** has head and neck with long, streaked, black-and-buff plumes; back and long scapular plumes fawn-brown with a pinkish blush. Overall brighter, more uniform buff above, wings and tail white, wing-coverts buff. Lore blue at first becoming yellow; bill bright greenish blue with black tip, legs and feet pink-orange. **Immature** similar to non-breeding adult but slightly darker, with neck and breast more heavily streaked, belly greyish and wing-coverts and primaries tinged with brown shading.

VOICE Usually silent, but may utter a shrill, harsh 'rrra' especially at dusk. When flushed, may call 'kek-kek-kek'.

BEHAVIOUR Normally prefers the dense cover of reeds or bushes alongside fresh water, but in early morning will feed more out in the open. Roosts in trees, often with other egrets and herons.

Solitary outside breeding season. Flight rapid and more fluttering than egrets, white of wings, rump and tail contrasting dramatically with buffy areas of plumage. Food consists mostly of freshwater insects and their larvae, but will take frogs, small fish, lizards, molluscs and worms.

RANGE Breeds throughout eastern Europe, Turkey, Iraq and Iran, wintering in sub-Saharan Africa. Also breeds across much of sub-Saharan Africa and in Madagascar where it is mainly sedentary but may cross between Madagascar and East Africa. Monotypic.

STATUS Uncertain; there are unconfirmed reports from Aldabra and the granitic islands.

SIMILAR SPECIES In breeding plumage similar to **Cattle Egret** which has longer neck but heavier jowl and much whiter, less buffy plumage (buff restricted to crown, breast and lower back). In other plumages, similar to **Madagascar Pond-heron** non-breeding, which has darker sepia brown upperparts and crown heavily streaked buff and heavier, darker streaked head and underparts; also appears stockier, more bull-necked with more powerful bill. **Indian Pond-heron** non-breeding lacks residual mane and is dumpier with shorter, more rounded wings. It is duller and darker on mantle (which is unmarked grey-brown), has more intense streaking and often shows a pale line above the lore between eye and bill.

INDIAN POND-HERON *Ardeola grayii*   Plate 11

Other names: Paddybird, Indian Pondbird   **French:** Crabier de Gray   **Creole:** Gas Endyen

DESCRIPTION Length 42-45 cm; wingspan 75-90 cm. A small heron, very similar to other *Ardeola*, but with dark unstreaked mantle. **Adult non-breeding** has head and neck dark brown streaked buff and drab grey-brown mantle and scapulars. Often shows a pale line from eye to bill above the lore (unlike Squacco Heron). Underparts white with breast heavily streaked dark brown. In flight shows pure white wings and tail, with grey smudging on webs of two outermost primaries. Basal two-thirds of bill blue or yellow-orange, black at tip. Legs and feet green. **Adult breeding** has head and neck unstreaked pale yellowish buff with long white nape plumes; mantle and scapulars deep reddish brown. **Immature** has the back brown, with buff streaking on scapulars as in non-breeding adult, but wing-coverts with buff and brown spotting, breast with strong, thrush-like buff and brown spotting, rather than the continuous wavy streaks of the adult. Dark shafts to primaries, the two outermost smudged grey and the next three with grey tips.

VOICE A high, harsh squawk or a rapid 'kwok-kok-kok'. May utter a deep croak when disturbed.

BEHAVIOUR Frequents margins of fresh or brackish water. Stands motionless or moves slowly. Less skulking than Squacco Heron and probably more tolerant of humans. Flight fast and agile. Feeds day or night, diet including small fish, frogs, insects, vegetation and baby turtles.

RANGE Breeds throughout the Indian subcontinent from the Iranian coast to the Maldives. Breeds November to February in southern India and may then disperse according to local droughts. Mainly sedentary but moves in response to rains and recorded as a vagrant to Saudi Arabia, Oman and Maldives. Monotypic.

STATUS Vagrant to granitic islands and Bird mainly September to November; not recorded elsewhere in entire Afro-Malagasy region.

SIMILAR SPECIES Squacco Heron non-breeding has a dull buff-brown mantle and retains suggestion of breeding mane. **Madagascar Pond-heron** non-breeding has more boldly streaked sepia brown mantle.

MADAGASCAR POND-HERON *Ardeola idae*   Plate 11

Other names: Malagasy Pond-heron, Malagasy Squacco Heron
French: Crabier blanc   **Creole:** Gas Malgas

DESCRIPTION Length 45-48 cm; wingspan 84-92 cm. Unmistakable in breeding dress (likely only at Aldabra) with pure white plumage and pinkish legs. Non-breeding birds are dark sepia brown above with thrush-like spotting on underparts. **Adult breeding** is pure white sometimes with a buff wash. Long, white nuptial plumes on crown, upperparts and breast give a fluffy appearance, the back plumes extending beyond the tail. Eye yellow, lore greenish. Legs pinkish, feet duller and greenish. Heavy, broad bill is greenish blue with a black tip. **Adult non-breeding** similar to other *Ardeola* but white wings and tail contrast with dark sepia brown mantle and scapulars with broad buffish or creamy stripes. Crown and nape brown with heavy blackish streaking, bright yellow-ochre face and upper breast with bold black-brown streaks, becoming thrush-like spots on lower throat. Bill greenish grey tipped black, legs greenish grey or yellowish. **Immature** similar to non-breeding adult, but with brown streaks on wing-coverts, more cream streaking on back and scapulars. Brown-grey smudging on webs of outer and tips of inner primaries and tail feathers, pale green eye and dark tip to orange bill.

VOICE A nasal, rather musical, rattled 'burr' on the approach of other species[1].

BEHAVIOUR Generally keeps to cover usually close to small inland water bodies, or (at Aldabra) on lagoon islets, but will feed in more open grassy areas at times. Secretive and solitary, more a forest species than its relatives and has a tendency to fly up into a tree when disturbed. Flight rapid. Feeds mainly inland, diet including lizards and insects.

BREEDING Breeds with Dimorphic Egret and Cattle Egret at eastern Aldabra, notably on Ile aux Aigrettes, where the colony virtually covers this islet. Nests are located at other islets in this vicinity. Nests are smaller than Dimorphic Egret, 20 cm in diameter, built about 2 m height under a canopy in bushes and trees. Material is mainly twigs laid around a fork in a branch with a softer lining. Clutch size is normally three and eggs are pale blue with indistinct white mottling. Breeding commences around October to November. Incubation lasts about three weeks, chicks having buffy yellow down at first. Fledging takes a further 2-3 weeks.

RANGE Breeds throughout Madagascar and Aldabra and migrates to East and central Africa between May and October. First-year birds are thought to remain in Africa. Vagrant to Comoros. May have an earlier Asian, not African origin[2]. Monotypic.

STATUS Breeds Aldabra (probably fewer than 100 birds[1]). There are few observations during May to August and some or all birds may migrate to Africa (in common with breeding birds on Madagascar). Colonisation may have been very recent. Abbott failed to record it during his three-month visit September to December 1892[3]. It was perhaps this species noted in 1897 as medium-size herons white with black teguments[4] but not recorded with certainty until 1967[1]. Vagrant to granitic islands November to March.

THREATS AND CONSERVATION Globally near-threatened and the small population of Aldabra must be vulnerable. Conservation measures by SIF include prohibition of tourist visits to breeding sites.

SIMILAR SPECIES Virtually unmistakable in breeding dress, though could possibly be confused with **Cattle Egret** in flight, which lacks two-tone bill and whose legs do not project beyond tail. Very similar to other *Ardeola* in non-breeding plumage. **Squacco Heron** non-breeding has paler more uniform dull buff-brown mantle and scapulars with paler crown, paler, finer streaking on underparts, more slender bill and smaller head appearing less bull-necked. **Indian Pond-heron** non-breeding has unmarked grey-brown (not dark brown) mantle.

References: 1. Benson and Penny (1971), 2. Snow (1978), 3. Ridgway (1895), 4. Voeltzkow (1897).

GREEN-BACKED HERON *Butorides striatus*   Plate 11

Other names: Striated (or Mangrove or Little Green or Green) Heron
French: Héron vert or Héron strié   **Creole:** Manik Ordiner

DESCRIPTION Length 40-48 cm; wingspan 52-60 cm. The only resident small heron of Seychelles, generally dark, stocky and mainly coastal. **Adult** is not obviously green (despite name), rather dark slate-grey on the back, darker on the crown, the wing-coverts and tail iridescent dark green with rufous buff scaling and flight feathers smoky grey-black. White on throat, sides of breast dark chestnut brown, rest of underparts grey with heavy black and cream spotting on the underwing. Legs pale yellowish green seen when standing; but in flight the most obvious feature is bright orange soles of the feet showing up clearly on a bird flying away from the observer. At the onset of breeding legs become brighter, eye colour deepens to orangey, bill becomes all-black and the crown feathers may be longer and crest-like. Bill black, with a yellow streak at the base of the lower mandible. Lore and eye yellow. Sexes similar, male averages larger and female has sides of neck and chest washed brown and more distinct spotting on throat. Race **degens** (granitics) has green-black back, wing-coverts edged rufous buff, grey-brown sides to upper breast with heavier spotting. Race **crawfordi** (Amirantes to Aldabra) has purple blue-grey back, wing-coverts broadly edged creamy white, ash-grey hindneck and sides of upper breast, with only faint spotting on upper breast. **Juvenile** has back dark brown and wing-coverts dark brown with white and rufous streaking. Flight feathers are smoky blue-grey with white tips. Neck and underparts creamy white, heavily streaked and blotched dark brown throughout.

VOICE A rasping 'korak' when disturbed. This alarm call is well represented by the Creole name 'manik'. Gives a variety of chuckling calls at nest.

BEHAVIOUR One of the typical birds of Seychelles to be seen on almost every coast, even on the smallest islands. Fiercely territorial, individuals patrol their own favourite stretch of shoreline or mangrove. Stance typically crouched and horizontal either perched on the edge of rock pools when low tide exposes the coral reefs or stalking through mangroves looking for mudskippers. Strong, direct flight, head appearing heavy. When at rest adopts a more upright stance and prefers to rest and preen in low trees near the shore. May be active throughout the day. Food includes small fish, frogs, skinks, grasshoppers, crickets and dragonflies. Small crabs may be caught also and may take eggs and chicks of small birds.

BREEDING In courtship display, constantly raises crest; also engages in short aerial chases and occasional duetting. Breeds throughout northwest monsoon, eggs noted from October to March, but may breed in any month. Nesting is generally solitary, occasionally in small numbers. Nest is an untidy platform of twigs with a shallow bowl, in a tree close to or over water. Two or four eggs are laid, pale green or blue and incubation is around 23 days. Within a week of hatching young are able to climb around in the nest tree, fledging in about 30 days.

RANGE Many races (30 commonly recognised, though some controversial), breeding mainly tropical and subtropical latitudes across the globe.

STATUS Breeds throughout Seychelles. Two endemic races occur, *degens* in the granitic islands and *crawfordi* from Amirantes to Aldabra. Studies suggest *crawfordi* to be Asiatic in origin whereas *degens* is African[1]. This overlap is perhaps a fascinating reminder that Seychelles' avifauna is a crossroads for two continents.

THREATS AND CONSERVATION Despite the prevalence of rats on so many islands, this is the commonest heron throughout. Every island, large and small, seems to possess at least one bird patrolling its beach and reef.

SIMILAR SPECIES Streaky immature birds may recall a miniature bittern and have sometimes been assumed to be **Yellow Bittern**, which is much paler buff with contrasting pale patches on wing-coverts and is confined to inland marshes where it is far more secretive. Immature **Black-crowned Night Heron** is larger with conspicuous pale spots on wings, mantle and scapulars.

Reference: 1. White (1951).

BLACK-CROWNED NIGHT HERON *Nycticorax nycticorax* Plate 12

Other names: Night Heron, Black-capped Night Heron **French:** Bihoreau gris **Creole:** Manik Lannwit

DESCRIPTION Length 55-65 cm; wingspan 105-112 cm. A chunky heron with large head, short thick neck and short legs, much smaller than Grey Heron and much larger than Green-backed Heron. **Adult breeding** has black crown with long, white head plumes, back and crown black with a blue-green gloss, white forehead and front cheeks. Tail and wings uniform slate grey, paler grey on underparts. Large round wings obvious in flight, toes projecting slightly beyond tail. Shortish legs are pinkish red, brightest for short period during courtship. Blue-black bill and lore, red eye. **Adult non-breeding** lacks plumes, the cap, back and bill duller and legs yellowish. **Juvenile** has the crown and back dark brown with buff streaks and spots, wings sepia brown with large white spots and underparts grey streaked with brown. Bill dirty yellow with black tip, eye brown at first becoming orange during first year, legs dirty yellow-green. **Second-year** similar to adult in general pattern but crown greyish black, plumes absent, back grey-brown and underparts washed pale grey-brown; may retain some streaking. Eye red-brown, bill blackish, yellow at base of lower mandible and legs yellow.

VOICE A crow-like croak 'quark', heard at dusk or after dark. Often calls in response to disturbance and in flight.

BEHAVIOUR Favours margins of freshwater wetlands, migrants sometimes coastal. Arboreal when roosting or nesting, remaining deep within vegetation during the daytime, more active at dusk or during the night. May be solitary or gregarious, often seen in small numbers in trees during late afternoon, moving to feeding areas around 6 pm. Distinctive, compact flight silhouette with rapid wingbeats of short, rounded, owl-like wings appearing rather laboured, legs barely projecting. Feeds in fresh, brackish or saltwater, alongside pools, streams in mangroves, or dry pastures, either standing motionless or walking slowly. Diet as varied as other herons, including fish, amphibians, insects, molluscs and small birds.

BREEDING In Africa, breeds during height of rains and is perhaps likely to breed December to April in Seychelles (young birds with down seen May). Courtship may involve an erect stance, standing on one foot then the other, ending hunched, bill pointed downward. Breeds colonially often with other herons or egrets. A platform nest is built in trees or bushes in which 3-5 pale blue-green eggs are laid. Incubation lasts *c*22 days, fledging *c*45 days.

RANGE Nominate race breeds from Spain east to Japan with scattered populations in Africa and Madagascar. Disperses widely after breeding, most European birds migrating south to sub-Saharan Africa from August to March. Other races breed North America, South America and Falklands. Vagrant as far afield as Amsterdam Island, Indian Ocean.

STATUS Resident Mahé, Praslin, Cousin and Silhouette. Invasion is very recent, with first confirmed vagrant report only in October 1992, a first-year bird on Mahé. Further vagrant records followed on Mahé in March 1993 and May 1994. Continuously present Mahé since November 1995 and Silhouette since August 1996. Downy juveniles observed on Mahé since 1996 and Silhouette since 1997, indicate probable commencement of breeding around this time. First nests discovered on Cousin by F. Payet, March 2000, a colony of at least six pairs at nest-building stage.

THREATS AND CONSERVATION It remains to be seen whether the toehold in Seychelles recently gained by this species will be consolidated. As with other waterbirds, this will depend in part on the conservation of freshwater wetlands, which are under pressure from development.

SIMILAR SPECIES Adult unmistakable. First-winter birds easily separated from **Green-backed Heron** which is much smaller and lacks white spots on wings and mantle.

YELLOW BITTERN *Ixobrychus sinensis* Plate 12

Other names: Chinese (Little) Bittern, Long-nosed Bittern
French: Blongios de Chine **Creole:** Makak Zonn

DESCRIPTION Length 30-40 cm; wingspan 42-47 cm. A small, skulking pale buff heron, superbly camouflaged among dry reed and grass stems but with strikingly patterned wings in flight. **Male** has a foxy orange mane on the neck with reddish brown face and hindneck and black crown. Upperparts and wing-coverts pale buff, tail black and underparts creamy with darker streaks on the breast. In flight, black flight feathers contrast sharply with pale buff wing-coverts and back. **Female** is darker chestnut brown on the back with diffuse buff margins to feathers of scapulars and mantle, faint rufous streaking on the black crown and chestnut streaks on pale buff foreneck. In both sexes legs yellow, eye yellow, lore yellowish green and bill yellow with a black culmen. Bill brightens to reddish pink in the breeding season. **Juvenile** has bold rusty streaking on the neck and breast; crown, back and scapulars are heavily mottled dark brown.

VOICE In breeding season calls a soft, repetitive 'oo-oo-oo' recalling Barred Ground Dove, but more monotonous and continuous. May utter a harsh 'kak-kak' in flight.

BEHAVIOUR Frequents margins of freshwater marshes where reeds provide cover. A solitary hunter, mainly crepuscular but sometimes seen during the day. Often seen flying over reedbeds early morning and late afternoon, flight low and rapid with occasional glides, legs dangling. Adept at climbing reed stems, moving from one to another. When disturbed it may raise its neck and 'freeze' in the manner of its larger cousins. When feeding moves forward slowly, occasionally pausing with neck extended. When standing motionless the bird has a unique tail-flicking action, flaring tail feathers to one side of the body, which may serve to distract or flush fish within range of its bill. Prey includes fish, frogs and aquatic insects.

BREEDING Breeds during northwest monsoon. In courtship, male raises crest feathers and wags tail from side to side. Nest is fairly bulky, built of dead reeds and other vegetation 30 cm to 1.5 m above the water. Normal clutch contains four eggs, with a bluish or greenish white colour. Incubation lasts c21 days. Both sexes share nest duties, feeding young by regurgitation. Chicks need only another two weeks before they can follow their parents for food.

RANGE An Asian species, breeding mainly India and Indo-China to Japan. Also breeds Seychelles and some remote Pacific islands in Micronesia. Northern populations migrate south to Philippines, Indonesia and New Guinea. Vagrant to Australia and Christmas Island, Indian Ocean. Monotypic.

STATUS Breeds Mahé, Praslin and La Digue, total population less than 100 pairs[1]. Confined to coastal freshwater areas on Mahé (e.g. the marshes at Intendance, Police Bay, North-east Point and Plantation Club), La Digue (the extensive network of marsh and freshwater pools on plateau) and north Praslin (marshes of Amitie to Lemuria Resort). Has also occurred Aride[2] and Curieuse. Seychelles is the most westerly population and the only one outside the Asian-Pacific region. Birds of Seychelles have no special characteristics and it has been presumed colonisation is recent event on the basis of similarity to Asian birds[3]. However, given the long distance migrations of some populations there may also be some continued inward gene flow from Asian vagrants.

THREATS AND CONSERVATION Now critically endangered in Seychelles, the continued survival of the Yellow Bittern is a cause for concern. Because it is neither endemic nor globally threatened it has been given a low conservation priority. Nevertheless, the Seychelles population is of considerable biogeographic interest and it is an outstanding indicator species for the health of marsh ecosystems[4]. The most serious threat remains the systematic drainage and clearance of the marshes in which it lives for agriculture, building and tourism developments. Much of the coastal plateau of Praslin was once similar to the waterlogged forest of La Digue, covering up to 18 ha historically, but today little more than 10% remains[5]. Marshes on Mahé have suffered even greater declines. For example the northwest plateau marshland has been reduced from 65 ha historically to a single pool of 0.11 ha, while the marsh at North-east Point, once an important bittern stronghold, has been progressively in-filled. Birds can survive in a network of ditches and small pools as on La Digue, but these too are steadily disappearing. Pollution of remaining marshes may also be a problem. Since the 1990s, marshes have been choked by introduced water hyacinth and water lettuce which reduces food supply by killing off fish, amphibians and other aquatic fauna[6,7]. Climate change and rising sea levels may pose a threat to some marshes. Several marshes have been breached by high tides in recent years, which may reduce food supply by, for example, eliminating amphibians. Habitat management might improve conservation prospects, notably on Silhouette where there is extensive marshland but this currently lacks suitable reedbeds[4].

SIMILAR SPECIES Cinnamon Bittern is more uniform cinnamon chestnut with no contrast in upperwing and no black on the crown or back and pale spotting on the mantle and back feathers. **Little Bittern** *I. minutus* (L33-38 cm, WS49-58 cm) is a potential vagrant. Nominate race breeds Europe to Siberia and northeast India and is a long-distance migrant mainly to southern half of Africa as far as Cape Province. Madagascar race *podiceps* (smaller and darker) may migrate to Africa during the dry season. Race *payesii* breeds sub-Saharan Africa, and is mainly sedentary. A small yellow-buff heron. Male has contrasting jet black crown, mantle, back, tail and flight feathers. Female duller and more similar to Yellow Bittern, but wing-patches more buff-brown and flight feathers less contrastingly black, more dark brown. Yellow bill with blackish upper tip and grey-green legs. Immature is tawny buff with dark bold steaks, similar to Yellow Bittern, wing-patch streaked. **Black Bittern** *I. flavicollis* (L54-66 cm, WS80 cm) breeds from India through to northern Australia and migrates south into Malaysia. Vagrant to Maldives and Christmas Island, so may be a potential vagrant to Seychelles. All-black, save for cream on neck and buff streaking on chin and upper breast. Female dark brown, juvenile mottled rufous to dark brown. Legs dirty yellow.

References: 1. Watson (1980), 2. Bowler and Hunter (2000), 3. Benson (1970d), 4. Gerlach and Skerrett (in press), 5. Gerlach and Canning (1996), 6. Gerlach (1996b), 7. Gerlach (1997b).

CINNAMON BITTERN *Ixobrychus cinnamomeus*   **Plate 12**

Other name: Chestnut bittern   **French:** Blongios cannelle   **Creole:** Makak Kannel

DESCRIPTION Length 40-41 cm; wingspan 58-62 cm. A small, uniformly chestnut bittern, slightly larger and stockier than Yellow Bittern, with more rounded and uniform wings, showing no contrast on wing-coverts or tail. **Male** is uniformly dark rufous on crown, hindneck, back, wings and tail with a prominent white malar-stripe and a darker stripe on the foreneck (sometimes indistinct or absent). Underparts warm rufous buff with a scattering of black spots on the sides of the breast. Legs yellowish green, lore yellow and bill greenish orange. **Female** tends to be a little darker and browner above and duller below, with dark brown streaks on the foreneck and breast and dark brown and buff markings on scapulars and upperwing-coverts. **Immature** is similar to female with dark brown crown and back, heavy brown streaking on the breast and belly and with prominent buff fringes to scapulars and wing-coverts giving a mottled appearance.
VOICE A territorial booming call but this is unlikely to be given by a vagrant. In flight may call 'krek-krek-krek'.
BEHAVIOUR A bird of freshwater swamps, reedy marshes and overgrown ditches. Solitary, shy and skulking, often freezing in an upstretched stance if disturbed. In flight, wingbeats are rapid and jerky. Most active at dawn and dusk but also often feeds in middle of day (more so than Yellow Bittern). Diet mainly frogs, small fish, eels, shellfish and insects.
RANGE Breeds in India and Indo-China as far east as Taiwan and Indonesia and Philippines. Though mostly sedentary, birds in the north of its range make seasonal movements in response to heavy rainfall, moving south into Malaysia. Monotypic.
STATUS Vagrant to granitic islands October; not recorded elsewhere in the entire Afro-Malagasy region[1]. This extraordinary record of an Asian bird arriving in Seychelles is a fascinating insight into how Yellow Bittern must first have colonised these remote islands.
 SIMILAR SPECIES Yellow Bittern is slightly smaller with paler buff plumage and contrasting black crown, black tail. In flight, pale wing-coverts contrast sharply with black flight feathers
Reference: 1. Lucking (1995).

EURASIAN BITTERN *Botaurus stellaris*   **Plate 12**

Other names: Great (or Common) Bittern   **French:** Butor étoilé   **Creole:** Gran Makak

DESCRIPTION Length 70-80 cm; wingspan 125-135 cm. A stocky, bull-necked, heron-like, skulking bird with the cryptic plumage of an owl. **Adult** has an extraordinary patterned plumage of tawny buff ground colour stippled and barred with black. Crown black, contrasting with the tawny buff neck and throat, finely barred with black. Mantle and scapulars have heavily black-barred centres, similar heavy barring also on the primaries and primary coverts. Bill greenish yellow, black at the base, with a black moustache extending down the neck from the gape. Sexes similar, though male averages larger. **Immature** similar to adult but generally has a paler cap and moustache, streaked brown not black and a less defined pattern to upperparts.
VOICE Gives a variety of croaking and cackling calls, but mostly silent outside breeding season. Famous for its loud, territorial booming, giving rise to name *Botaurus* meaning 'oxen-bull'.
BEHAVIOUR Favours freshwater marshes with extensive reedbeds, sometimes in damp fields. Rarely coastal, but can tolerate brackish water. A very wary, solitary bird, keeping to thick cover at all times and normally only taking flight when disturbed, quickly dropping again to cover. Mainly crepuscular and nocturnal but sometimes active by day. Despite size, clambers easily among reeds, often perching high up, head and neck visible above the vegetation. Rises awkwardly, neck protruding. Flight owl-like, long neck drawn in, legs dangling, wings broad and rounded. In reedbeds it waits motionless, or stalks slowly through cover, preying on fish and amphibians. Will freeze if disturbed, long neck extended and bill pointing skyward. Preys on large insects, lizards, mice and small birds.
RANGE Nominate race breeds throughout the temperate zone, from Britain through Asia to Japan. Only a partial migrant but first-year birds undertake extensive movements. Russian birds tend to winter in Iraq, Iran, Pakistan and northern India. Vagrant Sudan, Eritrea and Kenya. Race *capensis* of South Africa (darker on back) is mainly sedentary.
STATUS Vagrant to granitic islands; recorded October.
SIMILAR SPECIES Immature **Black-crowned Night Heron** is smaller and stockier and has darker upperparts with prominent pale spotting.

GREATER FLAMINGO *Phoenicopterus ruber*   **Plate 9**

French: Flamant rose   **Creole:** Flaman Roz

DESCRIPTION Length 120-145 cm; wingspan 140-165 cm. Large and unmistakable with very long thin neck and legs, pinkish white plumage and a bill like a broken nose. **Adult** has entire body from head to tail white, flushed with pink. Wing-coverts are bright crimson and flight feathers black, both these distinctive colours hidden when wings are folded. Legs pink with darker joints and in flight, extend as far beyond the tail as the neck stretches beyond the body. Bill pink with a black tip, eye yellow. Sexes similar. **Juvenile** has back streaky brown, head and neck grey-brown with underparts and tail dirty white. Legs are dark, dirty brown and bill grey with a black tip. In flight wing-coverts are brown and white and flight feathers blackish brown. Birds mature after four years, plumage becoming whiter and pinker and legs brightening with each succeeding season.

VOICE Feeding birds keep up a constant gabbling conversation with their 'nga-nga' calls, but make a louder, goose-like honking in flight.

BEHAVIOUR At Aldabra, resident birds frequent open, shallow inland pools; elsewhere vagrants are likely to be coastal. Highly gregarious, though vagrants may be solitary. In flight, wingbeats are rapid; outstretched head and neck giving a distinctive outline. On take off takes a short run, wings flapping and also runs several steps after touch-down. Flocks may fly in V-formation or more loosely. Normally feeds in tight flocks, walking in a slow and dignified manner, with head down and bill swinging from side to side through the water to sift tiny invertebrates and algae. The special filter in the bill traps crustaceans, molluscs and insects up to 2 cm in size.

RANGE Two races, *roseus* breeding at scattered localities around the Mediterranean and Turkey to India, Sri Lanka and Africa (sometimes considered distinct from the nominate race of Galápagos to Caribbean). Some populations are partially migratory or dispersive, moving in response to rains. Kenyan birds have been known to move to Botswana in extreme conditions and flocks appear in southwest Madagascar periodically, presumably irruptions from southern Africa.

STATUS Breeds Aldabra, almost wholly confined to the eastern end of the lagoon, in the Bras Cinq Cases-Bassin Flamant area (the latter named after the bird). May once have been much more common, Abbott reporting flocks of 500-1,000 in 1892[1]. The 1967 Royal Society Expedition recorded a maximum of 55 birds[2]. Later estimates have been somewhat lower. Recorded as a vagrant to Farquhar in October and to the granitics October-May.

BREEDING The remarkable discovery of breeding on Aldabra was only made as recently as April 1995 when staff on the atoll explored a remote pool in the Cinq Cases/Takamaka area of Grande Terre and observed a small grey chick following adults[3]. Four days later they returned to check for nests and found three complete and three partially complete nest mounds. This news was the first time breeding had been confirmed on any coral atoll world-wide and indeed the only other oceanic breeding site for this species is in Galápagos. An earlier indication of possible breeding came in September 1967 when a single fresh egg was found at Takamaka[2]. However, there were no nest mounds discovered on this occasion and the egg was not generally accepted as evidence of breeding, particularly in view of reports from other parts of the world that flamingos frequently drop eggs away from nest sites. Breeding was again confirmed in 1996, with four juveniles sighted in a group of 16 and three nest mounds containing egg fragments located[4]. It remains to be proven whether breeding at Aldabra is annual or sporadic. Flamingos elsewhere frequently favour some sites one year and not the next. Sometimes, Greater Flamingo breed at a new site in Africa and even if successful, do not return to breed in subsequent years. Flamingos form a strong pair bond. Much time is given to group displays months before breeding, continuing after the season. In these, birds carry out a series of ritualised postures including stretching the neck and bill skyward and turning the head from side to side, cocking the tail and spreading wings to display the strikingly patterned upperwing, or marching in one direction in dense flocks, suddenly turning 180° and marching back again. Birds will also suddenly turn the neck and head backward to preen or perform a display in which the wing and leg are stretched and extended backward on one side. Displays help to stimulate colonial breeding in response to local conditions (mainly water level and food supply). Colonies of flamingos elsewhere rarely breed with fewer than 50 pairs, yet there are no reports in recent times of groups this large at Aldabra. An interesting exception to this rule is the world's only other oceanic site, the Galápagos, a small, sedentary population, where groups as low as three pairs may breed. This may be an example of island populations adapting to their isolation by breeding in smaller groups. Nests consist of a mound of mud, both sexes helping with construction, working mud and stones with the bill to form the mound. A single chalky white egg (sometimes with a faint bluish hue) is laid, nest building continuing after laying. Sometimes nests may be reused in subsequent seasons. Incubation elsewhere lasts about four weeks (figures unknown for Aldabra). Adults sometimes assist hatchings by removing chunks of eggshell. Chicks are covered in grey down which is soon replaced by a darker downy plumage and are forced to leave the nest at 5-12 days after hatching, at which point they can swim and walk. Elsewhere, chicks often form large crèches, watched over by a few adults. Parents feed chicks with crop milk, a high protein secretion produced only by flamingos and pigeons. Non-breeding adults often feed chicks and even chicks of merely seven weeks in age may produce crop milk and feed another chick. Chicks can feed themselves by around five weeks old, by which time the straight bill with which they were born has acquired the typical kink of the adult. Nevertheless adults may continue feeding until chicks are 10-12 weeks old.

THREATS AND CONSERVATION The small population at Aldabra and the susceptibility of the species to disturbance must make it vulnerable. Its best protection may be the remote location and inhospitable nature of the terrain that it inhabits in eastern Aldabra.

SIMILAR SPECIES Lesser Flamingo *P. minor* (L80-105 cm, WS95-120 cm) is a potential vagrant; it breeds mainly in East Africa, moving large distances in response to changes in conditions at feeding grounds and is an erratic visitor to Madagascar. It is much smaller, adult uniform dark pink (generally brighter than Greater) with a deeper red bill having more extensive black tip, short base and more abrupt bend, and a dark red eye and surrounding skin (can appear black at distance, always darker than Greater). Juvenile (unlikely in Seychelles) slightly greyer than juvenile Greater Flamingo, with darker purple-brown bill having a smaller black tip. Immature paler, almost white, but retains similar differences in bill. At all ages, in flight, shows shorter neck and heavier body.

References: 1. Ridgway (1895), 2. Benson and Penny (1971), 3. Bergeson and Rainbolt (1995), 4. Bergeson (1996b).

WHITE STORK *Ciconia ciconia*   Plate 9

Other name: European White Stork   **French:** Cigogne blanche   **Creole:** Sigonny Blan

DESCRIPTION Length 100-102 cm; wingspan 155-165 cm. A very large, white, heron-like bird with red legs and bill and black wings. **Adult** has head, neck, underparts, back and tail white. Lesser and median wing-coverts white above and below, contrasting with black greater coverts and flight feathers. Long red legs and feet project well beyond tail in flight. Bill long, straight and bright red. Eye accentuated by black eyeliner. Sexes similar though male averages larger. **Immature** similar to adult but with duller, browner wing-coverts and duller legs. Bill duller than adult with black tip.
VOICE Normally silent outside breeding season.
BEHAVIOUR Favours open grassland or shallow marshes. Gregarious, migrating in large flocks. Flies with slow wingbeats, neck outstretched and slightly lowered, legs stretching well beyond tail. Soars on huge widespread wings, normally dependent on thermals to gain height and then glide long distances. On the ground, walks sedately. At rest often stands on one leg. Long bill and neck give a lengthy reach when feeding and the long legs allow wading into quite deep water in search of prey. Feeds on beetles, locusts, grasshoppers, lizards, frogs, fish, even snakes and mice.
RANGE Two races, nominate *ciconia*, which breeds from Europe to Turkey and winters from sub-Saharan Africa to South Africa, and *asiatica*, which breeds in Turkestan and winters from Iran to India. Either race may be possible in Seychelles, though the nominate is far more likely, both in terms of numbers and its East African flight path, and one specimen record was a bird of this race.
STATUS An irregular vagrant occurring at intervals of several years, but then in small numbers scattered throughout the islands, e.g. influxes occurred in 1983 and 1995 with 6-10 birds recorded on each occasion. Only the nominate race has been recorded. Migrating birds may pause in Sudan for a month or two, moving south again in November: this is when most Seychelles records have been, from December to January, a time of steady northwest winds and sudden rain storms. Such vagrants have favoured open areas such as grassy margins of airstrips.
SIMILAR SPECIES Based on known migration routes and movements in Africa, Black Stork *C. nigra* and Abdim's Stork *C. abdimii* are potential vagrants. **Black Stork** (L95-100 cm, WS165-180 cm) breeds in central Asia and migrates south to winter in South Africa, Pakistan and north India. Easily separated, adults mainly black with iridescent green-purple gloss and contrasting white belly. Bill, eye-ring and legs red. Immature brown where adult is black, with dirty grey-green bill and legs. **Abdim's Stork,** (L81 cm), breeds Senegal to Eritrea, migrating to eastern and central and southern Africa. Similar to Black Stork but smaller, with all-dark, black-bronze back, white belly, lower back and rump, dark grey-green bill, blue and red at base and legs dark with pink joints. Immature is similar in pattern but duller brown.

SACRED IBIS *Threskiornis aethiopicus*   Plate 9

Other name: Sand Ibis   **French:** Ibis sacre   **Creole:** Ibis Afriken

DESCRIPTION Length 65-89 cm; wingspan 112-124 cm. Similar to Madagascar Sacred Ibis, slightly larger with black trailing edge to wing and dark eye. **Adult** is white with a black head and neck, black legs and a brown eye, ringed deep red. Bill appears thicker than Madagascar Sacred Ibis (though not obvious on a lone bird). In flight there are, in addition to the black primaries on a predominately white wing, black tips to the secondaries, which form a black trailing edge. Bill black at base, horn at tip. Legs black. **Immature** similar to Madagascar Sacred Ibis, but with a much darker head and neck. Eye brownish black. Legs grey-black.
VOICE Usually silent away from breeding grounds. Sometimes gives a harsh croak in flight.
BEHAVIOUR Gregarious, though vagrants likely to be solitary. Otherwise similar to Madagascar Sacred Ibis.
RANGE Occurs throughout sub-Saharan Africa. Migrates several hundred kilometres in response to rains. Monotypic, if Madagascar Sacred Ibis is recognised.
STATUS Vagrant to Aldabra; recorded February.
SIMILAR SPECIES Madagascar Sacred Ibis has eye blue to white and pure white wings (lacking the black trailing edge of Sacred Ibis).

MADAGASCAR SACRED IBIS *Threskiornis (aethiopicus) bernieri*   **Plate 9**

French: Ibis de Madagascar or Ibis malgache   **Creole:** Ibis Malgas

DESCRIPTION Length 70-85 cm; wingspan 112-124 cm. A large mainly white bird with entirely white primaries and secondaries and a pale eye (in adult). **Adult** has feathers of back, underparts and wings pure white, save for long, lacy black plumes over the tail (modified tertials) and all-white wings, often with narrow black tips to the outer primaries. Head and neck are black and completely bare of feathers. Eye colour may vary from intense Wedgwood blue to pale blue to white, always paler than Sacred Ibis. Bill stout, long and curved like a curlew and wholly black; legs black. Sexes similar; male has a longer bill. **Adult breeding** has a patch of bright red skin in each armpit (which is otherwise dull brown or purple outside the breeding season) and more prominent, glossy plumes.
Juvenile has dirty white feathers flecked with black on the head and neck, milky coffee brown tertials, brown tips to primaries and secondaries and brown eye.
VOICE Call described as a 'weary groan, more squeaky when alarmed'[1]. May utter a harsh croak in flight.
BEHAVIOUR Feeds alone or in small groups favouring inland pools, also moving to the shores of the lagoon to feed at low tide and sometimes in open grassy areas. It is thought most of the population roosts together in trees. Flies with rapid wingbeats, neck outstretched. Feeds on crabs, extracted from soft muddy sand, snails in and around pools, and in dry terrain catches crickets, beetles and lizards. Will also scavenge, feeding on carcasses of giant tortoises or other carrion.
BREEDING Nests during northwest monsoon. In courtship, pairs bow to each other, necks stretched forward and downward. Male defends nest site aggressively. Untidy nests of twigs and sticks often bearing green leaves lined with dead leaves and grass are built in low bushes over or close to water. Eggs are laid November to December; with clutch up to three, (but more typically two) large white eggs, which are incubated for just under a month. Chicks have white down over most of the body and black down on the head and neck, mirroring adult plumage. Young fledge after five or six weeks, rarely more than one per nest. Low number of eggs and fledglings reflects survival and adults are known to live at least 21 years.
RANGE Confined to western Madagascar from the coast to 150 m altitude (nominate race) and Aldabra (race *abbotti*). Formerly included as races of Sacred Ibis and requires further study to establish status.
STATUS Resident Aldabra (100-250 pairs[2]), mainly at east end of Grande Terre (the sole breeding site) and to a lesser extent on Polymnie, Picard and Malabar. There is evidence of some increase in range since 1967, for birds are now resident (but do not breed) on Picard and Malabar where they were formerly absent.
THREATS AND CONSERVATION If recognised as a full species, it is probably Globally Threatened, with Aldabra as one of the last strongholds. Because of the remote location of Aldabra and its huge area, it is impossible to protect this population adequately. There is a small reserve staff based on Picard Island, but the breeding and roost sites are all at the other end of the lagoon in the southeast corner. At least the kind of routine persecution carried out by island labourers in the past seems to have ceased. Heavy human predation continued until relatively recently. Some killing of birds nesting near Cinq Cases continued into the 1990s until SIF took action to end this. Though tame, birds are highly susceptible to disturbance. Predation continues in Madagascar.
SIMILAR SPECIES Sacred Ibis has a black or dark brown eye and in flight, a black trailing edge to the wing.
Glossy Ibis *Plegadis falcinellus* (L49-66 cm, WS80-95 cm) is a potential vagrant, breeding discontinuously from North America to Australia, including Madagascar and East Africa; it is migratory and dispersive. Dark brown, duller in non-breeding plumage, neck streaked whitish. Immature has oily green sheen, head and neck browner with variable white to forehead, throat and foreneck.
References: 1. Penny (1974), 2. Rocamora and Skerrett (1999).

WHITE-FACED WHISTLING DUCK *Dendrocygna viduata*   **Plate 13**

Other name: White-faced Tree Duck   **French:** Dendrocygne veuf   **Creole:** Sarsel Figir Blan

DESCRIPTION Length 46-48 cm; wingspan 54-56 cm. An unmistakable, long-necked duck with a striking black-and-white head. **Adult** has back and breast dark chestnut, flanks finely barred black and white, appearing grey at a distance. Rump, tail and belly black. Neck long and dark, with a white neck-spot and diagnostic head pattern, black at the rear, white face with black eye and dark grey bill. In flight, wings are black above and below with chestnut leading edge to inner upperwing. Legs bluish grey. Sexes similar, though female has the white of the face tinged rusty brown. **Immature** is duller with head and neck mostly buff grey and the breast paler, acquiring the white head of adult at 3-4 months.
VOICE Noisy, the call a high-pitched trisyllabic whistle. Very vocal in flight.
BEHAVIOUR Normally occurs in flocks, frequenting lagoons and inland pools or lakes, but vagrants may be solitary or in small numbers. Flight slow and heavy on noticeably rounded wings. Walks with an upright posture and neck also held erect when alarmed. Feeds at night or dawn and dusk, on aquatic weeds and grasses taken while wading or by diving. Rests by day on sandbanks or islands.
RANGE Breeds throughout tropical South America, Africa as far south as the Cape, Madagascar and Comoros. Undertakes long-distance movements in times of drought, and presumably this is the stimulus

that causes small numbers to turn up in islands of the Indian Ocean. Monotypic.
STATUS Vagrant to Aldabra group; recorded April to June and September.
SIMILAR SPECIES None.

RUDDY SHELDUCK *Tadorna ferruginea*   Plate 13

Other name: Brahminy duck   **French:** Tadorne casarca   **Creole:** Sarsel Rouye Pal

DESCRIPTION Length 61-67 cm; wingspan 121-145 cm. A large, bulky duck with uniform rusty brown
body, white head and black-and-white wings. **Adult** has back, neck, breast and belly uniform bright rusty
brown, contrasting with whitish head (with black eye), black bill and legs and green speculum. In flight,
rusty brown body contrasts with black rump and tail and with wing pattern of black flight feathers and
prominent white upperwing-coverts and underwing-coverts. Sexes similar, but breeding male may show a
narrow black neck-collar (lost for most of winter) and female has a more buff head and neck contrasting
with cleaner white face. In worn plumage may appear much paler. **Immature** similar to adult female, but
duller with pale sandy head and drab buff back and tinged grey on forewing. Male acquires black collar
by February, but upperwing-coverts and scapulars are washed grey.
VOICE Noisy by nature. Male has a loud whooping call, female a louder, deeper, nasal 'ngah-ha-ha'.
BEHAVIOUR Frequents a wide range of inland sites including lakes, marshes and streams though vagrants
might be coastal. Generally in family parties outside breeding season, but vagrants are likely to be solitary. Strong flight with slow, heavy wingbeats. Mixed diet includes grasses, vegetables, locusts, shrimps,
fish and even frogs. Feeds on land or by upending in water.
RANGE Breeds from Morocco through to eastern Europe to China. The Asian population is migratory, wintering in India, Burma and southern China and at scattered sites down the Nile. Monotypic.
STATUS Vagrant to granitic islands; recorded December.
SIMILAR SPECIES None.

COMMON TEAL *Anas crecca*   Plate 13

Other names: Eurasian Teal, Green-winged Teal, Teal   **French:** Sarcelle d'hiver   **Creole:** Pti Sarsel

DESCRIPTION Length 34-43 cm; wingspan 58-64 cm. A very small duck with a green speculum bordered
with white. **Adult** in flight shows white bars either side of glossy green speculum, the inner wingbar broader than the trailing edge and broadening towards the outer wing. Underwing whitish in centre with dark
edges. **Male breeding** has chestnut head with a green gusset bordered in buff, from round the eye back
and down to the hindneck. Flanks and back grey, which seen at close range are made up of fine black
tracery. A triangular creamy yellow patch on side of rump is framed in black. Breast and belly grey, whiter
in the centre. Small grey bill, often orange at the base and dark grey legs. **Female** is mainly uniform buff
brown, mottled and streaked in darker brown with a thin, pale eye-stripe but no cheek-stripe as in
Garganey. When swimming, a pale horizontal line is visible along the side of the tail base. **Male in eclipse**
similar to female, darker above, the eye-stripe less distinct. **Immature** similar to female, but underparts
spotted brown and flanks striped darker compared to the clean white belly and scaly flanks of adult
female.
VOICE Male has a high-pitched whistling 'krit', female a high-pitched quack.
BEHAVIOUR Favours small pools and lakes, but in winter may occur on estuaries and mudflats. Highly
gregarious, with pair bonds formed in winter flocks, though vagrants may be solitary. Swift to take wing,
can spring almost vertically when disturbed. Diagnostic flashing flight with rapid wingbeats. Feeds on
seeds, water plants and insects.
RANGE Breeds throughout North America and Asia, from Europe to the Bering Strait. Winters southward,
nominate Palearctic race reaching Kenya, India and South-East Asia. Other races breed in Aleutians and
North America.
STATUS Uncertain; there are a number of unconfirmed reports from the granitic islands.
SIMILAR SPECIES Garganey is the only duck of comparable size, but the female has broad, pale stripes
over the eye and cheeks, while the male has striking dark chestnut head with a white streak from eye to
nape. Both sexes have a heavier all-grey bill.

MALLARD *Anas platyrhynchos*   Plate 13

French: Canard colvert   **Creole:** Sarsel Latet Ver

DESCRIPTION Length 50-65 cm; wingspan 81-98 cm. A medium-sized duck with a yellow-orange bill
and purple-green speculum. **Male** has metallic green neck and head separated from chocolate brown
breast by distinct white neck ring and ashy grey back and flanks. Tail-tip is grey with black central feathers that form tight curls. Bill greenish yellow, legs orange. In flight shows whitish underwing-coverts.
Purplish speculum prominently edged black and white. **Female** is uniform mottled buff-brown with pale

supercilium and dark eye-stripe, whitish sides to tail, brown-orange bill and orange legs. **Male in eclipse** similar to female but heavier, paler greyer face, blackish crown, warm brown breast with less streaking, more uniform above and uniform greenish yellow bill. **Immature** similar to female but duller, underparts more streaked, bill dull reddish brown.

VOICE Female has a loud quack, male a quieter, nasal 'veeb'.

BEHAVIOUR Will settle on the sea, but prefers freshwater lakes or ponds. Highly gregarious, though vagrants may be solitary. Flight swift and strong. Dabbles in shallow water, upending to feed on the stems of water plants, but will take seeds or insects on the ground.

RANGE Breeds throughout North America, Europe and Asia, from North Africa to Japan. Asian population winters further south, from Egypt across northern India to southern China. Nominate Palearctic race is the only one likely in Seychelles.

STATUS Possibly a vagrant to the granitics. Records are difficult to assess because of possible escapes.

SIMILAR SPECIES Drake is unmistakable, but female could possibly be confused with female **Garganey** (much smaller with distinct pale head-stripes), female **Northern Shoveler** (very long bill and grey panel on the wing-coverts) or female **Northern Pintail** (long thin neck and a blue-black bill). Eurasian Wigeon *A. penelope* and Gadwall *A. strepera* are potential vagrants which should also be considered. **Eurasian Wigeon** (L45-51 cm, WS71-85 cm), breeds throughout northern Europe and Asia and winters Nile valley, Middle East, India to southern China and Japan. Drake has squarish chestnut head with creamy yellow forehead, blue-grey bill, white belly and white wing-coverts contrasting with black flight feathers. Female and immature are dark pink-brown, with mottled back and short grey bill. Legs are dark grey to brown. Male in eclipse similar to female, but rich chestnut breast and sides contrast with white belly and forewing; white line along wing at rest. **Gadwall** (L46-58 cm, WS78-90 cm), breeds from Europe across Central Asia to China; northern birds move south in September, wintering East Africa, Middle East and India. Drake has a square, buff-grey head and breast, dark grey back and flanks and black rump. Female and immature are browner with darker mottling on breast and back. Both sexes have grey wing-coverts contrasting with white speculum that shows as white flash on swimming birds and have pale orange legs. Male in eclipse similar to female, less orange on bill and less coarsely marked.

NORTHERN PINTAIL *Anas acuta*   Plate 13

Other name: Common Pintail   **French:** Canard pilet   **Creole:** Sarsel Lake Pwent

DESCRIPTION Length 50-66 cm; wingspan 80-95 cm. An elegant duck with a slim neck, pointed body shape and narrow wings, giving a distinctive flight silhouette. **Male** is distinctive: a chocolate-coloured head and hindneck with white of breast extending up the sides of the neck to a fine point. Back and flanks finely stippled grey, the back covered in long black-and-white scapular plumes. White patch at the sides of rump. Long, pointed tail is black, edged white. Speculum bronzy green, with buff border above and white border below. Bill blue-grey, legs grey. **Female** similar to female Mallard, but with more uniform warm brown head, rounded head shape and all-dark slim, grey bill. Speculum duller than male but white rear border prominent at long range. **Male in eclipse** similar to female but has greyer upperparts. **Immature** similar to adult female, but with plainer, darker upperparts, cinnamon-coloured head and neck and more spotted below.

VOICE Male has a long 'greee' call and female a deep quack.

BEHAVIOUR Most commonly associated with shallow, wide open pools and lakes, but regularly uses coastal areas in winter. Highly gregarious for most of year. Flight swift with rapid wingbeats. Feeds mainly on water plants and seeds, taking aquatic insects when abundant. A dabbling duck, frequently upending to feed in shallow water.

RANGE Nominate race breeds throughout North America and northern Asia, wintering sub-Saharan Africa, India, Burma and Japan.

STATUS Vagrant to granitic islands November to February, occasionally appearing in two or threes.

SIMILAR SPECIES Female and juvenile are only likely to be confused with **Mallard**, which is a heavier duck, square-headed with orange on the bill and legs. Female **Gadwall** is smaller, short-necked and also square-headed.

GARGANEY *Anas querquedula*   Plate 13

French: Sarcelle d'ete   **Creole:** Sarsel Ordiner

DESCRIPTION Length 37-41 cm; wingspan 58-69 cm. A small duck with rapid flight, revealing distinctive pale grey forewings and green speculum bordered white. **Male** has a distinctive face pattern with a thick white stripe from the eye to the nape, dark brown crown and rusty brown cheeks and neck. Breast and back are mottled brown, flanks finely stippled grey. Long black-and-white ornamental scapulars over back, tail and rump mottled brown. In flight, the white trailing edge is wider than the white bar on the greater coverts and the upper forewing is pale grey. Both sexes have metallic green speculum with white borders. Bill and legs dark grey. **Female** is uniformly mottled dark brown, but with distinct facial pattern of two pale stripes above and below the eye, alternating with two dark stripes, one through the eye, the other across the lower cheek and a pale spot at the lore. The throat is often creamy white. **Male in eclipse** similar to female but brighter with whiter throat and belly and retains a pale grey forewing. **Immature** very similar to adult female, but with belly and lower breast covered in brown spots and streaks. In flight, the white trailing edge to secondaries is narrower than in adult.

VOICE Not very vocal. Drake has dry, rasping rattle, female a low quack.

BEHAVIOUR Feeds in shallow flooded fields or small ponds and lakes, preferring water with some thick cover round the edges. Will use lagoons or even settle on the sea. Often occurs in small groups of two to three, or occasionally larger flocks of half a dozen or more. Rises with agility from water, flying with rapid wingbeats. Feeds on worms, insects, seeds and plants that it takes by dabbling at the surface, ducking the head under the water, or, rarely, upending.

RANGE Breeds from Europe through Asia to the Bering Strait. Unusual among ducks in being a full migrant, flying south to winter in East Africa, the Middle East, India and South-East Asia. Monotypic.

STATUS Annual throughout Seychelles, reported mainly in the granitic islands in small numbers October to April.

SIMILAR SPECIES The only duck of similar size is **Common Teal,** which in the female has a more uniformly pale face and white edge to the tail. In flight shows darker wings, especially dark coverts on the forewing. The speculum is dark metallic green, also bordered with white, but the white wingbar is broader than the trailing edge, compared with Garganey in which the white trailing edge is the broadest. Often shows an orangey base to the smaller, grey bill.

NORTHERN SHOVELER *Anas clypeata*   Plate 13

Other name: Shoveler   **French:** Canard souchet   **Creole:** Sarsel Labek Kuiyer

DESCRIPTION Length 44-52 cm; wingspan 70-84 cm

A medium-sized, flat-headed duck with a long spatulate bill and bluish forewing. **Adult** of both sexes have distinctive bill, 20% longer than a Mallard and flared at the end, orange legs and feet and in flight shows diagnostic combination of bluish forewing and green speculum (grey-blue forewing and green speculum in Garganey). **Male breeding** (January onward) has metallic green head and yellow eye, dark mantle and long, black, white-edged scapulars, white breast, chestnut flanks, white patch on side of rump and black rump. Upper forewing is pale blue, speculum green with white border at front edge, which broadens toward wingtip. **Female** is mottled brown and buff, head paler. Upper forewing dull pale grey, speculum grey-green. In flight, dark brown belly contrasts with whitish underwing, upperwing with no white trailing edge. Grey-brown bill, orange at sides and base. **Male in eclipse** similar to female, darker brown on head and darker edges to feathers of breast and flanks and retains pale blue upper forewing. **Juvenile** similar to female but more uniform and duller, more streaked below and lacking the green speculum and white wingbar. Eye brown at first, later becoming yellow.

VOICE Mainly silent. Drake has a quiet 'tok-tok' and the duck, distinct quacking notes.

BEHAVIOUR Occasionally coastal, but prefers freshwater pools or lakes. Gregarious outside breeding season, though vagrants usually solitary. Flight rapid, large bill giving front-heavy silhouette, rather pointed wings set well back. Swims with neck outstretched, dabbling the surface to filter small items with the bill, or with head and neck immersed. Does not normally upend or dive like other ducks. Feeds on seeds, vegetation, molluscs, insects and tiny crustaceans.

RANGE Breeds throughout Canada, northern Europe and northern Asia, wintering in northeast Africa, India and southern China. Monotypic.

STATUS Vagrant throughout Seychelles October to March, with most records from the granitic islands in October or November.

SIMILAR SPECIES Noticeably larger than **Garganey**, which has a distinct head pattern and dark legs. Might be confused with **Mallard**, but this has a shorter, shorter, less flared bill, brownish forewing and blue speculum.

FERRUGINOUS DUCK *Aythya nyroca*   Opposite plate 13

Other name: White-eyed Pochard   **French:** Fuligule nyroca   **Creole:** Pti Sarsel Rouye

DESCRIPTION Length 38-42 cm; wingspan 63-67 cm. A small, uniform dark brown diving duck with rounded head, steep forehead, broad white wingbar and white undertail. **Male** has dark chestnut head, breast and flanks, black back, white under the tail and diagnostic white eye. Belly white but usually not visible on swimming birds. **Female** duller and browner, also with white undertail-coverts and belly but a brown eye. **Male in eclipse** similar to female but more chestnut on head and breast and retains white eye. In flight, both sexes appear very dark, the plumage contrasting with broad white wingbar (which runs from the secondaries right to the tip of the wing), white belly and undertail-coverts. Both sexes have slate grey bill with paler grey edges and black tip and dark grey legs and feet. **Immature** similar to female but browner and more uniform, pale grey undertail-coverts, barred with brown, pale edges to feathers of flanks and back, paler buff on foreneck and side of head, darker bill and grey eye.

VOICE Male makes a 'gek' call and some whistling notes, the female a quiet 'krrr' note.

BEHAVIOUR Prefers shallow freshwater pools or marshes, most active early morning and late afternoon, resting during the middle of the day. Secretive, diving at the first sign of danger; hides in dense vegetation rather than take flight. Usually solitary or in very small numbers outside breeding season. Rises from surface easily, almost springing vertically. Feeds by both dabbling and diving, mostly on seeds and roots of water plants. Sometimes coastal on migration.

RANGE Breeds from eastern Europe across southern Russia as far as the borders of Mongolia, with small outlying colonies in southern Spain, Turkey and Iran. Winters Nile valley and upper reaches of the Indus and Ganges rivers. Scarce migrant in Kenya December to March. Monotypic.

STATUS Vagrant; recorded Mahé in May[1], but probably more likely December to March as in Kenya.

SIMILAR SPECIES Most likely to be confused with female or immature Tufted Duck *A. fuligula* or Common Pochard *A. ferina*, both potential vagrants. **Tufted Duck** (L40-47 cm, WS65-72 cm), breeds from Europe as far east as Kamchatka; Russian birds winter south to Sudan, vagrant Somalia and Kenya, Asian birds winter from Pakistan and India to China. Head is more rounded, less peaked, bill blue-grey, shorter than Ferruginous Duck with broader black tip and body longer. Drake unmistakable all-black with white flanks and tuft on the head, female dark brown throughout (darker than Ferruginous, especially above) with a short head tuft and white blaze at base of bill. Some females may show a white undertail (particularly in autumn), but never as pure or as extensive as in Ferruginous Duck. Both sexes with bright yellow eye (grey-brown in immature) and broad white wingbar (but shorter and less prominent than in Ferruginous). Male in eclipse browner or greyer on sides and belly, crest shorter. **Common Pochard** (L42-49 cm, WS67-75 cm), breeds North America and Europe to Russia, wintering southward, vagrant Kenya. A diving duck, both sexes with broad grey wingbar and white belly. Drake has red-brown head, red eye, silver back, black breast and tail; female with dark brown head, grey-brown back and warm brown breast. Male in eclipse similar to female but with greyer body, darker on breast and reddish eye.

Reference: 1. Skerrett (1999d).

OSPREY *Pandion haliaetus*   Plate 14

Other name: Fish Hawk   **French:** Balbuzard pêcheur   **Creole:** Leg Peser

DESCRIPTION Length 55-58 cm; wingspan 145-170 cm. A virtually unmistakable, large long-winged, rather short-tailed raptor, dark brown above with white crown emphasised by black eye-stripe, white below with black carpal-patches. **Adult** appears very dark brown-black above save for white crown and paler bars on the tail, and the head appears all-white save for dark eye-band. White throat, breast, belly and wing-coverts contrast with black carpal-patches, primaries and secondary coverts and to a lesser extent with buffish brown breast-band. In flight, long narrow wings with only four 'fingers' and short square tail give a unique silhouette. From below, secondaries and tail are smudged grey. Bill black and feet blue-grey, eye yellow. Female larger than male, with broader breast-band. **Immature** similar to adult with paler tips to the back and upperwing-coverts, pale trailing edge to wings and tail, less distinct breast-band and less pure white underwing-coverts, which have an ochre stain, and eye is orange.

VOICE Mainly silent outside breeding season. Sometimes gives a shrill 'chook-chook-chook'.

BEHAVIOUR Hunts over water including coasts, bays and lagoons. Generally solitary outside breeding season. In flight, has powerful wingbeats, wings held angled recalling a huge gull. Patrols at 20-30 m above the water, occasionally hovering prior to stooping with outstretched talons to strike fish near the surface, sometimes submerging.

RANGE Nominate race, the only one likely in Seychelles, breeds across Europe and Asia, wintering sub-Saharan Africa and round the shores of the Arabian Sea to South-East Asia. Other races breed in the Caribbean, North America and Australia. First-summer birds of the nominate race remain in Africa.

STATUS Vagrant to granitic islands; recorded September.

SIMILAR SPECIES Pale morph **Booted Eagle** lacks entirely brown upperwing and back and has white uppertail-coverts, white underwing-coverts, blackish flight feathers and a different jizz due to longer tail and broader wings. **Short-toed Eagle**, a potential vagrant (see under Booted Eagle), has broader wings, lacks dark carpal-patches and wingbars, has more heavily marked underparts and a dark hood. **Western Honey-buzzard** pale forms have an obvious dark trailing edge to underwing, wings shorter and broader.

WESTERN HONEY-BUZZARD *Pernis apivorus*   Plate 14

Other name: European Honey-buzzard   **French:** Bondrée apivore   **Creole:** Papang Manzer Mousdimyel

DESCRIPTION Length 52-60 cm; wingspan 130-150 cm. A slim-tailed, small-headed hawk. Highly variable plumage, sometimes very pale but more typically barred rufous or greyish, sometimes blackish. All plumages have black carpal-patches on the underwing, tail with a broad black band at the tip and two dark bars at the base of the tail. **Male** typically has ashy grey head and upperparts. Upperwing has a broad

black trailing edge, absent in female. Breast and underwing-coverts covered in chestnut blotches. Bars form a striking series of concentric rings on underwing-coverts. Plumage very variable, dark forms with completely dark brown chest and underwing-coverts, pale forms with all-white chest and wings, save for distinct black carpal-patches. Underwing of both sexes shows a broad black trailing edge, but the male has outer primaries with a more clear-cut black-and-white division and fewer, shorter, but broader dark bars on the base of the flight feathers. The wingtip has five prominently fingered primaries. Tail pattern diagnostic: broad bar at tip, two narrow bars at base. Tail long, usually tight-folded in flight, showing rounded corners. At close range, small head, narrow black bill, dark grey cere and yellow eye characteristic. Legs bright yellow. **Female** is darker brown, upperparts more uniform unbarred dark brown. More extensively dark on fingers than male, flight feathers more barred. Breast and underwing-coverts covered in blotching forming more distinct bars. Little or no grey on head. **Immature** is more buzzard-like, has a darker tail with three or four tail-bands and four or five bars on flight feathers. Eye dark, surrounded by dark patch, cere yellow. Chest and wing-coverts may be dark brown, rufous or pale. Above, shows pale patches at base of primaries and base of uppertail.
VOICE A clear call, either 'peee-ya' or 'pee-he-oo'.
BEHAVIOUR Frequents forests with clearings and more open country on migration. Solitary and secretive, perching on branches within trees. In flight, soars on flat wings held straight out from the body with tail fanned. When gliding, wings arch down slightly, wings sweep back and long tail is folded closed. Walks easily on the ground and can even run. Distinctive horizontal posture with head held forward when perched. Renowned for preying on nests of wasps and bees, but will catch insects in flight or on the ground and take reptiles, amphibians, chicks, small mammals and fruit.
RANGE Breeds from Spain east into Russia and as far south as the southern rim of the Caspian Sea. Winters in the more wooded parts of sub-Saharan Africa. Uncommon in East Africa. Monotypic.
STATUS Vagrant to granitic islands and Amirantes September to October and March to May.
SIMILAR SPECIES Booted Eagle from below has paler tail and from above, a pale patch across upperwing-coverts and scapulars. **Black Kite** and **Yellow-billed Kite** are more uniformly dark, lacking any obvious barring on the underwing, have pale diagonal bands across upperwing-coverts and forked tails. Female **Western Marsh Harrier** has longer tail, more slender body and wings and is mainly dark with paler head and leading edge to wing. **Crested Honey-buzzard** *P. ptilorhyncus* (L52-68 cm, WS135-150 cm) is very similar and a potential vagrant. Breeds Siberia east to Japan, with resident southern populations from India to Borneo. Northern populations move south in September to winter in Indonesia and South-East Asia. Broader-winged and shorter-tailed, with a rufous underwing, no dark carpal-patches and six-fingered wingtip. A horseshoe mark on the throat, contrasting dark and light chest-bands and a single broad bar at the base of the tail are diagnostic. **Eurasian Buzzard** *Buteo buteo* (L50-57 cm, WS110-132 cm), also a potential vagrant, breeds across Eurasia, with some races migratory notably *vulpinus* breeding Scandinavia to central Asia, wintering sub-Saharan Africa, from East to South Africa and southern Asia. Broad-winged, dark brown above contrasting with rufous tail, rufous-brown breast and underwing-coverts with pale chest-band and broad black border to wings. Tail shorter than wing width with straight sides and sharp corners (tail longer in Western Honey-buzzard with slightly convex sides and rounded corners).

BLACK KITE *Milvus migrans*   Plate 14

Other name: Pariah Kite   **French:** Milan noir   **Creole:** Papang Nwanr

DESCRIPTION Length 55-60 cm; wingspan 135-155 cm. An all-dark raptor with the long, angled wings of a kite and a long, shallow-forked tail. **Adult** is dark brown throughout, relieved only by greyish diagonal bands on the upperwing, paler area near the hand on underwing and a paler, greyish crown. Paler secondary coverts on upperwing contrast with darker feathering on secondaries. At close range, a rufous wash to the underparts is visible. In flight, forked tail separates from all except Yellow-billed Kite but when spread can appear only slightly concave or even straight-edged. Cere yellow, bill black and feet and legs yellow. Sexes similar, female larger than male. **Immature** is paler throughout, with light brown to buff feathering on breast and upperwing-coverts, noticeable pale bases to underwing primaries, pale rump, pale tail base and a narrow pale band across greater coverts of the underwing.
VOICE A loud 'keee-ee-ee' like a young gull, also a buzzard-like mew.
BEHAVIOUR Frequents a wide variety of habitat including woodland with clearings, cultivation and margins of lakes or marshes. Solitary or in small numbers. Soars and glides with wings held flat or slightly arched, never raised in a 'V' like a buzzard or harrier. Tail is twisted constantly in flight, a typical feature of kites. Loose, slow wingbeats with head turned down give an impression of buoyant, unhurried flight. Very varied diet, which has made it one of the most abundant birds of prey in the tropics. May catch insects in the air, small mammals on the ground and snatch items such as injured fish from the surface of water. Associates freely with man, scavenging at refuse tips or stealing food from tables and plates.
RANGE Breeds from Europe through Asia to Japan. Western birds (race *migrans*) winter in sub-Saharan Africa. Asian birds (race *lineatus*; larger, with pale buff belly and undertail and more prominent white patch on underside of primaries) winter in Iraq, southern India and South-East Asia. Race *govinda* of Indian subcontinent to Malay Peninsula moves in response to rains and food availability.

STATUS Vagrant to granitic islands; recorded December.

SIMILAR SPECIES Yellow-billed Kite is the most likely confusion species: adults distinguished by all-yellow bill; paler at all ages, more rufous on head, underparts and tail, with more deeply indented tail. **Western Marsh Harrier** immature female can also appear all-dark, but usually shows a lighter, cream-coloured crown, more rounded tail and lacks the pale covert panels on the upperwing. In flight, harriers usually hold wings in a shallow 'V' and do not angle the tail in the manner of the kites. Dark-morph **Booted Eagle** similar, but has larger head, unbarred rounded tail, pale upperwing-coverts and scapulars, pale uppertail-coverts, lacks obvious pale area on underwing and has white spots at junction of wing and neck.

YELLOW-BILLED KITE *Milvus aegyptius*   Plate 14

French: Milan à bec jaune   **Creole:** Papang Labek Zonn

DESCRIPTION Length 54-59 cm; wingspan 132-150 cm. A large all-dark kite with all-yellow bill, rusty wash on head and breast and a forked tail. **Adult** has wings similar to Black Kite, but very rufous head, throat and chest. Head lightly streaked darker. Tail more rufous than Black Kite with indistinct dark bars. Whitish primary bases on underwings form distinct 'windows'. Bill all-yellow. Legs dirty yellow. Female significantly larger than male. **Immature** is paler than adult with more dark streaking and buff spotting on head, throat and chest. Creamy streaks on belly, narrow creamy tips to many feathers of upperparts. Bill darker and duller than adult, only the cere yellow.

VOICE Shrill, high-pitched whinnying or trilling.

BEHAVIOUR Frequents a wide range of habitat and, as Black Kite, often associated with human settlement. Often gregarious, including vagrants in twos or threes. Long wings and slightly forked tail, often twisted in flight, giving a distinctive kite silhouette. Diet includes fish, small mammals, insects, birds, reptiles, amphibians and carrion.

RANGE Nominate race breeds Sinai, Egypt and both sides of the Red Sea; from November onwards tends to move southwards through Egypt. Race *parasitus* occurs in sub-Saharan Africa, from West Africa through to Madagascar and Comoros and may also make extensive movements during the rainy seasons. Sometimes included as races of Black Kite.

STATUS Vagrant to Aldabra mainly October to February; where it has been possible to assign race these have proved to be *parasitus*, which is far more likely than the nominate.

SIMILAR SPECIES Black Kite is the most likely confusion species: adults have a black bill with yellow cere, are darker brown throughout, with grey streaking on head and throat, appearing to be capped, and with only a very shallow indented tail. Dark morph **Booted Eagle** similar, but has unbarred rounded tail, pale uppertail-coverts, lacks obvious pale area on underwing and has white spots at junction of wing and neck.

WESTERN MARSH HARRIER *Circus aeruginosus*   Plate 14

Other name: Eurasian Marsh Harrier   **French:** Busard des roseaux   **Creole:** Papang Lanmar

IDENTIFICATION Length 48-56 cm; wingspan 110-130 cm. A large, long-winged, long-tailed slender raptor, all-dark brown in female, dark brown with grey wings and tail in male. **Male** acquires grey wings and tail over several seasons, finally having a grey crown, leading edge to the wing, secondaries and tail. Back, rump and wing-coverts are dark brown. Underparts are rufous, the chest streaked dark brown. Grey secondaries contrast with black primaries. **Female** by comparison is all-dark brown, with a creamy buff crown and throat and a creamy splash on the leading edge of the inner wing; slightly larger than male. Often shows a pale patch on the breast. **Immature** resembles female, though with less cream feathering on the wing, (seldom showing a pale leading edge to the inner wing unlike adult female) and with the crown fringed more rufous. At close quarters, pale tips to greater coverts may be visible. **Subadult male** of about one year old resembles adult female but has grey on wings and tail and a dark trailing edge to the wing.

VOICE Normally silent; occasional chattering 'kek-kek' or wailing 'pee-yu'.

BEHAVIOUR Hunts alone, flying low over reedbeds, marshes and grassland with a slow, rocking flight and wings held in a shallow 'V', pausing before dropping suddenly onto prey. Vagrants could turn up in almost any habitat. Food includes small mammals, birds and frogs.

RANGE Two races, the nominate breeding from western Europe to Mongolia. Migratory in north of range, wintering sub-Saharan Africa and Indian subcontinent to Sri Lanka. Race *harterti* of northwest Africa is mainly sedentary. Replaced by Eastern Marsh Harrier *C. spilonotus* Siberia to Japan.

STATUS Vagrant to granitic islands January to February.

SIMILAR SPECIES Black Kite is the most likely confusion species. This is all-dark, with pale diagonal bars on the upperwing and a slightly notched tail which is constantly twisted in flight. **Booted Eagle** is similar but lacks all-dark wings and has a different flight silhouette, wings not held in a shallow V when gliding. **Western Honey-buzzard** has dark carpal-patches and a barred tail. There have been reports of unidentified ringtail harriers and if slow, rocking flight suggests a harrier, the following potential vagrants should be checked: **Pallid Harrier** *C. macrourus*, (L40-48 cm, WS97-118 cm), breeds from north of the Black Sea east to Kazakhstan, wintering south to Somalia, Kenya and Maldives. Slim, with light, tern-like flight. Male has unique whitish grey plumage throughout with a black wedge in longest primaries. Female with darker head and more contrast in facial markings (black eye-stripe and dark crescent round pale cheek), pale orange below and with dark smudging in innermost underwing secondaries, contrasting with paler outer primaries. Juvenile pale orange below with bold, dark neck-collar, edged cream. **Montagu's Harrier** *C. pygargus*, (L43-50 cm, WS96-116 cm), breeds from Spain east as far as Kazakhstan and winters in eastern Africa (Ethiopia to South Africa) and the Indian subcontinent as far south as Sri Lanka. Small, slim harrier, male with black line across grey upperwing and extensive black on wingtips. Female brown above with narrow white rump, rufous below, pale face and crown and uniform underwing. Juvenile with deeper rufous underparts and more rufous upperwing-coverts.

Harriers: top row Pallid Harrier, bottom row Montagu's Harrier

BOOTED EAGLE *Hieraaetus pennatus* Plate 14

French: Aigle bott **Creole:** Leg Lapat Plim

DESCRIPTION Length 45-53 cm; wingspan 100-121 cm. A small buzzard-like eagle which occurs in dark and light morphs, both with pale patches on upperwing-coverts/scapulars and pale uppertail-coverts (diagnostic). **Dark morph** resembles a buzzard, though with dark wings and paler brown panel on inner half of upperwing, pale patch at innermost primaries, whitish uppertail-coverts and pale tail contrasting with dark brown breast. A good feature, especially in dark morph, is small white patches at junction of neck and wing which show up well when seen head-on, like 'head lights' (sometimes also seen in Western Honey Buzzard, but less distinct). **Pale morph** is more common and distinctive, with all-dark flight feathers and tail, pale head and upperwing-coverts and dirty white underwing-coverts with distinctive spotting on the median and greater coverts. Brown throat markings contrast with a pale breast. **Intermediate morph** has underparts and underwing-coverts strongly washed rufous and a heavily streaked breast and belly. Eye yellow, orange or reddish. Bill grey-horn, blackish at tip, cere yellow. Feet yellow, claws black. Sexes similar, female larger. **Immature** is also polymorphic and similar to adult but with a rufous tint and streaked below and eye brown.
VOICE Generally silent outside breeding season. A high-pitched 'kee-kee' contact call.
BEHAVIOUR Frequents forests with clearings. Generally solitary outside breeding season. Flight rapid with deep, powerful wingbeats. Frequently glides, carpal joints forward and wingtips lowered, recalling Black Kite. Stoops to take lizards and small mammals on the ground but will also hunt birds through woodland canopy. Prey includes small to medium-sized birds, ranging in size from warblers to pigeons.
RANGE Breeds in Spain, North Africa, Turkey and east into China. Western birds move south into Africa, wintering from Ethiopia to South Africa. Eastern birds migrate south to Pakistan and India. Also breeds South Africa. Monotypic.

STATUS Vagrant to granitic islands; recorded November.

SIMILAR SPECIES Dark form may resemble **Black Kite** or **Yellow-billed Kite** These both have slower flight with shallow wingbeats, slimmer, longer, more angular wings, dark uppertail-coverts, longer, notched tails and always lack 'head lights'. Female and immature **Western Marsh Harrier** have a different silhouette, longer-winged with narrower outer wing, pale leading edge to forearm and dark uppertail-coverts. Pale morph may resemble a pale **Western Honey-buzzard**, but has pale centre to hand of underwing, contrasting sharply with dark secondaries and fingers, darker uppertail and upperwing-coverts/scapulars. The following are potential vagrants: **Short-toed Eagle**, *Circaetus gallicus* (L62-67 cm, WS162-178 cm), breeds from Spain through North Africa to Russia, wintering in sub-Saharan Africa. A pale, broad-winged raptor, white below with a dark 'hood', coffee-coloured above, narrow, dotted bands on wings and tail and narrow, square-cut tail. **Lesser Spotted Eagle** *Aquila pomarina* (L57-64 cm, WS143-168 cm), breeds from eastern Germany into Russia, wintering from Ethiopia south through East Africa to South Africa. Appears short-tailed, all-dark-brown below, paler buff head and wing-coverts and rump crescent contrasting with dark tail and flight feathers. Immature has pale tips to greater coverts, white 'windows' at base of inner primaries and a single, thin upperwingbar. **Greater Spotted Eagle** *Aquila clanga* (L60-70 cm, WS153-177 cm) breeds from Poland east through Russia to Korea, wintering Ethiopia, Indus valley and northern India. Very short-tailed, broad-winged eagle, dark brown with a black tail. From below dark brown body and wing-coverts contrast with grey flight feathers and tail. Immature dark with pale tail tip, white crescent on rump and two distinct white wingbars. **Steppe Eagle** *Aquila nepalensis* (L62-74 cm, WS165-190 cm), breeds from Black Sea through Tibet and Mongolia, wintering in Africa and India. Large eagle, varying from dark brown to buff or fawn, with paler patch on back, pale nape and lighter barring on underwing-coverts. Immature has broad white band along the underwing, two thin wingbars, white crescent on rump and white tips to flight feathers and tail feathers. **Eurasian Sparrowhawk** *Accipiter nisus* (male L29-34 cm WS58-65 cm, female L35-41 cm, WS67-80 cm), breeds from Europe east to Indochina, wintering from Kenya to India and Sri Lanka. Male slate grey above with reddish barring below, female and immature sepia brown above with fine dark barring below and darker tail-bands. Both sexes have yellow eye. **Levant Sparrowhawk** *Accipiter brevipes* (L30-37 cm, WS63-76 cm), breeds Russia, wintering sub-Saharan Africa, vagrants reaching Kenya. Noticeably sharper-winged than Eurasian Sparrowhawk. Male white below with black wingtips, washed pink-buff above. Female grey-brown above with dark wingtips, narrow brown barring below tinged pink-buff. Both sexes have red eye. Immature brown above, with large, thrush-like spots on breast, barred on underwing and tail.

LESSER KESTREL *Falco naumanni*   Plate 15

French: Faucon crécerellette   **Creole:** Katiti Latet Sann

DESCRIPTION Length 29-32 cm; wingspan 58-72 cm. A small falcon with silvery white underwing, slightly projecting central tail feathers and uniform, unspotted mantle in the male. **Male** has unspotted, uniform tan back, blue-grey hood, no moustachial-stripe, blue-grey greater secondary coverts and tail. Tail shows thick black subterminal band, white tips and slightly projecting central tail feathers. Breast buff with lines of fine spots. Underwing is silvery white. **Female** has a rusty red ground colour, heavily streaked and spotted throughout, the underwing pale. Both sexes have blue-black bill, yellow cere and feet and white claws, the latter feature diagnostic at close range. **Immature** similar to female, rump and undertail-coverts often tinged grey, moustachial-stripe indistinct. More spotting on underwing-coverts (especially in female), pale barring on flight feathers.

VOICE A diagnostic, trisyllabic, shrill 'kye-ki-ki' or 'tye-tye'.

BEHAVIOUR At breeding grounds, frequents open terrain; vagrants to Seychelles have favoured margins of airstrips and human settlement. One of the most gregarious falcons, though vagrants are likely to be solitary. Flight rapid with shallow wingbeats. Mostly insectivorous, hunting over open terrain for crickets, grasshoppers and beetles caught by stooping after a brief hover. Will catch flying insects such as termites and to a lesser extent lizards and small mammals.

RANGE Breeds from Spain and North Africa through southern Europe and Turkey, east to southern Russia and China. All Asian and European birds move south into Africa in October and November, wintering from Ethiopia to South Africa. Monotypic.

STATUS Vagrant to Bird and granitic islands; recorded December and April to May.

SIMILAR SPECIES Might be confused with the resident **Seychelles Kestrel** in the granitics, but this is smaller, has an unspotted breast and a dark moustachial-stripe. On Aldabra, confusion is possible with **Madagascar Kestrel**, also smaller, with darker underwing, whiter breast and belly and heavy spotting on flanks. **Common Kestrel** *F. tinnunculus*, (L31-37 cm, WS68-78 cm) is a potential vagrant; the nominate race breeds across Eurasia, resident in south and west of its range but northern and eastern populations highly migratory and widespread in East Africa, October to March. Male is distinctive, with dark moustachial-stripe, black spots on rufous back, heavier spotting on underparts darker underwing (but less dark at tip) and no bluish panel on upperwing. Females and immature very similar but with shorter wings and longer tail, wingtips falling well short of tail tip in bird at rest with darker ear-coverts and at close range, black not white claws. More boldly spotted below, underwing more barred and darker but less black at tip.

MADAGASCAR KESTREL *Falco newtoni*   **Plate 15**

French: Crécerelle malgache   **Creole:** Katiti Malgas

DESCRIPTION Length 25 cm; wingspan 45-50 cm. A small kestrel, bright rufous chestnut above, pale below with heavy spots on the flanks. **Adult** has rich chestnut brown back and wing-coverts with dark flight feathers and slate grey, barred tail. Head light rufous with fine dark streaking. Large, dark eye in pale face and small, dark moustache. Upper breast with a rufous buff wash and thrush-like black spots, heavier on the flanks. Lower belly and undertail-coverts white. Bill blue-grey with a yellow cere. Legs pale yellow. **Female** has head and tail browner, more heavily spotted both above and below; some females may have completely unmarked underparts. **Immature** similar to adult but the underparts are more heavily streaked and the flight feathers are edged buff.

VOICE A shrill 'ki-ki-ki-ki' like most other kestrels, from which it gets its Creole name.

BEHAVIOUR Usually conspicuous, perching on prominent vantage points such as bare tree branches or poles. Solitary or in pairs. Rapid flight, with shallow wingbeats interspersed with glides. Preys mainly on lizards and insects. Insects include grasshoppers and beetles, caught from low perches, and dragonflies and other flying insects caught by hawking. Abbott's day geckos *Phelsuma abbotti* are snatched from the trunks of trees and other lizards taken from the ground. Noted feeding on newly hatched green turtle young along the atoll beaches at Aldabra (R. Chapman pers. comm.) and on freshly killed Madagascar Turtle Dove[1].

BREEDING Breeds August to January, sometimes nesting in old nests of Pied Crows, large trees, palms and holes in trees or buildings. Also readily uses nest boxes. The abundance of such sites on Aldabra is very low which may severely restrict breeding[2]. Up to four reddish eggs are laid and like most kestrels the female incubates while the male brings food to her.

RANGE Occurs in Madagascar (where it is the most common bird of prey) and Aldabra only, of which the latter is sometimes considered a separate race *aldabranus*, slightly smaller than the Madagascar nominate race but otherwise indistinguishable. In Madagascar there is a rufous morph with very dark head and back and deep chestnut underparts. Vagrant to Comoros.

STATUS Resident Aldabra. Population estimated at 100 pairs[3] in 1967 but more recent data suggests a much smaller population of 20-50 pairs[4]. It is questionable whether there is a self-sustaining population.

THREATS AND CONSERVATION Pied Crows may be nest predators. Rats and cats compete for food and undoubtedly account for the relative scarcity of lizards on Aldabra (which made up 99% of prey brought to nest in one study at Madagascar[5]) compared to rat-free islands of Seychelles. The small population size, lack of breeding sites and remoteness of the atoll make it particularly vulnerable. For these reasons it must remain a priority for further research.

SIMILAR SPECIES Lesser Kestrel is a scarce vagrant, larger with whiter underwing and a buff wash on the breast.

References: 1. Schoeberl (1994), 2. Scoones *et al.* (1988), 3. Gaymer (1967), 4. Rocamora and Skerrett (1999), 5. Robenarimangason and Réne de Roland (1998).

SEYCHELLES KESTREL *Falco araea*   **Plate 15**

French: Crécerelle des Seychelles   **Creole:** Katiti Sesel

DESCRIPTION Length 18-23 cm; wingspan 40-45 cm. A very small kestrel with a dark head, rich rufous chestnut back and an unmarked breast; when perched appears small-bodied and short-tailed in silhouette. The only resident bird of prey in the granitics. **Male** has dark slate grey head, chestnut mantle and chestnut wing-coverts with dark crescents. Slate grey tail with dark bands, the broadest band at the tip. Chest and belly are pinkish buff and unspotted. Has a clear dark grey moustache, dark bill with yellow cere and yellow legs and feet. **Female** is larger and slightly paler, but also unmarked on the breast. **Juvenile** has underparts rufous with scattered dark streaks and spots, particularly on the lower breast, head dark blackbrown, streaked rufous, with light buff area round the eye and distinct dark moustache; first-year birds have a paler chestnut back with dark barring and a chestnut head with streaks. Flight feathers black with large rufous spots and narrow rufous buff tips, tail has a broad whitish buff terminal band.

VOICE Often heard before seen, calling noisily in flight or from a perch. A shrill, shaking 'shkea-shkea-shkea-shkea', or a more rapid 'kikikikiki', which gives it its Creole name.

BEHAVIOUR Occurs in a variety of habitats including forest, cultivation and around houses from the coast to the highest mountains, singularly or in pairs. Often associates with buildings to perch and roost. In flight has rapid wingbeats interspersed by gliding but does not hover. Hunts mainly from a perch, pouncing on prey. Takes mainly lizards, with over 92% of diet made up of geckoes and skinks, of which two-thirds are two species of green day gecko *Phelsuma spp*.[1]. Other items taken include insects, frogs, chameleons, rats and birds. May take Madagascar Fodies in the manner of sparrowhawk, panicking birds feeding on a lawn and snatching one of the scattering flock[2].

BREEDING Breeds August to October. Mean territory size around 40 ha[1]. In upland areas, most nest on cliffs, some in trees; below 200 m, coconut palms are favoured with cliffs, trees and buildings also used[3]. A loose nest is built of sticks. Clutch is 2-3 white eggs with brown markings, incubation lasting for 28-

31days, fledging 35-42 days and juveniles remain with adults for *c*14 weeks after fledging[3]. Cliff nests are most successful, then tree cavities, with buildings being half as successful as cliffs and palm trees the least successful of all possibly due to differences in predation pressure[3].

ORIGINS Seychelles was undoubtedly colonised from Madagascar and Seychelles Kestrel is most closely related to Madagascar Kestrel[4].

STATUS Endemic to granitic islands and resident Mahé (370 pairs), Praslin (ten pairs), Silhouette (30-40 pairs), North (five pairs) and some of the smaller satellite islands of Mahé and Praslin (10-15 pairs)[1]. IUCN Vulnerable. There appears to be little movement between islands. Probably once occurred throughout the granitic islands. In 1977, 13 birds were released on Praslin. This reintroduction was successful, for by 1987 there were possibly ten territories established on the island (pers. obs.). It may have reached 20 pairs by 1994[5] but has subsequently declined with several known territories abandoned[6]. The trend on Mahé and elsewhere since 1980 is unknown but there may also have been some decline[6].

THREATS AND CONSERVATION Past contraction of range was probably due to persecution and introduced predators. An early Creole name 'Manzer d'poul' means 'eater of chicken'. It would not be possible for this small kestrel to kill a chicken, but it may have snatched day-old chicks from the yards of settlements. In the past this species was also thought to bring bad luck and its presence near a house was said to indicate an imminent death. Such superstitions attracted widespread persecution and nests were also destroyed when found. Rats, cats, Barn Owls and Green-backed Heron may be predators of eggs and chicks, all introduced apart from the heron[3]. The greater success of remote cliff nests suggests that these are less prone to disturbance by introduced predators. There may be some nest competition from introduced Barn Owls and Common Mynas. The single most serious effect on the kestrels' ecology must have been the establishment of both cats and rats on all the larger islands, greatly reducing the main food supply, geckoes and skinks. Rarely moves between the islands, though there have been occasional sightings of lone birds on La Digue, which would appear to be suitable for a reintroduction. Nor has it been recorded on the smaller islands of Cousin and Aride, which being rat-free, have very high skink populations. As education improves, the days of persecution have receded and it is probably now limited by food supply or nest sites.

SIMILAR SPECIES There are no other resident falcons in Seychelles except **Madagascar Kestrel** which is confined to Aldabra. Vagrant falcons do occur, including **Eleonora's Falcon**, **Sooty Falcon** and **Eurasian Hobby**, but these are considerably larger and invariably appear longer-winged, darker and very fast by comparison. **Lesser Kestrel** has been recorded as a vagrant; this is much larger with a pale unspotted chestnut back, unmarked upperwing-coverts, silver white underwing-coverts and lines of spots on the breast.

References: 1.Watson (1981a), 2. Gerlach (1995), 3. Watson (1992), 4. Benson (1984), 5. Collar *et al.* (1994), 6. Rocamora (1997c).

RED-FOOTED FALCON *Falco vespertinus* Plate 15

French: Faucon kobez **Creole:** Katiti Lapat Rouz Was

DESCRIPTION Length 28-31 cm; wingspan 65-75 cm. A slate grey falcon with chestnut lower belly and undertail-coverts and red legs and feet. **Male** is blue-grey throughout with contrasting paler silver-grey flight feathers and darker, almost black tail contrasting with paler lower back. Chestnut lower belly, thighs and undertail-coverts contrast with blue-grey breast. Eye-ring, cere and legs and feet are all bright red-orange. Cheeks are uniform grey with no distinct moustachial-stripe (usually more obvious in Amur Falcon). **Female** is slightly heavier, with blue-grey upperparts, black flight feathers, narrowly barred tail and striking ginger orange breast, belly, underwing-coverts, crown and nape. Black moustachial-stripe and orange cere, legs and feet. **Juvenile** has the underwing heavily barred black and white with a black trailing edge. Upperparts greyish brown contrasting with paler greyish rump and tail, underparts white streaked reddish brown. Crown sandy brown, lacking grey tones of Amur Falcon, the moustachial-stripe shorter and untapered, often square-ended. **Immature** has flight feathers darker than adult, rufous tips to the lower back and wing-coverts and a white forehead. Immature male may show white around face and have more chestnut on the upper and lower breast. Immature female has a dark grey-brown back, paler chest with some streaking and fine streaking on the crown and nape. Dark mask over the eye and buff tips to the back and wing-coverts may still show from the juvenile plumage. Cere deep yellow to orange.

VOICE Both sexes may utter a 'kew kew kew' call in flight.

BEHAVIOUR Prefers more open ground, grassland, wetland margins or cultivated ground. Gregarious year round, but vagrants are likely to be solitary. Fast, agile flight with stiff wingbeats, sometimes gliding with wings swept back or soaring, tail spread and wings held straight. Hunts most actively at dawn and dusk. Almost entirely insectivorous, either hawking for insects in flight, watching from a vantage point, or hovering in open ground before dropping onto prey. Runs easily on the ground to catch insects. Though larger insects (crickets, grasshoppers and locusts) are preferred, will take frogs, lizards and even small birds at times.

RANGE Breeds from eastern Europe east through the steppes of Russia to Mongolia, migrating south through Turkey and Israel to winter in southern Africa. Monotypic.

STATUS Vagrant to granitic islands November to December.

SIMILAR SPECIES Adult male might be confused with dark **Eleonora's Falcon** or **Sooty Falcon**, but these lack the chestnut belly and thighs. Eleonora's Falcon also has uniformly dark upperwing and underwing-coverts darker than rear underwing. Adult Sooty Falcon from above has wingtip darker than rest of upper-wing and from below has underwing paler than body (reverse of Red-footed Falcon). Immature Red-foot-ed Falcon might be confused with immature **Eurasian Hobby**, but the latter has unbarred upperparts, heav-ily-spotted streaks on the breast, no trace of a rufous wash in the head or breast, entirely black central tail feathers and more uniformly barred, darker underwing. **Amur Falcon** male has pure white underwing-coverts. Male from above shows no contrast between pale grey tail and back (blackish tail of male Red-footed Falcon contrasting with grey back). Female below shows heavy black thrush-like spotting on the breast. Immature has grey-blue crown and upperparts with rufous wash (Red-footed Immature more sandy, lacking grey tones); moustache longer and more tapered; underparts streaked blackish (reddish brown in Red-footed).

AMUR FALCON *Falco amurensis*   Plate 15

Other name: Eastern Red-footed Falcon   **French:** Faucon de l'amour   **Creole:** Katiti Lapat Rouz Les

DESCRIPTION Length 28-30 cm; wingspan 65-70 cm. A dark falcon with clean white underwing-coverts in the male and white underparts spotted black in the female. **Male** is mainly dark sooty grey throughout except for obvious white underwing-coverts, darker moustachial-stripe, rufous thighs and undertail-coverts, orange-red cere, base of bill, eye-ring, legs and feet. From above, shows silvery grey primaries, as Red-footed Falcon, but the tail is pale blue-grey and lacks contrast with upperparts. When perched, a nar-row white line may be evident on the leading edge of folded wing. **Female** has a grey crown and nape, large black moustachial-stripe, breast whitish heavily spotted with black and underwing-coverts white with finer black spotting. Thighs and undertail-coverts are yellowish brown. Blue-grey upperparts are barred black. Tail white with black bands, the subterminal band widest. Bare parts similar to male. **Juvenile** has blue-grey crown with a rufous wash, long, tapered black moustache and grey upperparts washed rufous. Underparts and underwing-coverts are streaked blackish. **Immature** similar to female but with paler, finely streaked head and neck, rufous fringes to the feathers on the back and wing-coverts, fine bar-ring on the tail, buff unspotted 'trousers' and broad sepia-black streaking on the chest. Underparts and flanks may be streaked blackish, or heavily blotched.
VOICE A shrill flight call 'kew-kew-kew'.
BEHAVIOUR Frequents open country, but vagrants may turn up almost anywhere in Seychelles. Gregarious year round, but vagrants are likely to be solitary, sometimes in twos. Rapid flight, sometimes hovering with deep wingbeats, soaring on flat wings and gliding with carpal joint lowered and wingtips raised. More active in early morning and evening. Mainly insectivorous, preferring larger insects includ-ing locusts, grasshoppers and large beetles. When termites are swarming will hawk for these in flight, but otherwise commonly sits for long periods on a suitable perch, scanning open ground for moving prey. Will also hover and pounce in search of food, landing beside prey and may take frogs, small mammals and birds at times.
RANGE Breeds in northeast Asia, migrating late September across the Indian Ocean to southeastern Africa, where it arrives in November. Little is known of its return migration route, but this is thought to be along the East African coast and round the edge of the Arabian Sea and possibly across the western Indian Ocean. Monotypic.
STATUS Vagrant to granitic islands and Amirantes; more frequently recorded December to January but also in March.
SIMILAR SPECIES Red-footed Falcon is most similar, but the male from below has dark underwing-coverts and above a blacker tail contrasting with grey back. Female has unstreaked rufous crown, underparts and underwing-coverts, female Amur Falcon showing a similar tone only on thighs and undertail. Immature above is sandy brown lacking any grey tones and below has more obvious rufous wash on underwing-coverts. There might be confusion between immatures of Amur Falcon and those of **Eleonora's Falcon, Sooty Falcon** or **Eurasian Hobby** but these all lack the cleaner, whiter underwing flight feathers of Amur Falcon, distinctive in all plumages and quite unlike other species.

EURASIAN HOBBY *Falco subbuteo*   Plate 16

French: Faucon hobereau   **Creole:** Katiti Kannson Rouz

DESCRIPTION Length 28-36 cm; wingspan 69-84 cm. A long-winged, falcon with red thighs and a clear black-and-white face. **Adult** has dark slate grey back, wing-coverts and tail with black flight feathers. Crown and nape black forming a distinct two-lobed hood which contrasts with white cheeks and throat. Breast heavily streaked with black, underwing-coverts and tail finely barred and streaked black and white, giving a dark ground colour. Chestnut red 'trousers' show up as a clear red chevron in flight. Flight sil-houette resembles an enormous swift, with slim build, long, narrow, pointed scythe-like wings and a fair-ly short square-cut tail. Bill blue-black, cere and legs yellow. Sexes similar, female larger, broader-winged and often browner above. **Immature** has browner upperparts and pale buff tips to flight feathers, tail and wing-coverts giving a dark, mottled grey appearance above. At closer range, buff tips form a pale line at margin of greater coverts. Underparts have a pale rufous buff ground colour throughout, but with coarse brown-black streaking and trousers and undertail buffish rather than red with thick black bands under the tail and lighter, rufous bands in-between. Pale fringes on crown feathers give a paler cap, but face retains black mask and white cheeks. Bill blue-black and cere pale yellow. **Second-year** retains barred undertail but is otherwise similar to adult.

VOICE Rarely calls on migration; but may sometimes give a shrill 'ki-ki-ki-ki'.

BEHAVIOUR Frequents a wide variety of habitats wherever insect prey is abundant. Generally solitary. In flight, has stiff wingbeats, interspersed with long glides, wings held back. Often soars, wings held straight out and tail spread. Feeds almost entirely on the wing, either hawking over ponds and wetlands for dragonflies, catching large moths or bats at dusk, or, typically, running down smaller birds in aerial pursuit. Prey normally smaller passerines (sparrow size) but capable of taking small waders and even terns.

RANGE Nominate race breeds across Eurasia from Spain to Japan. Winters in southern Africa and northern India, arriving in November. Occurs as a regular migrant in Maldives, Somalia and Kenya.

STATUS Vagrant to granitic islands November to January.

SIMILAR SPECIES Peregrine Falcon has slower, more powerful flight and is larger, heavier in breast and wings with proportionately shorter, tapered tail and paler rump. **Eleonora's Falcon** has more languid flight and appears darker throughout with dark underwing-coverts contrasting with silvery flight feathers. **Sooty Falcon** also appears darker throughout, but with uniform underparts and no clear facial mask. **Red-footed Falcon** and **Amur Falcon** may also show rufous thighs but males are otherwise distinctive. Females have barred upperparts and tail, unlike unbarred Eurasian Hobby.

ELEONORA'S FALCON *Falco eleonorae*   Plate 16

French: Faucon d' Eléonore   **Creole:** Katiti Eleonora

DESCRIPTION Length 36-42 cm; wingspan 90-105 cm. A dark, long-winged, long-tailed falcon. Dark underwing-coverts contrast with paler flight feathers and usually shows a diagnostic coffee brown hue to upperparts. **Adult** occurs in two colour morphs, pale outnumbering dark about three to one. **Pale morph** is dark slate grey above with a grey, finely barred tail, slightly darker wingtips and outer tail and slightly paler rump/inner tail. Rufous brown below reaching to breast and heavily streaked. Most reliable feature is the underwing pattern where darker underwing-coverts contrast clearly with lighter, silvery bases to flight feathers and darker trailing edge. Black hood and long moustachial-stripe contrast with rounded, clean white cheek and throat. **Dark morph** mainly sooty black with coffee brown hue above and paler bases to primaries on underwing, more restricted than in pale morph. All adults when perched have wings the same length as tail. Eye dark brown, legs yellow with black talons. Female averages larger, heavier and broader-winged with pale blue cere and eye-ring (yellow in male). **Immatures** also occur in two colour morphs, though these are less clearly defined, differing mainly in pattern of underwing-coverts (mottled and barred in pale morph, entirely dark in dark morph) and undertail-coverts (pale buff and unbarred in pale morph, obvious dark barring in dark morph). Being shorter in wing and tail length, have an entirely different silhouette to adult, more similar to other falcons. Also generally browner than adult above, with pale tips to coverts giving a more scaly effect and more buffish below with buff rather than rufous tints to underwing-coverts and breast. Cere and eye-ring pale blue, bill bluish with blackish tip and lags greenish yellow. **Second-year** retains some worn feathers of juvenile plumage and has barred flight feathers and tail feathers but otherwise acquires more adult characteristics. **Third-year** similar to adult above but still with some barring on flight feathers and tail feathers.

VOICE Generally silent, but flight call a harsh 'kyer-kyer-kyer'.

BEHAVIOUR Almost entirely coastal, though outside breeding season also frequents inland wetlands, marshes and lakes. Gregarious, sometimes also associating with Eurasian Hobby, but vagrants are generally solitary. Very agile in flight with lazy, deep wingbeats, changing easily from a fast, flicking cruise into shallow, twisting dives. Outside breeding season (when feeds on small birds) takes mainly insects, large beetles, moths and dragonflies, captured on the wing (or occasionally by stooping onto the ground). Birds are caught in flight after a short chase or repeated stoops, ranging in size from warblers to pigeons.

RANGE Very restricted breeding range, from Canaries and Morocco, through scattered coastal sites in the Mediterranean east to Greece. Breeds late to feed young on migrating birds, abandoning colonies from mid-October and following the Red Sea and Somalia south to winter mainly in Madagascar, but also East Africa and Mascarenes. Monotypic.

STATUS Vagrant to granitic islands and Aldabra November to February.

SIMILAR SPECIES Sooty Falcon is the most likely confusion species, but this has larger head, narrower wings, shorter tail, a uniform underwing pattern without the contrasting coverts but less uniform upperwing showing contrast between darker primaries and silvery grey remainder of upperwing and a contrastingly darker tail tip. Immature Sooty is more diffusely streaked below, with less contrasting underwing and a broad dark subterminal tail-band. **Eurasian Hobby** is more uniformly streaked and barred beneath on whitish buff (not rufous) underparts, with red trousers in adult and a clear black hood often with a second lobe behind the eye, pale nape and white face and throat. **Red-footed Falcon** may appear similar, but is either all-dark with red thighs and no facial pattern (male) or with rufous (female) or buff (immature) underparts. Immature also has entirely barred uppertail and different head pattern. **Peregrine Falcon** is paler slate grey with distinct tail barring, which Eleonora's never shows, has heavier build, shorter, broader wings and different flight action. **Seychelles Kestrel, Madagascar Kestrel and Lesser Kestrel** are all-pale by comparison, with rufous upperparts and pale or lightly spotted underparts.

SOOTY FALCON *Falco concolor*   Plate 16

French: Faucon concolore **Creole:** Katiti Nwanr

DESCRIPTION Length 32-36 cm; wingspan 78-90 cm. A slim, long-winged uniform grey falcon, with large head and long central tail feathers, giving a wedge-shaped tail. **Adult** is uniform dark ashy grey below, with an all-grey head, slightly paler grey rump, darker grey flight feathers on upperwing and darker tail. Sometimes may show a paler chin or a darker moustache. When perched wings are noticeably longer than the tail. Flight silhouette diagnostic, with big head and long, narrow scythe-like wings appearing to reach behind short tail, underparts darker than underwing. Female slightly larger, heavier and darker than male with slightly less contrast in upperwing and lemony yellow cere and eye ring (bright yellow in male). Small, black bill and bright yellow legs in both sexes. **Immature** is browner above and below than adult, underparts with buff ground colour and darker streaking particularly on upper breast. Darker streaking shows up on underwing-coverts and also as fine dark bands on tail. Underwing lacks contrasting dark coverts of Eleonora's Falcon, but has a broad, dark trailing edge. Tail less obviously wedge-shaped than adult with broad terminal band below and dark, unbarred central feathers visible above and below. Shows a dark hood, with two lobes like Hobby, but cheeks and throat dirty brown and two pale patches on hindneck (absent in Eleonora's). Buff tips to tertials, greater coverts and primary feathers are typical of immature falcons. Cere and eye-ring bluish, legs pale greenish yellow.

VOICE Normally silent; a shrill 'kree-a, kree-a'.

BEHAVIOUR Favours dry, open country though absence of such habitat in Seychelles means it could turn up almost anywhere. Gregarious, though vagrants likely to be solitary. An agile, dashing hunter, taking larger prey from a shallow glide or stooping on items on the ground. Feeds mostly on insects (flying ants, termites or dragonflies), occasionally taking small birds and bats.

RANGE Breeds at scattered locations in Sudan, Egypt and Israel, with outlying colonies on the Red Sea and Persian Gulf. Main wintering range is Madagascar, where birds arrive in late October and depart again in April. Some appear to winter on the African seaboard, from Mozambique to South Africa. Monotypic.

STATUS Vagrant to granitic islands and Aldabra mainly November to December.

SIMILAR SPECIES Eleonora's Falcon and Hobby are the most likely confusion species. Size is little use to distinguish these in the field. **Eleonora's Falcon** is slightly longer and has more square-cut tail, but most reliably shows strong contrast on underwing between dark underwing-coverts and paler bases of flight feathers. Adult **Eurasian Hobby** has rufous trousers, but immature Hobby and Sooty Falcon may look very similar. Immature Sooty has heavier streaking on upper breast with buff ground colour, compared to uniform darkly blotched chest of Eurasian Hobby. The latter always shows clean white cheeks and throat compared to dirty brown face and throat of immature Sooty Falcon. Immature **Red-footed Falcon** and **Amur Falcon** are similar to immature Sooty, but with whitish rather than rufous flight feathers in the underwing.

PEREGRINE FALCON *Falco peregrinus*   Plate 16

French: Faucon pèlerin **Creole:** Katiti Pelren

DESCRIPTION Length 34-50 cm; wingspan 80-120 cm. A large, thickset slate grey falcon, with a deep, barred chest, broad wing bases and a relatively short, tapered tail. Classic 'crossbow' silhouette in flight. **Adult** is slate grey above with lighter grey rump, wings and chest uniformly fine barred beneath. Tail is relatively short, tapered in shape with dark bars, broader at square tip. Head black, dark hood and broad moustachial-stripe contrasting clearly with white cheeks, throat and upper breast. Bill blue-black, darker at tip. Cere, eye-ring and legs yellow. When perched, the wings are clearly shorter than the tail. In flight appears heavy, bull-necked and compact with broad bases to wings and tail. Female noticeably larger and heavier than male. **Immature** is dark brown-black above with paler tips to coverts, uniformly patterned underwing with black tips to primaries. Tail shows a clear buff tip. Underparts heavily blotched in vertical streaks, with broad spots and bars on flanks and barred undertail-coverts. Head pattern less distinct, with paler forehead and supercilium. Cere grey-blue, legs yellow.

VOICE A harsh, rasping 'kye-kye-kye-kye-kye' when agitated, but more usually silent on migration.

BEHAVIOUR A solitary hunter, mainly coastal, but also occurring in a variety of other open habitats. Purposeful, rowing flight; wingbeats more rapid and stiffer than large falcons, gaining height rapidly with swift soaring. Prey when sighted is pursued from a greater height, the falcon accelerating into a heart-stopping dive with wings close-folded to the body. Normal prey is pigeon-size birds, struck forcibly in flight. Prey ranges from birds as small as a finch and as large as a goose. When not hunting, birds spend long periods perched on high vantage points, observing and memorising local flight traffic.

RANGE The most widespread bird of prey in the world, breeding around the Arctic and northern Asia, Africa, India and China and from Tierra del Fuego on the tip of South America to Australia. Northern birds winter south to Africa, India and South-East Asia. Eurasian nominate race *peregrinus* is unlikely in Seychelles being only a partial migrant. Race *calidus*, breeding Lapland to central Siberia is more likely being a long-distance migrant; it is longer-winged, greyer above and less barred below. Also possible may be race *radama* of Madagascar and Comoros and *minor* of mainly sub-Saharan Africa.

STATUS Vagrant to the granitic islands November.

SIMILAR SPECIES There are many other falcons which are more likely in Seychelles and most are slim, long-winged and long-tailed. Peregrine Falcon has its own very distinct silhouette, deep-chested and short-tailed and is the only falcon with barred underparts. **Eurasian Hobby** is streaked on underparts and lacks paler rump/uppertail of Peregrine.

GREY FRANCOLIN *Francolinus pondicerianus* Plate 17

Other name: Grey Partridge **Creole:** Perdri

DESCRIPTION Length 33-35 cm; wingspan 48-52 cm. The only gamebird resident in the islands, a plump, finely barred grey partridge with a striking rufous undertail in flight. **Adult** has very fine grey and black barring on breast and belly and bolder white and grey barring on rump and tail. Mantle, back and wing-coverts browner with black-and-white streaking. Head rufous brown, with a pale streak behind the eye, dark crown contrasting with buff supercilium and yellow buff gorget bordered with black. Hindneck, rump and tail grey-brown, finely barred. In flight pale underwing contrasts with bright rufous undertail feathers. Bill dark grey. Sexes similar, but legs dark red with a spur in male, orange in female without spur. **Immature** similar to adult, but sides of neck and cheeks buff (lacking the grey cheek and gorget of adult), underparts buff with fainter vermiculations.

VOICE Contact call 'kila-kila-kila'. Most vocal at sunrise and sunset, with a distinctive call starting 'chek, chek, chek' building to a rapidly repeated 'kateeja-kateeja-kateeja'[1].

BEHAVIOUR In its natural range, adapted to arid conditions and a dry scrub habitat. On Desroches and Coëtivy, most frequently observed at margins of airstrip or paths. Gregarious, usually found in small groups. Keeps to cover at most times, but will stray out into open areas especially at dawn to feed. Prefers to run rather than fly, but when alarmed will take off with a loud whirr of wings, flying a short distance before alighting and running to cover. Diet mostly seeds, shoots and berries but also catches insects such as grasshoppers and termites.

BREEDING Little is known of the breeding season in Seychelles, though broods of chicks seen in early November[2] suggest they may breed during northwest monsoon. Nests on the ground, usually among grass or some denser cover, with a meagre lining of twigs and grass. Up to ten white eggs are laid and incubated for 18-19 days. In the manner of all gamebirds, the chicks can run within hours of hatching, following the hen bird as she scratches for food.

RANGE Resident from the coastal plains of Oman and Iran, through Pakistan and India to Sri Lanka. Three races (sometimes four) recognised, including the nominate in south India to Sri Lanka.

STATUS Resident Desroches and Coëtivy. Introduced as a gamebird to several islands around 1872 by Admiral Sir William Kennedy[3]. After visiting Farquhar in 1870 and enjoying an organised shoot of the 'several thousand' guineafowl which had multiplied to pest proportions in the maize crops of the island, he sent a 'cage of partridges from India' to the island, but the manager understandably declined the offer of another introduced pest! Kennedy released them instead on Félicité, where he heard later they were doing well. Around the turn of the century this species must have been more common, for it was reported present on 'practically all the islands between Seychelles and Madagascar'[4]. Penny reported two birds on the tiny sand cay of African Banks in 1971 but these had both disappeared on a later visit[5].

SIMILAR SPECIES None.

References: 1. Roberts (1991), 2. Johnson (1994), 3. Kennedy (1901), 4. Gadow and Gardiner (1907), 5. Penny (1974).

COMMON QUAIL *Coturnix coturnix* Opposite plate 17

Other name: Quail **French:** Caille des blés **Creole:** Kay

DESCRIPTION Length 16-18 cm; wingspan 32-35 cm. A tiny, chunky, almost tailless gamebird with streaked, brown back and paler buff underparts. **Adult** has mottled black and buff back, lightened by pale buff streaks from nape to rump. Rufous chest, with paler buff to cream belly. Bold cream darts edged in black along the flanks, more obvious in male. Streaked crown and thin buff supercilium. In flight, wings uniform mottled buff and brown. Bill grey, legs and feet pale yellow or brown. **Male** usually has a distinct head pattern with black 'bridle' from throat onto neck and cheeks and pale or rufous cheek-patches. **Female** is paler on the face and throat, without bold black lines or neck-collar. **Immature** similar to adult female, with no patterning on the cheeks, but with barring or spotting on the flanks rather than streaking.

Leg and bill colour as adult.

VOICE Male gives a rapidly spoken three note call likened to 'wet-my-lips' which may be repeated after a pause of a few seconds. Female often calls 'broo-broo'. When flushed, may give a low, trilling 'whree'.

BEHAVIOUR Runs along the ground, keeping to cover of short, dense vegetation such as grassland or cereal crops, but on migration will feed in open habitats, even semi-desert. Vagrants generally solitary and can be remarkably tame. When disturbed has weak, whirring flight on short wings, usually pitching down into cover at first opportunity. Over long distances flight is more starling-like: strong, rapid and direct. Feeds mainly on fallen seeds and small insects.

RANGE Nominate race breeds from Europe east into Russia and northern India, migrating south to winter in sub-Saharan Africa from Ghana to Sudan. Race *africana* (darker on underparts and upperwing-coverts) is resident in South Africa, with presumably the same race in Madagascar, Comoros and Mascarenes, and race *erlangeri* (much darker again) in East and northeast Africa. Birds of nominate race are a little paler and greyer towards central Asia. Birds of Comoros and Mascarenes are smallest.

STATUS Vagrant to granitics recorded December to May.

SIMILAR SPECIES There is potential confusion with Japanese Quail *C. japonica,* which is reared in captivity in the granitics and may possibly escape from time to time. **Japanese Quail** (L17-19 cm) is usually darker above, deeper rufous below, without any black bridle on throat of male. Male usually has rufous cheeks and sometimes a partial collar. Best distinguished in the hand on wing length of under 105 mm (Common wing 107-118 mm).

ALDABRA RAIL *Dryolimnas (cuvieri) aldabranus* Plate 17

Other name: Flightless White-throated Rail   **French:** Râle d' Aldabra   **Creole:** Tyomityo

DESCRIPTION Length 30-32 cm.
Unmistakable; the only resident rail of Seychelles and the last remaining flightless bird of the Indian Ocean, confined to Aldabra. **Adult** has dark brown upperparts tinged olive green and chestnut underparts, this colour extending onto the nape, crown and forehead. White barring on the flanks, though this is sometimes indistinct. Chin and throat white. Eye orange; legs blackish brown. Sexes similar, though male is usually a little larger with a very slightly larger bill. Female generally has a bright pink base to the upper mandible, which is dull red in male, the rest of bill dark grey but this is variable and not a reliable means of sexing. **Juvenile** is entirely dark brown with brown eye and black bill and legs.

VOICE Like many rails, has a large vocabulary. Ten categories of vocalisation have been identified[1]. Undisturbed birds, particularly lone ones, emit a quiet, low-pitched grunt, which may be a contact signal between pairs. When foraging in pairs, birds have a continuous low rasping call, barely audible to the human ear. Other calls include a throaty whistle, which follows the more usual grunt, and two-note alarm whistle a little like a wolf whistle. During the breeding season (early October to early March), the most commonly heard call (especially at sunset), is a duet which begins with one bird emitting the usual grunt, rising in intensity, tempo and pitch, followed by a series of rising notes in which the other bird joins in. A climax is reached with each bird producing simultaneous, shrill, loud notes. The head is tilted back and the bill open during these duets, which frequently are followed by copulation. The Creole name is imitative of the call.

BEHAVIOUR Tame and confiding, but capable of running fast over difficult terrain. Though flightless, can leap a metre or more vertically and may climb shrubs (possibly to investigate for insect prey). Naturally curious and will come to investigate any sound. The white throat may be displayed when pairs meet, facing each other motionless for a few seconds with bills held horizontal before separating again to forage. Very aggressive, implying some strong territorial behaviour. Young birds or males may leap up and kicking each other in a confrontation[2], birds 'flying at each other like game cocks'[3]. Feeds on invertebrates found in leaf litter, preferring shaded areas. Sometimes follows tortoises feeding in the disturbed leaf litter similar to the manner in which Cattle Egret may follow ungulates[4]. Also an active scavenger, with a varied diet that ranges from blowflies on tortoise carcasses, to food scraps around human campsites. At low tide moves through mangroves and limestone crevices, seeking out small snails and molluscs in the intertidal zone. Also eats crabs, turtle eggs where exposed and turtle hatchlings, many rails rushing to a nest that may erupt in daylight hours[5]. Giant tortoises on Aldabra will respond to light stimulation of their soft posterior from rails by adopting an erect stance enabling the birds to clean ectoparasites from the exposed skin; even the sight of a rail within one metre of a tortoise may produce this response[6]. This parallels co-operative behaviour between Galápagos giant tortoises and two species of ground-finches *Geospiza spp.* in which the finches clean ticks from the tortoises[7].

BREEDING Usually monogamous, forming exclusive territories within an area of common foraging ground. Breeds during northwest monsoon. Female defends the territory aggressively, with ritualistic display, male often playing a supporting role. A deep cup-shaped nest is built at or near ground level, usually well concealed among shrubs, using grass and casuarina needles. Three or four eggs are laid, coloured white and speckled with red. Chicks, reported around February, have shiny black down tinged olive. The down gradually becomes dull brown and the olive green tinge strengthens. Mortality appears high, with rarely more than one chick surviving to a few weeks old despite clutch size. There is a fairly lengthy period of post -fledging parental care, which may last until the next breeding season.

ORIGINS Undoubtedly from Madagascar and related to White-throated Rail *D. cuvieri*, which retains the ability to fly. The genus *Dryolimnas* is endemic to the western Indian Ocean. Its former range probably included Mauritius, where it is known from a single 1809 specimen, which might have been a vagrant, though sub-fossil remains suggest it was resident[8]. The earlier origin is unknown. Differs from birds of Madagascar in smaller size, significantly shorter wing (averages 121 mm versus 151 mm for *D. cuvieri*), slightly shorter tarsus and middle toe and slightly longer bill[9]. Also paler on lower flanks and belly, little or no black streaking on upperparts and white bars on belly less distinct.

STATUS AND DISTRIBUTION Endemic; confined to Aldabra atoll where most birds are found on Malabar. Also occurs on Picard, Polymnie, Ile aux Cedres and a few small lagoon islets, with a total population of c5,000-7,000 birds. It probably once occurred throughout the atoll. Birds of Ile aux Cedres are larger and paler than others on Aldabra and being isolated from these in the southeast corner of the lagoon, may be a relict group from an original Grand Terre population (R. Wanless pers. comm.). Flightless rails of closely allied forms were also recorded in the past on Astove, Cosmoledo and Assumption. On Assumption race *abbotti*, discovered by Abbott in 1892, was still 'very common' in 1908[10] but extinct by 1937[11]. This rail was distinct from the Aldabran form, with blackish streaks on the back and more distinct white bars on belly and flanks. William Stirling, shipwrecked on Astove in 1836 mentions 'a bird like a moorhen, numerous on the island...' He added in his diary 'one of the birds called by me a moorhen was caught today: it is about the size of a partridge; bill an inch long, head and breast, as far as the legs, reddish brown; chin, white; under the rump and wings and on the thighs, feathers barred with white;...the tail is short...and above darker like the back and wing-coverts, dirty green'[12]. Last recorded on Astove in 1908 (as *D. abbotti*)[10]. Recorded second hand by Abbott on Cosmoledo and said to survive on the atoll at South Island at least until 1908[10].

THREATS AND CONSERVATION Looking at this diminutive rail as it stalks through the undergrowth of Aldabra, it is difficult to appreciate its full significance. It is now the last surviving flightless bird in the Indian Ocean. The Dodo *Raphus cucullatus*, which is a part of our language, was neither the first nor the last flightless bird to become extinct in the area. Eggshells of the Elephant Bird *Aepyornis maximus* still survive in the sandy wastes of southwest Madagascar. Three metres high, the world's tallest bird was hunted by the first settlers. Visiting ships to the Indian Ocean in the 16 Century encountered a range of other flightless birds, all now extinct: the Dodo and the Red Rail *Aphanapteryx bonasia* of Mauritius (both gone by 1680), the Réunion Ibis *Threskiornis solitarius* (gone by 1700), the White Dodo *Victoriornis imperialis* of Réunion (extinct 1770) and Legaut's Rail *Aphanapteryx leguati* and the Rodrigues Solitaire *Pezophaps solitaria*, extinct around the same period. Abbott was concerned the fate of the Aldabra Rail would be similar. Writing in 1893, he said the birds were 'excessively tame and unsuspicious as well as inquisitive, they run up to inspect any stranger who invades their habitat, occasionally even picking at his toes. I am told that rails swarm upon the Cosmoledo and Astove atolls. I fear they are doomed to early extinction on Aldabra from the wild cats which will eventually reach the other islands of the group...'[3] It is ironic now that they survive on Aldabra, yet are gone from Assumption, Cosmoledo and Astove. Human predation was a serious factor in this decline. There are no known natural predators, though the well-concealed nest site suggests predation at some time and land crabs may possibly take young birds. Introduced rats may also do so, though adults have been seen to peck at and drive off rats from food scraps[2]. Indeed, when rails were first transferred to Picard and held in cages overnight, one pair killed a rat that burrowed into its cage (R. Wanless pers. comm.). Cats are probably a more important predator. Rails survive on Polymnie and Malabar in the presence of a higher density of rats than on Grand Terre, where there are cats present and no rails[13]. In the mid-1960s, if proposals for a military staging post on Aldabra had materialised, the rail population could have been devastated. A road linking Picard, Polymnie, Malabar and the eastern end of Grand Terre would have cut straight through the area with the greatest population density and enabled cats to spread throughout the atoll. Conservation measures suggested include the elimination of cats and perhaps rats on some or all of Aldabra's islands and the rehabilitation of Assumption where a wild population could be established once again. A translocation to Picard took place in November 1999, following the elimination of cats. Breeding commenced in January 2000[14], and was immediately much more successful than in a control area on Malabar. Captive breeding has also been recommended[15]. A genetic study may confirm full species status[14].

SIMILAR SPECIES: None.

References: 1. Huxley and Wilkinson (1977), 2. Frith (1977), 3. Ridgway (1893), 4. Penny and Diamond (1971), 5. Frith (1975a), 6. Huxley (1979), 7. MacFarland and Reader (1974), 8. Cowles (1987), 9. Benson (1967), 10. Fryer (1911), 11. Vesey-Fitzgerald (1940), 12. Stirling (1843), 13. Seabrook (1990), 14. Wanless (2000), 15. Hambler *et al.* (1993).

CORNCRAKE *Crex crex*   **Plate 17**

Other name: Land Rail   **French:** Râle des genêts   **Creole:** Ral Lezel Rouye

DESCRIPTION Length 27-30 cm; wingspan 46-53 cm. A mottled brown-and-black rail that stays in dense cover; when flushed it shows bright rust-red wings.**Adult** ground colour is tawny buff, streaked with rows of black spots from crown, down the back (where they are largest) to rump and tail. Flanks are grey-white with chestnut barring, and the face plain, with a blue-grey supercilium. The most striking feature is the wings, bright chestnut red when opened, forming a chestnut band on the side of the body when folded. Bill pink-brown and legs pinkish grey. Sexes similar, female less grey on side of head and neck. **Immature** similar to adult but more buff above and below, with little or no barring on flanks.
VOICE Silent in winter quarters.
BEHAVIOUR Tame but skulking, preferring to remain in dense cover at all times, unless moving from one patch of vegetation to another. A bird of dry grassland, only using damp ground or marshy sites when no other cover is available. Generally solitary outside breeding season. Short flights are weak and fluttering, legs dangling, but over long distances, legs are raised and flight action much stronger. Like many rails, the body is compressed laterally to allow it to squeeze between plant stems in dense vegetation. Runs or walks rather than flying, unless startled suddenly. Feeds on a variety of items, taking beetles, ants and worms from the ground, or pecking caterpillars and crickets off grass and leaves.
RANGE Breeds across Europe and Asia, from Ireland and France to Lake Baikal in Russia, wintering in grasslands of sub-Saharan Africa. Monotypic.
STATUS Vagrant to Bird and the granitic islands October to January.
SIMILAR SPECIES Spotted Crake is much smaller, green-brown with numerous white spots edged with black and buff undertail-coverts. **Striped Crake** is tiny, olive brown above, grey below, with a pattern of black-edged white stripes.

SPOTTED CRAKE *Porzana porzana*   **Plate 17**

French: Marouette ponctuée   **Creole:** Ral Pitakle

DESCRIPTION Length 22-24 cm; wingspan 37-42 cm. A small, dark brown rail covered in white spots, with a red-and-yellow bill and distinctive buff undertail-coverts. **Adult** is green-brown above, with black centres to feathers on crown, back and rump and white flecks throughout. Face and underparts slate grey, densely spotted with white, flanks olive green densely marked with short white and black bars. Flight feathers dark brown, with white outer web to first primary. Undertail-coverts pale buff, a diagnostic feature. Birds cock their tails as they slip away into cover, so this feature is often the 'last glimpse'. Bill yellow, red at base, legs olive green. Sexes similar, female slightly smaller with less grey below and more white spotting. **Immature** similar to female but with duller legs, browner rather than grey below and buff, unspotted face and white throat. Bill yellow-brown with yellow base at first, becoming olive with orangey base.
VOICE Normally silent in winter quarters, but occasionally male may make the typical call, a regular 'kwit, kwit, kwit' like dripping water.
BEHAVIOUR Prefers dense rush and sedge, rather than open wet sites. Generally solitary outside breeding season. Flies only reluctantly, keeping to the dense vegetation around marshes and wetlands. On migration, flight action is strong. Sometimes feeds with waders such as Wood Sandpiper. Occasionally swims over short distances. Feeds on insects, snails, earthworms, spiders and seeds or shoots of marsh plants.
RANGE Breeds in scattered localities across Europe and throughout Russia as far east as Lake Baikal. Birds move south in September, moving into East and South Africa (from Ethiopia to Botswana) and into northern India (from Pakistan to Bangladesh). Monotypic.
STATUS Vagrant to granitic islands December to February.
SIMILAR SPECIES Corncrake is much larger, with no white spots and striking rufous wings. **Striped Crake** is smaller, dark olive brown above with obvious white streaking on the back and dark rufous buff undertail.

STRIPED CRAKE *Aenigmatolimnas marginalis*   **Plate 17**

French: Marouette rayée   **Creole:** Ral Are

DESCRIPTION Length 18-21 cm; wingspan 36-39 cm. A tiny, short-tailed, dark olive brown rail, with white stripes on the back and rufous buff undertail-coverts. **Adult** most striking feature is white striping on the back formed by white feather margins. As it slips into cover when disturbed, the most useful field mark is deep rufous buff undertail, last thing seen as bird cocks its tail on retreating. In flight, white outer web on first primary shows up. Bill dirty green, legs and feet dirty greenish brown in both sexes. **Male** has forehead and hindneck dark brown, face buff, back and wing-coverts dark brown striped white, rump and uppertail black-brown streaked rusty, breast orange-buff, flanks olive brown with whitish edges and rufous thighs. **Female** has forehead and hindneck dark grey, face light grey, breast grey scalloped white, flanks grey edged white with back and rump as male. **Immature** similar to male, but unstriped above and with rufous tinge to face and breast and blue-grey legs.

VOICE Mostly silent in winter quarters. Grunts and growls reported and a 'chup' or 'yup' contact call.

BEHAVIOUR Favours grassy areas, mud and shallow pools rather than tall, dense vegetation and water much deeper than 20 cm. Highly secretive, most active late afternoon to dusk. Generally solitary outside breeding season. Flight similar to other crakes. Feeds on seeds, small snails, worms, small fish and insects such as small grasshoppers.

RANGE Breeds in West, East and southern Africa, but also an inter-African migrant, dispersal linked to regional rainfall but little is known of its movements outside the breeding season. Monotypic.

STATUS Vagrant to Aldabra recorded October.

SIMILAR SPECIES Spotted Crake is larger, with buff undertail-coverts and white spotting throughout. **Corncrake** is half as large again, with no white spotting or streaking and with rufous wings and pale barring on the flanks.

WHITE-BREASTED WATERHEN *Amaurornis phoenicurus*   Plate 17

French: Râle à poitrine blanche   **Creole:** Ral Pwatrin Blan

DESCRIPTION Length 32 cm; wingspan 45-54 cm. A black gallinule the size of a moorhen, with white face, breast and belly, cinnamon undertail-coverts and pale yellow-green legs. **Adult** has a white face, throat, breast and belly contrasting with black crown, hindneck, back, wings and tail. Back and wing-coverts have a slaty blue-black cast. Grey on flanks extends down to legs in irregular barring. The undertail-coverts and vent are bright cinnamon brown. Eye red, bill yellow-green with a red frontal shield, legs long and pale yellow-green. Sexes similar, female smaller and bill may be duller. **Immature** has upperparts dark olive brown, white of face and breast with greyish cast and may be flecked with black or brown and undertail-coverts duller than adult. Legs and bill duller and eye is yellow-green.

VOICE A variety of grunts, croaks and chuckles and a loud 'kee-wak, kee-wak, kee-wak-a-wak-wak' kept up for long periods. Silent outside breeding season.

BEHAVIOUR In breeding range frequents marshes, flooded grassland, thickets along watercourses and mangrove areas, but will also feed out in open habitats in the manner of a moorhen. In Maldives, feeds on beaches at low tide. Sometimes climbs bushes, swims and even dives. Not particularly shy, often feeding on open ground. Diet includes molluscs, worms, insects, seeds and vegetation.

RANGE Breeds throughout India south to Maldives, south China, Indonesia and the Philippines. The nominate race *phoenicurus* occurs from Pakistan to Japan, northern populations migratory, moving south within range and west to Arabia.

STATUS Vagrant to granitic islands recorded December.

SIMILAR SPECIES Common Moorhen is similar in size, but is brown-black throughout, has white undertail-coverts and a red bill. **Allen's Gallinule** is smaller, blue-purple below and olive green above with red bill and red legs.

COMMON MOORHEN *Gallinula chloropus*   Plate 17

French: Gallinule poule d'eau   **Creole:** Poul Do Ordiner

DESCRIPTION Length 32-35 cm; wingspan 50-55 cm. A black-and-brown gallinule with white undertail feathers, yellow-green legs and a red bill and shield. **Adult** head, neck, belly and flanks dark slate-black or slate-brown and back dark brown. Flanks demarcated by a row of thin white arcs; undertail feathers also white, flared as the bird raises and flicks its tail. Bill scarlet with a yellow tip, continued into a red 'shield' on forehead. Legs yellowish green with a red 'garter' at the top and very long toes. Sexes similar, male usually larger and brighter than female, with a larger frontal shield. **Juvenile** dull brown throughout with a white throat, no white flank line and dirty green bill, shield and legs. **First-year** acquires red bill and yellower legs but remains browner in the body than adult.

VOICE Strongest territorial call a loud, explosive 'prreuk!'. Alarm calls include a 'kittick' call and a repeated 'kerrk, kerrk, kerrk'.

BEHAVIOUR Usually associated with wetland margins, but also wanders far from water, feeding in open short grass areas and on the smaller islands in woodland. Strongly territorial and aggressive and encounters between different individuals often involve them 'mooning' at each other by cocking tails and showing off their white underpants! In extreme cases, these encounters can develop into fights, birds jumping high and using their feet like fighting cocks. Outside breeding season more gregarious, often in small groups. Flight rather weak and fluttering over short distances but strong over longer ranges. Scratches through leaves like chicken, taking invertebrates, pecking at fallen fruit, even scrambling into banana plants to peck at ripe bananas. Omnivorous, feeding on seeds, leaves, shoots, earthworms, insects and snails.

BREEDING May breed in any month: one pair observed over two seasons bred six times in the first and nine times in the second-year, raising 22 young[1]. In courtship, male chases female, both birds with neck outstretched. Female may bow, male nibbling feathers on back of head (sometimes female nibbles feathers of male). Pairs also greet each other by lowering head and flicking or sometimes fanning tail. Nests are saucer shaped, constructed from grass and reeds, often in dense mats of vegetation at ground level, but may be up to 2 m above ground on fallen tree trunks, etc. Clutch sizes studied on Aride varied from 4-13 eggs, the latter almost certainly the result of two females laying in the same nest. Eggs are sandy in colour, speckled or blotched with brown and purple spots. Incubation lasts around 23 days and chicks are led from the nest within hours of hatching. Though fed immediately by their parents, the chicks are also commonly fed later on by siblings from earlier broods. Chicks when new-born are fluffy and black with bright red-and-blue pates, these bright colours stimulating adults to feed them. Chicks raise their tiny, bald pink wings to beg food from the adult. Mortality is high, as despite large clutch sizes it is rare to see more than two successful chicks per nest.

RANGE Widespread, breeding on every continent except Australia and Antarctica including throughout eastern, central and southern Africa, Madagascar, Mascarenes and Seychelles. Has apparently recently colonised the Maldives (presumably from India) breeding since 1945. There is no evidence of vagrancy in the Indian Ocean, but Russian birds are thought to reach the Middle East and birds are known to cross the Sahara. The Seychelles population, formerly considered a separate race, is now generally considered to belong to *orientalis*, Asian in origin, occurring in the Andamans, south Malaysia, the Sundas and Philippines, with slaty-blue upperwing-coverts, less greenish wash and a larger shield. Race *pyrrhorhoa* breeds in Madagascar, Réunion, Mauritius and Comoros; race *meridionalis* occurs in sub-Saharan Africa.

STATUS Resident in the granitic islands, but only common on rat- and cat-free islands, notably Aride, Cousin and Cousine and at areas of open water elsewhere (e.g. Plantation Club, Mahé and the coastal plateau of La Digue). The population has increased on Aride, with only 4-5 pairs in 1978[2] but 98 birds in 1998[3]. First collected by Oustalet in 1878 and classified as an endemic race *seychellarum*. Birds measured in 1990 had mean wing lengths of 161 mm, compared to 180 mm in the nominate *chloropus*, though body weights were similar (318 g, cf. 315 g in *chloropus*). This might imply Seychelles birds are tending to flightlessness; wing length should be measured in any dead birds found, to check for possible long-distance vagrants. There is some movement between islands; c15 birds leave Cousine each year with recoveries of ringed birds on Praslin and Cousin, always immatures (P. Hitchins pers comm.).

THREATS AND CONSERVATION The greatest long-term threat to this species must be from the steady and inexorable loss of wetlands in Seychelles. Larger marshes have been drained for agriculture and the smaller ponds, which often sustain several pairs, are gradually filled in for building or spoiled by pollution and rubbish. The birds are productive and seem able to compensate for high losses of eggs or chicks; they need to, for on the larger islands feral cats and rats are common. They are still relatively common on La Digue where there are still large wetlands and scattered small pools. Birds are easily caught and were possibly fattened for the pot in the past, but this is not a serious problem today.

SIMILAR SPECIES None resident in Seychelles.

References: 1. Bullock (1989), 2. Warman (1979), 3. Bowler and Hunter (1999).

ALLEN'S GALLINULE *Porphyrula alleni*   Plate 17

Other name: Lesser Gallinule   **French:** Poule sultane d'Allen   **Creole:** Poul Do Allen

DESCRIPTION Length 22-24 cm; wingspan 48-52 cm. A small gallinule, blue-purple below and olive green above, with red bill, turquoise-blue shield and red legs. **Adult** is smaller and more slender than Common Moorhen, with blue-black head, deep olive green back and iridescent blue-purple-black underparts. Wings blue-green, flight feathers blue-black. Bill red with grey-blue shield (brighter, turquoise-blue in breeding plumage), eye brown (red in breeding plumage), legs dull dark red to crimson (brighter in breeding plumage). Combination of bare parts colours diagnostic. Sexes similar, male slightly larger than female and shield bright blue in male, green in female at beginning of breeding season. **Juvenile** is bright ginger-brown on face and sides of neck and breast, with white chin, throat and belly, buff undertail-coverts and dark brown mottled back. Bill reddish brown, legs brown, eye orangey. During the first year this is replaced with adult feathers, leaving **immature** with some buff flecking; bill duller than adult, with olive brown shield and paler, browner legs.

VOICE A guttural, frog-like 'gurr'. Also calls 'ke' in flight and clicks in alarm.

BEHAVIOUR Usually associated with open water, keeping to dense marginal cover by day and walking out onto floating vegetation to feed. Swims well, runs rapidly with head down and sometimes climbs reeds and bushes. Gait is more graceful than Common Moorhen. Vagrants are generally solitary. Fluttering flight over short distances, legs dangling. Mainly crepuscular, but sometimes active by day and on moonlit nights. Feeds mostly on shoots of water plants. Also takes seeds, fruits, worms, molluscs, insects and small fish.

RANGE Occurs in wetlands throughout sub-Saharan Africa and in the east from southern Sudan to eastern South Africa; also resident Comoros and Madagascar. Disperses or migrates as wetlands shrink in the dry season. Vagrant to Comoros and even Rodrigues (1500 km east of Madagascar). Monotypic.

STATUS Vagrant to Aldabra (recorded January to February) and the granitic islands (recorded March and July).

SIMILAR SPECIES Common Moorhen, which is common in the granitic islands is the only similar species known to occur; this has brown-black plumage and yellow-green legs.

PURPLE SWAMP-HEN *Porphyrio porphyrio* — Plate 1

Other names: Purple Gallinule, Purple Coot   **French:** Porphyrion bleu   **Creole:** Poul Ble

DESCRIPTION Length 45-50 cm; wingspan 90-100 cm. A large, heavy-bodied gallinule, with violet-blue head and underparts, blue-green back, massive, bright red bill, red eyes and long, red legs. **Adult** has head, neck, breast and belly iridescent violet-blue, with shiny olive green back and scapulars, rump and tertials. White undertail-coverts contrast with black tail feathers. In flight, broad, rounded wings are iridescent blue-green. Eye red with massive bill and bright red frontal shield, long, sturdy legs with long crimson red toes. Sexes similar. **Immature** is drabber than adult, with greyer tones on head, breast and flanks, paler on the belly, darker legs and bill blackish grey to dark red.
VOICE Not recorded from Seychelles. Elsewhere, fairly vocal, with a wide variety of grunts, cackles and shrieks. Alarm call a loud 'kree-ik'. Softer contact call similar to Common Moorhen but deeper.
BREEDING Details from Seychelles unknown. Elsewhere, a bulky nest of aquatic vegetation is built in thick growth over shallow water. Clutch is 3-5 buff or creamy eggs, blotched darker. Incubation 23-25 days; fledging 60-63 days.
BEHAVIOUR Frequents marshes and thick vegetation on the edge of freshwater wetlands. Generally solitary or in small groups outside breeding season. When alarmed will keep to dense cover, but comes out into the open to feed, taking mostly shoots, roots and seeds of waterside plants. Paces like Common Moorhen, using its long toes to walk over floating or matted vegetation and flicking its tail to reveal the white underside. Walks slowly, similar to a heron. Agile in climbing reeds and may swim or dive. Readily takes flight, showing broad wings and long dangling legs. Diet mainly aquatic vegetation, but will also take insects, molluscs, fish and other items.
RANGE Resident from the Mediterranean, through Africa, Middle East to India, South-East Asia and New Zealand. Race *madagascariensis* occurs from West to South Africa and Madagascar, sometimes considered a full species, Green-backed Swamp-hen *P. madagascariensis*; it has green upperparts and deeper blue head and neck. The race *poliocephalus* of India to Sumatra may also be a full species, Grey-headed Swamphen *P. poliocephalus*, differing in having a dark blue back.
STATUS Extinct. In truth no-one knows what species the 'Poul Ble' was. On size and general description something akin to Purple Swamp-hen seems most likely. No-one has yet unearthed bones which may be checked by zoologists, nor are there surviving skins or accurate biological descriptions. Today it survives only in place names, such as Anse Aux Poules Bleues on the southwest coast of Mahé, but this alone is sufficient proof of its abundance at one time. The first mention of it was by the Marion-Dufresne expedition of 1768, in the two ships La Digue and La Curieuse. An expedition party exploring the northwest of Mahé reported: 'There are in these rocks many giant tortoises, goats and also the "poules bleues", of which we killed one; they are like a large chicken, with blue feathers, the beak flat, wide and red and the same colour feet...' Two days later, following the northwest and north coast of Mahé, they report: 'We see always many gamebirds and 'poules bleues' which are much more 'farouches' than the other gamebirds'. The French word 'farouche' normally means 'savage', but can also mean shy or timid. However, the fierce nature of what would have been an extremely tame, fearless bird is implied in a second quote from Pierre Poivre, Governor of Mauritius, who, writing to the Commandant of Seychelles in 1770, said: 'If you could send me some pigeons, doves and other birds peculiar to your island, it would give me great pleasure, but don't send me any "poules bleues" for they are "malfaisantes' creatures", meaning, presumably, large and aggressive enough to make short work of any cage they were put in![1] Poivre's request is interesting given that there were also accounts of 'Oiseau Bleu' from Réunion between 1671 and 1730, described as 'dark blue' and 'about the size of a capon'. Authors[2] from this region also consider that this was probably a race of Purple Gallinule, but we may now never know for sure.
References: 1. Lionnet in Beamish (1981), 2. Diamond (1987).

CRAB PLOVER *Dromas ardeola* — Plate 18

French: Drome ardéole   **Creole:** Kavalye Loulou

DESCRIPTION Length 33-36 cm; wingspan 75-78 cm. A large avocet-like wader, white with black flight feathers, but with a massive black bill. **Adult** has pure white head and body, white wings with black flight feathers, black mantle and white rump and tail. Wings pure white beneath. In non-breeding plumage, crown to hindneck is streaked grey and scapulars, tertials and upperwing-coverts off-white. Legs long and bluish grey, eye black and the extraordinary powerful bill black, deep, sharp and massive. Sexes similar. **Juvenile** lacks black, with dirty smudging on wings and tail, a grey mantle and dark streaking on the hindneck. **Second-year** has blackish grey back but retains pale grey scapulars and upperwing-coverts of juvenile.
VOICE A harsh tern-like 'kurrk' and a barking 'ka-how', these calls mostly among flocks on the move.

BEHAVIOUR Frequents sandflats or lagoons at low tide, also patrols sandy beaches. Highly gregarious, often in large flocks. Flight strong and direct but not rapid. Walks slowly scanning for crabs, then runs them down before they can reach their burrows. Although molluscs, worms and mudskippers are taken, preferred prey is crabs including ghost crab *Ocypode spp.*, common on the coral sand beaches and the smaller, brightly-coloured mangrove crabs. The need to feed at low tide means that it is frequently active at night; the pure white plumage may help individuals in a flock locate each other after dark.

RANGE Breeds on islands from Somalia to Red Sea and Persian Gulf between May and July, later than other species which may compete with its specialist diet, Fewer breeding sites are known than can be accounted for by the numbers occurring at non-breeding sites. Outside the breeding season birds move south along the East African coast, occurring commonly on the coasts of Somalia, Kenya and Tanzania, (particularly between September and April) and reaching the Comoros and Madagascar. Monotypic.

STATUS An annual migrant throughout Seychelles, most common October to April. A few remain year round. Small flocks frequent the mud and sandflats off Victoria and Roche Caiman, Mahé. Largest numbers occur inside the lagoons of Aldabra and Cosmoledo (flocks of 1,000 or more recorded at each, with the atoll-wide population of Aldabra alone up to several thousand[1]) and to a lesser extent, St François (350 recorded). Perhaps as much as 10% of the world's population of c50,000 winters in Seychelles.

GENERAL With a body like an Avocet and a bill like a raven, the Crab Plover is strange enough. But even more remarkable is the fact that it excavates a burrow 1-2 metres into sandy ground close to the sea and incubates a single white egg underground! This may be protection against ground predators or from the heat of the sun, but whatever the reason, its very restricted range and incredible lifestyle make this wader one of the most extraordinary shorebirds in the world.

SIMILAR SPECIES None recorded though **Pied Avocet** *Recurvirostra avosetta* (L42-46 cm, WS67-77 cm) is a potential vagrant, breeding Europe to northeast China and locally in Africa and dispersing widely outside the breeding season. It has almost identical legs and plumage, but a very fine, upcurved bill instead of the Crab Plover's tomahawk.

Reference: 1. Betts (2000b).

EURASIAN OYSTERCATCHER *Haematopus ostralegus*   Plate 18

Other name: European Oystercatcher   **French:** Huîtrier pie   **Creole:** Attraper Zwit

DESCRIPTION Length 40-46 cm; wingspan 80-86 cm. A striking chunky, pied wader, black above, white below with a heavy, straight red bill and relatively short, pink legs. **Adult** has black head, back and tail contrasting with white underparts. In flight, black wings reveal a broad white wingbar and white rump contrasting with the black tail. Bill heavy, straight and bright red-orange. Eye and eye-ring red and legs pink. **Adult non-breeding** shows a white frontal neck band. Sexes similar, male with shorter, deeper bill and slightly broader tail-band. **Immature** has upperparts brownish black and shows pale fringes on back and wing-coverts, greyer legs and a larger white neck-collar. Eye reddish brown with eye-ring indistinct and dull orange. Base of bill orange-red, tip horn washed red. Legs grey washed pink.

VOICE A loud shrill 'kleep' which can become an insistent, accelerating 'peep-peep-pip-pip-ip-ip-ip' call.

BEHAVIOUR Away from breeding haunts a bird of open seashores, frequenting sandflats, sand bars or sites where freshwater meets the sea. At high tide roosts on higher, dry ground. Generally gregarious but vagrants are likely to be solitary or in twos or threes. Flight strong and direct with shallow wingbeats, lacking fluidity of many waders. Walks rapidly but lacks grace of long-legged waders. Probes deeply for worms in muddier sites, or uses the bill to stab open molluscs or crabs.

RANGE Nominate breeds from Iceland and Scandinavian coasts to western Russia, with race *longipes* to western Siberia and other races in northeastern Asia and New Zealand. Nominate winters from breeding grounds to West Africa, race *longipes*, winters Middle East to East Africa and western India. Race *longipes*, most likely in Seychelles, has a bill averaging 10 mm longer than the European race *ostralegus*, with a nasal groove extending more than half its length; also browner above, the back noticeably paler than the head and chest.

STATUS Vagrant to granitic islands and Aldabra; recorded in all months. Most recorded arrive October to November, sometimes remaining for the following summer (probably first-year birds and believed to be race *longipes*).

SIMILAR SPECIES None.

BLACK-WINGED STILT *Himantopus himantopus*   Plate 18

French: Echasse blanche   **Creole:** Bekir

DESCRIPTION Length 35-40 cm; wingspan 67-83 cm. A slender black-and-white wader, with long trailing pink legs and straight, needle-like bill. **Adult** has all-black wings and black back, sometimes with greenish gloss, white rump and tail. Head and underparts pure white save in the breeding season when some males have a grey tinge to crown and nape (others having a full black cap). Sometimes grey is more extensive on crown, nape and hindneck in non-breeding plumage. Eye red, legs pink, thin and very long, trailing well behind tail in flight and bill very fine, black and straight. Sexes similar, female smaller and duller brown above lacking gloss. **Immature** has upperparts and wings brownish with buff fringes to coverts, the mantle sepia brown, crown and hindneck dusky and white trailing edge to the wings. Eye yellow-brown and legs duller than adult.

VOICE A sharp 'kek', a yelping 'ke-ak' and a shrill, continuous 'ki-ki-ki-ki' of alarm.

BEHAVIOUR Prefers coastal areas and lagoons or the fringes of inland lakes. Generally gregarious but vagrants are likely to be solitary. Tame and approachable (more so than other waders) outside breeding season, bobbing head and scolding prior taking flight. Flight rapid and confident, showing long, narrow, pointed wings and nearly all of tarsus visible beyond tail tip. Walks with a graceful gait, feeding in water up to its knees searching for small insects, flies and even small fish. Probes mud or picks at surface, chases prey in water or adopts avocet-like scything action.

RANGE Nominate race breeds through southern Europe and Russia to Mongolia, wintering in tropical Africa. Also breeds throughout Central, East Africa, southern Africa and Madagascar, making periodic seasonal movements and throughout India and Sri Lanka. Three other races breed in America and one in Indonesia to Australasia. Northern populations are migratory. African population is mainly sedentary.

STATUS Vagrant to the granitic islands recorded September to October.

SIMILAR SPECIES None.

STONE CURLEW *Burhinus oedicnemus*   Plate 18

Other name: Thicknee **French:** Oedicnème criard **Creole:** Korbizo Gro Lisye

DESCRIPTION Length 40-44 cm; wingspan 77-85 cm. A thickset wader, with curlew-like plumage, heavy plover-like bill and a huge yellow eye set in a square head. **Adult** has pale, mottled buff-brown plumage reminiscent of a curlew, with fine streaking on head, neck and upper breast, heavier streaking on back and wing-coverts and a pale belly. Horizontal white bar across wing-coverts, paler in female, heavily bordered black in male. Head rather square with a prominent yellow eye set centrally and pale crescents above and below. In flight, shows bold wing pattern of pale brown inner wing and black flight feathers, with striking white patches. Bill heavy, yellow at base and black at tip. Legs pale yellow with prominent tarsal joint. **Immature** similar to adult but with rufous buff fringes to tertials and scapulars, less prominent bar on median coverts and more prominent white tips to greater coverts.

VOICE A whistling 'kur-lee' call, usually at twilight, but not normally heard away from breeding grounds.

BEHAVIOUR Normally a bird of open steppe, preferring dry, stony terrain (hence the name) at all times, or if not available, short open grassland. Generally gregarious on migration but vagrants likely to be solitary. Often remains inactive by day, relying on its camouflage and if disturbed, flies a short distance low over the ground with shallow wingbeats. Walks with stealth or runs with head and neck jutted forward. Over longer distances, flight is slow but powerful and fluid, trailing legs usually obvious. Feeds more actively in twilight or at night, taking beetles and crickets and occasionally larger items including lizards or even mice.

RANGE Breeds from Spain and North Africa through Turkey, southern Russia and Iran to India and Burma. Six races recognised including Asian race *harteri*, Indian race *indicus* and the North African race *saharae* (which breeds from North Africa to Egypt and the Middle East); all are mostly sedentary. Vagrants to Seychelles are more likely the nominate *oedicnemus*, a long-distance migrant which has reached East Africa to Tanzania.

STATUS Uncertain; a report from Aride, believed to be this species, pre-dated the formation of SBRC but details were lost.

SIMILAR SPECIES Curlews and **godwits** have similar mottled upperparts, but both have unmistakable long bills. There is no other wader of this size with these features. Only the **Pacific Golden Plover** or **Grey Plover** might be confused and these are half the size, slender, with finer black bills and legs.

COMMON PRATINCOLE *Glareola pratincola*   Plate 19

Other name: Collared Pratincole **French:** Glaréole à collier **Creole:** Pratinkol Lezel Borblan

DESCRIPTION Length 23-26 cm; wingspan 60-65 cm. A long-tailed pratincole with chestnut underwing-coverts and white trailing edge to the secondaries. **Adult breeding** is pale coffee brown above with dark flight feathers and black tail contrasting with clean white rump. Brown crown and upper breast contrast with distinct sandy buff gorget, bordered black. Lower breast and belly white, contrasting with dark underwing and dark chestnut underwing-coverts. Outermost tail feathers much longer, so that at rest, tail is equal in length or longer than wings. Deeply forked tail (more so than Oriental Pratincole) tipped black. **Adult non-breeding** has paler upperparts, head and chest feathers with buff streaking and throat spotted and streaked with an indistinct outline. Bill black with a red base and legs brownish black. Sexes similar, male slightly longer-winged and longer-tailed. **Immature** similar to non-breeding adult, but with buff tips and dark chevrons on back and wing-coverts and with buff, not white secondary tips. At rest, wings are longer than tail.

VOICE Utters shrill, tern-like calls resembling 'kikki' or 'kiri' notes.

BEHAVIOUR Frequents margins of wetlands, mudflats or open fields. Highly gregarious but vagrants may be solitary or in twos or threes. In flight appear like a dark tern, swooping over meadows, pools or along lines of trees to feed. On the ground look more like slender, long-winged, long-tailed plovers. Food mostly large insects, caught in flight, but will run and catch insects or leap up from the ground.

RANGE Nominate race breeds from southern Spain through eastern Europe to southern Russia, wintering sub-Saharan Africa and India. The smaller, darker northeast African race *erlangeri* makes local movements, crossing the Red Sea into Arabia. Race *fuelleborni* (also darker but almost as large as nominate) breeds over much of sub-Saharan Africa including Kenya and is nomadic.

STATUS Vagrant to granitic islands November to January.

SIMILAR SPECIES Black-winged Pratincole is darker, with all-black underwing and no white on the secondaries. **Oriental Pratincole** is darker, short-tailed with no white on the secondaries but similar chestnut underwing-coverts. Note that **Madagascar Pratincole**, *G. ocularis*, is also a potential vagrant (see under Oriental Pratincole SIMILAR SPECIES).

ORIENTAL PRATINCOLE *Glareola maldivarum*　　　　**Plate 19**

French: Glaréole orientale　**Creole:** Pratinkol Azyatik

DESCRIPTION Length 23-24 cm; wingspan 60-65 cm

A relatively short-tailed pratincole, with chestnut underwing-coverts and all-dark flight feathers. **Adult breeding** has throat with buff gorget bordered black. Dark brown above, darker flight feathers without white tips to secondaries. Underwing also dark with chestnut underwing-coverts. Flanks and lower belly more orangey buff than other pratincole species. Similar white rump and dark tail to Common Pratincole, but outer tail feathers barely longer than inner feathers, giving a short-tailed appearance with wings longer than tail at rest. **Non-breeding** has indistinct edge to gorget, greyer on neck and chest, breast and flanks with less obvious orangey buff suffusion. Bill black with a red base, legs brownish black. Sexes similar, male slightly longer-winged and longer-tailed with deeper tail fork. **Immature** has buff tips to back and wing-coverts and may show a very narrow buff-white trailing edge to the secondaries, but much less obvious than in Common Pratincole.

VOICE Sharp, tern-like 'kyit' or 'tyik' calls.

BEHAVIOUR When not feeding, roosts on the ground in either short grassland or ploughed fields. Gregarious, but vagrants solitary or in twos or threes. Like other pratincoles, has a swooping, swallow-like flight catching insects on the wing.

RANGE Breeds from Pakistan through India and Indo-China to southern China. Birds move south to winter in southern India, Indonesia, Philippines and northwest Australia. Recorded as a vagrant to Mauritius and Maldives. Monotypic.

STATUS Vagrant to granitic islands September to December.

SIMILAR SPECIES Common Pratincole has similar chestnut underwing-coverts, but a clear white trailing edge to the secondaries. **Black-winged Pratincole** has all-dark wings above and below, with black underwing-coverts. Note also that the **Madagascar Pratincole** *G. ocularis* (L23-25 cm, WS63-65 cm) is a potential vagrant, being a migrant from Madagascar to the East African coast and a vagrant to Mauritius and Réunion. Very dark above with a black crown and white eye-stripe, no throat gorget, rufous underwing-coverts and lower flanks, very short black-and-white tail and rump similar to a wheatear and short greyish black legs.

BLACK-WINGED PRATINCOLE *Glareola nordmanni*　　　　**Plate 19**

French: Glaréole à ailes noires　**Creole:** Pratinkol Lezel Nwanr

DESCRIPTION Length 23-26 cm; wingspan 60-68 cm. A dark-winged pratincole, with completely dark underwing. **Adult breeding** has buff gorget bordered black. Uniform dark brown upperparts with wing-coverts as dark as flight feathers and underwing-coverts similarly dark brown. White rump contrasts with dark tail as in other pratincoles. Tail-feathers midway between Common and Oriental in length. **Adult non-breeding**, as other pratincoles, has tawny feathering to head and neck, mottled breast and less distinct black border to gorget. Black bill with red base and brownish black legs. Sexes similar, male slightly longer-winged and longer-tailed. **Immature** very similar to immature Common, though with less grey on the upper breast. Shows the same buff tips and darker scaling to the back and wing-coverts. May show some chestnut tips to some underwing-coverts (but never the axillaries) and narrow buff tips to the secondaries, but otherwise still mostly dark brown underwing-coverts, a diagnostic feature.

VOICE Lower-pitched than Common; gives 'kikeek' and 'yup' flight calls in winter quarters.

BEHAVIOUR Prefers open ground, mown grass and meadows, but will feed over shrub vegetation along the shore. Gregarious but vagrants usually solitary. Has similar swallow-like flight to Common, snatching flying insects. Main prey crickets, grasshoppers, locusts and beetles, usually taken on the wing.

RANGE Breeds southern Russia as far east as the borders of Mongolia, migrating south in September and crossing over the Red Sea to winter in South Africa. The main passage is well inland, so that it is only a rare vagrant to the coasts of Somalia and Kenya.

STATUS Vagrant to granitic islands, Bird and Amirantes October to December and April.
SIMILAR SPECIES Common Pratincole has chestnut underwing-coverts and a clean white trailing edge to the secondaries. (This needs scrutiny, as chestnut underwing can appear dark brown unless seen clearly.) Beware immature birds that can show narrow buff secondary tips in both Black-winged and Oriental Pratincole. **Oriental Pratincole** has all-dark wings above, but chestnut underwing-coverts. Note that **Madagascar Pratincole** is a potential vagrant (see under Oriental Pratincole).

LESSER SANDPLOVER *Charadrius mongolus*   **Plate 20**

Other name: Mongolian Sandplover   **French:** Pluvier de Mongolie   **Creole:** Pti Plivye Ordiner

DESCRIPTION Length 19-21 cm; wingspan 45-58 cm. A small drab plover, brown above, white below, with more upright posture than Greater Sandplover, short legs set well back on body and showing very little tibia. **Adult non-breeding** appears drab grey-brown above and white below, with partial collar on sides of breast. Most useful features to distinguish from Greater Sandplover are bill length, leg length and jizz. In Lesser, legs appear shorter, showing little or no tibia and seem to be well back on body, so that bird appears unbalanced. Also, head is more rounded, less angular and bill is short, barely the same length as from bill base to eye. Bill black, legs olive grey to blackish (generally darker than Greater Sandplover). In flight toes do not extend beyond tail tip and shows a narrow, white wingbar, underwing white. **Adult male breeding** has broad rufous breast-band extending onto flanks, white throat and a thick black mask. Some individuals show traces of breeding plumage from late March onward. **Female breeding** has a brown mask, the rufous areas less intense and less extensive than male. **Immature** appears buff on chest, with pale buff margins to back feathers, tertials and wing-coverts (which disappear by midwinter).
VOICE A short, hard 'dirik' flight call, but may make softer, churring calls[1].
BEHAVIOUR Feeds on open sand and mudflats at low tide. Less common along sand beaches on the smaller islands, typically a bird of large flats or lagoons. Gregarious, often associating with other small waders including Greater Sandplover. Rapid flight action similar to cogeners. Runs quickly, dipping forward to take prey from the ground. Usually takes a much shorter run than Greater Sandplover, with up to four paces, (rarely more than eight) in each run pausing for 1-4 seconds[2]. Feeds on molluscs, crustaceans, worms and insects.
RANGE Breeds Tibet, southern China and the Kamchatka peninsula of Russia. Most likely race in Seychelles is *pamirensis*, which winters on coasts from South Africa to western India. Himalayan race *atrifrons*, which winters India to Sumatra, might also be possible; it appears darker grey above. Other races winter further east.
STATUS Annual migrant, occurring in small numbers throughout the islands, most commonly September to April, but a few remaining year round.
SIMILAR SPECIES Greater Sandplover is most likely confusion species, but this has longer, paler legs, set more centrally, more horizontal stance, more angular head and longer bill. **Caspian Plover** is a long-legged slender plover of grassland rather than mudflats, with broad white supercilium. **Common Ringed Plover** has short orange legs and black-and-white breast- and neck-collars.
References: 1. Hirschfield *et al* (2000), 2. Hockey (1993).

GREATER SANDPLOVER *Charadrius leschenaultii*   **Plate 20**

Other names: Large Sandplover, Geoffroy's Sandplover
French: Pluvier de Leschenault   **Creole:** Gran Plivye Ordiner

DESCRIPTION Length 22-25 cm; wingspan 53-60 cm. A small, drab plover, brown above and white below, with more horizontal posture and longer legs set centrally, showing more shin than Lesser Sandplover. **Adult non-breeding** is drab grey-brown above and white below, the brown from sides of neck forming an almost complete breast-band. Some birds may show traces of rufous at sides of breast-band and black mask either early or late in season. In flight, shows darker subterminal bar on tail, wings show narrow white wingbar, white underwing and toes project well beyond tail. Best distinguishing features from Lesser Sandplover are longer legs, showing much more of the tibia and set closer to the mid-point of the body giving a more horizontal posture in the standing bird. Also a more angular head with more sloping forehead and longer bill, its length greater than the distance from bill base to the eye. Bill black, legs variable from olive grey to black (usually paler than Lesser Sandplover). **Adult male breeding** has a narrow rufous breast-band, white throat and a broad black strap through eye. **Female breeding** has a brown mask, rufous areas less intense and less extensive than male. **Immature** has buff tinge to head and pale fringes on wing-coverts showing up as a pale oval on the folded wing.
VOICE Contact call a short, soft 'drreep', but may also make short, hard calls[1].
BEHAVIOUR Feeds on open sand and mudflats at low tide, moving in a series of short runs with pauses between. Foraging behaviour involves a longer run than Lesser Sandplover, with typically, more than four, very often more than eight paces, followed by a pause of 5-10 seconds[2]. Gregarious, often associating with other waders including Lesser Sandplover. Rapid flight action similar to cogeners. Picks small insects, crus-

taceans and worms and small crabs[1] from the surface of the mud.

RANGE Breeds from Turkey and Caspian Sea east to Mongolia, wintering along coasts of eastern half of Africa, Red Sea, Persian Gulf and India to Australia. Central Asian race *crassirostris* is most likely in Seychelles; it is longer-billed than other races. The shorter-billed race *columbinus*, which breeds Turkey to Afghanistan and winters south to Gulf of Aden is less likely but might be possible.

STATUS Annual migrant, occurring in small numbers throughout the islands, mostly September to April. Some first-year birds may remain year round.

SIMILAR SPECIES Lesser Sandplover has shorter legs (showing little or no tibia) set well back on the body, a more rounded head and shorter bill. **Caspian Plover** has more buff on head and wide, flared supercilium, fine bill, more upright stance and wingtips extending well beyond tail at rest. **Common Ringed Plover** has short, orange legs and black-and-white chest- and neck-collar.

References: 1. Hirschfield *et al* (2000), 2. Hockey (1993).

CASPIAN PLOVER *Charadrius asiaticus*   Plate 20

French: Pluvier asiatique **Creole:** Plivye Was Lazi

DESCRIPTION Length 18-20 cm; wingspan 55-61 cm. A small, elegant plover with a slender neck, fine, pointed bill and prominent supercilium. **Adult non-breeding** has a slender silhouette, with long, pointed wings and tail, small head with fine, tapered black bill and a clear white supercilium, flared behind the eye. Upperparts dark grey-brown with pale feather fringes in non-breeding plumage. At rest, wingtips extend well beyond the tail. Underparts and underwing are clean white (with dark grey underwing greater coverts). In flight, indistinct wingbar, more obvious at primary bases and toes extend well beyond tail. Bill black, legs yellowish brown to dull olive grey. **Adult male breeding** has broad rufous breast-band bordered black at lower margin, white supercilium, forehead and throat with narrow brown strap to eye. **Female breeding** has less intense rufous on breast-band that lacks black border (or has only a faint suggestion of a border). **Immature** is similar to adult, but with more buff in the breast-band and wing-coverts with darker centres and rufous buff or buff fringes.

VOICE A soft piping 'tik, tik' flight call and a sharper 'kuhit'.

BEHAVIOUR Prefers dry, short grassland areas (in contrast to the sandplovers) and often seen on open, mown grass sites such as airstrips. Gregarious, though vagrants to Seychelles are generally solitary, sometimes with other species including Greater Sandplover. Rapid flight with regular wingbeats and a long run upon landing. Hunts in the same 'run-stop' style as other plovers, but has a more upright stance than sandplovers. Diet almost exclusively insects, sometimes snails and grass seeds.

RANGE Breeds Caspian Sea to northwest China, wintering from South Africa to Red Sea. Occasional vagrant to the Indian subcontinent. Monotypic.

STATUS Vagrant, recorded mainly in the granitic islands but also south to Farquhar; most records October to March.

SIMILAR SPECIES Greater and **Lesser Sandplovers** are similar but occur typically on sand- and mudflats rather than grassland; they are shorter-tailed, shorter-winged (wingtips level with tail tip), show less neck, have an incomplete breast-band, less prominent supercilium and bills heavier at the tip. **Oriental Plover**, a very rare vagrant, is a more likely confusion species, but this is longer-necked, longer-legged, with distinctive upright stance, no wingbar and an all-dark underwing.

ORIENTAL PLOVER *Charadrius veredus*   Plate 20

Other names: Oriental Dotterel, Eastern Sandplover
French: Pluvier orientale **Creole:** Plivye Les Lazi

DESCRIPTION Length 22-25 cm; wingspan 61-67 cm. A small, elegant plover with an upright stance, long pale flesh-yellow legs, a long-necked stance, no wingbar and an all-dark underwing. Distinctive upright stance, accentuates long neck and legs. **Adult non-breeding** has grey-brown upperparts, the feathers fringed rufous or buff, broad buffish breast-band and clean white underparts. Dark crown and cheeks contrast with pale supercilium and white face. At rest, wings project noticeably beyond tail. In flight shows no wingbar on upperwing and distinctive dark grey-brown underwing contrasts with white belly, toes projecting beyond tail. Bill black and slightly tapered, legs pale flesh-yellow or flesh-orange. **Adult breeding** has pale orange breast-band with black lower margin and white face and sides of neck. **Immature** has more mottled breast-band, rufous buff face and supercilium and with more obvious pale fringes to back and wing-coverts, edged buff rather than rufous. Legs paler flesh-yellow.

VOICE Flight call a shrill 'tyip, tyip, tyip'. Also a short, piping 'klink' note.

BEHAVIOUR Frequents open short grassland areas, (e.g. island airstrips) but is sometimes found close to freshwater or on beaches. Gregarious, sometimes in association with Caspian Plover and Pacific Golden Plover. Runs swiftly and on the wing has a swift, erratic flight, climbing high. Feeds mainly on insects, but also sometimes on seeds and snails.

RANGE Breeds in dry steppes along the northern border of China, flying south after the breeding season to winter in Indonesia and northern Australia. Monotypic.

STATUS Vagrant to granitic islands, recorded October to November.

SIMILAR SPECIES Caspian Plover is most similar, but is slightly smaller, with shorter, darker legs and less upright posture. It also has faint wingbar and whiter underwing-coverts. **Pacific Golden Plover** is of similar size, but much darker above with bright yellow suffusion, smudged grey underparts and black legs.

Comparative sizes of waders 1: (1) Pacific Golden Plover, (2) Greater Sandplover, (3) Lesser Sandplover, (4) Caspian Plover, (5) Ringed Plover, (6) Curlew Sandpiper

COMMON RINGED PLOVER *Charadrius hiaticula*　　　　Plate 21

Other names: Ringed Plover, Great Ringed Plover
French: Pluvier grand-gravelot　**Creole:** Plivye Kolye Nwanr

DESCRIPTION Length 18-20 cm; wingspan 48-57 cm. A small, pale plover with black mask, white neck-collar. Pure white below with black breast-band and short orange legs. **Adult non-breeding** is grey-brown above, pure white below, with head and back separated by narrow white neck-collar. Black face-mask covers eye and extends onto ear-coverts, rounded at the base. Prominent white supercilium and no eye-ring. Narrow black band across upper breast runs up onto shoulder. Shorter tertials and longer primary projection than Little Ringed Plover. Prominent white wingbar in flight. Bill short and black with an orange base (sometimes all-dark). Legs short and dull orange. **Adult male breeding** has narrower breast-band, blacker on bill and brighter orange legs. **Female breeding** similar to male but less smart in appearance, often with much brown in the breast-band and face-mask. **Immature** has greyish face, white forehead and supercilium and an incomplete grey breast-band, becoming black by January, after which usually indistinguishable from adult. Wing-coverts with paler fringes than adult, bill all-black and legs dull orange or yellow.

VOICE Flight call a soft, two note 'cu-wip', rising in pitch.

BEHAVIOUR Prefers wide sandy tidal beaches and wet muddy areas, where it feeds with other waders. Several birds may gather with other waders at high-tide roosts until the water has fallen sufficiently to feed again. Rapid flight action, but slower than Little Ringed Plover, interspersed with gliding, long wings and long tail obvious. Typical plover feeding action of run, stop and peck to take small worms or crustaceans from surface sand or mud.

RANGE Breeds around Arctic shores, from Baffin Island to the Bering Strait. Race *hiaticula* of Canada to Europe winters in western Africa and Mediterranean; race *tundrae* of northern Scandinavia and Russia winters Caspian Sea, sub-Saharan Africa and Indian Ocean islands; it is slightly smaller and darker brown above.

STATUS Annual migrant in small numbers throughout Seychelles mainly October to March.

SIMILAR SPECIES The **sandplovers** have longer, darker legs and lack the white neck-collar. **Little Ringed Plover** is most similar, but this has dark legs, lacks a wingbar and seen close to, has a yellow eye-ring

LITTLE RINGED PLOVER *Charadrius dubius*   Plate 21

French: Pluvier petit-gravelot   **Creole:** Pti Plivye Kolye Nwanr

DESCRIPTION Length 14-17 cm; wingspan 42-48 cm. A very small plover with a black mask and yellow eye-ring, white neck-collar, single dark breast-band, pale flesh legs and no wingbar. **Adult non-breeding** is small and slim, uniform grey-brown above with obvious white neck-collar and buffish forehead and supercilium. A single brown band across upper breast separates white underparts from white throat. Obvious yellow eye-ring shows up in brown face-mask, which is pointed at the base. Longer tertials and shorter primary projection than Common Ringed Plover. In flight wings dark grey-brown and no wingbar. Bill black, legs dull pale flesh. **Adult male breeding** has black face-mask and collar and brighter eye-ring, clearer white forehead with white line and black frontal bar across the forecrown, leaving forehead white and pink legs. **Female breeding** has breast-band tinged brown and a narrower eye-ring than male. **Immature** is brownish, face dusky without obvious black-and-white mask and back and wing-coverts with buff scaling. Pale forehead does not extend as an obvious supercilium. Legs greyish yellow.

VOICE A distinctive 'peeu' call, falling in pitch and a shorter 'peep' or 'pip' note.

BEHAVIOUR Occurs along shorelines and on mudflats like Common Ringed Plover, but more likely to be found beside small ponds or other freshwater margins. Less gregarious than many waders. Rapid flight action, showing shorter, narrower wings and shorter tail compared to Common Ringed Plover. Also has a faster foraging action. Food includes beetles, insect larvae, flies, shrimps, spiders and some seeds.

RANGE Palearctic race *curonicus* breeds across Europe and Asia, from Britain and Spain through Russia, wintering sub-Saharan Africa and from the Indian subcontinent through to Indonesia. Race *jerdoni* mainly resident from India into South-East Asia.

STATUS Vagrant to granitic islands and Bird, recorded September to January.

SIMILAR SPECIES Common Ringed Plover has orange legs, two-tone bill (in adult), an obvious wingbar and no yellow eye-ring. **Greater** and **Lesser Sandplovers** are larger, with uniform sandy-grey upperparts, white below and with longer, dull grey-brown legs. **Kentish Plover**, *C. alexandrinus* (L15-17 cm, WS42-45 cm) is a potential vagrant. The nominate race breeds western Europe to northeast China, wintering sub-Saharan Africa to Indonesia, mainly north of the Equator but reaching Kenya coast. This is a similar small plover with greyish brown upperparts, white neck-collar, small dark breast markings (but no breast-band), white forehead and supercilium, greyish brown crown and nape (rufous in the breeding adult), fine black bill, white wingbar and black legs.

PACIFIC GOLDEN PLOVER *Pluvialis fulva*   Plate 21

Other names: Asian Golden, Eastern Golden, Asiatic Lesser Golden Plover
French: Pluvier fauve   **Creole:** Plivye Dore

DESCRIPTION Length 23-26 cm; wingspan 60-68 cm. An elegant plover, upperparts suffused golden yellow, underparts smoky grey and with long black legs. **Adult non-breeding** has upperparts, crown and hind-neck suffused golden yellow, prominent yellowish buff supercilium and a black eye set almost centrally in squarish head. Neck and breast smudged with dark grey from throat to belly. In flight, shows an indistinct wingbar, while uniform golden yellow flecking on back and wing-coverts contrasts with dark flight feathers and all-grey underwing contrasts with white belly. Bill black, tapering to fine point, legs black in all plumages. **Adult breeding** (rarely seen in winter quarters), has breast, throat and face jet black with a white border. **Immature** is brighter than adult, upper breast with a golden yellow wash finely flecked with brown and face showing clear yellow-buff supercilium.

VOICE A sharp two note 'tu-ee' and a melancholy 'kl-ee' call.

BEHAVIOUR Frequents short grassland areas such as airstrips, sand- or mudflats, beaches and lagoons. Fairly gregarious, often in small numbers with other waders. Rapid, confident flight with regular wing-beats. Feeding behaviour consists of typical plover short runs with pauses between, stooping quickly to snatch food. Diet mostly consists of worms and beetles, though some grass and seeds may be taken.

RANGE Breeds in high Arctic tundra from Kara Sea in Russia through to Alaska, wintering shores of Somalia, India, South-East Asia, Australia and the Pacific. Monotypic.

STATUS Annual migrant in the granitic islands, less frequent elsewhere recorded mainly October to April, a few sometimes remaining year round.

SIMILAR SPECIES Grey Plover occurs with greater abundance in similar areas. This is more bulky with longer legs and heavier bill. It is distinguished in all plumages by the dark grey upperparts and in flight by the whitish underwing with black 'armpits' (axillaries) and clear white wingbar and white rump.

GREY PLOVER *Pluvialis squatarola*   Plate 21

Other names: Black-bellied Plover, Silver Plover   **French:** Pluvier argenté   **Creole:** Plivye Sann

DESCRIPTION Length 27-30 cm; wingspan 71-83 cm. A large, plump, silver-grey plover; white underwing with black armpits, white wingbar and white rump. **Adult non-breeding** is a large, rather thickset plover, mottled grey and white above and below with white belly. Most striking in flight, when grey back and wing-coverts contrast with bold white wingbar, black flight feathers, white rump and grey tail. Underwing diagnostic in all plumages: clean white with black axillaries ('armpits'). Black eye set in a dark smudge through cheek, contrasting with white lore and supercilium. Bill large for a plover and broad at the tip. Legs black. **Adult breeding** has jet black face, chest and belly contrasting with grey back, rarely seen in winter quarters but vestiges remain in first arrivals of autumn. **Immature** is darker and browner above, with blackish centres and buff fringes to scapulars, giving the back a yellowish tinge. Underparts greyish with lines of fine streaking, more distinct than the grey mottling of adult.

VOICE Distinctive trisyllabic 'pleeooee' or shorter 'plooee'.

BEHAVIOUR Tends to be more exclusively coastal than Pacific Golden Plover, preferring to feed in muddy intertidal habitat, though sometimes seen singly along coral sand beaches or on reefs at low tide. Not as gregarious as many waders, though at high tide will associate with other shorebirds in grassland areas, standing in characteristic hunched posture. Powerful, rapid flight. Forages in typical plover 'run-stop-search' style, feeding on molluscs, crustaceans and worms.

RANGE Breeds in tundra of Russia to Canadian Arctic, migrating south in August to winter around coasts of Africa, India, South-East Asia, Australia and North and South America, making it one of the most widespread shorebirds in the world. Studies of museum specimens suggest that females travel further south than males. Monotypic.

STATUS Annual migrant, occurring in small numbers throughout the islands, mainly September and April. Many first-year birds remain year round.

Moulting adult Grey Plover

SIMILAR SPECIES Pacific Golden Plover is the most likely confusion species, but is slightly smaller with a more slender outline and finer bill with golden spangling on upperparts. Immature Grey Plover may show a 'golden' wash on the back, but is greyish on neck and breast and shows contrasting white lore and dark eye smudge on the face. In flight there can be no confusion: Grey Plover shows black armpits, bold white wingbar and white rump.

GREAT KNOT *Calidris tenuirostris* Plate 21

Other names: Greater Knot, Great Sandpiper **French:** Bécasseau de L'Anadyr **Creole:** Gran Bekaso

DESCRIPTION Length 26-28 cm; wingspan 62-66 cm. A very large, deep-chested calidrid, grey above and spotted grey below, with a heavy, straight bill. **Adult non-breeding** is plain grey above, with noticeable dark streaks in feather centres. Head, face and neck uniform with indistinct pale supercilium, streaked crown, nape and mantle a clear distinction from Red Knot in all plumages. Finely streaked grey, lower breast and belly white with heavier black spotting on upper breast sides and finer grey chevrons on flanks and lower breast. At rest, wingtips project beyond tail. In flight, upperparts appear mainly grey with a noticeable black carpal-patch; white wingbar and white rump contrasts with blackish tail. Bill is dark, straight and heavy but finely tapered at the tip, longer than Red Knot. Legs are dark slate- or greenish grey. **Adult breeding** is unmistakable, with bright rufous scapulars, dark mottled back and white underparts with large black spots. Sexes similar, though female averages larger and has less rufous in scapulars. **Immature** is much darker above, with dark covert centres having buffish fringes more reminiscent of Grey Plover and retained throughout first-winter. Head and neck are similar but upper breast has a buff wash with more obvious dark spotting and dark spots extending down onto flanks. Legs slightly paler than in adult, green-grey.
VOICE Mainly silent. May give a two-note flight call, 'nyut-nyut' with the accent on the first syllable.
BEHAVIOUR Away from breeding grounds, almost entirely coastal, preferring large expanses of intertidal mud or sand. Gregarious, solitary vagrants often associating with sandplovers and other shorebirds. In flight, wingbeats fairly slow. Feeds mainly by probing or pecking taking molluscs, crustaceans and sea cucumbers.
RANGE Very restricted breeding range in the high Arctic of northeast Siberia, close to the Bering Strait. In autumn migrates south to coasts of Pakistan, Bay of Bengal, South-East Asia and Australia. Monotypic.
STATUS Vagrant to granitic islands, recorded March and April.
SIMILAR SPECIES Grey Plover is superficially similar, especially in flight, but this has a squarer head, a short, blunt bill and different feeding action (walk and pause, not a steady, methodical plod). **Curlew Sandpiper** is the only other similar calidrid, but is whiter beneath with no spotting, has a finer, slightly downcurved bill and usually occurs in small flocks. **Red Knot** is slightly smaller and stockier with shorter legs, rather large head and heavier bill. Also has more distinct supercilium and wingbar and wider but less distinct grey rump.

RED KNOT *Calidris canutus* Illus. p. 205

Other names: Knot, Lesser Knot, European Knot **French:** Bécasseau maubèche **Creole:** Bekaso Sann

DESCRIPTION Length 23-25 cm; wingspan 57-61 cm. A large, stocky calidrid (only Great Knot is larger) with short, straight bill and short legs. **Adult non-breeding** is almost entirely grey, becoming white on belly and vent, best identified by size and jizz. Pale grey rump and tail lack contrast with mantle (Great Knot has contrasting narrow clear white rump). At rest, wingtips and tail tip are roughly equal. In flight, shows distinct white wingbar. Dark eye stands out in rather large head with white supercilium clearer than in Great Knot. Rather short olive green legs. Rather heavy bill. **Adult breeding** is virtually unmistakable, with brick-red face and underparts (less intense in female) and stocky build. **Immature** appears scaly due to whitish fringes and dark subterminal bands on feathers of wing-coverts, retained throughout first winter.
VOICE Mainly silent but may give a low, soft 'knot'. Alarm call is a 'kikikik'.
BEHAVIOUR Entirely coastal in winter quarters. Gregarious, often associating with other waders. More rapid flight action than Great Knot, wingbar more obvious. When feeding, does not walk then pause, like Grey Plover, but creeps slowly and steadily forward. Feeds mainly on molluscs, sometimes taking insects and crustaceans.
RANGE Breeds Arctic North America to Siberia and northeast Asia wintering southward. The nominate race of central Siberia reaches southern Africa to Australia and is most likely race in Seychelles.
STATUS Uncertain; some reports have been indeterminate between this species and Great Knot and have been accepted by SBRC as 'Knot sp.'.
SIMILAR SPECIES Great Knot adult non-breeding is very similar, a little larger and longer-legged with longer, more tapered bill. Appears less chunky, more tapered towards rear with longer neck and smaller head and has a streaked crown, nape and mantle in all plumages (unlike Red Knot). At rest, wingtips project beyond tail, supercilium and wingbar are less obvious and rump is purer white.

Adult winter Red Knot (left), and adult winter Great Knot

COMMON SNIPE *Gallinago gallinago* Plate 22

Other name: Fantail Snipe **French:** Bécassine de marais **Creole:** Bekasin Lanmar

DESCRIPTION Length 25-27 cm; wingspan 44-47 cm. A slim, straight-billed snipe, fast and sharp-winged in flight, with a broad white trailing edge to the secondaries. Identification of snipe species depends mainly upon differences in wingbars, underwing pattern, tail pattern, face pattern, general proportions of the body and flight. **Adult** has dark upperparts with a mixture of black, white, buff and chestnut mottling, contrasting with pure white belly. Creamy-buff supercilium and mid-crown-stripe contrast with broad black stripes over crown and with supercilium narrower than eye-stripe at the base of the bill. Back is mottled black and buff with broad buff stripes as far as the rump. In flight, shows a broad white trailing edge to secondaries, inconspicuous primary-covert wingbar and broad white barring on underwing-coverts. At rest, shows stripes over mantle and scapulars which are broad and creamy-buff and on closed wing, tertials are dark with narrow pale bars. Tail is longer than Pintail, extending well beyond primary tips at rest and is tipped rufous with very little white at sides. In the hand, 14-16 feathers in the tail. Bill is long and straight, mainly brownish and darker at tip. Legs short and variable in colour, either grey-blue or yellow-green. Sexes similar, female slightly longer-billed. **Immature** as adult but with wing-coverts more neatly fringed pale buff (but replaced in late autumn).
VOICE In flight gives a distinctive, rasping 'skee-ip' call.
BEHAVIOUR Frequents margins of marshes or where marshy ground is absent, will feed along the margins of streams or pools. Not very gregarious and vagrants are usually solitary. With such cryptic plumage, tends to freeze when approached, then take flight at the last minute, rising rapidly with a series of jinking turns and flying high and far until well distant. Bill is designed to probe for earthworms, but will also take insects, molluscs and some plant material including stems and seeds.
RANGE Several races breed right across northern hemisphere. Only the nominate *gallinago* is likely in Seychelles. It breeds from Britain across Europe and northern Asia as far east as Kamchatka and winters in sub-Saharan Africa, India and Indo-China.
STATUS Vagrant, mainly to granitic islands, recorded October to February.
SIMILAR SPECIES Pintail Snipe and **Swinhoe's Snipe** both lack the broad, white trailing edge to the secondaries, have less extensive white on belly, darker underwing, pale supercilium broader than eye-stripe at base of bill and have a less erratic escape flight. Pintail appears dumpier, with rounded wings, a shorter blunt bill and a short tail that barely projects beyond the wings at rest, toes projecting conspicuously in flight. **Jack Snipe** *Lymnocryptes minimus* (L18-20 cm, WS39-45 cm) is a potential vagrant, which breeds Scandinavia to eastern Siberia, wintering south to sub-Saharan Africa and the Indian subcontinent. It also has a white trailing edge to the wing but is distinguished from Common Snipe by smaller size, shorter bill, lack of crown-stripe and flanks streaked not barred. It also has a less erratic, twisting escape flight compared to Common Snipe.

PINTAIL SNIPE *Gallinago stenura* Plate 22

French: Bécassine à queue pointue **Creole:** Bekasin Lake Pikan

DESCRIPTION Length 25-27 cm; wingspan 44-47 cm. A pale, short-billed, short-tailed snipe, lacking a white trailing edge to the secondaries. **Adult** has tail barely projecting beyond wings; head appears small and bill shorter and slightly thicker than Common Snipe. At close range note supercilium is flared widely in front of the eye, tapering to nothing behind it (i.e. broader than the eye-stripe at bill base). 'V' stripes on mantle and scapulars, finer and whiter than Common Snipe, formed by a more complete crescent on each feather. In closed wing, the tertials almost completely cover tail and have broader pale bars giving a more obvious barred pattern. Pale bars of tertials are generally wider than the dark bars (about equal in Swinhoe's Snipe) and primaries are covered by tertials (Swinhoe's has short primary projection). In flight, feet project noticeably beyond the short tail which shows almost no white; wings appear rounded, lack broad white trailing edge to secondaries and have an obvious paler, greyish median-covert panel on upperwing; underwing appears dark, due to uniform dark grey barring on underwing-coverts. In the hand, easily identified with 26-28 tail feathers, outer 6-9 (generally eight) pairs of which are narrow and pin-shaped. Sexes similar, female averaging slightly longer-billed. **Immature** has upperwing-coverts with whitish fringes rather than the brown-buff feather tips of the adult, but generally indistinguishable from October onward. More worn primaries of first-winter compared to adult may be a more reliable characteristic.

VOICE Often calls when flushed, sometimes repeated, call more monosyllabic, throaty and slurred than Common Snipe, similar to the 'quack' of a Mallard[1].

BEHAVIOUR Feeds in damp ground typical of snipes but will forage in drier sites also, moving into dry grassland and grazed pasture. Vagrants are generally solitary. When flushed, rarely climbs as high as Common Snipe but has a less erratic escape flight. Feeding behaviour may include 'picking' as much as probing, with seeds, caterpillars and beetles taken.

RANGE Breeds in northern Russia, east of the Ural Mountains, wintering India, Indo-China and Indonesia. Monotypic.

STATUS Vagrant to granitic islands and Aldabra, recorded November and March.

SIMILAR SPECIES Great Snipe is larger and plumper with heavy barring beneath, white outer tail feathers and striking white bars on the greater coverts. **Common Snipe** is similar in size, but has a broad white trailing edge to the secondaries and the pale supercilium is narrow at base of bill. **Swinhoe's Snipe** also lacks the white trailing edge and has similar face pattern and wing pattern, but is larger, longer-billed and longer-tailed; in flight toes project less conspicuously and at rest shows a short primary projection and the pale and dark bars of tertials are roughly equal in width.

Reference: 1. Carey and Olsson (1995).

SWINHOE'S SNIPE *Gallinago megala*   Plate 22

Other names: Chinese Snipe, Forest Snipe   **French:** Bécassine de Swinhoe   **Creole:** Bekasin Swinhoe

DESCRIPTION Length 27-29 cm; wingspan 47-50 cm. A rather plain dark brown snipe with long bill, pointed wings, an obvious median-covert panel and dark underwing. **Adult** is slightly larger than Common Snipe but with supercilium broader than eye-stripe at base of bill. Also lacks white trailing edge to secondaries but has similar wing pattern to Pintail Snipe, with pale rufous median covert panel above and uniform dark barring underwing. Short primary projection, (tertials cover primaries in Pintail) with pale and dark bars of tertials equal in width (pale bars tend to be wider than dark in Pintail). Tail longer than in Pintail with more white visible at corners. In flight, toes still project slightly, but less conspicuously. In the hand, there are 20 feathers in the tail. Spring birds may appear dark in tone on the flanks, neck and face. Bill relatively long, brownish green at the base, dark brown at tip. Legs longer than other snipe, greenish yellow. Sexes similar, female averaging longer-billed. **Immature** similar to adult, may retain pale buff fringes to wing-coverts and tertials of juvenile, but this is rarely distinguishable in field. More worn primaries of first-winter compared to adult may be more reliable characteristic.

VOICE Typically silent, if it calls at all giving a single quiet 'chet' like Pintail Snipe, but thinner with less throatiness[1].

BEHAVIOUR Frequents similar habitat to other snipe, though sometimes found in drier locations. Generally solitary. Slow to take off when flushed and flight heavy and rather slow and direct. Mainly crepuscular and nocturnal. Feeds mainly on worms and insects.

RANGE Breeds in central Siberia and northern Mongolia, migrating south to winter in southern India, southern China, the Philippines, Indonesia, New Guinea and northern Australia. Has been recorded as a vagrant to Maldives. Monotypic.

STATUS Uncertain; some reports have been accepted by SBRC as indeterminate between Pintail and Swinhoe's Snipe.

SIMILAR SPECIES Common Snipe has broad white trailing edge to the wing, a longish tail, much white on belly, pale underwing and pale supercilium narrow at base of bill. **Pintail Snipe** has shorter bill and tail, no primary projection and has pale bars of tertials generally wider than dark bars.

Reference: 1. Carey and Olsson (1995).

GREAT SNIPE *Gallinago media*   Plate 22

French: Bécassine double   **Creole:** Gran Bekasin

DESCRIPTION Length 27-29 cm; wingspan 47-50 cm. A rather tubby snipe with slightly shorter bill than most other snipe, heavy barring on the belly and white outer tail feathers. **Adult** is slightly larger than Common Snipe, with more pot-bellied outline, shorter bill and longer legs. Most striking feature on ground is two rows of white formed by white tips to wing-coverts. In flight these show up as two white wingbars, in addition to white trailing edge to wing and distinct white outer tail feathers, unlike other snipe species. Diagnostic heavy barring on underparts right across belly and spotted, not streaked neck and breast. Bill shortish, brownish at base, darker at tip. Legs either grey-green or brownish yellow. Sexes similar. **Immature** is similar to adult, though white tips to wing-coverts are less obvious. Outer tail feathers tend to have more buff hue, but still show up as paler when tail flared. Similar to adult by December onward. **VOICE** Gives a low call 'ech' or 'ech-ech' when flushed.

BEHAVIOUR Frequently occurs in drier habitat than Common Snipe including short grassland lightly wooded terrain or ploughed ground. Crepuscular and nocturnal, being most active towards dusk. Vagrants are generally solitary. Has a much slower, heavier and more direct flight than other snipe; prefers a low escape route, landing again as soon as possible. Holds bill more horizontal than Common Snipe. Feeds mainly on worms and insects.

RANGE Breeds Norway, Sweden and northern Russia, wintering sub-Saharan Africa. Monotypic.

STATUS Vagrant to granitic islands, recorded October to December.

SIMILAR SPECIES Common Snipe is less bulky, longer-billed, unbarred on white belly, streaked rather than spotted on neck and breast, tail mainly rufous and with a broad white trailing edge to secondaries.

BLACK-TAILED GODWIT *Limosa limosa*   Plate 23

French: Barge à queue noire **Creole:** Limoza Lake Nwanr

DESCRIPTION Length 40-44 cm; wingspan 70-82 cm. A large, long-legged wader with a long, straight bill and striking flight pattern with black-and-white wingbars and black-and-white tail. **Adult non-breeding** appears uniform grey, paler off-white below. Crown feathers appear slightly darker, accentuated by distinct pale supercilium in front of eye. In flight reveals unmistakable pattern of broad white wingbars contrasting with black flight feathers and white rump above black tail; also white not barred underwing. Bill long and straight, pink at base, black at tip. Legs long and dark, tibia longer than Bar-tailed Godwit so that toes trail in flight much more conspicuously. **Adult male breeding** has vivid pink-orange head, neck and upper breast, contrasting with white belly, barred black on flanks and belly. Bill orange at base, dark at tip. **Female breeding** is much paler below than male with some white flecking, belly and undertail-coverts whitish, flanks barred, and retains much grey in scapulars. **Immature** has black-and-buff mottling on mantle, scapulars and wing-coverts and distinct bright rufous flush to neck and mantle. Similar to adult by December.

VOICE Flight call is a quiet 'kuk kuk'.

BEHAVIOUR Shows a distinct preference for muddy habitats in which to feed, (scientific name *Limosa* means 'of the mud') and is thus most often seen on sheltered coasts, river mouths and lagoons, feeding on intertidal mudflats. Vagrants are generally solitary. Strong flight, bill held horizontal and legs trailing. Probes deeply with its long bill for worms, but will also take insects and seeds.

RANGE Breeds throughout central Europe and central Russia, with outlying populations in eastern China. Nominate European and Asia race *limosa* winters mostly south of the Sahara, in the Nile valley and northern India; Siberian race *melanuroides* winters from Malaysia to north coast of Australia.

STATUS Vagrant, recorded from Bird to Alphonse, October to November.

SIMILAR SPECIES Most likely confusion species is **Bar-tailed Godwit**, with a similarly long, straightish but slightly up-tilted bill, shorter legs and in flight lacks strong black-and-white patterning on wings or tail.

BAR-TAILED GODWIT *Limosa lapponica*   Plate 23

French: Barge rousse **Creole:** Limoza Lake Are

DESCRIPTION Length 37-41 cm; wingspan 70-80 cm. A large, grey-brown wader with long dark legs, white rump and long, straight, slightly up-tilted bill. **Adult non-breeding** is dull grey-brown above with faint mottling and paler below. Dark crown and broad, greyish supercilium, in front and behind the eye. In flight, wings are uniform grey-brown with darker primaries, contrasting with a white wedge up the lower back and barred tail and toes projecting only slightly. Underwing may be white (race *lapponica*) or barred (*baueri*). Most distinctive feature is the long straight bill, pink at base, dark at tip and very slightly upturned. Legs dark, almost black, long but with shorter tibia than Black-tailed Godwit. **Adult male breeding** is dark brick red below, from crown and face through neck and belly to vent, mottled black and rufous above. Bill all-dark. **Female breeding** is considerably paler, generally lacking any rufous, the breast with some dark barring. **Immature** appears darker and brighter: back and wing-coverts have dark centres with buff spots, forming a mottled pattern similar to Whimbrel. Diagnostic buff wash to neck, scapulars and tertials. Similar to adult by January.

VOICE Flight call 'ki-biu' and alarm call a harsh 'ke-vik'.

BEHAVIOUR Feeds on coastal mudflats and shorelines, but with a distinct preference for more sandy areas, striding across ground between the tides and probing into muddy sand with its long bill. Gregarious, often in small numbers in Seychelles. In flight, head and neck retracted giving deep-chested appearance, legs barely visible beyond tail. Less graceful and erect stance compared to Black-tailed Godwit. Takes worms, crustaceans and even small fish.

RANGE Nominate race *lapponica* breeds round the shores of the Arctic Sea from northern Sweden and Finland, wintering round the coastline of East Africa, Iran and Pakistan, reaching Seychelles and Madagascar. Siberian race *baueri* (larger, with darker, spotted rump and barred underwing) winters South-East Asia and Australia.

STATUS Annual migrant, occurring throughout the islands mainly October to March. Most birds are nominate race but race *baueri* has been recorded on Frégate[1].

SIMILAR SPECIES Most likely to be confused with **Black-tailed Godwit**, which has a longer, straight bill, longer legs and a striking black-and-white pattern on wings and tail visible in flight. **Whimbrel** is similar in size and colour, particularly in flight, but has a strongly downcurved bill, paler blue-grey legs, close, fine streaking over the breast and belly and a dark striped crown. **Asian Dowitcher** *Limnodromus semipalmatus* (L33-36 cm, WS59 cm) is a potential vagrant. It breeds Central Asia, wintering eastern India through South-East Asia to northern Australia, with unconfirmed vagrants to Seychelles and Kenya. Smaller, with straighter, less tapering black bill; rump barred and underwing mainly white with fine black streaks and fine, black spotting on flanks.

Reference: 1. Lucking (1996b).

WHIMBREL *Numenius phaeopus*   Plate 23

French: Courlis corlieu   **Creole:** Korbizo Ordiner

DESCRIPTION Length 40-46 cm; wingspan 76-89 cm. A relatively large, mottled brown wader with a long, downcurved bill and two dark stripes over the crown. **Adult non-breeding** is similar to Eurasian Curlew but smaller, with the same closely patterned brown back and flight feathers and fine dense streaking on face, neck and breast. Crown pattern diagnostic: two dark stripes over top of head, with narrow pale line between. In flight, a white wedge up the lower back contrasts with barred brown tail and mottled brown wings with darker flight feathers. Both bill and legs are shorter than Eurasian Curlew, the dark downcurved bill slightly blunter and blue-grey legs appearing slightly short for bird's size. Sexes similar, female averages larger. **Adult breeding** indistinguishable from adult non-breeding. **Immature** may appear darker above than adult, with distinct buff hue below, coverts of back and wings with buff spots and fringes around dark brown feather centres until around December.

VOICE A distinctive seven note piping trill: 'pi-pi-pi-pi-pi-pi-pi'.

BEHAVIOUR At high tide may roost or feed in dry grassland sites inland from the coast, but at all other times feeds in intertidal areas. Gregarious, often with other waders. Flight light, direct and rapid. Probes for worms and molluscs, but will also stalk individually along beaches to catch, dismember and swallow the ghost crabs that are abundant on most coral sand shorelines.

RANGE Nominate race *phaeopus* breeds from Iceland through Scandinavia to northern Russia, but isolated populations occur in eastern Russia (*alboaxillaris*), Alaska (*variegatus*) and Arctic Canada (*hudsonicus*). Indian Ocean birds mostly of nominate race, but may include *alboaxillaris* (Steppe Whimbrel) which appear paler grey above, whiter on breast and belly and have all-white, unstreaked underwing.

STATUS Annual migrant in good numbers throughout the islands, mainly October to March, a few remaining year round.

SIMILAR SPECIES Eurasian Curlew is larger, with longer bill and legs but without the crown-stripes. **Bar-tailed Godwit** appears similar in flight, but has straight bill and darker legs. **Little Curlew** is possible: much smaller, has dark crown-stripes but with clear buff eye-stripe and much shorter, finer bill. **Slender-billed Curlew** *Numenius tenuirostris* is a slim, elegant curlew with short slim bill and white breast and belly patterned with diagnostic thrush-like black spots. Note Steppe Whimbrel (Whimbrel race *alboaxillaris*) also has an all-white underwing but is separated by other features including boldly striped head pattern.

SLENDER-BILLED CURLEW *Numenius tenuirostris*   Illus. p. 209

French: Courlis à bec grêle   **Creole:** Korbizo Labek Fen

DESCRIPTION Length 36-41 cm; wingspan 80-92 cm. A slim, elegant curlew with a short, slim bill and white breast and belly patterned with thrush-like black spots. **Adult** is the size of a Whimbrel, but with shorter bill, tapering to a fine tip. Crown dark but without pale mid-stripe (some may have a suggestion of stripe, but less distinct than in Whimbrel). White eye-ring is broken in front of the eye by a black loral-stripe. Diagnostic feature is breast pattern, small, thrush-like black spots on white ground colour and unmarked white belly. In flight shows almost pure white underwing-coverts. Tail appears whiter in flight, with spots rather than bars. Bill is tapered, brown with paler base, legs pale slate grey. Sexes similar, female longer-billed and averages larger. **Adult breeding** shows paler buff ground colour and more obvious spotting on breast and flanks. **Immature** similar to adult but with streaks rather than spots on neck and breast and flanks.

VOICE Like a shortened Curlew, 'ker-wee'.

BEHAVIOUR Feeds on wet grassland as well as coastal lagoons and mudflats. Often seen in small groups but vagrants are likely to be solitary. Wades more frequently than Eurasian Curlew, flight lighter and more rapid. Probes mud for molluscs and crustaceans or pecks in grass taking insects.

RANGE Very rare (IUCN Critically Endangered) and declining, known from only a handful of sites in central Russia and to winter in Spain, Morocco and Tunisia, with scattered unconfirmed records. Monotypic.

STATUS Uncertain; there have been published reports from Aldabra and Cosmoledo, not confirmed by SBRC[1].

SIMILAR SPECIES See under Whimbrel above.

Reference: 1. Gretton (1991).

Slender-billed Curlew (top), and Whimbrel

EURASIAN CURLEW *Numenius arquata*   Plate 23

French: Courlis cendré **Creole:** Gran Korbizo

DESCRIPTION Length 50-60 cm; wingspan 80-100 cm. A large, uniformly mottled brown wader with a very long, downcurved bill. **Adult non-breeding** from distance has undistinctive uniform dull brown or pale brown plumage but unmistakably long, downcurved bill, dark with pinkish base. Beautifully intricate plumage of dark spotting and streaking on buff-brown background. The dark eye stands out in a finely streaked face, neck and breast. No distinct flight pattern, save a white wedge on the rump and lower back contrasting with mottled brown wings and tail and grey markings in secondaries (unlike Whimbrel). Underwing pale, speckled with black in nominate race but mainly white in race *orientalis*. Long legs are blue-grey. Sexes similar, female longer-billed and averages larger. **Adult breeding** is brighter buff above, with heavier streaking below. **Immature** is shorter-billed than adult, paler below, with spots rather than streaks on flanks until around December.

VOICE Distinctive, drawn-out 'coor-liu' from whence its name.

BEHAVIOUR Strides slowly and deliberately across mudflats or grassland in search of prey. Gregarious, but generally solitary or in twos or threes in Seychelles. Direct rather gull-like flight, slower than Whimbrel with neck retracted, bill held parallel to ground. In wet ground the bill is thrust deep and twisted to locate worms or crustaceans below the surface. In dry habitats it may run to catch grasshoppers or beetles among the grass. Intertidal habitats are preferred for feeding and at high tide, birds may roost among mixed wader flocks on dry ground waiting for feeding areas to be uncovered by the falling tide.

RANGE Breeds throughout northern Europe (western race *arquata*) and east across Russia as far as the Amur River (eastern race *orientalis*, paler overall, whiter on underwing, belly, tail and uppertail-coverts with much longer bill and legs). Winters within west of breeding range and on coasts all round Africa, Red Sea to South-East Asia. Races mix in winter from Mediterranean to western India. However, only *orientalis* occurs in most of African region including Seychelles.

STATUS Annual migrant throughout Seychelles in small numbers, mainly October to March, outnumbered by the much more common Whimbrel.

SIMILAR SPECIES Whimbrel, the most likely confusion species, is smaller, with a shorter bill and distinctive dark and light crown-stripes. **Slender-billed Curlew** is an unconfirmed vagrant, but is very small, with short, fine-tipped bill and spotted rather than streaked breast. **Little Curlew** is half the size, with dark, striped crown and no white rump as in the other species. **Far Eastern Curlew** *N. madagascariensis* (L53-66 cm, WS110 cm) is a potential vagrant, breeding eastern Siberia and wintering south, mainly in Australia, vagrant west to Iran and Afghanistan. It is a huge curlew, darker overall and more buff-brown below than both races of Eurasian Curlew with no white on rump and no dark patch on lore

LITTLE CURLEW *Numenius minutus* Plate 23

Other names: Little Whimbrel   **French:** Courlis nain   **Creole:** Pti Korbizo

DESCRIPTION Length 29-32 cm; wingspan 68-71 cm. A tiny warm sandy buff curlew with dark brown rump and lower back, short, slender bill, less noticeably downcurved than other similar species, short, narrow neck and a small head. **Adult non-breeding** has similar mottled brown upperparts and streaked underparts to Eurasian Curlew, but at half the size (and three-quarters the size of Whimbrel), should stand out alongside. Dark crown similar to Whimbrel but with buff mid-stripe and a more distinctive buff eye-stripe and pale lore. Rest of plumage warmer, sandy brown than similar species. In flight, diagnostic all-dark brown upperparts with no white on rump or lower back as in other species, wings mainly dark with a pale central panel, flight feathers evenly black and unspotted with pale tips and underwing-coverts grey-brown. Shortish, bill, downcurved towards tip, dark at tip, pink at base and on most of lower mandible. Legs blue-grey. Sexes similar, female averages slightly larger. **Immature** is shorter-billed than adult, with white notches on tertials. Difficult to distinguish from adult in field by November onward.
VOICE A distinctive three note 'pe-pe-pe' rising slightly in pitch and alarm call a strident 'tew-tew-tew'.
BEHAVIOUR Less inclined to feed on shores than similar species, feeding rather in more open terrain, bare ground or short grassland. Often quite tame, allowing a close approach. Gregarious, though vagrants are likely to be solitary. Freer flight than similar species with shallow, languid wingbeats. Feeds by probing and picking, taking insects, spiders, berries and other plant material.
RANGE Breeds Siberia, wintering from New Guinea to northern Australia. Monotypic.
STATUS Vagrant to Bird recorded October to April.
SIMILAR SPECIES Whimbrel is larger, darker brown also with crown-stripes and with distinct white wedge on lower back and rump that all other similar species show. **Eurasian Curlew** is twice the size with a very long bill and has no crown-stripes. **Bar-tailed Godwit** is larger, with a straight bill.

COMMON REDSHANK *Tringa totanus* Plate 24

Other name: Redshank   **French:** Chevalier gambette   **Creole:** Kavalye Lapat Rouz

DESCRIPTION Length 27-29 cm; wingspan 59-66 cm. A medium-sized, grey-brown wader with orange-red legs and a straight bill with a red base. **Adult non-breeding** is uniform grey-brown above and paler beneath, with fine dark speckling on breast and belly. Obvious white eye-ring, indistinct supercilium and a dark line from bill to eye. In flight striking pattern of white on lower back and trailing edge of wings contrasts with black primaries and barred tail. Bill moderately long, mostly dark but orange-red at base. Legs diagnostic orange-red. Sexes similar, female averages longer-winged. **Adult breeding** shows brighter mottling on back and heavier spotting on face, breast and flanks. **Immature** with brighter buff fringes to dark brown back feathers, dark crown streaked buff and dense fine streaking on breast, fading to finer spotting on belly. Bill all-dark, legs orange-red. Difficult to age in field by November.
VOICE Loud yelping 'tew-hu' or 'tew-hu-hu' flight call.
BEHAVIOUR Outside breeding season almost exclusively coastal, feeding on mudflats, sheltered bays, lagoons and small inlets. Gregarious, though vagrants are likely to be solitary. Strong flight with rapid wingbeats. On ground, bobs up and down in excitement. Probes mud and sifts for worms, shrimps and tiny crabs.
RANGE Nominate race *totanus* breeds from Britain through to Russia, wintering tropical Africa, India to Indonesia. Race *ussuriensis* breeds from Urals to China wintering on shores from Red Sea and Persian Gulf to Indonesia and recorded annually in Kenya and Maldives; differs in breeding plumage, upperparts more cinnamon-brown.
STATUS Vagrant to granitic islands recorded February.
SIMILAR SPECIES The only other similar-sized waders with orange-red legs recorded in Seychelles are **Terek Sandpiper** (shorter orange-yellow legs and an upturned bill) and **Ruff** (dull orangey grey legs, scaly upperparts and short, dark decurved bill, pale at base). **Spotted Redshank** *T. erythropus* (L29-32 cm, WS61-67 cm) is a potential vagrant. It breeds northern Scandinavia to northeast Asia, wintering sub-Saharan Africa and western Europe, Mediterranean to South-East Asia. Also has orange-red legs, longer than Common Redshank (particularly the tibia). Wings lack broad white trailing edge. Adult non-breeding is paler, more uniform grey above, whiter below (similar to Common Greenshank), obvious white supercilium outlined by black loral stripe. Longer bill is entirely black on upper mandible, contrasting sharply with red base of lower mandible. Immature is browner above, barred grey below (unlike Common Redshank).

MARSH SANDPIPER *Tringa stagnatilis* Plate 24

French: Chevalier stagnatile   **Creole:** Kavalye Lanmar

DESCRIPTION Length 22-25 cm; wingspan 55-59 cm. A slender, grey-and-white wader with long, fine tapering bill and very long greenish legs. **Adult non-breeding** is pale grey above, with fine white scaling and faint dark streaking on the back and wing-coverts. Head and hindneck finely streaked and underparts uniformly pale. Prominent long, white supercilium gives a capped appearance. In flight, wings appear all-dark, contrasting with pale tail and a narrow white gusset that reaches up the lower back almost to the nape while toes project conspicuously. Bill dark, showing little grey at base, very thin and straight, tapered to tip. Legs grey-green, very long and thin in proportion to body. Sexes similar, female averages larger. **Adult breeding** has dark brown back with buff mottling and face, neck and chest with fine black streaking. **Immature** is slightly browner on crown and hindneck with pale buff fringes to back and wing-coverts. May be indistinguishable from adult by November.

VOICE Flight call a shrill 'kiu kiu kiu'; also a loud 'tyip tyip tyip' alarm call when flushed.

BEHAVIOUR Found on sheltered coasts, but prefers inland pools and wetlands, where it wades in shallow water, picking out insects and shrimps or sifting through fine mud for small snails with a delicate action. Gregarious, but vagrants are likely to be solitary or in twos or threes. Rapid, agile flight, legs trailing. Erect, elegant stance. In alarm sometimes bobs head or flexes wings and neck.

RANGE Breeds from eastern Ukraine across Russia to eastern Siberia wintering sub-Saharan Africa, India, South-East Asia to Australia. Monotypic.

STATUS Vagrant to granitic islands, recorded mainly October to January but also May.

SIMILAR SPECIES The most likely confusion species is **Common Greenshank** with similar grey-and-white plumage but it is larger and less graceful with a long but fairly stout slightly upturned bill, a broader white wedge up the back and legs projecting only slightly in flight.

COMMON GREENSHANK *Tringa nebularia*   Plate 24

French: Chevalier aboyeur **Creole:** Kavalye Lapat Ver

DESCRIPTION Length 30-34 cm; wingspan 68-70 cm. A medium-sized wader, uniform grey above and white below, with a slightly upturned bill and greenish legs. **Adult non-breeding** appears uniform grey above, though seen closer, mantle and wing-coverts are finely scalloped and streaked to give a scaly effect. Crown, neck and sides of breast finely streaked grey, contrasting with clean white breast and belly. In flight, uniform grey wings (with darker primaries) contrast with broad, pure white gusset up lower back and rump and tail with grey bars and toes project only slightly. Bill grey, darker at tip, slightly uptilted and thicker than in Marsh Sandpiper. Legs long and greenish grey. Sexes similar, female averages larger. **Adult breeding** is darker above, with bold black and grey mottling on scapulars and heavy streaking on crown and breast. **Immature** is darker than adult non-breeding, the mantle and wing-coverts mottled dark with pale edges and upper breast and neck with darker streaking. May be indistinguishable from adult by November.

VOICE A loud, ringing, three-note flight call, 'tew-tew-tew'.

BEHAVIOUR Found by pools and wet marshes, in mangrove swamps, or on reefs, lagoons or intertidal sand and mudflats. Fairly gregarious, sometimes solitary. Direct flight with deep wingbeats, revealing long, pointed wings. Feeds along the water's edge, wading into deeper water to search for food items, which include invertebrates and small fish.

RANGE Breeds from Scotland and Scandinavia through northern Russia to the Kamchatka peninsula, wintering sub-Saharan Africa, Arabia, Persian Gulf, India, South-East Asia and Australia. Monotypic.

STATUS Annual migrant in small numbers throughout Seychelles mainly October to April, sometimes remaining year round.

SIMILAR SPECIES Marsh Sandpiper has a similar pattern of plumage, but is smaller and more delicate, with a much finer, straighter bill, a long, white supercilium giving a capped appearance, a long narrow wedge up the back almost to the nape and long, spindly legs that project conspicuously beyond the tail in flight. **Wood Sandpiper** is smaller still, with a short straight bill and browner, more speckled back.

GREEN SANDPIPER *Tringa ochropus*   Plate 24

French: Chevalier cul-blanc **Creole:** Kavalye Deryer Blan

DESCRIPTION Length 21-24 cm; wingspan 57-61 cm. A small dark sandpiper with a bold white rump and blackish underwing. **Adult non-breeding** shows uniform dark green-brown back and wings and lighter grey crown, face and upper breast contrasting sharply with clean white underparts. Cheeks and neck are grey with faint streaking. A dark line reaches from eye to bill and above this is a wide pale supercilium in front of the eye, absent or indistinct behind eye. In flight above, dark almost black wings contrast with square white rump, only a few wide black bars on tail and toes project only very slightly. In flight below, black underwing contrasts with white belly. Bill straight and dark with slaty blue base. Legs also dark, shorter than Wood Sandpiper. Sexes similar, female averages larger. **Adult breeding** has more white flecking on back and dark streaking on crown, neck and breast. **Immature** similar to adult non-breeding but with much finer buff speckling on the upperparts and noticeably on the tertials and bill and legs duller green. Indistinguishable from adult November onward.

VOICE Call a clear 'pueet-wit-wit' when disturbed.

BEHAVIOUR Like other sandpipers, has a characteristic bobbing action with tail and hindquarters. Because of its frequent solitary occurrence and tendency to feed around small pools, ditches or stream sides, often not seen until disturbed, when it zigzags, climbs steeply and then may fall back to ground further on. Occurs on tidal inlets and lagoons, but generally prefers inland pools, flooded meadows or ditches. Feeds on insects, small crustaceans and some plant matter, taken by pecking, rarely probing.
RANGE Breeds Sweden through Russia to eastern China. Winters in sub-Saharan Africa, Arabia, India and South-East Asia. Monotypic.
STATUS Vagrant to granitic islands, recorded October to January.
SIMILAR SPECIES Wood Sandpiper is browner with more prominent pale flecking on upperparts, lacks clear-cut breast-band, has longer, paler legs and pale supercilium prominent in front and behind eye. In flight, underwing is pale and legs project more beyond tail tip, which is narrowly barred. **Common Sandpiper** is smaller and browner with clear-cut breast-band and distinctive white wingbars.

WOOD SANDPIPER *Tringa glareola*   Plate 24

French: Chevalier sylvain   **Creole:** Kavalye Dibwa

DESCRIPTION Length 19-21 cm; wingspan 56-57 cm. A long-legged, elegant sandpiper with a speckled brown back, streaked breast and pale underwing. **Adult non-breeding** has grey-brown back finely speckled with light spots, neck and breast smudged grey with faint streaking and white underparts. Crown darker and streaked, noticeable pale supercilium extending behind the eye to the rear of the ear-coverts. Streaked and barred breast and flanks do not form a breast-band, merging to white belly. In flight, wings are broader at base than Green Sandpiper. From above, uniform grey-brown upperwing contrasts with white rump and narrow, grey tail-barring, the entire length of toes projecting beyond tail. From below, white belly and paler wing-coverts contrast with darker flight feathers. Bill straight and dark, legs dirty yellowish green. Sexes similar, female averages slightly larger. **Adult breeding** has back with bold grey and white mottling and distinct streaking on neck and breast. **Immature** has upperparts a warmer brown than adult due to yellow-buff edging to coverts. Legs often paler and yellower than in adult. Outer primaries fresher and darker than adult in early spring.
VOICE In flight a clear 'jiff-if-iff'.
BEHAVIOUR Prefers to feed along the margins of freshwater pools, marshes or stream sides and is only rarely seen on the shoreline. More social than Green Sandpiper, but still often solitary. Sometimes swims. When disturbed, climbs quickly into the sky giving its clear flight call. Pecks and probes, sometimes sweeping bill through water taking molluscs, crustaceans, worms, small insects and even small fish and frogs.
RANGE Breeds from Scandinavia to the Kamchatka peninsula. Winters in sub-Saharan Africa, India, South-East Asia and northwest Australia. Monotypic.
STATUS Annual migrant, occurring throughout the islands mainly October to March.
SIMILAR SPECIES Common Sandpiper is smaller and shorter-legged, with a white wingbar. **Green Sandpiper** has a more 'black-and-white' appearance, dark above with fine mottling, more clear cut breast-band below and supercilium is indistinct behind the eye. In flight, black underwing contrasts with a clean white belly and legs project only slightly beyond tail tip, which is broadly barred.

TEREK SANDPIPER *Xenus cinereus*   Plate 25

French: Chevalier bargette   **Creole:** Kavalye Trakase

DESCRIPTION Length 22-24 cm; wingspan 57-59 cm. A small mud-grey wader with a long upturned bill and short yellowish legs. **Adult non-breeding** is uniform muddy grey above with a darker bar at the shoulder. Grey head, neck and upper breast form a clear boundary with the clean white lower breast and belly. In flight, darker primaries contrast with white secondaries, recalling wing pattern of Common Redshank, but narrower and less sharply defined, rump and tail paler, smudged grey-brown. Bill seems inordinately long for body size; yellowish orange at base and noticeably upcurved. Legs relatively short, yellow or dull yellow-orange. Sexes similar, female averages slightly larger. **Adult breeding** has black centre stripes to scapulars, which form a dark shoulder-stripe, obvious in flight and dark carpal bar. Grey breast streaked blackish. **Immature** is paler grey above with finer black streaking and scaling on scapulars and wing-coverts and a buff wash to breast and back feathers. Legs may appear yellower.
VOICE Call 'tyu-du-du', softer than Common Redshank, or a rolling 'trururrut'.
BEHAVIOUR Found on coastal mud- or sandflats, lagoons or beaches, where it walks or runs in search of food, even wading into shallow water. Frequently accelerates from brisk walk to mad dash, chest held low and legs seemingly set well back. May bob tail in the manner of Common Sandpiper. Often solitary except at mudflats near Victoria where often in small numbers. Flight direct at low level over mudflats. Pecks and probes, sometimes chasing prey. Diet includes insects, molluscs, crustaceans and worms.
RANGE Breeds from Finland, throughout northern Russia almost as far east as the Bering Strait, wintering around the shores of Africa, Arabia, India, South-East Asia and Australia. Monotypic.

STATUS Annual migrant in small numbers, mainly to the granitic islands October to April, a few remaining year round.
SIMILAR SPECIES No other wader has this combination of small size, yellowish legs and long upturned bill.

Comparative sizes of waders 2: (1) Ruddy Turnstone, (2) Terek Sandpiper,
(3) Common Sandpiper, (4) Little Stint

COMMON SANDPIPER *Actitis hypoleucos*   Plate 25

French: Chevalier guignette   **Creole:** Kavalye Bat Lake

DESCRIPTION Length 19-21 cm; wingspan 38-41 cm. A short-legged sandpiper, marbled brown above and white below, with a sharp-edged breast-band and thin white wingbar. **Adult non-breeding is** uniform olive brown above (finely marbled at close range) with brown head and upper breast, thin black eye-stripe and white supercilium. Underparts white, forming a sharp boundary with brown upper breast with a clear white wedge up onto the shoulder. Long tail projects well beyond wingtips at rest. In flight, uniform upperparts contrast with darker wings with clear white wingbar. Bill as long as head, straight and dark. Legs are greenish grey and relatively short. Sexes similar, female averages larger. **Adult breeding** is virtually identical but with the breast tinged browner, lightly streaked brown on neck and breast, and upperparts slightly glossy greenish brown with fine dark brown shaft streaks. **Immature** has buff tips and fringes on feathers of upperparts giving more scaly, scalloped plumage than adult.
VOICE Often reveals presence with a shrill, high-pitched 'seep-seep-seep' call which carries long distances.

BEHAVIOUR Found wherever there is water, and though it prefers the margins of ponds, ditches and streams, regularly occurs along rocky shorelines, beaches, lagoons and mudflats. Usually solitary, with typical crouched stance, head pitched forward and tendency to bob tail and hindquarters rhythmically, more exaggerated and more often than other waders that share this habit. Unique low-level flight alternately fluttering wings and gliding, with a glide and usually an abrupt turn accompanied by a final wing-flutter immediately prior to landing. Walks briskly, clambering over rocks and other obstacles with ease and sometimes swims. Pecks and stabs to take small invertebrates, mostly insects, molluscs, crustaceans, amphibians and some plant material.

RANGE Breeds throughout Europe and east through Russia as far as northern Japan. Winters throughout sub-Saharan Africa, the Arabian coast, India, South-East Asia and Australia. Monotypic.

STATUS Annual migrant throughout Seychelles mainly October to March.

SIMILAR SPECIES Green Sandpiper is larger and appears very dark above, with black underwing; **Wood Sandpiper** is longer-legged with more freckled upperparts. Neither has the white shoulder wedge nor the white wingbar of Common Sandpiper.

GREY-TAILED TATTLER *Heteroscelus brevipes* Plate 25

Other names: Polynesian Tattler, Siberian Tattler
French: Chevalier de Sibérie   **Creole:** Kavalye Lapat Zonn

DESCRIPTION Length 24-27 cm; wingspan 60-65 cm. A uniformly ashy grey wader with dark grey upperparts, dark straight bill and yellowish legs. **Adult non-breeding** is reminiscent of Common Redshank in size and shape, but is uniform ashy grey on back, wings and head, with paler grey wash on the face, breast and flanks, toning to paler belly and vent. Distinct white supercilium contrasts with darker lore. In flight, tail, rump and wings are all uniformly ashy grey, underwing grey also, contrasting with paler breast and belly. Bill quite long and straight, dark with yellowish base and with a nasal groove extending halfway down its length. Legs dull yellow, occasionally greyish. Sexes similar, female averages slightly larger. **Adult breeding** is darker above, more distinct speckling on neck and breast, grey barring on flanks, legs brighter yellow. **Immature** similar to adult, but with white spotting on back and wing-coverts (usually replaced by December) and fine greyish barring on breast and flanks.

VOICE A two note, rising 'tu-wip' call; also a ringing 'tlee-tlee'.

BEHAVIOUR Outside the breeding season frequents coastal areas, occurring on shorelines as well as reefs, sandflats and lagoons. Vagrants are generally solitary. Rapid flight with flicking action, revealing long, unmarked, grey wings. Can appear surprisingly tame and may crouch at first rather than take off. Makes similar bobbing movements to Common Sandpiper. Feeds mainly by pecking, on crustaceans, molluscs, insects and sometimes small fish.

RANGE Breeds northern Siberia and further east from the Lena River to the Bering Strait, wintering in Philippines, Indonesia, Papua New Guinea, Australia and New Zealand. Monotypic.

STATUS Vagrant to Bird Island and the granitics, recorded remaining at Bird year round.

SIMILAR SPECIES Only other comparable wader with yellow legs is **Terek Sandpiper,** but this is much whiter beneath with slightly upcurved bill.

RUDDY TURNSTONE *Arenaria interpres* Plate 25

Other name: Turnstone   **French:** Tournepierre à collier   **Creole:** Bezroz

DESCRIPTION Length 21-25 cm; wingspan 50-57 cm. A stocky wader with short orange legs, rufous upperparts and black-and-white chest collar. **Adult non-breeding** has mottled rufous brown upperparts contrasting with clean white underparts. In all plumages the black chest-band with small white lapels is distinctive. Head is dull brown, scapulars pale orange, wing-coverts dark brown with underparts contrasting pure white. Strikingly patterned in flight, a series of white stripes shows up on dark back and white wingbar and white rump contrasting with black flight feathers and tail. Bill short, black and slightly upturned. Legs short and orange. **Adult male breeding** has streaked black crown, white sides to head, black facial mask and black-and-white neck-collars. Mantle and wing-coverts become vivid chestnut-orange. **Female breeding** is not as smart, being less orange above, (often absent on wing-coverts) less white on head, more streaked on crown, less cleanly cut breast-band. **Immature** duller brown throughout, dark brown upperparts with buff fringes to mantle and wing-coverts (disappearing by December).

VOICE Quiet trilling 'ku-tu-tut' flight calls.

BEHAVIOUR A bird of rocky shorelines and though sometimes seen singly on rocky coasts, it is more often encountered in small groups feeding on exposed reefs at low tide. Flight slow, with powerful deep wing-beats, lacking the grace of similar-sized waders. Probes in rock crevices for small crustaceans, or flicks over stones (hence the name) to locate crabs, worms or rockhoppers.

RANGE Breeds Arctic shores of Eurasia (race *interpres*) and North America (race *morinella*). Outside the breeding season, birds may be encountered on almost any shoreline in the world, American race *morinella* wintering from southern USA to South America, the nominate race *interpres* wintering from Europe,

Africa through to Australia.

STATUS Annual migrant in large numbers throughout Seychelles mainly October to March, first-year birds often remaining year round.

GENERAL Adults colour-ringed on Cousin have apparently migrated and been re-sighted in subsequent winters. This highlights the probability that some non-breeding visitors to Seychelles are not wind-blown vagrants but actually regularly pinpoint the islands on their migrations. One ringed on Cousin in November 1982 was recovered in Kazakhstan in August 1986 and again recovered on La Digue in November 1986. There have also been ringing recoveries from Cousin in Iran and Dagestan.

SIMILAR SPECIES None.

RED-NECKED PHALAROPE *Phalaropus lobatus*   Plate 25

Other name: Northern Phalarope   **French:** Phalarope à bec étroit   **Creole:** Falarop Labek Fen

DESCRIPTION Length 18-19 cm; wingspan 32-36 cm. A tiny, slender wader, grey and white with a black mask through the eye and with a black, needle-like bill. **Adult non-breeding** is ashy grey above, with a scalloped effect formed by white edges to the feathers. Underparts, neck and head white save for black strap through eye and cheek. In flight, wings are dark with a distinct white wingbar and tail grey with white edges to rump. Bill black, straight, very thin and tapered. Legs black. **Adult female breeding** is dark sooty grey above, with a buff stripe over scapulars, chestnut neck, white throat and a small white spot above eye. Underparts are pale grey, with darker, bold mottling on flanks. **Male breeding** is duller than female with less clear-cut pattern and often a narrow pale supercilium. **Immature** is dark brown above with buff edges to coverts giving a 'tortoiseshell' effect. Crown and hindneck dark brown, large dark eye-patch and pinkish buff neck and upper breast contrasting with white throat and belly. In flight, as adult, dark with a white wingbar.

VOICE A sharp 'kitt' or 'kirrit' call in flight.

BEHAVIOUR Phalaropes are unique among waders in preferring to swim in search of food, using their slightly lobed toes to help them paddle. Away from the breeding grounds they gather in their thousands to feed in the open ocean, but vagrants are likely to be solitary. Flight action over long distance similar to a small *Calidris*, though over short distances is more flitting and erratic. Has a buoyant, tail-up stance when floating similar to Common Moorhen, paddling and spinning hither and thither to pick tiny invertebrates from the surface of the water.

RANGE Breeds alongside tundra pools in the far north, with a circumpolar distribution from Iceland through Scandinavia and Siberia to Arctic Canada. European and Siberian birds winter in a small area of northwest Indian Ocean from the Gulf of Aden to the coast of Oman; the eastern Siberian breeding population winters in a small area of the Western Pacific around the Indonesian and Papuan islands. Monotypic.

STATUS Vagrant to granitic islands and seas north of Denis, recorded November.

SIMILAR SPECIES No other wader recorded in Seychelles has a black mask through the eye kinked down on the ear-coverts, nor has the preference to swim at all times rather than walk or wade. However, **Grey (or Red) Phalarope** *P. fulicarius* (L20-22 cm, WS40-44 cm) is a potential vagrant, breeding on Arctic coasts and wintering mainly off west South America, western and southwestern Africa; recorded as a vagrant to Kenya with old unconfirmed records from Seychelles. It is larger, paler, more uniform grey, with heavier, thicker bill.

SANDERLING *Calidris alba*   Plate 26

French: Bécasseau sanderling   **Creole:** Bekaso Blan

DESCRIPTION Length 20-21 cm; wingspan 40-45 cm. A small, tubby wader, silvery grey above and clean white below, which runs and feeds energetically along the beach just above the wave line. **Adult non-breeding** is pale silvery grey above with bright white face and underparts. Black carpal-patch on wing shows up as distinct black smudge on folded wing. In flight, shows broad white wingbar bordered with black. Bill short, straight and black, legs short and black. **Adult breeding** shows rust-red face and red upper breast with black spotting, streaked crown and nape and back mottled black and chestnut. Male breeding is generally brighter than female. **Immature** has black centres to scapulars forming chequered pattern on back. Pale buff wash to sides of breast and mantle. Crown and face with fine, black streaking, giving head a darker appearance than in non-breeding adult. Similar to adult non-breeding by December onward.

VOICE Flight call a loud 'twik, twik'.

BEHAVIOUR Prefers sand beaches, often in small groups, with a diagnostic habit of running only inches ahead of the lapping waves. Rapid flight, wings flickering to reveal striking wing pattern. Feeds on tiny invertebrates stranded on the shoreline, picking or probing in the wake of the receding tide wash, then nimbly scampering ahead of the next incoming wave. No other wader feeds so actively, running to and fro within reach of the tide.

RANGE Breeds on high-Arctic tundra, from northern Canada to Greenland, Spitsbergen and Russia.

Wintering birds can be found anywhere round the coasts of North and South America, Africa, Arabia, India, South-East Asia and northern Australia. Monotypic.

STATUS Annual migrant, recorded throughout the islands mainly October to March.

SIMILAR SPECIES Little Stint also has black bill and legs, but is smaller, more delicate with a finer bill, shorter legs and dark grey centres to scapulars in non-breeding plumage. **Red Knot** is larger, with longer bill, spotting on breast and flanks and greenish legs. **Curlew Sandpiper** is similar size but with longer, downcurved bill and buff wash on upper breast.

LITTLE STINT *Calidris minuta*   Plate 26

French: Bécasseau minute   **Creole:** Pti Pti Bekaso

DESCRIPTION Length 12-14 cm; wingspan 34-37 cm. A tiny, black-legged stint, immature birds with clear white 'braces' on the chestnut back and adult with dark grey back and white underparts. **Adult non-breeding** is dark grey above, the scapulars with dark central streaks, crown, cheeks and sides of breast grey and finely streaked. White on face, breast and belly give a generally paler appearance to all other small waders except for Sanderling. Short tail does not project beyond wingtips. Short, straight, black bill, fine at tip. In flight, shows white sides to rump, grey sides to tail and a long, narrow wingbar. Legs short and black. **Adult breeding** shows a tortoiseshell pattern above, more intense in male, with back mottled black, rufous and white. Face, nape and breast-sides rufous with dark, streaked crown. **Immature** with clear white 'braces' forming a distinct V on the chestnut-orange back; rust-coloured fringes to lesser coverts and tertials. Note that it appears 'red-necked' but the brighter lesser coverts and tertials distinguish from immature Red-necked Stint. First-winter has darker centres to scapulars than adult and buff fringes to wing-coverts. May be distinguishable from adult until around January.

VOICE A sharp 'tip' or 'tit' flight call.

BEHAVIOUR Feeds on open mudflats, sandflats, in tidal creeks or the edges of inland pools. Gregarious, though often solitary on Seychelles. Flight free and light with rapid wingbeats. A rapid feeding action, pecking from the surface rather than probing, though will occasionally wade into shallow water also. Diet mainly insects, small crustaceans and molluscs.

RANGE Breeds round the shores of the Arctic Sea, from northern Norway into Russia as far east as New Siberian Islands. Birds winter from Mediterranean, to sub-Saharan Africa, coasts of Arabia and India and along the river valleys of the Indus and the Ganges. Monotypic.

STATUS Annual migrant, most commonly in the granitic islands from October to April.

SIMILAR SPECIES Red-necked Stint actually appears greyer than immature Little Stint, having pale grey lesser coverts and tertials. At all times Red-necked is longer-winged, shorter-legged and bill is slightly thickened (not fine) at tip, but non-breeding birds may only be separable on measurements. Wing/tarsus ratio is 5.0 or more in Red-necked, 5.1 or less in Little Stint[1]. **Sanderling** is larger, tubbier and altogether cleaner, silver-grey above and white below.

Reference: 1. Hayman *et al.* (1986).

RED-NECKED STINT *Calidris ruficollis*   Plate 26

French: Bécasseau à col roux   **Creole:** Bekaso Kolye Rouz

DESCRIPTION Length 13-16 cm; wingspan 35-38 cm. A very small, tubby, black-legged stint, with short black legs and rufous scapulars contrasting with grey lesser coverts and tertials. **Adult non-breeding** is light grey above with dusky tertials and clean white underparts. Head with dark streaked crown and a dark line through lore and eye, accentuating the paler supercilium. Bill short and black, slightly thickened at tip, legs also very short and black. **Adult breeding** has face and upper breast rufous, scapulars mottled black, rufous and white, lesser coverts grey, mantle and upper breast-sides with dark flecking on a white background (unlike Little Stint). Sexes similar. **Immature** may be distinguishable until around November, appearing mostly grey above and paler below, with a faint rufous wash on crown and mantle and rufous fringes to black-centred scapulars contrasting with pale grey wing-coverts and tertials. May show a faint V on the mantle, but this is never distinct. Breast-sides are smudged grey with a hint of warmer buff, but never 'red'.

VOICE Flight call a coarse 'kreep' or 'keeep'.

BEHAVIOUR Frequents tidal flats, sand bars and the edges of small pools. Vagrants likely to be solitary. Does not tower when flushed, flight rapid and free. Almost hovers prior to landing. Moves busily over wet ground, pecking rapidly at the surface to find food. Diet includes worms, crustaceans, seeds and molluscs.

RANGE Breeds in Arctic from central Russia east to the Bering Strait and Alaska. Winters round shores of Bay of Bengal, through Philippines, South-East Asia and the Pacific islands to Australia and New Zealand. Recorded as a vagrant on the coasts of Somalia, Kenya and Mozambique. Monotypic.

STATUS Uncertain; there are several unconfirmed records.

SIMILAR SPECIES Little Stint is the most likely confusion species, which appears much more rufous in the immature, with distinct pale 'braces', compared to the two-tone back of the Red-necked, in which rufous scapulars contrast with grey wing-coverts. Little also has slightly finer bill and different flight call.

TEMMINCK'S STINT *Calidris temminckii*   **Plate 26**

French: Bécasseau de Temminck   **Creole:** Bekaso Temminck

DESCRIPTION Length 13-15 cm; wingspan 34-37 cm. A very small stint with very short yellowish legs, short, slightly downcurved bill, dull brown back and tail just longer than wingtips. **Adult non-breeding** appears uniformly dull grey-brown above and white beneath, with very short, yellowish grey legs and short, dark bill slightly downcurved at tip. Wings when folded, just shorter than long tail. In flight, short, narrow white wingbar not extending onto primaries and bright white outer tail feathers show up clearly (diagnostic). **Adult breeding** has rufous tints to crown and mantle and distinctive black-centred scapulars with broad grey tips. **Immature** is distinguishable until around November, showing slightly more buff-grey on upperparts, the back scaly with fine black crescents on upper scapulars and warm buff scalloping on lower wing-coverts. Bill blackish, paler at the base. Legs yellowish and appear too short for the slightly elongate body, giving distinctive 'sawn-off' silhouette.
VOICE A trilling 'tirrrit' given in flight.
BEHAVIOUR Prefers to feed on muddy ground in sheltered inlets, wet fields or small wetlands, less in open areas. Vagrants are generally solitary. When flushed, has a tendency to climb swiftly, unlike low flight of other stints, jinking from side to side as it rises. Picks beetles, small flies or tiny worms from surface mud or vegetation, rarely if ever probing.
RANGE Breeds round the shores of the Arctic from Norway to the Bering Strait. Winters Mediterranean, Nile valley, sub-Saharan Africa and coasts of Arabia, India and South-East Asia to Taiwan. Monotypic.
STATUS Vagrant to granitic islands and Bird, recorded September to December.
SIMILAR SPECIES In flight, other stints have greyish not white outer tail feathers. **Long-toed Stint** also has yellowish legs and pale 'braces' on mantle, but has longer legs, longer neck, more upright posture and darker, more mottled plumage. **Little Stint** is similar in size but with black legs, shorter tail and longer wingbar. Plumage pattern and shape can recall **Common Sandpiper**, but latter is larger

LONG-TOED STINT *Calidris subminuta*   **Plate 26**

French: Bécasseau à longs droigts   **Creole:** Bekaso Ledwa Long

DESCRIPTION Length 12-13 cm; wingspan 33-35 cm. A small, well-patterned, long-necked, long-legged stint with pale yellowish legs, 'braces' and a slightly downcurved bill. **Adult non-breeding** is dark grey above, with dark centres to scapulars and wing-coverts. Broad pale supercilium, often split over the eye, contrasts with dark, streaked crown and cheeks, sometimes appearing rufous-capped. Pale grey streaking on upper breast and underparts white. Downcurved dark bill with obvious paler base to lower mandible. Long tibia, legs pale yellow with noticeably longer central toe and all toes, including hind toe longer than in other stints, toes projecting beyond tail in flight. **Adult breeding** has rufous tints to crown, cheeks and upper breast, bright rufous edges and black centres to scapulars and tertials. Sexes similar. **Immature** is most likely plumage to encounter, clean white underparts contrasting with bright rufous mottled back. Clean white V on edge of mantle, scapulars and tertials dark-centred with bright rufous edging. Crown streaked rufous, breast-sides buff with fine dark streaking. Bill dark, paler at base and slightly decurved. Legs yellowish or yellowish orange.
VOICE Flight call is a soft 'chirrrup', reminiscent of a House Sparrow's chirp.
BEHAVIOUR Feeds on the edge of pools or wet ground, like Temminck's Stint, preferring sheltered fresh-water sites to open tidal flats. Vagrants are generally solitary. Quite different posture from other stints; the longer neck, shorter tail and tendency to 'sit up' give it a more forward, upright stance rather like a tiny Ruff. When flushed, tends to 'tower' like Temminck's Stint, gaining height rapidly. Diet includes insects, crustaceans and molluscs.
RANGE Breeds in scattered areas from the Ural Mountains east to the Bering Sea. Winters eastern India, South-East Asia, Philippines, Indonesia to western Australia. Monotypic.
STATUS Vagrant to granitic islands, recorded November.
SIMILAR SPECIES The white 'braces' on the immature bird also occur on immature **Little Stint**, but this has black legs and a shorter bill. **Temminck's Stint** is the only other possible confusion species with pale legs, but this is duller, with buff fringing to scapulars, never the bright rufous wash extending from crown to tertials.

PECTORAL SANDPIPER *Calidris melanotos*   **Plate 27**

French: Bécasseau à poitrine cendrée   **Creole:** Bekaso Pwatrin Sann

DESCRIPTION Length 19-23 cm; wingspan 42-49 cm. A drab, grey-brown sandpiper with a slightly downcurved bill, dark yellowish legs and a distinctive dark chest bordering a white belly. **Adult non-breeding** has grey-brown upperparts, mantle and wing-coverts having uniform dark centres. The crown is streaked brown, contrasting with a paler streaked supercilium. Face, hindneck and whole of upper breast with dark, regular streaking to give a fine 'corduroy' effect. This pattern ends abruptly on the lower breast, forming a diagnostic border against the uniform white belly. Shows a long primary projection. In flight, has a pale, indistinct wingbar but rump has black centre with white edges. Bill is mainly dark, pale at base and slightly downcurved at tip. Legs are dark yellowish grey. Non-breeding sexes similar, but male always larger. **Adult male breeding** has black centres and narrow rufous fringes to scapulars and distinct dark grey-brown breast and neck with white spotting. **Female breeding** has a yellowish wash to breast and some streaking on flanks. Mantle, scapulars and tertials dark brown with paler fringes. **Immature** has rufous buff fringes to mantle, scapulars and tertials retained to November give a warmer brown coloration, feathers black in centre. Prominent white V-shapes on mantle and scapulars. Creamy supercilium with fine streaking.

VOICE Flight call a harsh 'chrrrt'.

BEHAVIOUR Occurs with other waders on coastal mudflats and intertidal areas, but also has a tendency to feed around grassy pools, in wet grassland or on mown grass areas. Erratic snipe-like escape flight. Relaxed feeding action, taking insects, seeds and crustaceans.

RANGE Breeds in coastal tundra in the far north, from Siberia through Alaska to Canada, wintering South America and Pacific as far south as New Zealand. Monotypic.

STATUS Vagrant to granitic islands and Bird, recorded October.

SIMILAR SPECIES Sharp-tailed Sandpiper is a possible confusion species, but this has shorter bill, more distinct white supercilium flared behind eye outlining brighter rusty cap and warm ochre flush to the lower chest, streaking continuing onto flanks and undertail-coverts.

SHARP-TAILED SANDPIPER *Calidris acuminata*   Plate 27

Other name: Siberian Pectoral Sandpiper
French: Bécasseau à queue pointue   **Creole:** Bekaso Lake Pwent

DESCRIPTION Length 17-20 cm; wingspan 42-48 cm. A brownish sandpiper with a prominent supercilium, distinct brown 'cap', rather short, slightly decurved bill, and no sharp demarcation on lower breast. **Adult non-breeding** appears drab grey-brown above with dark centres to feathers of back and wing-coverts. The only distinctive feature is a dark chestnut brown 'cap' more noticeable because the prominent white supercilium contrasts with the darker crown, particularly behind the eye. White eye-ring is more prominent than in Pectoral Sandpiper. Upper breast is greyish, fading to a whitish lower breast and belly (not sharply divided) with small markings extending as streaks or chevrons (arrowhead shapes) onto flanks and undertail-coverts, which are unmarked or only very finely streaked in Pectoral Sandpiper. No distinctive pattern in flight save white edgings to the rump and slightly more obvious wingbar than Pectoral Sandpiper. Bill dark grey. Legs dirty greyish green. **Adult breeding** has less dense pectoral markings which are less vertically aligned compared to Pectoral Sandpiper, but these extend as spots or chevrons to flanks and undertail. Sexes similar. **Immature** is much brighter than the adult, with chestnut fringes to the mantle, scapulars and tertials, bright chestnut cap and very distinct white supercilium. Warm ochre-buff flush to the breast, streaking confined to upper breast and sides of lower breast. Grey-olive to yellowish legs. May remain distinguishable from adult until December.

VOICE Twittering flight call 'trrt-wheep-wheep' recalling Barn Swallow.

BEHAVIOUR Will feed on open mud and sandflats often with other waders, but prefers brackish lagoons or small freshwater pools with some short vegetation for cover. Vagrants are generally solitary. Less upright stance than Pectoral Sandpiper, but similar flight action. Pecks and stabs, diet includes insects, snails, crustaceans and worms.

RANGE Breeds northern Siberia in exposed tundra, wintering mainly in the Pacific from New Guinea to New Zealand. Monotypic.

STATUS Vagrant to granitic islands, recorded October to February and July.

SIMILAR SPECIES Pectoral Sandpiper has a distinctive chest pattern and less distinct cap and supercilium. **Curlew Sandpiper** is even longer-billed with clean white rump and blackish legs.

CURLEW SANDPIPER *Calidris ferruginea*   Plate 27

French: Bécasseau cocorli   **Creole:** Bekaso Korbizo

DESCRIPTION Length 18-23 cm; wingspan 42-46 cm. A largish, elegant sandpiper with distinctive white rump, longish black legs and very long, downcurved black bill. **Adult non-breeding** appears uniform pale grey above and clean white below. Head and face finely streaked, upper breast white with very fine, sparse streaking. Shows a prominent white supercilium. In flight, bold white wingbar shows up against black flight feathers and broad white rump contrasts with black tail. Bill black, long and downcurved, tapered from a broad base. Legs quite long and black. **Adult male breeding** is vivid chestnut-orange from face through to lower belly, back spangled black, white and chestnut with grey wing-coverts. **Female breeding** is paler on underparts, many feathers flecked white and has dark brown bars on the belly. **Immature** has plumage darker brown than adult non-breeding with a scaly pattern over the back and wing-coverts, retained throughout first winter, more streaking over head and upper breast and buff flush to crown, back and particularly upper breast, giving individuals a warmer, slightly tanned look.

VOICE Flight call a quiet, trilling 'tirrilip'.

BEHAVIOUR May feed in either sandy or muddy habitats around the coast, usually in small flocks. Rapid, agile flight. At high tide small feeding flocks regularly fly inland to feed on short grass areas. Birds jab and probe like starlings for food (insects, shrimps and small worms) and will wade into shallow water to feed also.

RANGE Breeds in Arctic Siberia, wintering mainly on coasts of sub-Saharan Africa, Arabia, India, South-East Asia and Australia. Monotypic.

STATUS Annual migrant in good numbers throughout the islands mainly October to March, a few remaining year round.

SIMILAR SPECIES The only other small waders with a distinctive square white rump are **Green Sandpiper** and **Wood Sandpiper**, which have uniform upperparts, no obvious wingbar, shorter, straight bills and paler legs. **Red Knot** is a similar grey-and-white colour, but plumper, with shorter, straight bill and light barring onto flanks. **Broad-billed Sandpiper** is smaller, thicker-billed, shorter-legged, with a strong crown pattern. **Dunlin** *C. alpina* (L17-21 cm, WS32-36 cm) is a potential vagrant, breeding on Arctic shores and wintering southward, sometimes reaching Kenya. It is similar in non-breeding plumage, more dumpy and has shorter tarsus; darker grey on breast with less conspicuous supercilium and longish black bill downcurved only at tip. In flight, lacks white on lower rump and uppertail-coverts. Immature is buffish, with some rufous on crown and back and dark blotches on flanks.

BROAD-BILLED SANDPIPER *Limicola falcinellus* Plate 27

French: Bécasseau falcinelle   **Creole:** Bekaso Labek Laz

DESCRIPTION Length 16-17 cm; wingspan 37-39 cm. A small sandpiper with a distinct 'kink' at the tip of the curved bill, adult grey and white and with snipe-like immature plumage. **Adult non-breeding** is mostly grey above, with darker shafts to back feathers and white fringes to wing-coverts, giving a scaly effect. Sometimes shows a noticeable dark carpal-patch at the shoulder. Head, neck and upper breast-sides grey, finely-streaked. Head with noticeable paler crown-stripe, broad white supercilium and darker eye-stripe. In flight, dark leading edge contrasts with uniform grey back and upperwing and white wing-bar. Bill shape diagnostic: broad and straight for most of length, with slight downward kink at tip, dark with brownish base. Legs dirty greyish yellow-green. **Adult breeding** is very dark above, with heavy black centres to back feathers, boldly streaked neck and upper breast and strong black crown-stripes. Sexes similar. **Immature** plumage may be retained until January; it is reminiscent of a snipe with very dark centres to the back and wing-coverts, chestnut fringes to mantle, white fringes to wing-coverts and distinct tram-lines over back. Black crown with paler crown-stripes and broad, pale split supercilium contrasting with darker streaked head. Upper breast spotted rather than streaked. Belly and flanks white and unstreaked.

VOICE Flight call a buzzing trill 'chreeet'.

BEHAVIOUR In wintering grounds favours soft, muddy tidal flats. Vagrants are generally solitary, but associates with other waders. Low short, escape flight; over distance, recalls Curlew Sandpiper but with shallower wingbeats. Feeds mainly on worms, with insects, crustaceans and seeds also taken, picking from side to side, sometimes probing, sometimes making short runs.

RANGE Nominate race *falcinellus* breeds in northern Scandinavia and northern Russia, wintering in East Africa, southern Red Sea and along the coasts of Iran and India. East Siberian race *sibirica* winters South-East Asia to Australia; it has a broader lower supercilium and narrower upper supercilium and brighter rufous fringes to feathers of upperparts.

STATUS Vagrant to granitic islands, recorded October.

SIMILAR SPECIES There are no other small waders with a crown pattern like this, but **Little Stint** immature also has tramlines over back, but more rufous back, shorter bill and black legs. **Long-toed Stint** immature also has 'V' on back, but this too has a short, fine-tipped bill.

BUFF-BREASTED SANDPIPER *Tryngites subruficollis* Plate 27

French: Bécasseau roussâtre   **Creole:** Bekaso Pwatrin Rouye

DESCRIPTION Length 18-20 cm; wingspan 43-47 cm. A distinctive wader with small pigeon-like head, short straight bill, warm plain buff plumage and yellowish legs. **Adult non-breeding** has diagnostic unstreaked buff face, prominent dark eye surrounded by a pale eye-ring, head pigeon-shaped and fine streaks on the crown. Upperparts scaly with fine spotting from hindneck to mantle, scapulars with dark brown centres broadly fringed in buff and less well-marked wing-coverts. Underparts sandy buff unmarked except for neat black spotting at side of breast. In flight, upperparts and rump uniform scaly brown, underwing white with a blackish crescent on the primary coverts, no obvious wingbar in upperwing. At close range, note feathering on lower mandible extends further than on upper (unlike in Ruff). Bill short, slender, straight and all-black, legs yellowish. **Adult breeding** similar to non-breeding plumage. Sexes similar, but male always larger. **Immature** similar to adult, but with darker centres to wing-coverts and whitish

fringes, giving an even more scaly appearance. This plumage may be retained throughout first winter.
VOICE Normally silent; sometimes a low 'krrrt' flight call.
BEHAVIOUR Rarely seen by water. Most often feeds in short grass areas including airstrips where it can be approached closely. Gregarious, though vagrants are likely to be solitary. Escape flight erratic and snipe-like, legs sometimes trailing. Flight over longer distances freer and more plover-like. Moves with a high-stepping gait, picking at insects and sometimes seeds in the grass or on the ground.
RANGE Breeds on the Arctic shores of Alaska, western Canada and a few sites in eastern Siberia. Birds winter in Argentina, but occur as vagrants to Europe, North Africa, Egypt and Kenya. Monotypic.
STATUS Vagrant to Bird and Platte, recorded October to November and March to April.
SIMILAR SPECIES Most likely to be confused with **Ruff**, a more frequent vagrant. Ruff is larger, with longer legs and bill and in juvenile plumage has more rufous fringing to the back and wing-coverts, more rufous crown and faint streaking on the face and neck.

RUFF *Philomachus pugnax*　　　　　　　　　　　　　　　　**Plate 27**

Other names: Reeve (female only)　**French:** Combattant varié　**Creole:** Bekaso Gro Vant

DESCRIPTION Length 26-30 cm (male), 20-25 cm (female); wingspan 54-58 cm (male), 48-52 cm (female)
A medium-sized wader with distinctive 'scaly' back pattern, yellowish legs, white tail sides and a straight bill slightly decurved at the tip. **Adult non-breeding** is grey-brown above with solid, dark feather centres, giving a distinctive scaly appearance. Distinctive silhouette at all times, longish neck, small head and slightly pot-bellied outline. Whitish underneath with grey mottling on breast, pale throat and pale sides to face. In flight, no distinctive pattern save for white sides to tail. Bill dark with paler base, mostly straight, but noticeably downcurved towards the tip. Legs vary from dirty yellowish green to dull orange. Sexes similar in winter plumage but female considerably smaller than male. **Adult male breeding** has elaborate ruff of head and throat plumes, black, white, rufous or barred. Legs brighter orange-red. **Female breeding** with bold dark chevrons on back, breast and flanks. **Immature** is brighter than adult, with very dark centres and rufous fringes to back and wing-coverts, giving scaly 'tortoiseshell' effect. Crown rufous and finely streaked, neck and breast with warm sandy buff flush. Bill dark, legs dull yellowish green. May be inseparable from adult from December onward.
VOICE Usually silent. Sometimes gives a quiet 'ker' or 'kuk'.
BEHAVIOUR In wintering areas prefers brackish sites to tidal shores, though it does occur on sand or mudflats. A tendency to frequent wet or dry grassland sites, particularly for roosting. Highly gregarious but vagrants are generally solitary or in pairs. Lazy, sometimes erratic flight, frequently gliding including just prior to landing. Upright stance, head raised, tail lowered. Probes in mud or picks at surface, taking insects, seeds and aquatic plants.
RANGE Breeds at scattered sites through Europe, mainly in Scandinavia and western Russia and as far east as Siberia. Outside the breeding season, birds move south into Mediterranean, sub-Saharan Africa and to a lesser extent the Persian Gulf and the coasts of India. Monotypic.
STATUS Vagrant to granitic islands and Bird, recorded August to November.
SIMILAR SPECIES There are few waders of this size with dirty yellowish legs and warm buff breast: only **Buff-breasted Sandpiper** is at all similar. This has similar scaly upperparts and same warm sandy buff flush to breast, but a uniform pale buff wash over face without any hint of eye-stripe or streaking, giving an open, honest look. Rather pigeon-like head shape and shorter, straight bill also distinct. At close quarters (and this species allows close approach, unlike Ruff) note also that the feathering on the lower mandible extends noticeably further onto the bill than that on the upper mandible. Immature **Sharp-tailed Sandpiper** has similar leg and breast colour, but chestnut tints to upperparts and combination of chestnut crown, white supercilium and dark cheek give a quite different appearance

SOUTH POLAR SKUA *Catharacta maccormicki*　　　　　　　　**Plate 28**

Other name: McCormick's Skua　**French:** Labbe de McCormick　**Creole:** Asasen McCormick

DESCRIPTION Length 51-54 cm; wingspan 125-135 cm. A medium-sized, heavily built polymorphic skua. Most show some contrast between dark upperparts and paler underparts, are greyer and colder than Antarctic Skua, with black bare parts and usually a (diagnostic) pale nape. **Adult** is structurally similar to Antarctic Skua, but a little smaller and slimmer, with flatter forehead, more rounded head, slightly shorter tail (wingtips project beyond tail, unlike Antarctic) and slimmer, more pointed wings (but differences are very subtle). Paler area at base of bill, especially the upper mandible may stand out against a dark mask or hood. Side of neck sometimes streaked yellowish (not as pronounced as Antarctic), but usually less mottled more uniform above than Antarctic. Cold dark brown tail is very short. Base and shafts of primaries white, forming a wing-flash prominent on both underwing and upperwing (particularly the latter). Rapid primary moult may create large gaps in the wing (involving up to four primaries at one time, compared to only 1-2 in Antarctic Skua). Eye dark brown. Legs black. Bill black, less hooked and less powerful than

Antarctic. Sexes similar. **Pale morph** is readily identifiable by sharp contrast between pale buff-grey head/underparts and cold dark brown upperparts with blackish underwing-coverts; sometimes underparts are faintly streaked or mottled. **Dark morph** (much rarer) is uniform dark grey-brown, colder than Antarctic, lacks any rufous tone (though sometimes with a pinkish cast) and usually has a paler nape which sometimes extends onto the mantle. **Intermediate morphs**, common between the two extremes, show some degree of contrast between paler underparts/head and buff-brown upperparts. **Adult breeding** has yellow hackles on pale nape unlike Antarctic Skua. **Juvenile** is distinctly greyer than adult, with no brown tones and showing much less variation. Head/underparts vary from pale to mid-grey, upperparts cold dark grey with finer paler grey feather tips. Smaller white primary flash than adult. Legs and feet bluish. Bill two-toned: blackish or pale blue-grey with a black tip, but may become all-black by October. By **second-year**, similar to adult. Accurate ageing at sea difficult.

VOICE Normally silent away from breeding grounds.

BEHAVIOUR Pelagic and solitary or in small parties outside the breeding season. Flight low, direct and more agile than Antarctic Skua. Piratical, harrying seabirds up to the size of boobies with short, direct attacks but less prone to kleptoparasitism than Antarctic Skua, feeding mainly by plunge-diving for fish. May follow fishing vessels and may take carrion.

RANGE Breeds mainly on Antarctic coast. Disperses northward February to April as far as Alaska and Greenland. Some appear to migrate clockwise in the Pacific and possibly the Atlantic. Indian Ocean records are poorly documented. May occur regularly in the Red Sea, Arabian Sea and Persian Gulf, which might suggest it is also regular in Seychelles waters[1,2]. Adult probably has more limited movements than immature.

STATUS Vagrant; recorded granitic islands around July but might occur throughout Seychelles.

SIMILAR SPECIES Antarctic Skua appears, larger and fiercer, has uniformly dark plumage and heavier bill without pale 'nose-band'. **Pomarine Skua** is less bulky, less hunched, has more angular wings and less white in wings, especially in upperwing. Immature Pomarines are usually separated by barred rump, axillaries and underwing-coverts (except in the darkest individuals).

References: 1. Hollom *et al.* (1988), 2. Shirihai (1996).

ANTARCTIC SKUA *Catharacta antarctica*   **Plate 28**

Other names: Brown (or Southern, or Southern Great) Skua
French: Labbe antarctique   **Creole:** Asasen Antarktik

DESCRIPTION Length 61-66 cm; wingspan 150-155 cm. The largest skua, with heavy build, broad wings and a malevolent demeanour. Lacks contrast between warm, dark brown, yellowish-streaked upperparts and underparts. **Adult** is generally dark brown, often with rufous tones (unlike South Polar Skua). Some have a slightly paler head and underparts. Upperwing and tail blackish brown, underwing dark brown, white bases and shafts of primaries forming a prominent wing-flash. Race *lonnbergi* (most likely race in Seychelles), has distinct pale edges to feathers of mantle and scapulars; also blackish brown underwing-coverts contrast with paler body. Heavily marked with pale streaks and mottling above, worn birds showing a pale-blotched mantle and scapulars, contrasting with darker head and wings. Large, angular head lacks a contrasting dark cap. Rump and tail short and broad (unlike Pomarine Skua) but not as short as South Polar, wingtips being about equal to tail length (wingtips project in South Polar). Eye light brown; bill black (sometimes with paler base), large and hooked; legs black. Sexes similar; no seasonal variation. **Juvenile** similar to adult, but more uniform (lacking streaking), with darker, warmer brown upperparts more cinnamon underparts, less distinct wing-flashes (sometimes absent) and less streaking. Legs dark grey, usually with pale spots on the tarsus. Bill dark grey, sometimes tipped black. **Second-year** and subsequent years prior to adult plumage are similar but darker, more uniform.

VOICE Normally silent away from breeding grounds.

BEHAVIOUR Pelagic and normally solitary in Seychelles waters. Flight direct and low over ocean surface. Highly predatory, feeding mainly on other seabirds including terns and shearwaters; seen to take Fairy Tern and Lesser Noddy in Seychelles. May follow fishing vessels and may take carrion.

RANGE Taxonomy is complex and somewhat controversial. Three races are recognised by most authorities: *antarctica* (Falkland Skua; Falklands, southern Argentina), *hamiltoni* (Tristan Skua; Gough, Tristan da Cunha) and by far the most widespread and most likely to occur in Seychelles waters *lonnbergi* (Sub-Antarctic Skua; circumpolar including Antarctic Peninsula, sub-Antarctic islands). Differences are slight, but *lonnbergi* is the largest race, with more powerful bill than other races, sometimes considered a separate species. Breeds Southern Ocean islands to Antarctic Peninsula; fledging and dispersal around February, some as far as tropical latitudes. In the Indian Ocean has been recorded as far north as India and Oman.

STATUS Annual throughout Seychelles mainly May to September.

SIMILAR SPECIES South Polar Skua has variable plumage but almost all show some contrast of paler head, nape and underparts with dark upperparts/underwing and are colder, with greyish tinge to head and body. Also, less fierce in appearance with less powerful bill, smaller, more rounded head and may show a pale 'nose-band'. **Pomarine Skua** is smaller with proportionately longer tail and less obvious wing-flashes.

Great Skua *C. skua*, sometimes claimed from Seychelles, generally has a capped appearance but is otherwise similar. However it is unlikely to reach the Indian Ocean and the timing of most reports (the austral winter) would suggest sightings are probably Antarctic Skuas.

POMARINE SKUA *Stercorarius pomarinus*   Plate 28

Other names: Pomarine Jaeger, Twist-tailed Skua   **French:** Labbe pomarin   **Creole:** Asasen Pomarin

DESCRIPTION Length 46-51 cm (including tail up to 19 cm); wingspan 125-135 cm. A large, powerful, polymorphic skua with broad rounded head and a heavy, hooked, two-tone bill. Adult has diagnostic spoon-like twisted central tail feathers, though less obvious, shorter or even absent in non-breeding plumage. **Pale morph** (over 90% of total population) is generally darker than Arctic Skua, has a more shaggy appearance to vent, more obvious breast-band, larger wing-patches (most extensive on underwing) and more barring on flanks. Tawny brown upperparts, diffuse brown cap (all-black in breeding plumage), whitish chin, cheeks, throat and hindneck. Underparts are usually white with a dusky vent (sometimes entirely white), but are highly variable. Shafts and bases of outer primaries white, most extensive on underwing, forming a prominent crescent; smaller pale patch in front of this is visible at close range on the underwing. Underwing-coverts are paler than the body (generally opposite in South Polar). Tail blackish, paler below towards the base and on outer feathers. May show some indistinct barring on uppertail- and undertail-coverts. **Dark morph** is mainly dark tawny brown, cap and flight feathers usually darker with yellowish or buff cheeks and hindcollar. All morphs have a dark brown eye; dark grey legs and long, deep bill which is usually grey, darker at the tip with a prominent gonydeal angle. Sexes similar. **Immature** is grey-brown or black-brown above, with a pale hindcollar following moult in winter quarters and no head streaks (unlike Arctic Skua). Shows buff tips to feathers on back and upperwing with underwing more barred than adult. Most easily identified by double white wing-patches on underwing (the inner one crescent-shaped). Breast and underparts are evenly barred, strong and pale on the uppertail- and undertail-coverts, lower belly, vent and sometimes, flanks. Tail projections are short and rounded or even absent. Legs pale grey with whitish or bluish tinge, tibia and most of foot black. Bill pale grey with black tip, more obviously two-tone than adult. By **second winter** resembles adult, but with paler barring on axillaries and underwing-coverts and less contrasting cap.
VOICE Normally silent away from breeding grounds. Alarm call is 'yek-yek'.
BEHAVIOUR Normally pelagic and solitary outside breeding season, though small flocks may follow ships or migrate together. Strong, direct flight, more laboured than Arctic Skua, with slow, gull-like wingbeats, putting on bursts of speed when chasing gulls and terns. Piratical, living largely by this means outside breeding season. May kill smaller seabirds, often attacking petrels in winter quarters.
RANGE Breeds in Arctic tundra, wintering mainly north of Equator but small numbers regular off South Africa. Present year round off Arabian coast, more common in winter migrating across Eurasia. Also recorded Mozambique Channel north to East Africa, mainly October to March. Monotypic.
STATUS Uncertain; probably a vagrant, but reports to date have failed to rule out confusion species.
SIMILAR SPECIES **Arctic Skua** and **Long-tailed Skua** are less heavily built (particularly Long-tailed) with pointed central tail feathers (when present) and less white in wings. In absence of tail-streamers, size, broad-based wings, barrel chest, thick neck and large head of Pomarine rule out Arctic or Long-tailed Skua. **Antarctic Skua** is larger with broader wings and a proportionately shorter tail. **South Polar Skua** is bulkier, more hunched, has less angular wings and more white in wings, especially the upperwing, and never shows a paler patch on uppertail-coverts unlike juvenile Pomarine.

ARCTIC SKUA *Stercorarius parasiticus*   Plate 28

Other names: Parasitic Jaeger, Richardson's Skua   **French:** Labbe parasite   **Creole:** Asasen Arktik

DESCRIPTION Length 41-46 cm (including tail up to 18 cm); wingspan 110-120 cm. A small, elegant, dark-winged, polymorphic skua, with short, pointed central tail feathers, intermediate in build between Long-tailed Skua and Pomarine Skua. Flat crown and slender bill give triangular shape to head. Underside rounded; wings long and slender and tail fairly long, length accentuated by pointed central tail feathers. **Dark morph** is entirely tawny brown or grey-brown with uniform dark hindneck and neck in winter, slightly darker cap and flight feathers and 3-4 white primary shafts on the upperwing (usually only two in Long-tailed Skua, with which confusion is most likely). White wing-panel may be indistinct above but obvious below. Underwing greyish, upperwing often showing contrast between dark grey-brown coverts and blackish flight feathers. Brown upperparts show little if any contrast (unlike Long-tailed) between secondaries and wing-coverts. Tail blackish brown, sometimes with a black terminal band. **Pale morph** has diffuse, brown cap, indistinct in non-breeding plumage, with pale yellow or whitish cheek, more dusky in non-breeding birds. Underparts whitish (but highly variable), vent dusky or white and a brown breast-band (variable in size, sometimes incomplete). **Intermediate morphs** occur, highly variable between light and dark morphs. Non-breeding birds mantle, back and scapulars edged white or buff, uppertail- and undertail-coverts barred brown and white. Pointed central tail feathers are shorter and less obvious in winter when may also show indistinct barring on uppertail- and undertail-coverts, duller cap and scruffier under-parts. All morphs have eye brown, cere whitish, and bill and legs blackish. Sexes similar. **Juvenile** plumage, retained until February, is variable (more so than Pomarine Skua) from black to brown to grey-ish, barred brown, with warmer more rusty ground colour than Pomarine. Buff edges to tips of closed pri-maries (lacking in Long-tailed), primary shafts and bases as adult; generally distinguishable by blue-grey or pink-grey bill with black tip and barred underwing-coverts and axillaries. Usually has pale hindneck (unlike Pomarine) and some head streaking. Gradually loses pale feather fringes and rusty tinge. **Subadult** also has barred underwing and features intermediate between adult and immature.

VOICE Normally silent away from breeding grounds.

BEHAVIOUR Pelagic, forming flocks on usual migration routes, though vagrants are generally solitary. Piratical and aerial with falcon-like dash and glide, low-level flight, attacking seabirds, forcing them to dis-gorge their catches. Sometimes follows fishing vessels.

RANGE Breeds Arctic south to cool temperate zones, light morphs further north than dark, dispersing August reaching up to 30-50° south in Atlantic and Pacific. Very common off southwest Africa by late October though much less frequent off southeastern coasts. Monotypic.

STATUS Vagrant; recorded Bird and granitic islands January to February, but probably equally likely any-where in Seychelles.

SIMILAR SPECIES Long-tailed Skua is similar in size but slimmer in build, with narrower wings, longer tail and more rounded head giving a more tern-like appearance than other skuas. Very long tail-streamers (when present), blunt not tapered, smaller bill and little or no white in upperwing, black trailing edge to wing. Immature is generally greyer than Arctic, with creamy-white, barred upperparts giving a scaly appearance, little white on upperwing and a large white patch on heavily-barred underwing. **Pomarine Skua** is more heavily built, body deepest at the belly, with heavier head and bill; also has spoon-shaped central tail feathers (when present) and more extensive white in broader wings.

LONG-TAILED SKUA *Stercorarius longicaudus*   Plate 28

Other names: Long-tailed Jaeger, Buffond's Skua
French: Labbe à longue queue   **Creole:** Asasen Lake Long

DESCRIPTION Length 48-53 cm (including tail up to 29 cm); wingspan 105-115 cm
A small graceful skua, adults with very long narrow tail-streamers. Polymorphic, though dark morph is very rare; immature colour is highly variable. **Adult** has upperparts cold grey, scaled buff or whitish in win-ter. Upperwing-coverts grey-brown, contrasting with black primaries and secondaries, latter forming a nar-row trailing edge (a contrast lacking in Arctic Skua). Lacks prominent wing-flashes, having little or no white in the upperwing (usually has only two white primaries, whereas Arctic Skua has three or more) and no white in underwing, which is dark grey and may appear blackish. Underparts to vent grey, often white on lower breast with a solid breast-band. Smaller, more rounded head than Arctic Skua, with neat, black cap, flecked whitish in winter. Tail blackish brown with long narrow central feathers, not tapered and thick at the base as in Arctic Skua, but these may be shorter or even absent in winter. Slimmer body and narrow-er wings than Arctic Skua, giving a more tern-like jizz. **Dark morph** is extremely rare; it has grey-brown lower breast and belly. All morphs have bill blackish grey, smaller than Arctic Skua, less gonydeal angle and legs bluish black. Sexes similar. **Adult breeding** lacks breast-band and has longer central tail feathers. **Juvenile** can look chunkier than adult and plumage is highly variable, but is generally colder grey-brown than Arctic Skua, lacking rufous tones. Head yellowish white, grey on sides and nape, variable in tone. Upperparts show well-defined whitish barring, contrasting with grey-brown plumage, appearing scaly at a distance (Arctic Skua has more buff bars and browner plumage). Shows little white on upperwing, but a large patch on underwing and lacks buff edges to primary tips as in Arctic Skua. Underparts grey, often darker on breast, white immediately below. Tail-coverts barred (some Arctic Skuas have plain undertail-coverts). Underwing heavily barred, especially on axillaries (some dark Arctic Skuas have uniform under-wing-coverts). Tail projections blunt at tip, generally shorter than adult (Arctic Skua tapered, pointed and shorter). Bill blacker at tip (40-50% compared to only about 25% in Arctic), black extending past gony-deal angle. Juvenile plumage is moulted in first-winter and adult characteristics gradually acquired. **Immature** similar to juvenile, underparts cleaner often emphasising dark hood and breast-band, scapulars and upperwing-coverts grey-brown; pale hindneck and sometimes shows traces of black cap. Similar to adult by **second winter** but with barred underwing.

VOICE Silent at sea.

BEHAVIOUR Highly pelagic and usually solitary outside breeding season. Has more continuous flapping flight than Arctic Skua with occasional glides. Flies higher than other skuas, but may also hug waves and often settles on sea. Snatches fish from surface. May parasitise small terns, but generally with less force or persistence than other skuas.

RANGE Breeds circumpolar, nominate *longicaudus* from Scandinavia to Siberia and *pallescens* (in which white of breast continues to belly) from eastern Siberia to Arctic North America and Greenland. Winters south of the Equator to sub-Antarctic waters, both races recorded off South Africa. Large numbers arrive off Namibia late September, building up through October to November, in small numbers east of Cape of

Good Hope to Mozambique Channel, Kenya, Madagascar.

STATUS Uncertain; some reports of small skuas are indeterminate between this and Arctic Skua. Scattered reports around Indian Ocean and pelagic habits suggest possibility.

SIMILAR SPECIES Arctic Skua is heavier, with broader wings lacking darker trailing edge, plumage less cold grey or grey-brown and shorter central tail feathers (when present) and has less buoyant flight.

HEUGLIN'S GULL *Larus heuglini*   Plate 29 & 30

Other name: Siberian Gull   **French:** Goéland de Heuglin   **Creole:** Golan Siberyen

DESCRIPTION Length 60-70 cm; wingspan 142-160 cm. Similar to Lesser Black-backed Gull, but larger and heavier with bigger head and more powerful bill. **Adult** has dark slate-grey upperparts, white tail and white underparts. Head white, heavily streaked on crown and neck, especially lower hindneck in winter to early spring. Outer primaries gradually shade blacker, showing less contrast than in Lesser Black-backed Gull race *graellsii* and are tipped white with a small white mirror on the outermost primary. Only the outermost primary is black (in *graellsii*, outermost two are black). Shows a white trailing edge to the wing. Eye pale yellow, eyelids dark red. Legs long and yellowish or pinkish. Bill powerful and yellow with red gonydeal spot (sometimes dusky or even a dusky subterminal band) and prominent gonydeal angle. Sexes similar, female smaller. **First-winter** has whitish head and underparts. Generally paler than Lesser Black-backed Gull with paler mantle and, at rest, shows dark diamond shapes on scapulars and dark-centred tertials with white fringes may be obvious. Solid blackish terminal tail-band typically covers about one-third of visible tail/rump area. Long pinkish legs and heavy black bill with a prominent gonydeal angle. **Second-winter** plumage is greyer above; bill may have a yellowish base and eye may be pale. Immature moults later than adult so second-winter plumage possibly best combined with second-summer and called second-year[1]. **Third-year** acquires much grey in upperparts, bill usually with obvious subterminal band and pale eye, but retains wide black tail-band and yellow or pinkish legs. **Fourth-year** is slate-grey above, with a whitish eye, banded bill and pinkish yellow or dull yellow legs. Resembles adult by fifth year but may even then retain some immature features especially in colour of bare parts.

VOICE A loud 'kyew-kee-kee'.

BEHAVIOUR Generally coastal and gregarious, though vagrants may be solitary or in pairs. Flight strong and buoyant. Feeds on molluscs, worms and crustaceans.

RANGE Breeds northern Russia from about 45-80° east, wintering south to Arabian Peninsula and East Africa. Regular Kenya coast, sometimes in substantial numbers November to March[2], where both the nominate *heuglini* and race *taimyrensis* occur; the latter is larger, longer-legged, paler with little head streaking, duller grey upperparts and often has pink legs. Taxonomy is controversial and some authorities group with Lesser Black-backed Gull, others with Herring Gull *L. argentatus*.

STATUS Vagrant to granitic islands; recorded February.

SIMILAR SPECIES Lesser Black-backed Gull is smaller, less heavily built and has a less powerful (but proportionately longer) bill; first-winter has darker back contrasting with pale rump and less well-marked scapulars. See also **Kelp Gull**, a potential vagrant, under Lesser Black-backed Gull.

References: 1. Harris *et al.* (1996), 2. Zimmerman *et al.* (1996).

LESSER BLACK-BACKED GULL *Larus fuscus*   Plate 29 & 30

French: Goéland Brun   **Creole:** Golan Ledo Nwanr

DESCRIPTION Length: 52-60 cm; **wingspan:** 135-150 cm. A large but elegant gull, slightly smaller than Heuglin's Gull with slimmer build. At rest, long wings give tapered appearance towards tail. **Adult** has upperparts and most of upperwing greyish black (black, often with a brownish tinge in race *fuscus*), showing little or no contrast with black wingtips; broad white trailing edge, faint white leading edge and rump and tail white. Head whitish with heavy dark streaking. Underparts white with brown streaks on the sides of the breast and back of the head/neck. Usually shows a white mirror on the outermost primary and often a smaller mirror on the second outermost primary. Underwing white with broad dark grey subterminal trailing edge and black tips. Bill yellow with a red spot near the gonys. Eye pale yellow with red eye-ring; legs yellow. Sexes similar; male averages larger. **Adult breeding** is similar to non-breeding, but head is pure white and wing has a white mirror on the outermost primary only. **First-winter** has brownish white head, darker ear-coverts and whiter chin and throat. Greyish white underparts with heavy brown bars on sides of the breast, flanks and undertail-coverts. Upperparts are dark brown edged white, giving a scaly pattern, scapulars dark but weakly marked compared to Heuglin's Gull. White rump contrasts with black tail-band. Bill black, legs dull flesh, eye brown. **First-summer** is similar to first-winter, with head and underparts whiter and upperparts and upperwing mainly uniform brown. By **second winter** the bill becomes pale, tipped black and the tail-band is smaller. Adult characteristics are gradually acquired by fourth winter, though even at this age some traces of immaturity may be retained.

VOICE A deep, loud 'yew' or 'keyi-kee-kee'.

BEHAVIOUR Favours sheltered coasts and harbours. Flight graceful and buoyant. Gregarious, though rarely associating with other species and vagrants usually solitary. Takes fish by dipping to surface or surface-plunging; feeds on shoreline on molluscs, worms and crustaceans. Also may take insects in the air, scavenge at dumps and occasionally steal from other birds.

RANGE Race *fuscus* (has the blackest upperparts and is sometimes recognised as a full species, Baltic Gull) breeds Scandinavia to White Sea wintering Africa (regular Kenya, vagrant to South Africa) to southwest Asia. Race *graellsii* breeds western Europe, wintering southward, vagrant Kenya. Another race *intermedius*, winters mainly western Europe to West Africa. Taxonomy is complex and not fully understood.

STATUS Vagrant throughout Seychelles, December to February. Only race *fuscus* has been recorded with certainty, though *graellsii* may possibly occur. SBRC categorised the species as annual and did not collect records until this status was changed in 1996 when it became evident Heuglin's Gull was also occurring[1].

SIMILAR SPECIES Heuglin's Gull is larger and more powerful with heavier bill, some recalling Great Black-backed Gull *L. marinus* (though the latter is a vagrant only as far south as North Africa, so is very unlikely to reach Seychelles). First-winter Heuglin's has whiter head and a paler mantle than Lesser Black-backed Gull, showing less contrast with rump, solidly dark tertials and bold diamond-shaped marks on the scapulars. **Kelp Gull** *L. dominicans* (L54-65 cm, WS128-142 cm) is a potential vagrant and should be considered. Breeds South Africa where mainly resident but non-breeders reach Mozambique and are vagrant to Kenya. Slightly larger, more heavily built and more compact than Lesser Black-backed Gull, wings extending only slightly beyond tail at rest, larger head, heavier bill, thick, olive (not yellow) legs and usually a dark eye.

Reference: 1. Skerrett (1996c).

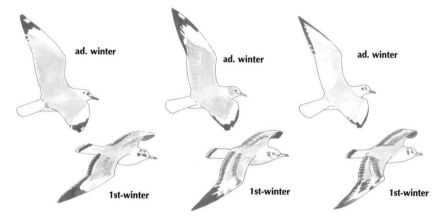

Gulls in flight, left to right: Brown-headed Gull, Grey-headed Gull, and Black-headed Gull

GREAT BLACK-HEADED GULL *Larus ichthyaetus*   Plate 29 & 30

Other names: Pallas' (or Greater Black-headed) Gull
French: Goéland ichthyaète   **Creole:** Gran Golan Latet Nwanr

DESCRIPTION Length: 57-61 cm; **wingspan:** 150-170 cm. A large deep-chested gull with long, sloping forehead accentuating length of powerful, banded bill; white eye-crescents at all ages. **Adult non-breeding** head is mainly white with dark mask highlighting white eye-crescents. Dark streaking over crown to hindneck. Upperparts grey; underparts, rump and tail white. In flight, wing pattern distinctive, broad wings showing mainly white primaries, the outer six having black patches forming a subterminal crescent. Flight profile appears front-heavy due to deep chest, long neck/head and heavy bill. Bill pale yellow, banded black, paler at tip. Long, greenish yellow legs. Sexes similar; male averages larger. **Adult breeding** is the only large gull with complete black hood (often acquired as early as January) and a banded bill. **First-winter** usually reaches winter quarters after post-juvenile moult, with grey mantle and scapulars (unlike any other large gull) and mainly white underparts. Dusky white eye-crescents. Hindneck and nape with brown spots that often extend onto the breast and flanks. Mantle and scapulars grey, unlike any other large gull of similar age and underparts mostly white. Wings blackish brown with pale midwing-panel visible in flight. Rump and tail white with broad black subterminal band. Bill greyish pink, tipped black, often becoming yellow with black subterminal band. Legs brown, greyish or pink. **Second-winter** has greyer wings with extensive pale patches on the outer wing, a slender tail-band and bare parts similar to adult. **Second-summer** similar, but with partial black hood and further reduced tail-band. **Third-winter** as adult with incomplete remnants of tail-band and less white in tips of primaries. Resembles adult by third summer. **VOICE** Normally silent away from breeding grounds. Call a loud crow-like 'krar-a'. **BEHAVIOUR** Frequents large lakes and coasts. Usually solitary or up to three individuals, sometimes with other gulls. Flight slow, heavy and lazy. Follows fishing vessels and scavenges at dumps. Steals food from other gulls and terns. Normally very timid. Feeds mainly on fish, with some crustaceans, birds, eggs and even reptiles taken.

RANGE Breeds on lakes from Black Sea to Mongolia. Migrates southward to east Mediterranean, Red Sea, west coast of India. Vagrant to Kenya December to April (regular at Malindi). Population is increasing and number of vagrancy records is rising. Monotypic.

STATUS Uncertain; large gulls, almost certainly this species, have been reported on Mahé prior to formation of SBRC, but no confirmed records.

SIMILAR SPECIES First-winter **Heuglin's Gull** has a whitish head, a black bill and black centres to scapulars.

BROWN-HEADED GULL *Larus brunnicephalus*   Illus. p. 225

Other name: Indian Black-headed Gull   **French:** Mouette de Tibet   **Creole:** Golan Latet Bren

DESCRIPTION Length: 41-43 cm; **wingspan:** 110-118 cm. A medium-sized gull, similar to Black-headed but broad black wingtip breaks white leading edge of upperwing and dark grey leading primaries on underwing. **Adult non-breeding** has head mainly white; ear-coverts, crown and nape brownish grey (darker than Grey-headed Gull); brown hood acquired by March. Upperparts grey, underparts, rump and tail white. Upperwing mostly grey, with a white oval shape on outer primaries and primary coverts and broad black wingtips broken by a prominent white mirror visible above and below. Underwing is mostly grey, darker than Black-headed Gull, leading primaries dark grey. Eye whitish or yellow, eyelids bright red and obvious white eye-ring. Bill bright red, tipped blackish; legs bright red. Sexes similar; male averages larger. **Adult breeding** has head paler brown than Black-headed Gull with more obvious black rim at rear. **First-winter** has head whitish, crown, nape and ear-coverts brownish grey. Upperparts grey, some feathers with pale brown edges. Rump and tail white with brownish subterminal band (broader than Black-headed or Grey-headed). Upperwing grey with broad dark brown or blackish trailing edge and sooty brown primaries, white at base. Underwing darker than Black-headed, only white on inner coverts. At rest shows a brown carpal bar. Eye dark, becoming whitish during first winter or spring; legs and bill yellow or orange-red, bill tipped blackish. Develops partial hood and brighter bare parts by first summer with full adult characteristics by second-winter.

VOICE A raucous, crow-like 'ceeaw'. Usually silent away from breeding grounds.

BEHAVIOUR Breeds on inland waters, wintering mainly in sheltered coastal regions. Gregarious, often associating with Black-headed Gull where they occur together. Strong flight with leisurely wingbeats, gliding effortlessly. Scavenges refuse discarded from ships and follows fishing vessels. Takes fish, prawns, crustaceans and when inland feeds on insects, worms and other invertebrates.

RANGE Breeds central Asia, dispersing south to Persian Gulf, Indian coast and South-East Asia. Rarely ventures far offshore. Monotypic.

STATUS Uncertain; see comments under Black-headed Gull.

SIMILAR SPECIES Black-headed Gull has more extensive white leading edge to wing and a darker eye.

GREY-HEADED GULL *Larus cirrocephalus*   Illus. p. 225

French: Mouette à tête grise   **Creole:** Golan Latet Sann

DESCRIPTION Length 39-42 cm; wingspan 100-115 cm. A medium-sized gull, larger and longer-necked than Black-headed Gull, darker mantle and upperwing with white edge from carpal area. **Adult** has sloping head with pale grey hood, complete in breeding plumage, otherwise shading to white on forehead and chin and darker on ear-coverts, recalling pattern of Black-headed Gull but with a much paler ear-spot, absent in some birds. Upperparts grey, rump and tail white. Underparts white, sometimes with pink flush in breeding season. Upperwing grey, wingtips black with white mirrors on both webs of outer two primaries and a white wedge on bases of outer primaries to carpal area. Underwing dusky, black at tips with white mirrors. Eye pale; bright red legs, longer than Black-headed. Bright red bill, thicker than Black-headed, with blackish subterminal band near tip. Sexes similar; male averages slightly larger. **First-year** similar to non-breeding adult, but eye brown, bill yellowish with pink tip; sides of breast brownish grey, crown mottled grey-brown, ear-coverts dusky. Head markings paler than Black-headed with pale grey nape and darker grey upperparts lacking white edges to scapulars; brownish diagonal bar across wing-coverts. Tail white with narrow black subterminal band (narrower than Brown-headed; probably best distinction from this species). Develops partial hood by end of first year and full adult characteristics by third year.

VOICE Noisy; similar to Black-headed Gull. A repetitive 'kaar' during disputes and a crow-like 'caaw-caaw'.

BEHAVIOUR Inland and coastal; often gregarious. Flight less agile than Black-headed Gull, with slow, shallow wingbeats and more gliding, appearing bulkier and showing broader, straighter wings. Feeds mainly on fish either on surface or while in air. Also scavenges, may rob terns or take chicks. Very upright stance.

RANGE Two races: *poiocephalus*, the only race likely in Seychelles, breeds West Africa, Upper Nile Valley south to South Africa and Madagascar, vagrant Comoros. Nominate *cirrocephalus* breeds South America. STATUS Uncertain; see comments under Black-headed Gull. There is a published report from Farquhar, the closest island of Seychelles to Madagascar, though no details survived to the date SBRC was formed[1]. SIMILAR SPECIES Black-headed Gull is smaller and less bulky, has more slender bill, darker eye (in adult), paler underwing, paler mantle and upperwing and narrower more angled wings with extensive white leading edge. Non-breeding adult has much darker ear-spot.
Reference: 1. Feare and Watson (1984).

BLACK-HEADED GULL *Larus ridibundus* Plate 29 & 30

Other names: Common (or European, or Northern) Black-headed Gull
French: Mouette rieuse Creole: Golan Latet Nwanr

DESCRIPTION Length: 34-37 cm; wingspan: 100-110 cm. By far the most likely small to medium-sized gull to be seen in Seychelles. All plumages distinguished by extensive white leading edge to upperwing and whitish underwing. Adult non-breeding has head white with a blackish ear-spot. Mantle and most of upperwing pale grey, with prominent white leading edge, outer primaries tipped black. Underwing pale grey to whitish (but can appear darker in field), outer primaries black with white inner webs and secondaries. Tail and underparts white. Eye brown or crimson brown. Bill dull red with black tip; legs dull red. Sexes similar; male averages larger. Adult breeding similar to non-breeding but with chocolate brown hood and brighter bare parts (often acquired in winter quarters). First-winter has head white with black ear-spot. Mantle mainly clear grey and underparts white. Upperwing grey with white leading edge, blackish brown trailing edge and brown carpal bar. Rump and tail white with narrow, black subterminal band. Bill dull yellowish flesh with extensive dark tip, legs dull yellowish flesh to orange. By first summer the bare parts become orange and develops a partial brown hood flecked white; carpal bar faded pale brown.
VOICE A harsh 'krrya', 'kwar' or 'kraa'. May be vocal at all times of year.
BEHAVIOUR Generally coastal in Seychelles, favouring muddy shores, often on inland waters elsewhere. Gregarious and aggressive. Agile flight, with steady deep wingbeats, wings appearing pointed. May dip to surface or surface-plunge to feed. Also feeds while walking, making fluttering pounces, or paddle-feeds. Sometimes a scavenger, sometimes piratical or may follow fishing vessels. Prey includes fish, marine invertebrates, insects and scraps at dumps.
RANGE Breeds from Iceland across most of Europe and Asia to Mongolia, migrating southward; vagrant South Africa. Numbers increasing and annual to East Africa since 1971. Usually considered monotypic, though some authorities recognise race *sibiricus* of northeast Siberia (slightly larger billed).
STATUS Near-annual in granitic islands; most records December to January of singles or up to three birds.
SIMILAR SPECIES The only small gull recorded with certainty. However, some reports to SBRC have failed to rule out other possibilities, in particular Brown-headed Gull and Grey-headed Gull[1]. In 1997, SBRC created a new category of record, 'Black-headed/Brown-headed/Grey-headed Gull', to classify such reports pending a clearer picture emerging. Brown-headed Gull is readily separated by broad black wingtips which break white leading edge of upperwing, prominent white mirror near wingtip (above and below), dark grey leading primaries on underwing and a pale eye with more obvious eye-ring. First-winter has broader sooty brown primaries, white at base and bold black trailing edge to wing. Grey-headed Gull is larger than Black-headed Gull with darker underwing, mantle and upperwing, broader, straighter wings having more extensive black tips, longer, heavier bill, longer legs, sloping head, pale eye (in adult), paler ear-spot and more upright stance. Potential vagrant small gulls include White-eyed Gull *L. leucophthalmus* (L39-43 cm, WS100-109 cm), breeds Red Sea, Gulf of Aden, vagrant Maldives and possibly Kenya, Sooty Gull *L. hemprichii* (L42-45 cm, WS105-113 cm), breeds Red Sea to Kenya, wandering south to Tanzania/Mozambique border, Slender-billed Gull *L. genei* (L42-44 cm, WS90-102 cm), breeds southern Europe to Pakistan; vagrant south to Kenya and Franklin's Gull *L. pipixcan* (L32-36 cm, WS81-93 cm), breeds North America, vagrant South Africa and Mozambique.
Reference: 1. Skerrett (1997).

WHISKERED TERN *Chlidonias hybridus* Plate 32

Other names: Marsh (or Black-fronted) Tern French: Guifette moustac Creole: Dyanman Moustas

DESCRIPTION Length 23-25 cm; wingspan 74-78 cm. Distinguished from other marsh terns by larger, stockier form. Longer wings than White-winged Tern, with heavier, stubby bill. Structure and generally pale plumage means far more likely to be confused with Common Tern, but has shorter tail with more shallow fork and broader, shorter, less angled wings. Adult non-breeding has white underparts, sometimes with a faint grey patch on the sides of the breast, black on head reduced to a line through the eye and across the nape. Crown flecked black, hindneck greyish. Upperparts entirely pale grey, tail grey, upperwing grey with paler bases and darker tips to outermost primaries. In transition may retain black crown while losing dark underparts. Bill black and stubby (though male is longer-billed than female); legs long,

dull red or black. **Adult breeding** is easily identified by entirely dark grey upperparts and underparts (dark-est on belly) contrasting with white cheeks and white undertail-coverts; black cap contrasts sharply with white cheeks and pale grey underwing contrasts with dark belly; bill and legs blood-red. **Juvenile** has hind-crown and nape dark brown streaked grey and the back brown, feathers edged whitish contrasting with grey wings, rump and tail. However, this plumage is soon moulted. **First-winter** has upperparts and rump mainly grey and underparts white; black on the head largely confined to the ear-coverts and crown, lobed at the rear, reaching almost to the mantle (more extensive than White-winged Tern); upperwing grey with darker trailing edge, bases of outer primaries white or light grey; tail grey edged white with faint dusky tip; bill and legs blackish. Similar to adult by first summer but underparts mainly white.
VOICE A loud grating 'uurhk'.
BEHAVIOUR Favours inland sites at breeding grounds, but also coastal mudflats and estuaries in winter. Vagrants are generally solitary. Steadier flight than White-winged Black. Takes insects and small fish, dip-ping to surface or hovering before making shallow plunge-dives. Diet also includes frogs, crabs and even mice.
RANGE Nominate *hybridus* breeds southern Europe and North Africa discontinuously to southwest Asia wintering Africa to southwest Asia. Race *sclateri* breeds Madagascar/southern Africa, *delalandii* (darkest race with narrow cheek-stripe) East Africa, *indicus* from eastern Iran to north India and *javanicus* in north-east India and (probably) Sri Lanka. Other races breed eastern Asia and Australasia. Resident and migra-tory throughout range. Races differ slightly in coloration of breeding plumage. Genus is sometimes merged with *Sterna*.
STATUS Vagrant to granitic islands (race or races unknown); recorded October and March/April, some-times in breeding plumage in April. No records from outer islands, almost certainly due to lack of observers.
SIMILAR SPECIES White-winged Tern has white (not grey) rump and a whiter hindcollar. **White-cheeked Tern** is larger with more deeply forked tail and grey (not white) vent. **Common Tern** has longer more deeply forked tail, narrower, more angled wings; non-breeding shows a blackish carpal bar, rump paler than tail, darker outer primaries and red legs.

WHITE-WINGED TERN *Chlidonias leucopterus*   Plate 32

Other names: White-winged Black (or Marsh) Tern, White-tailed Tern
French: Guifette leucoptère   **Creole:** Dyanman Lezel Blan

DESCRIPTION Length 20-23 cm; wingspan 63-67 cm. The only annual marsh tern; a small compact tern without a prominent tail fork. **Adult non-breeding** shows an obvious contrast between silver-grey upper-parts and white or pale grey rump. Black patch behind eye and grey-flecked crown, usually separated by a white supercilium, though some are white-headed. Dark central nape-stripe, visible from rear, broadens downward to white hindcollar. Upperparts pale grey with blackish bars on lesser coverts and secondaries. Paler rump may contrast with tail and back, becoming less obvious in late winter. Outer webs of outer tail feathers white. Usually retains some black on the underwing often forming a diagnostic black bar; also a black smudge is visible at the base of the forewing on the breast. Tail white or pale grey. Shape is useful in identifying non-breeding birds: rounded head and delicate bill give a gentle appearance. Bill black and distinctly shorter than head; legs light red or orange. Sexes similar (but male averages darker). **Adult breed-ing** is unmistakable: black body and underwing contrast with silver-grey upperwing, white carpal bar and white rump and tail; bill dull red; legs red. **Juvenile** shows sharp contrast between dark back, pale wings and white collar and rump. **First-year** similar to non-breeding adult, but with narrow, blackish carpal bar and secondary tips forming a faint trailing edge. Bill black; legs pink or reddish.
VOICE A loud 'kerr', 'kwek' or 'krek'.
BEHAVIOUR Prefers inland locations though also seen along coasts. Often present in small numbers. Flight swift, agile and direct. Hunts with regular, slightly stiff wingbeats over water, dipping to snatch food from the surface. Hovers less than Whiskered and does not plunge-dive. Will hunt over grassland for grasshoppers, hawk for insects or snatch flies and caterpillars off vegetation. Occasionally takes small fish.
RANGE Breeds Ukraine to south China, wintering mainly in sub-Saharan Africa (Chinese population win-ters in South-East Asia) usually on inland lakes and rivers. Monotypic.
STATUS Annual throughout Seychelles, usually present in small numbers from October to April. Birds in full breeding plumage often present in April.
SIMILAR SPECIES Whiskered Tern is larger and stockier with longer wings and heavier, stubby bill. **Black Tern** *C. niger* (L22-26 cm, WS56-62 cm) is a potential vagrant. Occurs on West African coasts south to the Cape, a few reaching Natal; vagrant Kenya. Adult non-breeding and immature show a dark smudge on side of breast and extensive solid black cap.

GULL-BILLED TERN *Gelochelidon nilotica*

Plate 31

French: Sterne hansel   **Creole:** Dyanman Gro Labek

DESCRIPTION Length 35-38 cm; wingspan 100-115 cm. A large, heavy, whitish tern with broad head, short, slightly forked tail (less forked than most terns). Combination of head markings and stubby black bill diagnostic. Broad wings, heavy body, short bill and longer legs give a distinctive gull-like jizz. **Adult non-breeding** shows head white with a blackish eye-crescent and fine black streaking on crown and hindneck (latter may be difficult to see in field). Uniform silver-grey upperparts and rump (Sandwich Tern has white rump and tail) and white underparts. Dark trailing edge to upper and under primaries (Sandwich lacks a dark edge on upper primaries and has less obvious grey edge to under primaries). Underwing white, show-ing little contrast with upperwing. In strong sunlight, shows a slight translucence to secondaries and inner-most primaries. Heavy, black bill with marked gonydeal angle; long, thick, black legs. **Adult breeding** has a black cap extending to the hindneck and no crest, (unlike Sandwich Tern). **First-winter** similar to adult non-breeding, whiter with brownish wash to crown and hindneck and brown edges to the feathers of the scapulars and wing-coverts gradually lost with age; darker grey primaries. Lacks dark crown of first-win-ter Sandwich, but has a black wedge-shaped patch behind the eye; brown tips to white tail (gradually lost with age) and brown legs.

VOICE An unmistakable deep, low 'ger-vick'.

BEHAVIOUR Usually solitary in Seychelles, hunting along coastlines or at rest on tidal flats. Gull-like slow steady flight on long, broad wings. Flight silhouette distinguishes from Sandwich Tern, bill merging even-ly into broad head and elliptical body. Generally does not plunge-dive but drops to the surface to take food; sometimes hawks for insects.

RANGE Six races breed North and South America, Europe, Asia to Australia. Nominate *nilotica*, breeds Europe to northwest China wintering southward and is the only race recorded in Africa. The genus is often merged with *Sterna*, but behaviour (intermediate between *Sterna* and 'crested' terns) and unique combi-nation of characters support separate treatment.

STATUS Annual migrant in small numbers throughout Seychelles, mainly October to December.

SIMILAR SPECIES Sandwich Tern has a more slender, longer, black bill, usually tipped yellow, appearing 'stuck on', not an extension of the head as in Gull-billed; less powerful, more elegant with lighter flight; more angled, more translucent wings and white rump and tail. Adult has dark leading edge to outer wing (lacking in Gull-billed) and a shriller call.

CASPIAN TERN *Hydroprogne caspia*

Plate 31

French: Sterne caspienne   **Creole:** Dyanman Zean

DESCRIPTION Length 47-54 cm; wingspan 130-145 cm. The largest tern, identified by size, bulk, large dagger-like blood-red bill with black tip and pointed wings; blackish outer primaries of underwing con-trast with pale grey upperwing.

Adult breeding has a black cap with a small crest on the nape. Upperparts pale grey, underparts and rump white. Upperwing pale grey, slightly darker on inner webs of outer primaries. Underwing white with blackish outer primaries decreasing in extent inward (diagnostic). Tail short with a shallow fork. Bill blood-red ringed black at or near the tip. Legs long and black, so stands taller still than other terns. **Adult non-breeding** similar to adult breeding, but forehead and crown have fine white streaking; outer primaries sometimes darker; bill duller with a broader black ring. **Juvenile** similar in build to adult; bill shorter and duller orange-red, legs yellow-brown. Brown feather tips on mantle, scapulars, tertials and tail give a scaly appearance. Cap is dark brown, spotted white or pale brown on the forehead and crown. Rump uniform whitish grey. Faint grey-brown carpal bar. Brown tips to wing-coverts, dark secondaries and grey tertials (with two brown bands at tip). Tail white, barred grey and brown at the tip. Bill orange-red, with a broad-er black band than the adult sometimes covering the tip. Legs pale at first, darkening to black within a few months. By end of **first year**, similar to adult non-breeding but primaries and primary coverts darker grey; some faint brown tips to other wing-coverts and tail tipped brown. Similar to adult non-breeding by **second year**, some retaining brown-tipped tail.

VOICE A diagnostic loud 'krowk' or 'kaar-arr', similar to a heron. Juveniles begging call (at rest or in flight) 'klee-aee'.

BEHAVIOUR Coastal, rarely being seen far from land. Not very gregarious. Flight powerful and direct with slow shallow wingbeats, the wings held stiffly alternately flashing dark primaries of the underwing and pale of the upperwing. Flight silhouette suggests a miniature booby. Hovers and plunge-dives in shallow water submerging completely to catch small fish. Often aborts dive at the last moment. Aerial skimming has been recorded, though not yet in Seychelles. Feeds on fish, prawns and crabs, rarely taking floating refuse or terrestrial food items. May take eggs or young of other birds.

BREEDING One or two greyish eggs with brown blotches are laid April to August. Nest is an unlined, shal-low scrape in sand at the edge of tidal flats, usually close to, sometimes below, high spring tide line. Minimum inter-nest distance at Aldabra is $c10$ m[1]. Both parents incubate. Incubation lasts about 21 days, fledging 47-49 days. Post-fledging parental care may be four months or more.

RANGE Breeds North America, north and central Europe eastward to Siberia, South Africa and Australasia. Disperses widely outside breeding season. Sometimes merged with *Sterna* but enormous bill marks out the species from all other terns. Monotypic.

STATUS Resident and breeds Aldabra, the world's only known oceanic breeding site. Small numbers resident Aldabra (up to 10 pairs) and breed on Iles Moustiques. Breeding also recorded on Esprit (in 1968[1]), Picard (first noted 1995[2]) and has probably bred Ile Michel[1]. Highest recorded count is 30 birds on Iles Moustiques, May 1972[1]. May breed Astove, where there are regular sight records of up to five birds feeding in the lagoon, or at Cosmoledo where there have also been sightings. Recorded as a vagrant to the granitic islands in November.

THREATS AND CONSERVATION Such a small population as exists on Aldabra must be considered highly vulnerable despite the protected status of the atoll. During 1969, when the population was studied closely, the majority of egg losses were caused by high tides, though predators, especially black rats, may have contributed[1]. On Astove feral pigs pose an additional threat to possible breeding attempts.

SIMILAR SPECIES Size, heavy build and bill colour/shape distinguish from all other terns.

References: 1. Diamond and Prys-Jones (1986), 2. Augeri and Pierce (1995).

GREATER CRESTED TERN *Thalasseus bergii*   Plate 31

Other names: Crested (or Swift or Yellow-billed) Tern, Great Crested Tern
French: Sterne huppée **Creole:** Dyanman Sardin, Golan Sardin

DESCRIPTION Length 46-49 cm; wingspan 125-130 cm. A very large tern, with large, drooping, yellow bill and long, narrow, pointed wings. Large, angular head with a shaggy crest. **Adult breeding** has a black cap separated from bill by a narrow white band and crested at the nape. Grey upperparts, paler on rump and tail, white on underparts. Upperwing mainly grey, primaries whitish with darker inner webs and secondaries tipped white, forming a pale trailing edge. Underwing white with dark grey tips and shafts to the outer primaries. Heavy, decurved, cold yellow bill, sometimes with a greenish tint towards the base. Legs black and long. **Adult non-breeding** similar to adult breeding but bill paler and duller, forehead and forecrown white, a narrow white eye-ring and the hindcrown and nape black with white streaking; worn outer primaries are dark grey. **Juvenile** is darker grey above than adult, mottled black and white, the crown blackish brown with feathers edged white; bill darker and often has a blackish tip, a shorter crest and the sides of face streaked brownish. Some brown tips to feathers on the scapulars, the rump paler (though occasionally as dark as upperparts). Tail grey, with contrasting blackish outer tail feathers and often a black tip. Primaries blackish and secondaries dark brown edged white; pale wing-panel showing at rest a faint carpal bar on the upperwing. **First-year** birds are similar to adult non-breeding, but may retain some brown tips on the scapulars, have darker primary coverts and a faint carpal bar.

VOICE In flight a soft 'whep-whep'. Feeding birds give a continuous 'keek-keek' and a harsh 'krow'.

BEHAVIOUR Coastal, often seen in small parties at water's edge. May fish alone or in small groups. Elegant, bounding flight. Plunge-dives, usually in depths of less than 1 m, but occasionally dips to surface taking fish, squid and other aquatic prey including turtle hatchlings. Regularly bathes at water's edge.

BREEDING One or occasionally two eggs are laid on bare earth, the site sometimes lined with a few small stones. Eggs vary in colour from pale creamy with almost no spots to dark and heavily spotted. Breeds in dense colonies (average distance between nests about 50 cm), on small coral islands or, at Aldabra, on small lagoon islets. On Aldabra there are two laying periods: December to January and June to August[1]. Incubation lasts 25-30 days, fledging 38-40 days. Given the extended post-fledging dependency of chicks (perhaps at least four months[2]) and moult time, successful breeders probably miss the next breeding period and return to breed after a 12-month cycle[3]. Elsewhere breeding has been recorded on African Banks in July[4] and November[5], and on Etoile May to September[6].

RANGE Breeds Indo-Pacific and Red Sea, Persian Gulf. Six races generally recognised with others sometimes proposed. Race *thalassinus* breeds Seychelles, Rodrigues, Tanzania and Chagos; this is the smallest race and has a broad white band on the forehead. Race *velox* breeds Red Sea, Arabian Sea and Bay of Bengal; bigger, darker on upperparts (almost as dark as Sooty Tern), larger-billed and shows a white U-shape at the tips of the secondaries. Race *enigma*, the palest race, breeds on islands off Madagascar and Mozambique. Genus is often merged with *Sterna*.

STATUS Breeds Aldabra (Iles Chaland, Table Ronde, Ilot Déder, Grand Mentor, Ilot Marquoix, Champignon des Os and Sesame), Cosmoledo and Etoile. Probably also breeds Farquhar and Providence and formerly bred African Banks but no recent records. One study suggested the total Aldabra breeding population may be about 150 birds[7], but this could be too low if birds are not resident all year, or do not breed every six months as seems likely[1]. Present at breeding sites year round and also a regular non-breeding visitor throughout Seychelles. Breeding race is *thalassinus*, but *velox* has occurred as a vagrant to Bird September-October and *enigma* might be possible.

THREATS AND CONSERVATION Colonies are prone to human disturbance and will desert eggs and small chicks if approached. Disturbance may have been the cause of extinction on African Banks. Even efforts by scientists at Aldabra to study colonies have resulted in eggs and small chicks being deserted[1].

SIMILAR SPECIES Lesser Crested Tern is smaller and more elegant in appearance; has less angular head, straight orange bill and a front-heavy appearance.

References: 1. Diamond and Prys-Jones (1986), 2. Feare (1975b), 3. Ashmole (1965), 4. Ridley and Percy (1958), 5. Vesey-Fitzgerald (1941), 6. Stoddart (1984), 7. Diamond (1971).

LESSER CRESTED TERN *Thalasseus bengalensis*   Plate 31

Other names: Little (or Small) Crested Tern
French: Sterne voyageuse   **Creole:** Pti Dyanman Sardin, Pti Golan Sardin

DESCRIPTION Length 36-41 cm; wingspan 95-110 cm. A large elegant tern, smaller than Greater Crested Tern, with an orange bill, more rounded head and neater crest. **Adult** has black cap, reduced in non-breeding plumage: forehead and forecrown to beyond peak is white, mottled greyish. Whiter-headed than any stage of Greater Crested Tern, showing a black spot in front of the eye, a black band behind the eye to the nape and a white eye-ring. Underparts, sides of face, chin and throat white. Pale bluish grey upper-parts, rump and tail paler with some tail feathers whitish. Outer webs of primaries silver-grey, inner webs white. Underwing white with a dark trailing edge on the outer wing. At rest, wingtips project beyond the tail. Bill orange (brighter in breeding adult) and straight. Legs black and shorter than Greater Crested. **Juvenile** is scalloped brown on scapulars, mottled brown on leading edge of wing with brownish flight feathers, primary coverts and corners of tail. Legs dull orange, becoming dusky. **First-winter** as non-breeding adult, except bill pale yellow, faint carpal bar (which fades rapidly with age), darker dusky grey tail corners and flight feathers with the outer primaries brown and indistinct brown tips to wing-coverts forming dark bars. By **second year** similar to adult non-breeding but with dark secondary bars, outer primaries and sides of tail.
VOICE A high-pitched, rasping 'kik-kirek' or 'kriik'. Also a 'kee-kee-kee'.
BEHAVIOUR Coastal and gregarious, often associating with other terns. Flight fast and graceful, vertically plunge-diving to catch fish, usually submerging completely. Hunts with bill held horizontal. Will also dip to surface and settle on water.
RANGE Nominate *bengalensis* breeds Red Sea, Pakistan, Laccadives and Maldives and migrates southward throughout the western Indian Ocean outside the breeding season. Race *torresii*, darker than nominate, breeds Persian Gulf and Indonesia to Australia. A third race breeds North Africa, wintering West Africa. Genus is often merged with *Sterna*.
STATUS Annual throughout Seychelles, mainly December to April. Most or all are likely to be race *bengalensis* but *torresii* may be possible.
SIMILAR SPECIES Greater Crested Tern is heavier, with more angular head, longer-winged, heavy drooping yellow bill and less elegant flight. **Sandwich Tern** has black bill, usually tipped yellow, which is held downward when hunting, (not horizontal as in Lesser Crested)

SANDWICH TERN *Thalasseus sandvicensis*   Plate 31

French: Sterne caugek   **Creole:** Gran Dyanman Eropeen

DESCRIPTION Length 36-41 cm; wingspan 95-105 cm. A large, graceful, whitish tern with angular head, flat crown and slender, long, yellow-tipped black bill. **Adult non-breeding** has white forehead, black crown streaked with black and a short, black crest at the nape; rest of head white except for small black spot in front of eye. Upperparts pale grey; underparts rump and tail white. Long, slender, angled wings, the sec-ondaries and inner primaries appearing translucent in strong sunlight. Upperwing pale grey with inner webs of the outermost primaries darker. Underwing white except for darker webs and tips of outermost primaries. Shortish tail but with elongated outer tail feathers. Eye dark brown; long, slender, straight bill, which is black with yellow tip; legs black. **Adult breeding** has complete black cap and a more prominent crest. **Juvenile** has a paler tip to bill and narrow black scaling on mantle and scapulars (lost by early winter). **First-winter** head resembles adult non-breeding but browner, streaked grey-brown on the crown. Scapulars grey, underparts and rump white; tail grey-white with black tip. Upperwing white with blackish V-shapes on the wing-coverts and primary tips (reduced with age). Bill entirely black or with a restricted yellow tip, shorter than adult.
VOICE A shrill 'kirrick', inflected upward. Also a 'gwit' alarm call.
BEHAVIOUR Coastal and pelagic. Gregarious, often associating with other tern species. Flight strong and swift, with deep wingbeats, the head and neck protruding, head and bill pointing downward. Plunge-dives from a height of up to 10 m above the surface, submerging for longer than other smaller terns with which it associates. Also dips to surface. Food mainly small pelagic fish.
RANGE Nominate *sandvicensis* breeds Europe and North Africa and migrates southward, often well off-shore, mainly to western and northern seaboard of Africa; rare off Kenya, vagrant to Natal. Other races occur in North and South America. Genus is often merged with *Sterna*.
STATUS Vagrant to granitic islands October to December.
SIMILAR SPECIES Gull-billed Tern has uniform pale grey upperparts and tail (Sandwich Tern has white rump and tail). Calls are very different, Gull-billed being deep and low, Sandwich guttural and shrill. Gull-

billed has shorter, broader, less pointed wings and an all-black bill (not tipped yellow); tail shorter, less forked. Unlike Sandwich, it rarely plunge-dives, is heavier in build with shorter thicker neck, no shaggy crest and longer legs.

COMMON TERN *Sterna hirundo*   Plate 32

Other name: Black-billed Common Tern   **French:** Sterne pierregarin   **Creole:** Dyanman Eropeen

DESCRIPTION Length 31-35 cm; wingspan 77-85 cm. A medium-sized tern, all plumages (except juvenile) having a blackish wedge on the primaries of the grey upperwing and a whitish underwing with a broad, diffuse blackish trailing edge. **Adult non-breeding** has forehead and forecrown white, rear crown and nape black. Dark wedge in upperwing is usually present except in fresh moult, but some black is likely on the inner primaries. Underwing whitish, with broad, dusky trailing edge to the primary tips. Broad, dark carpal bar evident at rest. Upperparts grey with a grey wash to the rump and tail and white underparts. No tail-streamers and at rest tail length approximately equals wingtip length. **Adult breeding** (reported from Seychelles March onward) has glossy black cap extends to nape with white wedge between the cap and bill (broader than in Arctic Tern); grey upperparts contrast with white tail in which the outer web of outer feathers is black, moderately long tail-streamers and greyish white underparts. Bill black, sometimes with a dull red base, becoming scarlet with a black tip in breeding plumage from egg laying to near time young fledge. Legs dull red, brighter in spring. Sexes similar. **First-winter** similar to adult non-breeding structurally, a little smaller with blunter wingtips. Grey upperwing with thick, dark carpal bar and secondary grey bar narrowly tipped white. Grey rump and white underparts. Forehead whitish, crown, nape and ear-coverts blackish brown. Usually has a pale orange base to bill (Roseate Tern has all-black bill). Gradually acquires adult characteristics by third summer.
VOICE An extensive vocabulary including a drawn-out 'kree-eer' or rapid 'kek-kek-kek'.
BEHAVIOUR Mainly coastal in winter. Gregarious and commonly in the company of other tern species. Rather languid flight. Plunge-dives, sometimes hovering first, catching mainly small marine fish, sometimes crustaceans and insects.
RANGE Nominate *hirundo* breeds North America and Europe wintering mainly south of Tropic of Cancer, *tibetana* (slightly darker grey upperparts with shorter bill) in Turkestan and southwest China wintering mainly eastern Indian Ocean, *minussensis* in central Asia wintering mainly northern Indian Ocean, and *longipennis* in northeast Asia wintering mainly South-East Asia to Australia.
STATUS Annual throughout Seychelles in small numbers October to April, most frequent on passage. It is not known which race or races occur; in East Africa *hirundo* is most common, but *tibetana* may also occur.
SIMILAR SPECIES Roseate Tern has extremely long tail-streamers and no black on the underwing primaries, more protruding head, longer bill and narrower wings; never shows contrast between back and rump and has quicker, shallower wingbeats. **Whiskered Tern** can recall Common Tern but has a very shallow tail fork and uniform grey rump/tail. **White-cheeked Tern** is smaller, with shorter, more slender wings, shorter legs, longer bill (drooping towards tip), darker on the upperparts with whitish primaries; at rest shows broader black carpal bar and in flight underwing has a pale mid-wing-panel visible at long range, flight more elegant and bouncing. Potential vagrants might include **Arctic Tern** *S. paradisaea* (L33-35 cm, WS66-77 cm) and **Antarctic Tern** *S. vittata*. (L35-40 cm, WS74-79 cm). Both occur as non-breeding visitors in the Indian Ocean, but generally south of Madagascar.

ROSEATE TERN *Sterna dougallii*   Plate 32

Other names: Rosy (or Dougall's or Graceful) Tern   **French:** Sterne de Dougall   **Creole:** Dyanman Roz

DESCRIPTION Length 33-38 cm; wingspan 72-80 cm. A medium-sized, whitish, elegant tern with long tail-streamers and shortish, narrow wings. **Adult breeding** has a black cap with a straight line at the base (not curving behind the eye), rest of head white, the nape having a greyish tinge. Pale grey upperparts lack contrast with white rump and tail. Underparts white, with a rosy tinge noticeable in all months when present at breeding colonies. At rest, long tail-streamers stretch well beyond the wingtips. Pale grey upperwing with three outermost primaries blackish and pure white underwing. Against strong light, the secondaries and innermost primaries are translucent. On arrival at breeding colonies, bill is black and legs red. Bill becomes red from around time of egg laying to fledging; by time of departure legs are usually dull red and bill blackish. **Adult non-breeding** has crown and forehead white, shorter tail-streamers and no rosy tinge to breast. **Juvenile** has dark brown crown, nape and ear-coverts, the forehead soon becoming white and chin, throat and hindneck white. Underparts whitish with clear pink in the breast feathers. Upperparts pale grey with golden brown scaly mantle and scapulars, becoming clearer grey with age. White underwing with faint greyish primary tips. Dark carpal bar fades during the **first-year**. Wings lack dark tips to the under primaries, unlike immature Common Tern or Arctic Tern. Bill and legs black, gradually lightening to orange-red.
VOICE Extremely vocal at breeding colonies. Call a harsh 'chiw-ik', year round. Alarm call a loud rasping 'krak'. When defending its territory a rapid, repetitive 'kek-kek-kek'.

BEHAVIOUR Mainly pelagic outside the breeding season. Elegant flight with rapid wingbeats, the narrow, centrally situated wings giving an apparent stiff flight action. Often travels long distances to feeding sites from breeding colonies. Highly gregarious year round. Generally hunts over deep water, sometimes along shores, feeding mainly by plunge-diving at speed at an acute angle (similar to a booby) from 2-10 m to catch small pelagic fish. Also dips to the surface to catch fish driven there by larger predators.

BREEDING Breeds in dense colonies, arriving later and leaving earlier than other terns. Activity is strongly synchronised. On Aride, arrives towards end of April and lays mid- to end-May (sometimes early June) in sub-colonies between 25 m and 80 m above sea level[1]. One or two eggs, creamy with dark brown spots, are laid in a shallow scrape (very occasionally three or even four eggs have been recorded[2]). Number of nests with two eggs is usually low, but may increase dramatically in some years, probably in response to food supply (e.g. on Aride nearly 50% in 1996[2] but only 4% in 1997[3]). Incubation takes about 17-24 days, both parents sharing in duties though female does most work, and fledging 25-30 days[4]. Post-fledging parental care may last several months. In Seychelles, this species is the first to desert when difficulties such as poor food supply are encountered. Generally leaves breeding colonies by the end of August. Complete breeding failures have often been recorded on Aride.

RANGE Breeds discontinuously between 56°N and 35°S along coasts of North and South America, Europe, Africa, India and Australasia. Five races recognised including Western Indian Ocean race *arideensis* (named after Aride), sometimes included in *bangsi* which also breeds in the western Pacific. Geographical variation of races mainly involves bill colour. As *arideensis* has the least amount of black on the bill of any race during the breeding season (becomes entirely red, compared to *bangsi* which has only 50% red), there may be a good case for recognition of this race.

STATUS Breeds May to August on Aride (*c*1,250 pairs[5]), African Banks (*c*150 pairs) and Etoile (*c*150 pairs[6]). May also breed Goëlettes, Farquhar and on Bancs du Providence. Has disappeared as a breeding bird from Mamelles, Ile Sèche, Récif, Ile aux Vaches Marines, Bird, and probably elsewhere. In 1936, Vesey-Fitzgerald noted Roseate Terns on Mamelles and recorded hearsay reports of breeding[7], but Ridley and Percy made no mention of them in 1955[8]. Ile Sèche and Récif are mentioned in many references as breeding sites though without documentary support. Ile aux Vaches Marines is mentioned as having 'a large colony of Roseate Terns' in 1955[8] but breeding ceased at some time during the mid-1980s (last record 1982[9]). Away from the granitic islands, a colony on African Banks numbered 250-300 pairs in 1966[10]. In 1997, there were 82 pairs (P. Constance, R. Nolin pers. comm.). Used to breed on Goëlettes, Farquhar in 1897[11] but there have been no reports for a century until 50 birds were seen on the island in 1999[12]. This lack of sightings may be due to a lack of observers. Also used to breed on Bancs Providence at least until the 1970s, when large numbers of chicks were observed being taken by Grey Herons (G. Savy pers. comm.). Aride holds possibly the last viable breeding population in Seychelles being the only wardened colony, yet despite protection the population has declined here too. In 1973, there were estimated to be 2,500 pairs[13] and in 1975 4,300 to 4,800 pairs[14]; this declined to 900-1,000 pairs in 1988[15] and a low point of 426 pairs in 1994[4], rising to around 1,250 pairs each year 1996-99[16].

THREATS AND CONSERVATION The future of Roseate Tern as a breeding species in Seychelles is a cause of great concern. It is prone to human disturbance, especially early in the breeding season when entire colonies may desert. Human predation has eliminated the species from much of its former range, especially due to egg collecting. Rats and other introduced predators may also have contributed to this decline. Problems with food supply appear to be the main factor affecting breeding success[17], which could be linked to over-exploitation of inshore fishing stocks, or changes in water temperature and ocean currents. Introduced Barn Owls have preyed on Roseate Terns on Aride, accounting for 4% of the breeding population in 1993[4], but following implementation of control measures in 1996, losses have been virtually eliminated[16]. Pisonia seeds can stick to feathers, preventing birds from flying, and on Aride staff check colonies for fallen seed piles prior to the nesting season. One reason for the decline elsewhere in the world, highlighted by ringing recoveries, has been trapping and killing of birds at fishing villages along coast of West Africa, a habit not practised in Seychelles. Wintering grounds of Seychelles birds are unknown and the possibility of similar exploitation outside the region remains. A ringing programme to ascertain the wintering grounds of Seychelles birds, a survey of prey items and a survey of all Indian Ocean breeding sites have been recommended[14].

SIMILAR SPECIES Confusion is unlikely during the breeding season when no similar species is likely in Seychelles. At other times is whiter, especially on the underwing than **Common Tern** with proportionately longer bill and stiffer flight.

References 1. Warman (1979), 2. Maul (1997), 3. Ramos (1998), 4. Ayrton (1995), 5. Bowler and Hunter (1999), 6. Skerrett (1995), 7.Vesey-Fitzgerald (1936), 8. Ridley and Percy (1958), 9. Skerrett (1994c), 10. Ridley and Percy (1966), 11. Farquhar (1900), 12. Skerrett (1999b), 13. Procter (1974), 14. Warman (1977), 15. Bullock (1988), 16. Bowler and Hunter (2000), 17. Ramos (2000).

WHITE-CHEEKED TERN *Sterna repressa* **Plate 32**

French: Sterne à joues blanches   **Creole:** Dyanman Lazou Blan

DESCRIPTION Length 32-34 cm; wingspan 75-83 cm. A medium-sized tern, smaller, shorter, longer-billed than Common Tern with narrower wings, jizz more like a marsh tern. **Adult non-breeding** has white forehead shading to greyish with streaks on the crown, black on the hindcrown, nape and ear-coverts, having more black on the head than Common Tern. Upperparts, rump and tail grey, contrasting with paler primaries and white underparts; no tail-streamers. Sides of rump are whiter than the adjacent belly and rump (visible at long range). New whiter inner primaries during December to April moult contrasting with darker outer primaries (visible close range only). Underwing diagnostic: dark secondaries and primary tips contrast with whitish median coverts and greater coverts. Black or blackish red bill, long and drooping at tip giving a Roseate-like appearance. Legs black. **Adult breeding** has dark grey underparts and upperparts contrasting with prominent white cheeks and black cap; moderately long tail-streamers. Long bill is red with black tip and legs red. **First-year** differs from first-year Common and from marsh terns in that grey rump and tail lack contrast with upperparts. Upperparts may retain juvenile brown scaling and broad, blackish subterminal crescents to scapulars and tertials. Darker, more extensive blacker carpal bar than Common though may be less distinct due to dark grey upperwing. May show dark brown tip to tail. Blackish outer primaries of juvenile are retained until at least December, with inner webs pale grey and secondaries dusky grey with brown tips. White underparts and underwing mainly white, primaries and secondaries pale grey. Bill black sometimes with a pale orange base.
VOICE A loud grating 'kee-eerr' with accent on second syllable, similar to Common Tern.
BEHAVIOUR Mainly coastal, occurring up to 10 km off land. Single or gregarious, often in flocks of mixed tern species. Graceful, bouncing flight, feeding by dipping to surface, landing but not submerging. Diet mainly fish and some marine invertebrates.
RANGE Breeds Red Sea, Persian Gulf, Laccadives and coasts of northwest Indian Ocean to western India, moving south along coastlines August, to southern India, East Africa; vagrant southward to South Africa. Monotypic.
STATUS Uncertain; there are unconfirmed reports from the granitic islands and Aldabra.
SIMILAR SPECIES Common Tern is less compact with longer bill, different underwing pattern, paler grey upperparts, white rump and less graceful flight. Both **Saunders' Tern** and **Little Tern** are smaller with shorter tail, straighter bill, more rapid flight; paler above with less distinct carpal bar and less black in primaries.

BLACK-NAPED TERN *Sterna sumatrana* **Plate 34**

French: Sterne diamant   **Creole:** Dyanman Likou Nwanr

DESCRIPTION Length 30-32 cm; wingspan 61 cm. A small, whitish tern with black bill, black legs and a narrow black band running from eye to eye which is broadest at the nape. **Adult** has head white except for black eye-stripes that join behind the nape. Very pale grey upperparts can appear almost pure white in bright sunlight, lacking contrast with white underparts. Outer web of the outermost primary is black, visible at close range. Bill and legs black, contrasting sharply with white plumage. **Juvenile** has head whitish with poorly defined, greyish brown eye-stripe. Mantle sepia brown mottled grey-brown, rump whitish, tail pale grey and underparts white. Bill dull yellow at first, gradually darkening to black with age. By end of **first year**, similar to adult but with dusky carpal bar on the upperwing, dark grey primaries and the black eye-stripe not as distinct. May retain some brown streaking on the crown and some brownish tips to scapulars.
VOICE A high-pitched 'kee-ik' repeated up to four times. Also a deeper, sharp 'cheet-cheet-cheeter' alarm call.
BEHAVIOUR Frequents coral atolls, feeding mainly in small numbers inside the lagoon or up to 1 km offshore. Usually dips to surface to catch small fish or marine invertebrates, but occasionally plunge-dives or submerges from a swimming position. Feeds mainly on small fish (4-8 cm), generally alone, sometimes in twos and threes.
BREEDING Nests on the ground on small predator-free islands. At Aldabra, nests almost entirely on small islets within lagoon close to channels connecting the lagoon to the ocean; eggs usually laid January to February but recorded in most months[1]. Elsewhere, may breed at other times of year (recorded July to August on African Banks and September to November St Joseph Atoll). Nest sites are bare rock sometimes within tussock grass occasionally lined with fine rock grains, small twigs and leaves. Most Aldabra islets support only one pair with a maximum of three pairs on a single islet and a minimum inter-nest distance of 4 m[2]. One or two eggs are laid, greenish blue with dark spotting. Incubation lasts 21-23 days. Chicks fledge at 24-28 days and associate with parents for at least one month after fledging[2].
RANGE Two races known from the tropical Indo-Pacific. Race *mathewsi* breeds in western Indian Ocean, including Seychelles to Maldives and Chagos.
STATUS Breeds Aldabra (50-70 pairs[3]), Cosmoledo (c50pairs[4]), African Banks (c10 pairs[4]), St Joseph Atoll, Farquhar and Bancs Providence. May be resident in the vicinity of breeding sites year round but also recorded as vagrant to granitic islands and Bird June to October.

THREATS AND CONSERVATION An uncommon breeding seabird in Seychelles effectively protected only at Aldabra, where a small population is confined mainly to tiny islets (it has been recorded at the extreme eastern tip of Malabar). It was probably once more widespread and its present distribution is restricted by the presence of rats and cats elsewhere.

SIMILAR SPECIES At a distance or against sunlight could be mistaken for **Fairy Tern**, adults of which are pure white, have no black eye-stripe and have deep blue base to the black bill.

References: 1. Gaymer (1967), 2. Diamond and Prys-Jones (1986), 3. Betts (2000b), 4. Rocamora and Skerrett (1999).

BRIDLED TERN *Sterna anaethetus*   Plate 33

Other names: Brown-winged (or Smaller Sooty, or Panayan) Tern   **French:** Sterne bridée   **Creole:** Fansen

DESCRIPTION Length 34-36 cm; wingspan 77-81 cm. A small, slim-winged tern; smaller, paler, browner and less uniform above than Sooty Tern, with white forehead extending behind the eye as a narrow, white supercilium. **Adult** is pale sepia brown above, grey-brown on the mantle and white below, greyer on the belly. Crown and nape are black, sometimes separated from greyish upperparts by a narrow, pale collar (often difficult to see). White supercilium accentuates black line through the eye more so than Sooty Tern and the black loral-stripe continues to the gape with even width. Also smaller and shorter-winged, though still relatively long-winged compared to other terns. Flight feathers are a little darker than upperparts (most noticeable at rest). Tail grey-brown with white outer web to outermost feathers. Tail projects slightly beyond the wings at rest. White inner webs of the three outer primaries form a white fan shape at the leading edge of the wing contrasting with grey inner primaries. Contrast on the underwing is less distinct compared to black-and-white pattern of Sooty Tern. Long, black bill is almost equal in length to the head. Legs strong and black. Non-breeding somewhat paler and less distinctly patterned; the crown and lore speckled paler and less demarcated from white forehead, while pale edges to the feathers of mantle and sometimes scapulars adds to less smart appearance. **Juvenile** has upperparts paler than adult, broadly tipped white or cream and underparts mainly whitish, greyer on the flanks and sides of breast; head pattern similar to adult but the crown is brownish with fine white streaking and little contrast between the head and the mantle. Tail shorter and entirely greyish brown. **First-year** similar to adult, outer feathers tipped pale grey, but lacking white outer webs of adult.

VOICE A yapping 'whup-whup' when at breeding colonies. At evening communal roosts, excited barking and trilling calls. At sea, a variety of grating cries.

BEHAVIOUR Though not as pelagic as Sooty Tern may be sometimes encountered out of sight of land. One ringed as a nestling on Cousin in May 1973 was recovered off Pemba May 1974[1]. Gregarious and will associate with other terns. Rarely swims, but readily perches on driftwood, boats etc. Flight buoyant and graceful with deep wingbeats, usually very low over the sea. Generally hovers and dips to surface but will occasionally plunge-dive without submerging. Feeds mainly on small (less than 6 cm) fish, occasionally squid and even aquatic insects.

BREEDING In Seychelles it is unique in breeding synchronously in an eight-month cycle[1]. On Aride, between 1988 and 1997 the cycle varied from six to 9.5 months, averaging eight[2]. Nest site is usually shaded by a rock or vegetation. Clutch is a single egg, greenish speckled brown and purple, laid in a shallow scrape with little or no lining. Incubation, carried out by both sexes, lasts 28-30 days; fledging 55-63 days. Juveniles will leave the colony after about 35 days and are possibly independent of parents by this age. Though synchronous within one colony, colonies on different islands are not synchronous. For example, eggs have been present on Récif when Aride birds were fledging[3], and the cycle on Cousin has been observed to be up to ten weeks ahead of Aride[4].

RANGE Breeds throughout tropical seas, with six races recognised, though these are not separable at sea. Race *antarctica* breeds Indian Ocean including Kenya, Seychelles, Madagascar, Mascarenes and Maldives to Andamans. African population breeds in an annual cycle, unlike Seychelles.

STATUS Breeds in granitic islands on Aride (*c*100 pairs), Cousin (*c*600 pairs[5]), Cousine, Zavé, Booby, Mamelles, Récif (1,000 pairs) and Ile Sèche. Also on Bird, recorded breeding for the first time in 1994. Common in Amirantes, though not known to breed except in small numbers on Desnoeufs. Less common further south and further west, where still breeds on small rat-free islets of Cosmoledo[6] and there are old records of breeding on Farquhar and Providence. Population trends not well known except Aride where 1,100-1,500 pairs were present in 1988[7], but only 141 pairs in 1995[5] and 79 pairs in 1997[2]. The largest colony may be Récif, where 3,000 birds have been counted (J Nevill pers. comm.).

THREATS AND CONSERVATION Though not particularly common, it is not regarded as threatened. Its main breeding colonies in Seychelles are protected and eggs are not generally favoured by egg collectors. Rats may have caused extinction on some islands. Barn Owls pose a threat and pellets examined on Aride showed a high proportion of Bridled Tern remains[2].

SIMILAR SPECIES Sooty Tern is black (not grey and brown) above, larger and lacks white supercilium behind eye. Immatures easily distinguished, Sooty having blackish (not greyish white) underparts.

References: 1. Diamond (1976), 2. Betts (1998), 3. Skerrett (1994c), 4. Carty and Carty (1996), 5. Burger, Lawrence and Davis (2000), 6. Mortimer and Constance (1999), 7. Bullock (1989).

Bridled Tern (left), and Sooty Tern

SOOTY TERN *Sterna fuscata*   **Plate 33**

Other name: Wideawake Tern   **French:** Sterne fuligineuse   **Creole:** Golet

DESCRIPTION Length 36-39 cm; wingspan 82-94 cm. A medium to large tern, blackish above, white below. Long-winged with very long tail-streamers, white at tip. Underwing shows strong contrast between blackish flight feathers and white underwing-coverts. Rounded white forehead-patch is restricted to area in front of eye.

Adult has black cap contrasting sharply with white forehead, separated from white chin and throat by black loral-stripe which narrows from the eye to the base of the bill; edges of this white patch are rounded in front of the eye, (extends as a white supercilium behind the eye in Bridled Tern). Black cap extends continuously to entire black upperparts, tail, rump and upperwing. Outer tail feathers of long streamers white, obvious in flight. Underparts also white with a slight grey tinge to the belly, especially obvious in rear views. When flying low over clear tropical seas often appears turquoise blue below due to reflection from the surface. Underwing has striking black-and-white pattern, blackish grey flight feathers sharply contrasting with white wing-coverts. Non-breeding birds may be more faded and have a few white edges to feathers of upperparts. Bill black; legs black. **Juvenile** has upperparts mainly blackish brown tipped cream, most heavily on scapulars with white belly-patch. Underparts also blackish brown tipped whitish on the lower belly with whitish undertail-coverts and no tail-streamers. Adult plumage is achieved at 1-2 years though may retain some dark spots on the underparts up to five years old. Sometimes retains smoky grey underparts up to time of first breeding.

VOICE A nasal 'er-wakey-wake' (interpreted 'wideawake' giving other name). Breeding colonies are extremely noisy day and night. Alarm calls include a prolonged 'kree-aah' and a short low 'krork'.

BEHAVIOUR Highly pelagic, returning to land only to breed. Highly gregarious, often seen in flocks and with Bridled Terns and Brown Noddies. Strong, bounding flight. Avoids landing on water being unable to swim. Usually dips to surface, rarely plunge-diving, feeding on squid and small fish (mainly 6-8 cm, sometimes larger) including many deep-water species that only ascend to the surface at night.

BREEDING Nests in huge, dense colonies during southeast monsoon. Nest densities are greatest in open areas with only 30-50% vegetation cover[1]. First birds arrive at colonies towards end February or early March, numbers building up strongly into April. High-altitude flight over breeding colonies is a common activity April to May. In courtship, both sexes hold wings vertically over the back with head thrust forward. In ground display, one circles the other, wings drooping, neck arched, head turned slightly away taking short runs; this may be observed throughout the breeding season, especially when changing nest duties. During display the outer bird may break off to attack neighbours in adjacent territories. One egg is laid, usually at end May to early June, in a shallow scrape with little or no lining. Egg is white spotted with various shades of brown. For unknown reasons laying appears to commence earlier on African Banks than elsewhere in Seychelles (recorded from second week in May[2]). Laying is seasonally synchronous and usually diurnally synchronous with 94% of eggs laid between midday and 18.00 hrs[3]. Each member of a pair usually attends to incubation on alternate days though some adopt a two-day strategy and others a less regular regime[1]. Incubation shifts lengthen towards end of laying season. Incubation usually 28 days[4]. Hatching success may vary from 75% in the centre of colony to only 10% at the extreme edges[4]. Most losses are due to predation. Predators include rats on some islands, cats on Cosmoledo, Cattle Egrets and Ruddy Turnstones. Sometimes there may be mass desertion of eggs or young caused by heavy tick infestation[5,6]. If eggs are lost, birds may re-lay 12-13 days later, sometimes laying a third egg if the second is lost; however the survival rate for late eggs is poor[1]. Fledging lasts 56-70 days, young birds remaining near the colony for a further 14-21 days[1]. Juveniles fly over the water erratically, dipping to the surface to collect seaweed and imitating the darting flight of adults over shoals of fish. Post-fledging parental care may last about two months. Studies on Bird show that many return to the same colony each year and chicks return as adults to breed at their natal colony[7,8]. However, in 1995 one Sooty Tern arrived on Aride that had been ringed as a chick on Kermadec Islands, New Zealand in 1961 (a distance of 13,350 km from Aride) and still in good breeding condition at 34 years of age. Also, colour-ringed birds from Bird have been recorded at other Seychelles colonies including Desnoeufs and Farquhar and birds ringed African Banks have turned up at Bird. Egg collection may disrupt the tendency to return to the same colony each year[9].

RANGE Several races not separable at sea, breed throughout tropical seas. Race *nubilosa* (which has pale grey tinge to belly) breeds Red Sea and throughout the Indian Ocean to Philippines. Non-breeding birds are frequent visitors south to Natal. Vagrants reach east of Cape of Good Hope and north to Britain. Ringed birds from Bird Island have been recovered from India and Australia.

STATUS Breeds Bird (*c*700,000 pairs[9]), Aride (*c*360,000 pairs[10]), Récif (*c*10,000 pairs), L'Ilot Frégate (*c*1,000 pairs), African Banks (*c*5,000-10,000 pairs), Desnoeufs (*c*500,000 pairs), Etoile (*c*5,000 pairs), Goëlettes, Farquhar (*c*260,000 pairs[9]) and Wizard (Grand Ile), Cosmoledo (*c*1,100,000 pairs[9]). Some colonies have been wiped out by egg collection, others substantially reduced in size. In 1955 there were breeding colonies on Cousine, Ile Sèche (1,000 pairs), Ile aux Vaches Marines and Rémire that no longer exist today[11]. The same 1955 study found on Mamelles a broken egg but no other evidence of breeding on an island that was once reported to have a large colony. On Cousin, nesting reportedly ceased 1947 or earlier[11], while on Rémire, eggs were harvested ruthlessly and though it was possible to collect 135,800 eggs in 1959 by 1966 the colony was extinct[12]. Today, the colonies on Récif and African Banks are protected areas, though subject to heavy poaching pressure and populations have declined. On African Banks, the colony was estimated at 43,300 pairs in 1955[11], 34,000 in 1966[12] and 20,300 in 1974[13]. On Desnoeufs the current population compares with earlier estimates of 1,121,000 pairs in 1955[11], 1,831,000 in 1966[12] and 1,195,000 in 1979[14]. Eggs are legally collected from roughly two-thirds of Desnoeufs but the actual area protected has not remained fixed in all years. Legal egg collection for export to Mahé was carried out on Cosmoledo only in two seasons (1990 and 1991) though former residents of the atoll also collected in the past, prior to the abandonment of the settlement[15]. On islands where effective protection has been given, the increases in population have been dramatic. On Aride, the population in 1966 was estimated at 60,000 pairs while on Bird, the population rose from 18,000 in 1955 to 120,000 by 1966 during a period when eggs were not seriously cropped[12].

THREATS AND CONSERVATION This remains the most common seabird in Seychelles, though over-collection of eggs and killing of adults and chicks has led to extinction on some islands (e.g. Cousin, Cousine, Ile Sèche, Mamelles, Rémire and Platte) and a drastic decline on others. Colonies on Booby, L'Ilot Frégate, Récif, African Banks and Etoile may be in danger of extinction (despite all except for L'Ilot Frégate having legal protection). The Ministry of Environment protects colonies within the granitics as far as resources permit but no protection is given to outer islands due to logistics and expense. On Aride, poaching is a problem, though vigilant wardening and assistance from Ministry of Environment minimises losses. On Cosmoledo, the level of poaching is unclear but reports include some entering illegally from Comoros. This might threaten the colony, the largest in Seychelles. Loss of habitat may be a threat on some islands, including the spread of introduced plants such as Zepi Ble on Desnoeufs and Manz Tou on Aride. Introduction of rats and cats to some islands has probably had some impact on numbers (e.g. both rats and cats are present on Grand Ile, Cosmoledo[15]) though the habit of breeding in dense, noisy colonies may intimidate some predators and restrict losses to the periphery. On Bird, crazy ants, which may carpet the ground, pose a threat. However, man has had the greatest impact on numbers, a problem recognised as early as 1941[16]. A major study recommended eggs should be collected from only two islands – L'Ilot Frégate and Desnoeufs – and should be restricted to about 20% of the Seychelles population east of the granitic islands[4]. Collection generates an income to Ministry of Environment through a 15% commission on sales and is hailed by some as a model of sustainable exploitation with conservation benefits. However, others question this position. Proponents of egg collecting argue that: 1. A sensible controlled harvest is better than a free-for-all. If colonies had no economic value they would be abandoned to poachers who would take even higher numbers of eggs. 2. This is a traditional harvest important to local culture that should be respected. 3. In some colonies, nesting is limited by the available area so that vegetation management and controlled collection can increase productivity. If eggs are not collected nesting sites might become overgrown and productivity would fall. 4. Conservation must pay for itself if it is to be sustainable and a genuine effort is being made to conserve the species by providing an income to support conservation at sites such as Récif. Opponents counter that: 1. Collection is not economic but must be supported by the State, which is motivated by political, not financial, or conservation considerations. 2. Collection has been responsible for extinction of many colonies and now threatens the existence of others. 3. Seychelles has become a middle-income country with a rich and varied diet; it strongly embraces environmental protection and promotes itself as an ecotourism destination. Continuation of exploitation may be perceived as incompatible with this position. 4. The roots of this harvest as a tradition are questionable. In the past, millions of egg yolks were barrelled and exported for uses such as paint manufacture. 5. Far from dampening poaching pressure, legal collection may encourage demand, support high consumption above what is sustainable and may fuel poaching.

SIMILAR SPECIES Bridled Tern is smaller, shorter-winged, browner above, with narrow, more tapered eyebrow stretching behind the eye as a white supercilium. Immature may resemble a **Brown Noddy** at a distance but noddies are uniform in colour with wedge-shaped tails and stiff-winged flight whereas Sooty Tern has upperparts flecked with white, the tail forked, and more buoyant flight.

References: 1. Feare *et al.* (1997), 2. Skerrett (1995), 3. Feare (1976c), 4. Feare (1976d), 5. Feare (1976a), 6. Feare (1976b), 7. Feare and Gill (1994), 8. Feare and Gill (1995a), 9. Feare (1999), 10. Bowler and

Hunter (2000), 11. Ridley and Percy (1958), 12. Ridley and Percy (1966), 13. Feare (1979b), 14.Wilson and Chong-Seng (1979), 15. Mortimer and Constance (2000), 16. Vesey-Fitzgerald (1941).

LITTLE TERN *Sterna albifrons*   Plate 33

French: Sterne naine   **Creole:** Pti Dyanman

DESCRIPTION Length 22-24 cm; wingspan 51-56 cm. A very small, white, highly active tern virtually indistinguishable from Saunders' Tern except in breeding plumage. **Adult breeding** has a black cap with a triangular white patch on the forehead reaching to or behind the eye (Saunders' forehead-patch is shorter, squarer and never reaches behind the eye). Black loral-stripe, broader in male, widens towards the eye. Upperparts are grey, paler on the rump with contrasting white tail (no obvious contrast in Saunders'). Black primaries form a black leading edge on the outer wing (less extensive than Saunders', typically only one or two outer primaries black, smudged grey). Bill yellow with a black tip. Legs orange-yellow, brighter than Saunders'. In **adult non- breeding** from August onward, the crown whitens, the rear crown remaining black but flecked with white. The black mask is also flecked whitish, and there is a black area in front of the eye. Bill darkens from yellow to black and legs become grey. Tail is greyer, lacking any obvious contrast with upperparts. **First-winter** is similar to adult non-breeding with dark head and crown distinctly spotted dark, scaly upperparts and dusky carpal bar. Bill black with a dirty yellow base.
VOICE A sharp 'ket' or 'ket-ket'.
BEHAVIOUR At breeding grounds, favours inland sites unlike coastal preference of Saunders' Tern. However, non-breeding birds found along coasts. Gregarious, often in huge flocks. Energetic, swift and purposeful flight similar to Saunders' Tern. Feeds mainly on small fish (up to 6 cm), occasionally molluscs and aquatic insects.
RANGE Six races recognised, breeding West Africa and across Europe and to northwest India and from eastern and southern Asia to Australasia. Nominate *albifrons* of Europe to Asia is most likely to occur in Seychelles. Winters southward as far as South Africa.
STATUS At least a vagrant, recorded in the granitics in breeding plumage April, but could be more frequent than is currently known, due to the difficulty of separation from Saunders' Tern in winter plumage. Huge flocks have been reported from Aldabra claimed as this species, but these have not ruled out Saunders' Tern. Any dead Saunders'/Little Terns should be retained for confirmation of identification and preservation.
SIMILAR SPECIES Adult breeding **Saunders' Tern** has smaller, more square white forehead-patch, paler upperparts lacking obvious contrast with rump and tail, duller legs and more extensive black on primaries, including primary shafts. Non-breeding birds are almost identical.

SAUNDERS' TERN *Sterna saundersi*   Plate 33

Other name: Black-shafted Little Tern   **French:** Sterne de Saunders   **Creole:** Pti Dyanman Saunders

DESCRIPTION Length 21-23 cm; wingspan 50-53 cm. The northwest Indian Ocean equivalent of Little Tern, from which it is doubtfully separated. It is only safely separable in adult breeding plumage, rarely seen in Seychelles. A very small, white, active tern. **Adult breeding** distinguishable in the field from Little Tern in first to arrive, August to early September and last to leave in April. At this time, smaller white forehead-patch extends only to the eye where it is square cut or rounded, not extending into a white supercilium. Black loral-stripe is broad in front of the eye, (broader still in male). Upperparts distinctly lighter grey than Little lacking any contrast with rump and tail (Little has rump paler than back). Short tail, with tail-streamers often shorter than Little. Narrow, pointed wings, with outermost primaries blacker than Little, with black (not white) shafts and the black area more extensive and contrasting more sharply because of the paler upperparts (typically at least three outer primaries jet black contrasting with whiter wings and back). Long, slender bill is yellow tipped black and legs shorter, darker and browner than Little.
Adult non-breeding has upperparts darker grey than adult breeding and slightly darker than Little; forehead white, similar in extent to Little, the black mask slightly broader than Little. Bill black and legs grey or brown, sometimes with a yellowish tinge. **First-winter** similar to adult non-breeding, head dark with crown distinctly spotted dark, scaly upperparts and dusky carpal bar. Bill black, sometimes with a dirty yellow base; legs orange-red. By first summer similar to adult non-breeding, bill and legs with a yellowish tinge and head pattern often similar to adult breeding.
VOICE Calls 'ket-ket', similar to Little, possibly not as sharp.
BEHAVIOUR At breeding grounds, prefers the coast more than Little Tern. In Seychelles invariably seen along coasts, beaches and mudflats, generally in small groups. Flight is energetic, swift and purposeful, sometimes hovering with head pointing downward. Hovers, sometimes for lengthy periods, then plunge-dives to the surface, taking off again quickly. Feeds on small fish, sometimes molluscs and insects.
RANGE Breeds Red Sea, Persian Gulf to northwest India, dispersing southward into the Indian Ocean from about August to April. Monotypic.
STATUS Annual throughout Seychelles September to April.

GENERAL Small terns on Bird, African Banks and Cousin were identified in the 1970s as Damara Tern *S. balaenarum* and led to many spurious claims of this species, which almost certainly referred to sightings of Saunders' Tern or perhaps Little Tern[1]. Damara Tern does not occur in the Indian Ocean[2].

SIMILAR SPECIES Little Tern is almost identical in non-breeding plumage but at other times might be distinguished by larger and less square white patch on forehead, less black on wingtips, primaries having white not black shafts and darker grey upperparts contrasting with paler rump.

References: 1. Penny (1974), 2. Feare and Bourne (1977).

BROWN NODDY *Anous stolidus*   Plate 34

Other names: Common (or Greater) Noddy   **French:** Noddi brun   **Creole:** Makwa

DESCRIPTION Length 38-40 cm; wingspan 77-85 cm. An all-dark sooty-brown tern with a whitish cap and wedge-shaped tail. Larger and stockier than Lesser Noddy with heavier bill and pale-centred underwing. **Adult** has greyish white forehead and crown sharply demarcated from black lore (merging into grey at rear crown and nape) giving a more scowling appearance than Lesser Noddy. Upperparts warm dark brown; underparts dark greyish brown. Shorter, broader wings than most terns, the upperwing blackish brown with wing-coverts browner, underwing dark brown with a greyish centre. Broad, blackish brown wedge-shaped tail appears pointed when closed and during moult can even appear forked for a short time. Eye dark brown with white crescents above and below the eye. Bill powerful, thick and black, upper mandible obviously curved; legs dark brown. Sexes similar, male averages larger. **Juvenile** similar to adult, somewhat darker, pale crown less distinct and more restricted, the forehead more greyish, the rear crown browner (almost uniform with upperparts) and more poorly demarcated. Indistinct pale fringes to feathers of upperparts and wing-coverts.

VOICE A harsh, crow-like 'crawk' uttered in flight and at rest, more frequent at night. Also has several cackling and purring calls given during courtship.

BEHAVIOUR Pelagic and gregarious, feeding in flocks, often with other terns, both by day and in moonlight. Flight purposeful and direct, sometimes appearing falcon-like or skua-like, with broader wings and slower wingbeats than Lesser Noddy, emphasising heavier appearance. Hovers, then snatches prey from the surface. May sweep low to drink or bathe, though seldom settles on water for very long. Birds seek out shoaling baitfish, especially where herded by feeding tuna shoals. Though most commonly takes small fish and squid, may take larger items up to 16 cm.

BREEDING Main breeding season in granitic islands, Amirantes and Farquhar group is March to October. However on Aldabra and Cosmoledo the season is September to March, with peak laying in December and January. The reason for this difference is probably seasonal movements of ocean currents and the associated migration of tuna[1]. In courtship, adults nod their heads to each other (hence the name noddy). Most nest in leaf bases of coconut palms where available on islands such as Aride. Also nests in rock or coral cavities and ledges, particularly where trees are few or absent, in small loose colonies. The first eggs are usually laid early May in granitic islands. Invariably a single egg is laid, white with a little faint purple and brown speckling. Adults cool eggs by dipping their breast feathers in the sea before incubation and prevent chicks overheating by standing over them in the heat of the day. There is a striking difference in the colours of chicks, which may be white (one in five), or dark brown with a few fawn intermediates[2]. It may be that in situations where there is predation, one or other colour morph could help camouflage chicks from predators. Adults are very aggressive towards intruders near the nest site. Incubation lasts 31-34 days and fledging may take 37-47 days, varying with food availability. Fledging success is similar at both rock and most tree sites, though losses are much higher in palms where storms can easily dislodge nests[2]. Young birds remain at nest site for several weeks after their first flights, parents continuing to feed them. Late young may still be at nest sites in the granitic islands in October, but typically there is a noticeable exodus at this time. They may move further south to follow the fish shoals, roosting on the nearest island at night. In the granitic islands, a small minority nest out of season (mainly December to February) when studies show they take up to two weeks longer to fledge their chicks[2]. This strategy may be of some benefit if food supplies fail during the main breeding season.

RANGE Five races usually recognised, breeding throughout the tropics. Race *pileatus* breeding Indian Ocean and most of Pacific is slightly longer-winged and longer-tailed than the nominate.

STATUS Breeds Aride (*c*8,000 pairs[3]), Cousin (*c*1,000 pairs), Cousine (*c*900 pairs[4]), Ile aux Vaches Marine, Frégate, Ile Sèche, Mamelles, Récif, L'Ilot Frégate, Zavé, Booby, Coëtivy, Platte, Bird (*c*10,000 pairs[5]), Denis, African Banks (*c*4,000 pairs), Rémire, D'Arros, St Joseph Atoll, Marie-Louise (*c*2,000 pairs[5]), Etoile (*c*1,000 pairs[5]), Desroches, Desnoeufs (several thousand pairs), Alphonse, Goëlettes, Farquhar (10,000 pairs[5]), Cosmoledo (a few hundred pairs[5]) and Aldabra (*c*3,500 pairs[5])[4]. Largest Aldabra colony is Noddy Rock (*c*600 pairs); others include Table Ronde and Pink Rock.[1]

THREATS AND CONSERVATION Eggs are sometimes taken. It seems likely birds once bred on most islands but were wiped out by egg collection, killing of adults and chicks by man and by introduced rats and cats. On Ile du Lys, Iles Glorieuses a mass attack by rats was once reported in which 50-100 roosting birds were killed in a four-hour period though remarkably nearby eggs and chicks were apparently untouched[6]. Natural predators include Pied Crow and Grey Heron.

SIMILAR SPECIES Lesser Noddy is more slender, more narrow-winged, blacker, with more extensive grey crown which merges evenly at the lore, the bill is more slender and proportionately longer than Brown Noddy and the underwing is all-dark. Immature **Sooty Tern** has upperparts flecked with white, a forked tail and more buoyant flight.

References: 1. Diamond and Prys-Jones (1986), 2. Bullock (1989), 3. Bowler and Hunter (1999), 4. Wright and Passmore (1999), 5. Rocamora and Skerrett (1999), 6. Van der Elst and Prys-Jones (1987).

Adult Lesser Noddy (left), and adult Brown Noddy

LESSER NODDY *Anous tenuirostris* Plate 34

French: Noddi marianne **Creole:** Kordonnyen, Kelek

DESCRIPTION Length 30-34 cm; wingspan 60-70 cm. A small to medium all-dark tern, with grey crown, more extensive and less sharply demarcated at the lore than Brown Noddy. Long slender bill. **Adult** has forehead whitish grey, the crown and nape grey merging evenly; there is no sharp demarcation with lore, giving a gentler appearance than scowling Brown Noddy. Upperparts very dark brown with a slight greyish cast and the tail blackish brown. Underparts are charcoal-grey and the underwing evenly dark (unlike Brown Noddy). Lighter in build with narrower wings than Brown Noddy, black bill finer and proportionately longer. Eye dark brown with white crescents above and below the eye; legs black. **Juvenile** similar to adult, pale cap more contrasting and the plumage browner.

VOICE Usually silent at sea, but at colonies gives a variety of purrs and rattles.

BEHAVIOUR Pelagic, but some may roost at breeding sites year round. Gregarious, often feeding in dense rafts about 2 m above surface. Flight more buoyant with more rapid wingbeats than Brown Noddy and less likely to settle on sea. Hovers and dips to snatch small fish and squid driven to the surface by tuna and other predatory fish. Occasionally splashes but does not plunge-dive. Outside breeding season large flocks may roost on outlying islands such as Frégate, Bird and Silhouette, presumably to be closer to feeding grounds than their breeding sites.

BREEDING Breeds during southeast monsoon. Pairs undertake courtship flights in which both birds fly rapidly, close to each other, in long zigzags. Often seen to attempt copulation in a tower of up to five birds. Unlike Brown Noddy nests exclusively in trees. Nests are constructed from damp leaves, especially Pisonia where available, which is soft and malleable when damp, hardening as it dries out. Seaweed is occasionally used by birds nesting near the coast. Prior to egg laying, around April, birds gather in large numbers (usually at dawn) on the beach to consume coral fragments as a source of calcium[1]. Eggs may be laid April to June depending on food supply and the arrival of fish shoals associated with the change in monsoon winds around this time. A single egg is laid, white with a little purplish speckling. Incubation takes 30-36 days, fledging a further 42-47 days[1]. Young birds remain close to the nest site, departing early December. Adults begin drifting back to the colony late February, with a strong build up in April.

RANGE Nominate *tenuirostris* breeds Seychelles, Cargados Carajos, Réunion and Maldives. Another race, *melanops*, breeds western Australia. May possibly be conspecific with Black Noddy *A. minutus*, which replaces Lesser Noddy in the Atlantic and Pacific; in western Australia the two breed alongside each other.

STATUS Breeds Aride (c170,000 pairs[2]), Cousin (c80.000 pairs[3]), Cousine (c60,000 pairs[4]), Frégate (c7,500 pairs[5]), Bird (c300 pairs), Denis, Récif, Rémire and Marie-Louise (c3,500 pairs[6]). Formerly bred African Banks[7]. In 1955, Lesser Noddy was less numerous than Brown Noddy on Aride (fewer than 20,000 pairs), probably because Pisonia, their favoured nesting tree, was cut annually at this time. In 1973 when Aride was purchased by Christopher Cadbury for RSNC the population was 90,000[8]; by 1988 it had risen to 170,000 pairs[1] and has since remained around this figure. Vagrant to Aldabra.

THREATS AND CONSERVATION The three largest colonies in the world (Aride, Cousin and Cousine) are protected, wardened islands. Elsewhere where the bird breeds it is not seriously threatened by egg collectors. Rats probably pose the greatest threat and colonies are largely confined to rat-free islands.

SIMILAR SPECIES Brown Noddy is heavier, browner, broader-winged, heavier-billed with a more restricted grey crown sharply demarcated at the lore and a pale-centred underwing.

References: 1. Bullock (1989), 2. Bowler and Hunter (1999), 3. Burger and Lawrence (1999a), 4. Wright and Passmore (1999), 5. Burger and Lawrence (1999b), 6. Rocamora and Skerrett (1999), 7. Stoddart and Poore (1970a), 8. Warman and Todd (1984).

FAIRY TERN *Gygis alba*   **Plate 34**

Other names: White Tern, White Noddy   **French:** Gygis blanche   **Creole:** Dyanman Blan, Golan Blan

DESCRIPTION Length 28-33 cm; wingspan 70-87 cm. The world's only pure white tern. A graceful, delicate tern with translucent wings and midnight blue base to its black bill. **Adult** has entirely white plumage, unique among terns. Black eye-ring, gives appearance of a large black eye. Tail shorter than most terns with a shallow fork. Slightly uptilted black bill is blue at the base and appears stuck-on. Legs slate-blue to blackish with reduced webs and well-developed claws. **Juvenile** has brown mottling on the nape, upperparts washed grey with brownish or grey scalloping, a black spot behind the eye and the upperwing very pale grey with blackish shafts to the outermost primaries.

VOICE Emits a low purring chuckle when perched. Also a mechanical clicking and a distinctive, buzzing 'byowp' when excited.

BEHAVIOUR Present at breeding islands year round, but also encountered well out of sight of land. Flies at low altitude above the sea, often in pairs, twisting, turning and sometimes hovering. Tame and inquisitive at breeding sites though can be disturbed if approached too closely. Snatches small fish from the surface, occasionally taking squid and crustaceans. May carry to chick up to six small fish, which are neatly arranged in bill.

BREEDING In the granitic islands, may breed year round though most eggs are laid January to March. At Aldabra there is a definite season between early September and early April (by contrast in Chagos most breed May to August). In courtship pairs bill and coo at nesting sites 2-3 weeks before laying a single egg on a bare branch (sometimes on a bare rock), usually in a pit or fork for support. Egg is greyish white with dark grey, black or reddish brown spots denser at the blunt end. Parent sits behind the egg, fluffing out feathers of the lower breast to incubate. Incubating birds sit tight to prevent egg loss, but if disturbed, fall away backwards to avoid dislodging egg. Abandoned eggs are soon taken by predators such as Wright's skink *Mabuya wrightii* or Seychelles Fody. Incubation lasts around 21 days, fledging 50-60 days[1]. Newly hatched chicks have well developed claws to cling to the branch where they were born.

RANGE Four (sometimes more) races are recognised, spread throughout the tropics with race *candida* breeding Seychelles and Mascarenes to Marquesas (Pacific Ocean). Seychelles/Mascarenes populations have sometimes been ascribed to a separate race, *monte*, but this is not separable from *candida* (this distinguished by blue base to bill and dark primary shafts). Other races occur in south Atlantic and Pacific.

STATUS Breeds throughout Seychelles, though only in good numbers on islands where there are no cats or black rats, notably Aride (c1,700 pairs during southeast monsoon and 5,600 pairs during the northwest[2]), Cousin (c.1,200 pairs[3] during southeast monsoon and c3,600 pairs during northwest[4]), Cousine (1,000-1,500 pairs[5]), Frégate (2,000-4,000 pairs[6]), St François (1,000-1,500 pairs[7]) and Marie-Louise (2,000-4,000 pairs[7]). Population figures represent birds present at one point in time and year round figures are much higher. At Aldabra (100-400 pairs), breeds mainly in mangroves surrounding the lagoon, especially the northern rim, with smaller numbers along the northern coast and on islands of West Channels.

THREATS AND CONSERVATION Rats and cats have probably significantly reduced numbers and the largest populations are on islands free of these predators. Additionally, introduced Barn Owls have been blamed for the decline in the granitic group. Certainly many terns were taken on Aride prior to control measures introduced in 1996 where studies showed remains of Fairy Terns in 80% of owl pellets. Yet Aride retains the largest population in Seychelles. Also, breeding Fairy Terns made something of a comeback on Mahé during the 1990s indicating some other factor must have contributed to an earlier decline here and on other islands. A reduction in direct human interference is probably an important reason for the reversal in the species' fortunes. Older Seychellois recall how airguns were popular prior to the 1977 coup d'état; any bird was considered a target and a stationary white bird on a bare branch would have been easy prey. With regard to the more modern human activity of tourism, a study on Cousin found no significant difference in the number of breeding attempts or breeding success within 5 m or from 5-10 m of tourist paths[8]. If there was any discernible trend it was towards greater breeding effort and success close to tourist paths. A later study on Aride also found no significant difference between breeding success within 10 m of tourist paths and success beyond this limit[9].

SIMILAR SPECIES Unlikely to be mistaken for any other seabird except at a distance. **Black-naped Tern** which has very pale grey upperparts, black eye-stripe and an all-black bill.

References: 1. Bullock (1989), 2. Bowler and Hunter (2000), 3. Burger and Lawrence (1999a), 4. Burger (2000b), 5. Wright and Passmore (1999), 6. Burger and Lawrence (1999b), 7. Rocamora and Skerrett (1999), 8. Wilson (1982), 9. Haysom (1995).

FERAL PIGEON *Columba livia*   **No colour plate**

Other names: Common (or Domestic, House, Roof or Street) Pigeon
French: Pigeon biset   **Creole:** Pizon Domestik

DESCRIPTION Length 29-36 cm; wingspan 63-70 cm. One of the most familiar birds of city centres across much of the northern hemisphere, now well established in Seychelles. **Adult** plumage is variable, but usually blue-grey with a purple and green sheen on the neck, a white rump and two black wingbars. Some are predominately white, black or piebald. Feral birds usually have shorter, stouter bills than their wild ancestors. Bill grey, cere whitish; legs red. **Immature** is duller and darker, with markings less well-defined compared to adult. Little gloss on neck.

VOICE A monotonous 'oo-roo-coo' frequently repeated.

BEHAVIOUR Common around buildings, grain warehouses and airstrips. Highly gregarious, usually found in small flocks. Flight strong and direct. Diet mainly seeds.

BREEDING May breed in any month. In courtship, male performs a display flight with slow deep wing-beats, some wing clapping followed by gliding with wings raised and tail spread. Male displays to female by spreading the ruffling head and neck feathers and bowing deeply while continually crooning. Untidy nest built of twigs, often colonially, frequently on ledges of larger buildings such as Central Police Station, Victoria. Away from human settlement may nest in coconut palms. Two white eggs are laid. Incubation takes 17 days and fledging about 35 days, this short period making it possible to raise several broods annually. Both sexes share nesting duties. Chicks are fed on pigeon milk, a rich secretion regurgitated from the crop of the adult.

RANGE Original stock bred mainly on coastal cliffs and other rocky outcrops from Europe to India. This wild range has contracted, while feral stock has spread throughout Europe and Asia, sometimes mixing with wild forms. Introduced and common over much of North and South America, (as far north as Alaska, south to Tierra del Fuego), especially in large cities and farmland. Also West Indies, Pacific Ocean islands, Australia, New Zealand, St Helena, South Georgia and elsewhere. Probably introduced to Mauritius from Europe around 1715 and now found throughout the island[1].

STATUS Introduced and resident Mahé, Praslin, La Digue and Silhouette. It is unknown when introduction took place. This may be a fairly recent event; as in the early 1970s the only wild birds known in Seychelles were a small population on Frégate. Birds also existed on Alphonse in the 1950s until destroyed on the advice of a visiting doctor as rain water was collected from roofs for drinking (D Gendron pers. comm.). On Mahé, numbers greatly increased during the 1990s, especially around the port area and grain warehouses. It is likely to spread further unless controlled as it can cross between islands as demonstrated by one bird reaching Aride in 1996. On Assumption, a pair was brought from Aldabra around the mid-1980s, having been kept in captivity by a member of staff on Aldabra (contrary to regulations). This pair were released and multiplied to 69 birds by 1994[2]. An attempt at eradication in 1995 reduced population to five birds[3]; these were eradicated the following year.

SIMILAR SPECIES None in Seychelles.

References: 1. Rountree *et al.* (1952), 2. Skerrett (1994a), 3. Skerrett (1996b).

EUROPEAN TURTLE DOVE *Streptopelia turtur*   **Plate 35**

Other names: Turtle Dove, Common (or Western, or Isabelline) Turtle Dove
French: Tourterelle des bois   **Creole:** Tourtrel Eropeen

DESCRIPTION Length 26-28 cm; wingspan 47-53 cm. A small, slim, graceful dove, adult with a black-and-white patch on the side of the neck. Smaller, slimmer and paler than Oriental Turtle Dove. **Adult** neck pattern rules out all but Oriental Turtle Dove: 3-4 wide bars on the side of the neck, usually against a whitish background (4-6 bars in Oriental Turtle Dove, usually against a bluish background). Breast is tinged lilac, the belly and undertail-coverts whitish. Back and rump blue-grey, often with some brown areas. On upperwing, chestnut or orange-buff coverts have clear-cut black centres with a blue-grey panel on the inner wing. Diamond-shaped tail is dark grey to black, edged white (as in Oriental race *meena* though race *orientalis* tail edged grey). Yellow eye, black at periphery of the eye. Red eye-ring, more obvious than in Oriental. Legs dark pink. Bill blackish with pale tip. **Immature** similar to adult but less distinctive and browner, lacking neck-patch. Brownish breast does not extend to the belly (as it may in Oriental), sometimes with diffuse, pale ochre feather fringes (orange-pink when present in Oriental) giving a scaly appearance; brown primaries have diffuse pale edges.

VOICE A monotonous purr. Mainly silent in winter quarters.

BEHAVIOUR Favours open woodland, feeding on the ground and roosting in trees. Highly gregarious on migration, flocks often numbering many thousands but Seychelles vagrants have been solitary. Flight rapid and agile, wings held well back, with bursts of wingbeats in which the tail pattern is revealed. Often glides and spreads tail on landing, again revealing pattern. Feeds mainly on the ground. Diet mainly seeds (sometimes snails); often found near farms feeding on grain.

RANGE Breeds western Palearctic to about 85°E. Nominate *turtur* breeds western Europe to Siberia; race *arenicola* breeds Morocco to northwest China. Two other races breed within Africa. Most populations winter in Sahel region but *arenicola* (slightly smaller and paler) has been recorded wintering India. Vagrant to Somalia, Kenya, South Africa and Maldives.

STATUS Vagrant to granitic islands, Bird and Aldabra November to December. Race unknown but probably the nominate or possibly *arenicola*.

SIMILAR SPECIES Oriental Turtle Dove is darker, larger and heavier (recalling Feral Pigeon in shape rather than a slim European Turtle Dove) with brownish nape contrasting with grey crown, five or more black bars on neck-patch, larger dark grey centres and narrower rufous-fringes to wing-coverts and darker brownish pink breast. In flight, shows two narrow pale wingbars with smaller, darker blue-grey wing-panel and shorter tail may have a grey terminal band (race *orientalis*).

ORIENTAL TURTLE DOVE *Streptopelia orientalis*   **Plate 35**

Other names: Rufous (or Eastern, or Mountain) Turtle Dove
French: Tourterelle orientale   **Creole:** Tourtrel Oriental

DESCRIPTION Length 33-35 cm; wingspan 53-60 cm. A large, heavily built, generally dark turtle dove, longer and heavier than European Turtle Dove. **Adult** has forehead and crown grey contrasting with vinaceous-brown nape, hindneck and mantle (nape and hindneck of European is greyish, as is rest of head, contrasting with brown mantle). Neck with 4-6 dark bars on the side (3-4 in European), normally with a bluish (sometimes whitish) tone to pale surrounding patch; (this ground colour is merely an indicator, not diagnostic). Breast brownish pink, this colour more extensive towards the vent than the breast colour of European, which is also more lilac-tinged. Undertail-coverts of race *orientalis* are grey, though *meena* has white undertail-coverts (as European). Large, dark grey centres to wing-coverts with diffusely defined rufous margins give a scaly appearance. Often shows a double wingbar both at rest and in flight (wingbars virtually absent in European) and a smaller, darker blue-grey wing-panel compared to European. Tail is grey-black, edged grey in *orientalis* (white in *meena*) and shorter than European. Outer web of outermost tail feathers is dark, variable in extent, more restricted in *meena* (always pure white in European). Back and rump usually uniform blue-grey (European usually shows some brown). Outermost tail feathers usually blackish on the outer web (not usually so in European). Eye-ring variable but usually red, generally more limited and difficult to observe compared to the obvious red eye-ring of European. Bill dark grey, purple at cere and base; legs purplish. **Immature** similar to adult, generally less distinctly marked sometimes lacking neck bars, with darker wing-coverts tipped pale grey forming more obvious pale wingbars. Breast darker than European and breast colour more extensive (as in adult), sometimes with orange-pink feather fringes (pale ochre if present in European). Generally a browner nape and bluer rump than European. As adult, often shows more solid dark centres to wing-coverts and scapulars than European, black primaries with well-defined rufous edges. Adult characteristics acquired during first winter.
VOICE A hoarse coo, usually of four syllables. Also calls 'doo doo-doo, hoo-hoo hoo haw' repeated monotonously. Very different to the purr of European.
BEHAVIOUR Favours open woodland. Forms small parties in winter, but vagrants likely to be solitary. Flight strong, less agile than European, showing broader wings. Usually very wary. Feeds on seeds and berries.
RANGE Breeds east, central and southern Asia, west to Urals. Northern populations migrate to India and South-East Asia, occasionally reaching Maldives. Vagrant westward to Britain. Six races generally recognised including nominate *orientalis*, which breeds from central Siberia to Japan and south to Himalayas wintering South-East Asia, and *meena* of central Siberia to Iran and Kashmir wintering India. Other races occur but are mainly sedentary.
STATUS Uncertain; SBRC has accepted some reports of turtle doves as being indeterminate between Oriental and European.
SIMILAR SPECIES European Turtle Dove is smaller and more slender with a black subterminal band to the longer white-rimmed tail, has a browner rump, a larger paler blue-grey wing-panel and no wingbars. Nape is grey (not vinaceous-brown) concolorous with forehead and crown (not contrasting) and has 3-4 (not 4-6) neck bars. Dark centres to wing-coverts are smaller, more jet black and more clearly defined from broad orange-brown fringes.

MADAGASCAR TURTLE DOVE *Streptopelia picturata*   **Plate 35**

Other names: Painted (or Red Turtle) Dove   **French:** Tourterelle de Madagascar or Pigeon de Madagascar
Creole: Tourtrel Dezil

DESCRIPTION Length 25-26 (*rostrata*), 31-32 (*picturata*), 29-30 cm (*coppingeri*); wingspan 37-38 (*rostrata*), 44 (*picturata*), 42-43 cm (*coppingeri*)
A sturdy, medium-sized, mauve ground-feeding dove. **Adult** of race *rostrata* is characterised by smaller size, dark plum-purple head and blue-grey lower breast (pink-purple in other races). Dark pink-purple upper chest, mauve, sky blue and black necklace, grey belly and white vent. Purple-brown mantle merges to dark brown on the lower back and scapulars with wing-coverts dark blue-brown and a dark brown rump. Race *picturata* has blue-grey head, mauve-and-black necklace, purple-pink breast, paler pink-buff belly and white undertail-coverts; tail pale grey below. Mantle, scapulars and wing-coverts purple-brown. Wings are brownish, the tail grey-brown with ash grey outer tail feathers tipped white. In the Aldabra group race *coppingeri* occurs, having dark purple-mauve head and dark pink-purple breast merging to greyish belly and white vent, the necklace mauve and black; upperparts and rump deep blue-brown. Eye dull yellow with orange eye-ring. Bill blue-grey, darker at base. Legs reddish. **Immature** lacks the mauve colour of adult, generally more dark brown. Upperparts are dull sepia brown-grey, wing-coverts brown

with rufous tips, chestnut-orange shafts to flight feathers, bright rufous tips to secondaries and tertials and grey-brown tail. The breast is dull brown-grey, buff tips on feathers of lower breast forming a buff wash. Throat very pale buff-grey and only one or two dark spots on the neck. Belly whitish in race *picturata*, grey in race *coppingeri*, the latter being generally paler and duller. Eye dark brown. Bill dirty yellow with paler base. Legs dull reddish at front, pale grey at rear.

VOICE A repetitive 'oo-coo-rrr', usually uttered while perched in trees. Song 'worp-oo-oo, worp-oo-oo'.

BEHAVIOUR Frequents open woodland, woodland margins, roadsides and clearings. May be solitary, in pairs or in small parties. Strong, rapid flight, usually snapping wings two or three times on take off. At Aldabra, birds frequently cross between islands of the atoll via the channels. Perches and roosts in dense vegetation. Diet mainly seeds, with some fruit and occasionally insects and small snails.

BREEDING Breeds mainly October to February, but may also lay outside these dates. In courtship the male follows the female dragging the tail and sometimes wings, the head raised, occasionally accompanied by bubbling coos and an intensification of tail- and wing-dragging. When the female strides away from the male, the latter flutters and hops to catch up. Males sometimes fight, charging with head lowered and beak held out until one retreats. A loose nest of twigs is constructed in a tree or bush, usually at a fork. Usually two eggs are laid. Chicks, covered in bright orange down at first, are fed on 'pigeon milk', the regurgitated cellular lining to the adult's crop

ORIGINS Endemic to the western Indian Ocean. Originally included in the genus *Columba*, some authorities consider it may be linked ancestrally to Lemon Dove *Aplopelia larvata* of Africa. It is longer-legged and bulkier than other *Streptopelia* species, which may be simply an adaptation to terrestrial life. Race *picturata* is endemic and abundant throughout Madagascar, except for the High Plateau. Races also in Comoros (race *comorensis*) and Diego Garcia (race *chuni*). Nominate race introduced to Diego Garcia, Reunion and Mauritius. Remarkably, Iles Glorieuses has apparently been invaded from Aldabra by race *coppingeri*.

STATUS Four (possibly five) races: one endemic to the Amirantes now extinct, one endemic to the granitics, one generally presumed introduced and present in the granitics and Amirantes, one endemic to Aldabra and one at Cosmoledo possibly the same race as at Aldabra or possibly a hitherto undescribed race. Two races occur in the granitic islands, the endemic *rostrata* and the nominate *picturata*, which is thought to have been introduced. The plum-purple-headed *rostrata* has been assumed to have bred with the grey-headed *picturata* so that today just a few *rostrata*-types survive on Aride, Cousin, Cousine and Bird. Even on these islands it has been suggested there may be few or no pure forms left. However, there has been some suggestion that birds in the granitics may actually be reverting to the red-headed form following sightings of such birds on Silhouette and even Mahé. Alternatively, this form may have survived at higher altitudes and is now spreading to the lowlands following a reduction in persecution[1]. In one survey on Aride, 10% of the population of 200-400 birds showed features of *rostrata*[2]. The origin of Bird's turtle doves is interesting as none was noted in 1970s and 1980s and they only became common in 1990s; birds were probably always present in small numbers and may be genetically closer to *rostrata* than any elsewhere[3]. Race *picturata* is resident on the main granitic islands where woodland exists, and on Denis. Race *picturata* has usually been considered introduced. Newton visiting Mahé in 1865 related that it '...was introduced, it is said, some years ago by a late Inspector of Police. It is not very common and I saw it nowhere else'[4]. However, in 1767, prior to settlement, two types of pigeon were reported, red and grey suggesting both *rostrata* and *picturata* may have been present even before permanent settlement of the islands[5]. Race *coppingeri* is resident throughout Aldabra. It has been assumed this same race may be the one also resident on South Island, Cosmoledo[6,7]. However, differences in plumage and ecology suggest the latter may be an undescribed race[6,7]. It was long thought to be extinct on Cosmoledo but was rediscovered in 1982[8]. An attempt at translocation to Menai in the late 1980s apparently failed[9]. Race *coppingeri* is extinct on Assumption (it was probably this race last noted in 1908[10] until three were seen in 1977[11], again illustrating the ability of the species to cross open water). Probably once occurred on Astove, where Stirling in 1836 found 'a bird which seemed to be between a pigeon and a dove'[12]. In the Amirantes, race *saturata*, also mauve-headed, was resident on Alphonse, Poivre, Desroches and almost certainly elsewhere in the group but is now extinct probably due to human influences. Survived on Alphonse until the early 1950s when it was deliberately exterminated as a pest on drying copra (D Gendron, pers. comm.). Today, race *picturata* exists in the Amirantes on D'Arros, St Joseph Atoll and Rémire, though the origins of these birds – introduced or natural colonisers – is not known. They were first noted in the Amirantes when Parker collected two specimens on St Joseph in September 1967. These have been assumed to be introduced, but without supporting evidence, except that the birds were smaller-winged than typical *S. p. picturata* and were assumed to have hybridised with *S. p. rostrata*[13]. They are powerful fliers, have crossed open ocean to Assumption recently and even reached Iles Glorieuses historically so it is not impossible that race *picturata* arrived in Amirantes naturally perhaps even subsequent to the extinction of *saturata* and did not wipe out this race through inter-breeding, as is sometimes assumed. It is likely that Madagascar Turtle Doves of one race or another occurred throughout Seychelles prior to the arrival of man. For example, there are references to birds, almost certainly this species, on Farquhar, where Commander Farquhar found 'several kinds of doves are resident: one, a very small, short-winged species (*Turtur rostratus*) is abundant'[14]. Also it was probably this species reported at one time from Providence as 'une sorte de pigeons bruns'[15] and St Pierre as 'pigeons bruns'[16].

THREATS AND CONSERVATION Dupont recorded that on Aldabra it was 'being used as an article of food and there is some chance of this being destroyed entirely'[17]. In the 1970s it was much more common in eastern Aldabra then western, possibly due to past exploitation[18]; today, it appears once again to be common. Abandonment of human predation has undoubtedly led to an increase in numbers both at Aldabra and the granitic islands. A genetic study of birds on Cousin, Aride, Cousine and Bird could determine whether any pure *S. p. rostrata* survive. If so, it may be possible to reduce or prevent further hybridisation only on Bird, where efforts might usefully be concentrated. Similar work might also determine whether the birds on Cosmoledo are a distinct race and would help to establish the origins and relationships of all the races in Seychelles, perhaps leading to conservation implications.

SIMILAR SPECIES None

References: 1. Gerlach (1999), 2.Betts (1998), 3. Feare and Gill (1995b.), 4. Newton (1867), 5. Gerlach (1998), 6. Skerrett (1999b), 7. Rocamora and Skerrett (1999), 8. Mortimer (1984), 9. Mortimer and Constance (2000), 10. Fryer (1911), 11. Prys Jones *et al.* (1981), 12. Stirling (1843), 13. Benson (1970c), 14. Farquhar (1900), 15. Unienville (1838), 16. Froberville (1848), 17. Dupont (1907), 18. Benson and Penny (1971).

BARRED GROUND DOVE *Geopelia striata*   Plate 36

Other names: Zebra (or Peaceful, or Placid or Barred) Dove
French: Géopélie zébrée or Colombine zébrée   **Creole:** Tourtrel Koko

DESCRIPTION Length 20-23 cm; wingspan 24-26 cm. A very small, long-tailed ground-feeding dove. The only small dove resident in Seychelles. **Adult** has upperparts grey, tinged brown with heavy black-and-white barring on the sides of head, neck and flanks. Rest of underparts are pinkish, paler on belly. Face is grey, contrasting with brown hindcrown and pale blue exposed skin around the eye and lore and at base of bill. Tail long, slender and graduated, outer feathers broadly tipped white. Wings have bright tangerine webs on primaries, obvious on take off and more noticeable beneath. Blue-grey coverts on underwing form a paler bar at base of flight feathers. Bill bluish grey; legs pinkish with purple scaling. **Juvenile** similar to adult except paler, scruffier and shorter tail. Lacks dense barring on flanks, pale buff bars on coverts, no blue eye-ring or bill.

VOICE A rhythmic 'popitu-pop-po-pop' and a high-pitched trill followed by cooing. In courtship, with raised spread tail exposing white tips of outer tail feathers, a harsher 'caaw-caaw-caaw'.

BEHAVIOUR Frequents open areas from the coast to high hills. Solitary, in pairs or occasionally small flocks. Tame and confiding especially when attracted to food around houses, hotels and restaurants. Takes off vertically, flight direct. Feeds on seeds, mostly of grasses, sedges and herbs, so mainly found in open areas. Also takes insects and small invertebrates. Around human habitation feeds on rice, crumbs from tables etc. where these are made available.

BREEDING Breeds year round, but mainly October to April, song rendered throughout this period. In courtship male chases female periodically bowing and raising tail vertically, spreading tail feathers to reveal white tips. Builds nest of casuarina needles, grass and other materials on top of a rough platform of twigs, often high in trees, sometimes low down on rat-free islands. Usually two white eggs are laid, both parents sharing incubation and rearing duties. Generally only one chick survives. Incubation lasts 13 days, fledging 11-12 days. Chicks can fly at 21 days.

RANGE Resident South-East Asia to Australia. Introductions are widespread in the Indian Ocean, to Mauritius, Réunion, Rodrigues, Iles Glorieuses, Chagos and Seychelles; elsewhere introduced to Tahiti, Hawaii, St Helena and parts of South-East Asia and Indonesia outside the species' natural range. Newton in 1865 found them common in coastal areas of Seychelles and claimed they were introduced from Mauritius[1]. In Mauritius they had been introduced by 1768[2]. Birds of Chagos were introduced from Seychelles in 1960[3]. Monotypic.

STATUS Introduced; common resident on all main granitic islands, Bird, Denis, Coëtivy, D'Arros, St Joseph, Desroches, Farquhar and Assumption. Introduced on Assumption from Mauritius in 1976.[4]

SIMILAR SPECIES None, but can resemble **Seychelles Kestrel** from a distance when seen in silhouette perched on wires or posts.

References: 1. Newton (1867), 2. Bernadin (1773), 3. Loustau-Lalanne (1962), 4. Prys-Jones *et al.* (1981).

COMORO BLUE PIGEON *Alectroenas sganzini minor*   Plate 36

Other names: Sganzin's Wart (or Blue Wart) Pigeon, Comoro Fruit Dove
French: Pigeon bleu des Comores or Founingo des Comores   **Creole:** Pizon Olande Aldabra

DESCRIPTION Length 24-25 cm; wingspan 36-40 cm. The only blue-and-white pigeon at Aldabra, where it is the counterpart of Seychelles Blue Pigeon. **Adult** has deep blue upperparts, belly and tail and a silvery white head, neck and breast. Lacks wattle of Seychelles Blue Pigeon but has a red oval area around the eye. Silvery white plume feathers on back of the neck may have minute red tips, though these are rarely visible in field. Slight trace of red or purple in tail. Sexes similar, male slightly brighter silver-grey on neck and breast than female. Eye yellow; bill yellow; legs dark grey. **Immature** is duller than adult, eye dark brown with paler brownish or olive surround. More olive green tones above and below. Lacks red and blue of adult. Bare area around eye usually dull purple, occasionally dull olive.

VOICE A low coarse coo repeated four or five times.

BEHAVIOUR Frequents trees; in Comoros it is associated with evergreen forest while on Aldabra, is especially plentiful in mangroves (perhaps the nearest equivalent available) where it roosts and breeds[1]. Tame, often perching in the open. Solitary, in pairs or groups of up to 20 when feeding. Flight strong and swift often crossing channels between islands of Aldabra. Diet is fleshy fruits including those of ficus, bwa zil mowa and vouloutye[1]. Frequently flies between islands mainly across channels.

BREEDING Courtship similar to Seychelles Blue Pigeon, including similar display flight commonly seen September onward. Breeds mainly October to March. Loose platform of twigs built in a tree at 1-4 m above ground or in mangroves, generally concealed deep within vegetation. One white egg is laid. Incubation is probably about 28 days and fledging 21 days, as is the case for Seychelles Blue Pigeon.

ORIGINS Genus endemic to western Indian Ocean (see Seychelles Blue Pigeon). Differs from nominate *sganzini* of Comoros in slightly smaller size, though there may be some overlap.

STATUS Endemic race, found only on Aldabra (1,000-2,500 pairs[2]), never recorded with certainty elsewhere, but it may have been this species recorded in the 19th Century on Cosmoledo and on Astove where 'pigeon hollandais' was described as ' smaller and paler than those found at Seychelles'[3]. By 1907 it was extinct[4]. In addition, Captain Fairfax Moresby recorded in St Pierre and Providence, probably around 1822, 'a species of small blue pigeon are in great abundance and so seldom disturbed that they do not fly at man's approach, but are knock'd down with Sticks, we found them excessively good eating, these birds build and nest on the pisonia tree and other dwarf trees which cover the surface of the islands'[5]. Regrettably the exploitation of these birds was the probable cause of their extinction and nothing further is recorded of their existence.

THREATS AND CONSERVATION Once heavily exploited on Aldabra. Their future should be secured with Aldabra's designation as a nature reserve and World Heritage Site.

SIMILAR SPECIES None on Aldabra.

References: 1. Benson and Penny (1971), 2. Rocamora and Skerrett (1999), 3. Rivers (1878), 4. Dupont (1907), 5. Stoddart and Benson (1970).

SEYCHELLES BLUE PIGEON *Alectroenas pulcherrima*   Plate 36

Other names: Red-crowned (or Warty-faced) Blue Pigeon, Red-crowned Wart Pigeon, Seychelles Fruit Dove   **French:** Pigeon bleu des Seychelles or Founingo rougecap   **Creole:** Pizon Olande Sesel

DESCRIPTION Length 23-25 cm; wingspan 38-40 cm. A fairly large, sturdy, broad-winged pigeon, mainly dark blue with a whitish bib and red bare skin around the eye and top of the head. **Adult** has deep blue upperparts, wings, tail and belly, which may appear blackish or purple-black in some lights and primaries have a smoky grey cast. Blue areas contrast sharply with the greyish white breast, neck, sides of face and nape. Scarlet wattle in front of the eye extends as a cap on the forecrown and around the eye. Greyish plume feathers around the necks of adults may be crimson-tipped, though rarely visible in field. Male slightly larger with larger red wattles. Eye yellow; bill yellow; legs dark grey. **Immature** similar in structure to adult but with darker grey head and bib, many feathers narrowly edged green. Feathers of crown much darker still with obvious green edges. Primaries blackish grey with bluish cast and narrow, pale outer edges. Rest of wing and upperparts also blackish grey with bluish cast, primary coverts and greater coverts edged white, rest of wing-coverts, scapulars and upperparts edged green or yellowish green. Dark pink wattle around eye, eye yellow. Bill dark pink with grey tip.

VOICE A low hoarse coo, 'kwo-ko-ko-ko-ko-ko-ko-ko-kop', often uttered from dense vegetation.

BEHAVIOUR Frequents woodland from coast to high hills. May be solitary, in pairs and occasionally in small groups when feeding. Flight strong and swift. Diet mainly seeds and fruits, including those of takamaka, vouloutye, cinnamon and guava.

BREEDING Breeding may take place at any time of year but is concentrated in October to April. In courtship may fly high above trees, before plummeting downward at a steep angle, wings held rigidly forward and downward. This procedure may be repeated before landing in a tree. Also male bows and coos to female, raising plume feathers on the neck and head or struts up and down a branch plumes extended, turning head 180° then back again. Nest, built entirely by the female, is an untidy platform of twigs in a tree or shrub. Usually one white egg is laid, occasionally two. Incubation takes about 28 days, fledging about 21 days.

ORIGINS Genus is endemic to the western Indian Ocean where represented by Madagascar Blue Pigeon *A. madagascariensis*, Comoro Blue Pigeon *A. sganzini* (race *minor* at Aldabra and nominate *sganzini* in Comoros) and formerly Mauritius Blue Pigeon *A. nitidissima* of Mauritius (extinct c1826). Strong contrast between pale areas and blue areas of plumage and blue tail (not red as in Madagascar Blue Pigeon), suggest it is closest to Comoro Blue Pigeon[1]. May have arrived from Comoros by an earlier colonisation than that of Aldabra[1]. *Alectroenas* may have evolved in Madagascar from Asian and Australasian *Ducula* and *Ptilinopus*[2]. Described by Sonnerat and given the name 'Pigeon violet à tête rouge d'Antigue', after Antigue, the port of Panay, Philippines, presumably due to confusion in his collections.

STATUS Endemic to granitic islands. Numbers were reduced by persecution, but now making a comeback. In 1980, was rarely seen below 200 m, but now occurs to sea level. Fairly common on Mahé, Praslin, La Digue and other large wooded granite islands. Introduced to Cousin 1990 and now breeds successfully.

Became extinct on Aride (previously known only from three specimens in the American Museum of Natural History, collected in 1905), but has re-invaded by natural means; first recorded October 1990 when five birds were noted[3] and has subsequently bred in small numbers. Also re-established on Curieuse during 1980s in line with increase in woodland.

THREATS AND CONSERVATION Once taken for food, but this practice has now virtually ceased. In 1906 it was said to be '...readily caught by means of a noose fastened to a long stick, it falls an easy victim to natives, by whom the flesh is greatly esteemed'[4]. It was also shot by farmers as an alleged pest. Closely related Mauritius Blue Pigeon was persecuted by hunting parties, encouraged from 1775 onward by a bounty paid for the killing of all 'vermin' (vermin being interpreted very liberally) and finished off by introduced monkeys. Seychelles escaped introduction of monkeys but introduced rats and cats may take a heavy toll. Habitat conservation is important to the survival of the species because blue pigeons have muscular gizzards, which break down seeds to aid digestion. Consequently, unlike other pigeons, their droppings contain no viable seeds.

SIMILAR SPECIES None in granitic islands.

References: 1. Benson (1984), 2. Goodwin (1967), 3. Castle and Mileto (1991), 4. Nicoll (1909).

GREY-HEADED LOVEBIRD *Agapornis canus* Illus. below

Other names: Lavender-headed (or Madagascar) Lovebird
French: Inséparable à tête grise **Creole:** Pti Kato Ver

DESCRIPTION Length 15 cm; wingspan 22-23 cm. A small green-and-grey parrot introduced early in the 20th Century, now extinct. **Male** is mainly green; more yellow-green below and brighter green on rump with grey head, nape and upper breast; black underwing-coverts and black band on green tail. **Female** is entirely green, paler on the head. Bill bluish white on upper mandible, pale pink on lower; eye dark brown; legs pale grey. **Immature** similar to adult, male's grey head suffused greenish. Bill yellowish, black at base of lower mandible.

VOICE Highly vocal, particularly in flight. Gives a sparrow-like chirp or a quiet chatter and sometimes squawks or whistles.

BEHAVIOUR Frequents woodland margins and the vicinity of human habitation. Gregarious, seen in flocks of up to 50 at one time on Mahé[1]. Fast, direct flight. Feeds on seeds.

BREEDING Nests in tree holes. Lines nest with pieces of chewed leaves, bark and grass. Clutch in Madagascar usually three white eggs. Season in Seychelles was recorded as October to March[1]. Both sexes share incubation which lasts c23 days; fledging 43 days.

RANGE Endemic to Madagascar; nominate race in the east and west of the island and race *ablectanea* in the south. Introduced to Comoros; also introduced Réunion and Mauritius, where now extinct.

STATUS Introduced early in 20th Century to Mahé and Silhouette, now extinct. According to Dauban, was introduced around 1904-1906 (Loustau-Lalanne claims 1906[1]), during a visit by a vessel of 'Messageries Maritimes,' which used to call at Seychelles monthly en route to Marseilles. A cage full of Grey-headed Lovebirds was smashed one day during unloading at Victoria and the birds escaped[2]. Dauban recalled as a boy, seeing birds in Victoria's gardens from 1910 and later seeing large flocks in Gordon Square from 1925. Died out for unknown reasons, possibly due to inbreeding, or perhaps competition and nest predation from Common Myna and/or Seychelles Bulbul. By 1962, it had disappeared from the east coast and was confined to the area from Souvenir to Grand Anse then south to Anse à la Mouche[1]. No reliable sightings on Mahé since 1977. There is also a record of a single male on Assumption in 1977, of unknown origin; several species were introduced from Mauritius at this time[3], but not all established themselves, including for example c100 Budgerigars (S. Blackmore pers. comm.).

References: 1. Loustau-Lalanne (1962), 2. Dauban (1979), 3. Prys-Jones *et al.* (1981).

Grey-headed Lovebirds, male (left), and female

SEYCHELLES BLACK PARROT *Coracopsis (nigra) barklyi* **Plate 36**

French: Perroquet noir (or Cateau noir) des Seychelles **Creole:** Kato Nwar

DESCRIPTION Length 26-31 cm; wingspan 48-54 cm. The only surviving endemic parrot of Seychelles. Plumage entirely dark grey-brown, appearing blackish from a distance or against bright sunlight. Typical parrot in structure, with large head, large hooked bill, broad rounded wings and slightly rounded tail. **Adult** is dark grey-brown above and below except for greyish undertail-coverts. Outer webs of primaries and primary coverts may have a grey-green cast. Dark bill becomes slightly paler during the breeding season. Some bluish bare skin around the eye visible at close range. Brown legs. **Immature** similar to adult but duller and paler with pale undertail-coverts and yellowish tinge to bill. Also pale tips to wing-feathers and mustard flush to face, throat and upper breast.

VOICE A variety of unmistakable high-pitched whistles. Typically a trisyllabic 'weee-tooo-twee', lower on the second note. Common Myna may sometimes impersonate Black Parrot, but whistle is less strong.

BEHAVIOUR Frequents river valleys, woodland, gardens and cultivated areas wherever suitable food plants are available. May be seen early morning and late afternoon flying in and out of Vallée de Mai and down to sea level at Grand Anse and elsewhere. Fairly gregarious, often in pairs or small groups feeding in treetops. Flight powerful with slow wingbeats and intermittent gliding. Agile, acrobatic and graceful, often hanging upside down to take fruit or flowers. Picked fruit frequently held in one claw while feeding. Wary of humans but often approachable. Occasionally sunbathes with wings drooped. Known to have at least 20 food plants, of which 12 are introduced. Particularly fond of fruits of the endemic palm latagnen lat and the introduced bilenbi[1]. Diet also includes fruits, buds and flowers of introduced plants such as guava and pawpaw and endemic plants such as bwa rouz, bwa dou, bwa kalou, palmis and coco de mer (flowers).

BREEDING In courtship, pairs bow and gently touch bills. Breeds mainly throughout the northwest monsoon, October to April, with peak laying December. Nests in chamber in dead trees including albizia, casuarina, palmiste, Horne's pandanus and coco de mer at heights of 2.5-9 m. Artificial nest boxes made from hollow palm trunks with metal guards on the exterior to prevent rat predation are also readily used; these are about 1.5 m in height. Usually two or three white eggs are laid (if three, at least one chick always dies). Incubation lasts 15-18 days, fledging 59-61 days (V. Laboudallon pers. comm.). Post-fledging parental care may continue to June.

ORIGINS *Coracopsis* is a well-defined genus endemic to western Indian Ocean. Birds of Seychelles and Comoros (race *sibilans)* are very similar in size and plumage. Seychelles Black Parrot shows grey in outer webs of primaries (unlike *sibilans*) and undertail-coverts are sometimes paler than rest of body (*sibilans* always concolorous). Both are much smaller than Lesser Vasa Parrot *C. nigra* of Madagascar (which averages 35 cm, about 17% bigger) and both are grey-brown whereas birds of Madagascar are dark brown, lacking any grey tones. Possibly related to Pesquet's Parrot *Psittrichas fulgidus* of New Guinea, which has a similar distinctive flight. An earlier African origin is unlikely[2].

STATUS Endemic to Seychelles, where only recorded as breeding from Praslin, though possibly once occurred throughout Praslin group. It has sometimes been claimed it was introduced from Comoros but this is incorrect. 'Coffee-coloured parrots', evidently this species, were recorded by the Marion Dufresne expedition of 1768, two years prior to any human settlement. In 1976, 65% of birds counted were in the Vallée de Mai region[1], but since then there has been a spread into lowland areas, at least for feeding[3]. In addition to the Vallée de Mai entrance and Viewing Lodge, parrots may be observed around Britannia Hotel, Villa Flamboyant, Fond Boffay Forestry Office and elsewhere, early morning and late afternoon. Since 1988 has been regularly seen on Curieuse feeding on jamalac, bilenbi and pawpaw flowers at Anse Jose and Caimant[4]. There are specimens in Paris from Marianne (collected 1875 by de l'Isle) and at the American Museum of Natural History from Aride (collected 1907). First estimate in 1964-65 suggested a population of 30-50 birds[5]. Annual surveys by V. Laboudallon from 1982, suggest an increase in population of *c*40% between 1985 and 1996 with a 1997 estimate of 200-300 birds[3].

THREATS AND CONSERVATION Once persecuted by farmers due to liking for introduced fruits. Small boys also killed birds with catapults for 'sport'. However, today there is an increased awareness of the value of birds both intrinsically and as a tourism asset. In Madagascar the Lesser Vasa Parrot is often hunted for food, or captured as a pet[6], habits that have fortunately not spread to Seychelles. Possibly the greatest threat is introduced rats, common on Praslin. Nestboxes protected by rat guards have been erected at some sites, but in natural sites most nests are predated when chicks begin to call at three days old. In one study, 72 rats were trapped in the vicinity of a single nestbox during the period when chicks were vulnerable (V. Laboudallon pers. comm.). Habitat destruction and a growing human population may also put pressure on the species. Bush fires are frequently recorded on Praslin and have sometimes destroyed both breeding sites and food plants. Numbers may be limited by shortage of nest sites, dead trees often being cleared away and only in Praslin National Park are these generally left to provide nest sites. The planting of feeding trees such as latagnen lat, ficus and bilenbi would benefit the species[1].

SIMILAR SPECIES At close range unmistakable though optimistic visitors to Valleé de Mai frequently claim fleeting glimpses of **Seychelles Blue Pigeon** and **Madagascar Turtle Dove** to be Black Parrots.

References: 1. Evans (1979), 2. Benson (1984), 3. Rocamora (1997c), 4. Bullock *et al.* (1989), 5. Penny (1968), 6. Langrand (1990).

SEYCHELLES PARAKEET *Psittacula (eupatria) wardi*   **Plate 1**

Other name: Seychelles Green Parakeet   **French:** Perroquet vert (or Perruche) des Seychelles   **Creole:** Kato Ver Sesel

DESCRIPTION Length 41 cm**; wingspan**: 44 cm. A large, mainly green parakeet, now extinct. **Adult** of both sexes mainly bright green, with a slight yellow tinge below, yellow undertail and iridescent green on rump. Long, graduated tail, central feathers pale blue tipped yellowish green. Prominent red shoulder-patch (on secondary wing-coverts). Eye yellow; large, red bill; dusky grey legs. **Male** had nape tinged blue and a black collar. **Female** similar, without blue tinge to nape and no collar, somewhat duller and shorter tail feathers. **Immature** similar to female, with shorter tail.
VOICE Not known, but probably similar to Alexandrine Parakeet *Psittacula eupatria,* which has a variety of harsh, deep shrieking calls, often given in flight.
BEHAVIOUR Originally a bird of the Seychelles forests, it adapted to areas cleared for cultivation. Occurred in small flocks. Diet included fruits and habit of feeding on crops led to intensive persecution.
BREEDING Unknown. Probably nested in tree holes.
ORIGINS Possibly a closely related full species, or possibly a race of Alexandrine Parakeet *P. eupatria,* which occurs throughout the Indian subcontinent to Vietnam, Sri Lanka and the northern Andaman Islands.
STATUS Endemic race or perhaps a full species, extinct since around the turn of the 20th century. Exact original distribution is unknown, but included Mahé (first noted by Marion Dufresne expedition of 1768 and first described by Newton[1]), Praslin (first noted by Marion Dufresne expedition in 1768 and last reported 1871[2]) and Silhouette. Probably common before the arrival of man; Prior found 'a considerable number of green parrots' in 1811[3]. In 1876, it was noted that due to the clearing away of the natural forests, the planting of coconuts and the ruthless killing of the parrots 'there cannot be much doubt that they are doomed to extinction'[4]. Sadly, this prophecy proved correct. Marianne North painted two tame birds on Mahé in 1883 that were said to have come from Silhouette[5], the last record from the island. Last recorded on Mahé, March 1893, a specimen shot by Abbott[6]. Possibly survived to around turn of 20th Century.
References: 1. Newton (1867), 2. Pike (1872), 3. Prior (1820), 4. Newton and Newton (1876), 5. North (1892), 6. Ridgway (1895).

RING-NECKED PARAKEET *Psittacula krameri*   **Opposite Plate 36**

Other names: Rose-ringed (or Long-tailed or Green) Parakeet
French: Perruche à collier   **Creole:** Kato Ver Kolye Roz

DESCRIPTION Length 38-42 cm (including tail 19-28 cm); wingspan 42-48 cm. Unmistakable in Seychelles. A large almost entirely green parrot with a long, graduated, pointed tail and a deeply hooked red bill. **Adult** of both sexes appears big-headed, long and tapering with plumage mainly bright grass-green, often more yellowish green on the breast. Central tail feathers are turquoise, yellow at tip; the outer tail feathers yellow. Underwing-coverts yellowish green, noticeable in flight. Eye pale yellow; legs greenish yellow. Bill blood-red, sometimes orangey towards the hooked tip of the upper mandible and suffused blackish on lower mandible. Legs olive. **Male** (by third year) acquires a rose-pink collar with pale blue cheeks and nape (race *borealis*); a black line below the collar extends and widens under the chin to meet below the bill. **Female** lacks a collar (or collar is indistinct), lacks blue on the head and is slightly smaller than the male. **Immature** similar to adult female but has yellower tinge to plumage, shorter tail and pink bill tipped paler.
VOICE Highly vocal, often drawing attention by its loud, screeching raptor-like call: 'kee-ak....kee-ak'. Alarm call an abbreviated but accelerated 'ak....ak'.
BEHAVIOUR Favours gardens and lightly wooded areas often close to human habitations. Breeding pairs usually solitary, but gregarious at other times. Roosts communally, often with other species including mynas. Not very territorial except in the immediate vicinity of the nest hole. Flight powerful, rapid and direct with flickering wingbeats, the bull-head and long tapered tail giving a comet-like shape. Sometimes suddenly dives or turns, reminiscent of a wader. Diet mainly seeds (including bwa nwar), ripe fruit (including guava and star fruit), flowers and nectar. Sometimes feeds on the ground, walking with tail slightly raised.
BREEDING May pair for life. In courtship, male walks towards female, neck extended and head slightly to one side, then leaps backward. Male may also feed female by regurgitation, continued during incubation. Nests in holes in trees or sometimes buildings. In trees, both sexes assist in carving out the hole, which is guarded with vigour. Clutch size elsewhere is 2-6, but typically four. Eggs are smooth, oval and white. Incubation, performed entirely by female, takes 22-28 days, fledging 40-45 days. Young are fed by parents for about 14 days after fledging.
RANGE Natural range Afghanistan to Burma and Sri Lanka and sub-Saharan Africa north of the Equator. It is not certain whether populations in Iran, Arabia, Hong Kong and elsewhere have been established by natural range extension or by introductions. Certainly widely introduced including in Britain, other parts of Western Europe and in the Indian Ocean on Mauritius and Réunion. Four races recognised, including

borealis from Pakistan to southeast China, which is larger, bluer on head and greyer below.

STATUS Introduced and resident as a breeding bird on Mahé. Date of first arrival in Seychelles not known, though one parakeet, thought to be this species sighted on Mahé in 1974. Also noted throughout 1980s, with one seen regularly in St Louis/Victoria area[1]. One also resident Silhouette from 1995[2]. First sighting of more than one individual was in 1997 on Mahé, with up to five individuals present at Point Conan including a male of race *borealis*[3]. It has been suggested from the distribution pattern of wild parakeets and the frequency of Asiatic vagrants recorded in Seychelles the species could be a natural colonist[3]. However, it is known that at least one introduction was made in 1996, following the gift of two birds from a visiting Indian ship's captain to a local merchant, who reported that one died and one escaped[1]. Other birds possibly also arrived on Indian ships. In Asia is often considered an agricultural pest, for example in citrus orchards, red pepper crops and maize plantations. On the other hand it has been suggested if a viable population develops in Seychelles, it could restore a significant part of the natural seed dispersal and pollinator niche (lost with the extinction of the Seychelles Green Parakeet) essential to the islands' ecosystems[4].

SIMILAR SPECIES None in Seychelles.

References: 1. Skerrett (1998), 2. Matyot (1996), 3. Gerlach and Gerlach (1997), 4. Gerlach (1997a).

GREAT SPOTTED CUCKOO *Clamator glandarius*   Plate 37

French: Coucou geai   **Creole:** Koukou Pitakle

DESCRIPTION Length 38-40 cm (including tail 14-18 cm); wingspan 58-61 cm. A large, heavily built, crested cuckoo with long wings (broader and blunter than in Common Cuckoo) and a disproportionately long, narrow, wedge-shaped tail. **Adult** has dark grey upperparts and upperwing, boldly spotted white on the scapulars and wing-coverts, with secondaries and tertials broadly tipped white. Conspicuous silver-grey crown and ear-coverts with a small crest on the hindcrown and a black nape. Pale yellow on the shoulders, lower face, throat and breast, shading to creamy white on the underparts. Tail-feathers are boldly tipped white. In flight, shows three or four white wingbars and creamy underwing-coverts. Eye red or brown with red eye-ring; bill blackish, base of lower mandible and legs grey. **Immature** similar in build to adult but has upperparts uniform blackish, duller spots on the wing-coverts, narrower tips to tertials and secondaries, top half of head blackish with crest less obvious, throat and upper breast warmer buff and primaries rusty brown (conspicuous in flight and retained until first summer).

VOICE A whistling 'keeoow-keeeow-wow-wow'. Alarm call is crow-like.

BEHAVIOUR Favours dry open country, woodland clearings and cultivated areas with scattered trees, generally avoiding dense woodland. Shy and unobtrusive, but often perches in the open or tree-tops. Usually solitary or in pairs (vagrants are likely to be solitary), sometimes in small flocks on migration. Flight strong and direct, with slow wingbeats, long tail giving rakish appearance. Feeds on hairy caterpillars, termites, other insects and lizards, sometimes on the ground.

RANGE Breeds southern Europe, migrating mainly to sub-Saharan Africa. Also resident sub-Saharan Africa and an intra-African migrant, breeding South Africa September to March, wintering further north. Seychelles vagrants might potentially derive from either population, though long-distance European migrants are probably more likely. Monotypic.

STATUS Vagrant; recorded October on Bird only, but potentially might occur almost anywhere.

SIMILAR SPECIES Common Cuckoo and most other cuckoos likely in Seychelles are shorter-tailed, lack a crest and are not spotted on the upperparts. The only other long-tailed and crested cuckoo is **Jacobin Cuckoo**, but this is readily distinguished by pied plumage and smaller size.

JACOBIN CUCKOO *Oxylophus jacobinus*   Plate 37

Other names: Pied (or Black-and-White or Pied Crested) Cuckoo
French: Coucou jacobin   **Creole:** Koukou Zakoben

DESCRIPTION Length 31-33 cm (including tail 12-14 cm); wingspan 45-50 cm. A medium-sized, crested, long-tailed cuckoo similar in structure to Great Spotted Cuckoo, but readily distinguished by pied plumage and smaller size. **Adult** has upperparts black, sometimes having a bluish green sheen with a prominent pointed crest. Upperwings also black with a prominent white patch at the base of primaries conspicuous in flight. Underparts white or greyish white, sometimes washed buff. Tail long and graduated, feathers with prominent white tips. There is a dark morph (unrecorded in Seychelles), all-black except for a small white patch at the base of the primaries. Eye brown, bill black, legs dark grey. **Juvenile** similar to adult but upperparts dark brown and underparts washed buff; crest less developed, ash grey throat and buff spots on underside of tail (white in adult) which is tipped buff. However, this plumage is soon lost, and **first-winter** resembles adult, but with much dark brown in the upperparts.

VOICE A piping 'pew-pew' leading into a loud, laughing cackle.

BEHAVIOUR Frequents open woodland avoiding both dense forest and arid areas. More gregarious than other cuckoos, often in pairs or small groups. Flight straight and rapid. Perches in bush tops with a vertical posture. Frequently flicks tail and when alarmed raises crest and lowers tail. Feeds in trees and on the ground mainly on caterpillars and insects.

RANGE Race *pica* breeds Iran to north India and migrates to Africa departing September to October, possibly across the Indian Ocean. Race *jacobinus* is mainly resident south India/Sri Lanka, though migrants have been recorded Zimbabwe, Mozambique and Natal; it is shorter-winged and shorter-tailed than *pica*. Race *serratus* breeds in sub-Saharan Africa except for the lowland forests of West and central Africa and arid regions of southwestern Africa; this population migrates within Africa; it is pale grey on the underparts and faintly streaked on the throat and breast.

STATUS Vagrant to Bird and granitic islands, most records December or March, race unknown.

SIMILAR SPECIES Great Spotted Cuckoo is also crested and long-tailed, but is much larger and not pied.

COMMON CUCKOO *Cuculus canorus*   Plate 37

Other names: Grey (or Eurasian) Cuckoo   **French:** Coucou gris   **Creole:** Koukou Sann

DESCRIPTION Length 32-34 cm; wingspan 55-60 cm. Falcon-like, with a long tail and long, narrow, pointed wings; by far the most likely cuckoo to be seen in Seychelles. **Adult male** has upperparts, upperwing and throat slate grey, paler on the rump and uppertail-coverts, latter tipped white. Underparts below the breast are white with dark grey, wavy bars more widely spaced on the belly. Axillaries and underwing-coverts grey, barred white. Tail feathers are blackish-tipped and spotted white along the shafts and inner webs. Female appears in two morphs. **Grey morph female** is similar to male but tinged rusty brown on breast which may have some dark barring. **Hepatic morph female** (much less common) has upperparts, wings and tail rufous, barred blackish brown, and underparts are washed buffish. In both sexes eye is orange with yellow eye-ring. Bill blackish, base of lower mandible greenish; legs yellow. **Immature** plumage is variable, but wings and tail always heavily barred, sometimes resembling a kestrel. May be similar to a hepatic female but browner with narrow white feather fringes above and often a white patch on the hindcrown. May also may be grey to grey-brown above (still with white nape-patch) with some barring, feathers tipped whitish, or predominately brown with dark bars; crown feathers edged buff or white, feathers of upperparts tipped white or pale buff and tail blackish brown barred white and tawny.

VOICE Two-note 'cook-ooo' of breeding male not heard in winter quarters. Female has a bubbling chuckle sometimes heard in sub-Saharan Africa.

BEHAVIOUR Frequents woodland. Usually solitary but sometimes in small flocks on migration. Shy, but often perches in the open with rather horizontal posture. Flight direct and hawk-like but with rapid, shallow wingbeats below the horizontal, often followed by a long glide, head held high, appearing small in proportion to the body while white spots on dark tail of adult stand out. Feeds in trees on caterpillars, insects, centipedes, spiders and eggs, often swooping to the ground and landing clumsily. Sometimes feeds on the ground, where walks with a clumsy waddle.

RANGE Breeds across Europe and Asia south to China, entire western Palearctic population and much of Asian population wintering Africa south of 10°S to South Africa. East Asian populations winter South-East Asia. Four races are generally recognised, with the nominate *canorus* breeding western Europe to Korea and probably the most likely in Seychelles. Race *subtelephonus* (slightly paler with narrower black bars) breeding central Asia may also be possible.

STATUS Vagrant to near-annual throughout Seychelles. Most records are October to December or March to April suggesting passage possibly of birds between Asia and Africa.

SIMILAR SPECIES Asian Lesser Cuckoo is smaller, more compact and shorter-tailed, the dark tail and rump contrasting with paler back; has fewer bolder, well-defined, black bars on the underparts. The majority of reports of Common Cuckoo from Seychelles have been accepted by SBRC as '*Cuculus sp.*' because descriptions fail to rule out other possible vagrant species; indeed some reports appear to suggest other species, as yet undetermined, may occur. Potential vagrants might be African Cuckoo *C. gularis*, Indian Cuckoo *C. micropterus* and Oriental Cuckoo *C. saturatus*. **African Cuckoo** (L32-34 cm) is endemic to sub-Saharan Africa and is highly migratory within Africa, possibly moving in response to rains. Barring on flanks of male somewhat finer than Common and indistinct on lower belly. Undertail-coverts whiter with less obvious barring and no buff tinge. Outer tail feathers barred, not spotted as in Common. Female has grey on chest more restricted, sometimes washed buff with fine barring. There is no hepatic morph. Bill blackish at tip, yellow at base, heavier and broader than Common (8 mm : 7.25 mm). Immature similar to Common but greyer, darkest on the crown and mantle. Crown feathers tipped white and hindcrown white. Upperparts barred, tipped white. Underparts similar to Common, but throat often paler. Third spot from the top of the central tail feather averages 2.34 mm (Common 1.35 mm). **Indian Cuckoo** (L33-34 cm, WS52-54 cm) occurs from Pakistan south to Sri Lanka and east to east Asian islands. Generally browner than Common Cuckoo with bolder, more widely spaced bars on underparts. Feathers of mantle and upper breast have rufous margins, absent in Common Cuckoo. Brown tail has blackish subterminal band, probably the best field character. There is said to be no hepatic morph. Reddish brown (not yellow) eye. Bill heavier than Common. **Oriental Cuckoo** (L30-32 cm, WS51-57 cm) breeds across Russia, south to Japan, wintering south to Indonesia and Australia. Similar to Common Cuckoo, but darker, slimmer, slightly shorter-winged and shorter-tailed. Underwing dark with pale bar in centre (Common Cuckoo underwing pale overall). Male differs from Common mainly by more intense, broader bars on underparts, chest generally grey and upperparts and wings more blue-grey giving head a paler appearance. Undertail-coverts have heavy black blotches (not fine bars). Underparts have more yellowish or buff wash (especially on under-wing-coverts and undertail-coverts) and no barring on the leading edge of wing giving a mainly white patch below the carpal joint. All these characteristics variable and extreme care should be exercised in identification. Female grey morph has more buff on chest and neck and more coarsely marked than Common. Hepatic female is more distinctly different, with more strongly barred, blacker upperparts and underparts, especially on chest, back, rump, tail and large inner wing feathers. Lower mandible yellower than Common and legs more orange (sometimes tinged red). Both grey and rufous juveniles occur, all birds more boldly barred than Common, with warmer underparts and heavily marked chest-band and sides of breast.

Cuckoos: (1) African, (2), Common, (3) Indian, (4) Oriental, (5) uppertail of Indian, (6) uppertail of Oriental

ASIAN LESSER CUCKOO *Cuculus poliocephalus* Plate 37

Other names: Little (or Small or Asian Little) Cuckoo
French: Petit coucou **Creole:** Pti Koukou Azyatik

DESCRIPTION Length 28 cm; wingspan 38-40 cm. Similar to Common Cuckoo but smaller, more compact and shorter-tailed. Grey above with a grey breast and a white belly barred blackish; more boldly barred than Common Cuckoo, but with fewer bars. **Male** is uniform slate grey on the back contrasting with darker tail and rump (unlike Common). Underparts boldly barred black and white, the black bars broader, less wavy and more clearly defined than Common. White spots on tail near shafts, more spotted on outer tail feathers. Undertail-coverts white, washed pale buffish often heavily barred blackish (unlike Madagascar Lesser Cuckoo). **Grey morph female** similar to male with a brownish wash on the breast and sides of neck. **Hepatic morph female** is barred brown above often with unmarked rufous crown/nape and rump/uppertail. In both sexes, eye brown, with yellow eyelid and yellow eye-ring. Bill blackish, the base of each mandible yellow; legs yellow. **Immature** similar to adult female with crown feathers tipped brown, nape brown with white spots and flight feathers and scapulars spotted brown. Tail generally barred brown, undertail-coverts barred black.

VOICE A yelping 5-6 note whistle, the first two notes short, the third may be long or short and the last softest. Usually silent but calls in East Africa prior to spring migration.

BEHAVIOUR Frequents open woodland where may be secretive and solitary. Perches on lower branches often with a more upright stance than Common. Flight direct and rapid with shallow wingbeats and short glides. Diet mainly caterpillars and insects.

RANGE Breeds across central Asia from Afghanistan to Japan ,wintering India, eastern and southeastern Africa, possibly regularly crossing Indian Ocean. In East Africa there are few records before March, suggesting most arrive direct from the Indian Ocean south of Tanzania. On return journey often numerous in coastal Tanzania and Kenya late March to mid-April. In Africa often moves with Common Cuckoo (both species once recorded on Frégate together by authors). Monotypic.

STATUS Vagrant throughout Seychelles. Most records November to December but also April.
SIMILAR SPECIES Common Cuckoo is larger, longer-tailed, less compact with thinner, more wavy barring on the underparts. Some reports to SBRC of Asian Lesser Cuckoo fail to rule out **Madagascar Lesser Cuckoo** *C. rochii* (L28 cm), considered conspecific by some, which may only be distinguishable by voice, the lack of a hepatic phase and likelihood at a different time of year. Present Africa April to September (Asian Lesser present November to April). Breeds Madagascar September to April. A few remain year round but most migrate to East Africa where some overlap occurs in April, when Asian birds have fresh plumage but Madagascar birds are worn. Adult male slightly larger than Asian Lesser Cuckoo (Wing 162-174 mm: 142-162 mm) but female similar in size. Undertail-coverts pale cream or whitish buff, often unbarred or barred in centre and at tip, sometimes throughout (Asian Lesser Cuckoo has buff undertail-coverts heavily barred black).

Lesser Cuckoos: Asian Lesser Cuckoo (top), and Madagascar Lesser Cuckoo

MADAGASCAR COUCAL *Centropus toulou insularis* Plate 37

Other name: Black Coucal   **French:** Coucal malgache or Coucal toulou   **Creole:** Toulouz

DESCRIPTION Length male 38-41 cm, female 41-46 cm; wingspan 41-46 cm.
A large, skulking, blackish bird confined to Aldabra with chestnut wings in breeding plumage and a long, black, fan-shaped tail. **Adult breeding** is glossy black except for chestnut wings. Eye red: bill and legs black. **Adult non-breeding** is duller, more dark greyish brown, browner on the wings, streaked and spotted whitish on the throat and breast and the bill browner. Female is significantly larger than male. **Immature** has blackish upperparts, speckled cream on the mantle and crown. Wing cinnamon-rufous barred dusky-brown. Underparts dusky-brown, striated on the upper breast and throat. Rump and tail dark brown and finely barred, the tail having a green tinge.
VOICE Call a descending staccato 'co-co-co-co-co', likened to a water bottle being emptied. Male has a higher pitched voice than female[1]. Nestlings may hiss like a snake, a trait possibly inherited from African parental stock as an anti-predator device as there are no snakes on Aldabra, or possibly merely coincidental.
BEHAVIOUR Frequents woodland where moves through dense vegetation with great agility, running up trees and along branches. Generally solitary or in pairs, skulking but fairly tame. Often reluctant to fly and when does so, usually only glides a short distance with a few lazy wingbeats. Sometimes runs along ground. Most food is taken in low vegetation or on the ground. Diet mainly large insects, other invertebrates and lizards. Birds eggs and chicks might sometimes be taken.
BREEDING In courtship, both sexes hold bodies horizontal, waving their long tails from side to side. Male may offer female an insect prior to mounting and copulation. Loose cup-shaped nest is built mainly by the male in a tree or shrub, often in full view, about 1-2 m above the ground. Female may assist with construction and weaving, apparently doing the more intricate work. Male pushes material downward and moves around shaping the nest until one side is pushed up higher than the other[1]. Nesting material includes dead leaves, palm fronds and twigs with softer lining material such as casuarina needles and

smaller pieces of palm frond. After the first egg is laid, nesting material is worked from higher side to lower side creating a domed roof. Clutch is two sometimes three white eggs laid about nine days apart. Both sexes share incubation though the male does more. Incubation lasts 15 days, fledging 19 days[2]. Nestlings, if handled, may excrete a foul smelling sticky black liquid; this is probably an anti-predator adaptation. Young and adults may burst through the rear of the nest chamber to escape predators, or if disturbed by human approach.

ORIGINS Undoubtedly derived from the nominate race of Madagascar from which Aldabran race differs in having a longer tail and, in non-breeding plumage, having black restricted to the lower belly and under-tail-coverts[3]. There is also some variation in juvenile plumage, birds of Aldabra (and drier western Madagascar) having speckled plumage, whereas those of humid eastern Madagascar are barred[4]. Earlier origin is not very well known and it may possibly be of hybrid origin[3].

STATUS Endemic race found only Aldabra where it remains fairly common. Also formerly found on Assumption[5] (race *assumptionis*, a synonym of *insularis* though slightly shorter-tailed and darker on wings and mantle) and on Cosmoledo[6] now extinct.

THREATS AND CONSERVATION The future of this species should be secure given the status of Aldabra as a nature reserve and World Heritage Site. Young birds may be vulnerable to rats or cats, present on the larger islands of Aldabra Atoll.

SIMILAR SPECIES None.

References: 1. Woodell (1976b), 2. Frith (1975b), 3. Benson (1984), 4. Woodell (1976a), 5. Nicoll (1906), 6. Rivers (1878).

BARN OWL *Tyto alba* **Plate 38**

Other names: White (or Screech or Common Barn) Owl   **French:** Effraie des clochers   **Creole:** Ibou Blan

DESCRIPTION Length 33-35 cm; wingspan 85-93 cm. The only resident owl of Seychelles other than the much smaller, darker Seychelles Scops-owl. Readily identified by pale upperparts and underparts giving a ghostly appearance. **Adult** is pale golden buff above, flecked with dark grey and fine, white vermiculations. Underparts white with pale golden-brown wash, speckled darker brown, fairly large and broad pale rounded wings, flight feathers light grey, striped darker with a brownish wash. Short tail with central tail feathers light grey, crossed with four dark grey bands and white outer tail feathers. Large head, lacking ear-tufts, with heart-shaped white facial disc. Brown marks on internal side of eyes, about the same size as blackish eyes. Legs flesh or greyish, upper tarsus feathered white; bill ivory white. Sexes similar. **Juvenile** similar to adult, but upperparts darker grey, underparts washed golden brown. Adult plumage is assumed at about two months of age.

VOICE A shrill, eerie screech uttered in flight. At rest, may hiss or snore. Snaps bill if disturbed near nest site. Noisy when not actively hunting.

BEHAVIOUR Roosts in trees or among rocks and usually only found away from woodland when hunting at rubbish dumps. Normally sedentary and not territorial, though may have regular hunting routes. Nocturnal and crepuscular, rarely seen during the daytime unless disturbed at nest or roost. Flies low when hunting, taking rodents, lizards, insects and birds. On Aride, analysis of pellets indicates terns, both adults and young, are the preferred prey, particularly during the main tern breeding season (March to August); at other times, mice feature prominently with passerines forming a very low proportion of prey[1]. In another study of kills where remains were sufficient to make identification possible, prey species were predominantly Fairy Tern followed by Bridled Tern then Lesser Noddy; also taken are Roseate Tern, Brown Noddy, Audubon's Shearwater and house mouse[2].

BREEDING On Aride, the main breeding season is January to September[3]; details are lacking for other islands. Courtship chases are common in which the male screeches more frequently than the female. Male flies in and out of the chosen nest site and female stimulates male with behaviour similar to a young bird. There is much tongue-clicking and rubbing of cheeks prior to frequent copulation. Nests in natural cavities between granite rocks. No nest is built, but a scrape made on the cavity floor well back from the entrance. Female remains for long periods at the nest site prior to laying 4-8 white eggs[3] at intervals of 1-4 days. Incubation is carried out entirely by the female, though male may occupy nest site for brief periods while female perches nearby. Incubation averages 32 days. Newly hatched young have light, white down sparser on the underparts, replaced at around 14 days by longer more profuse down. Both parents feed young. By 30 days of age, facial disc is well developed and primaries begin to grow, down disappearing by 45 days of age when plumage is complete. Young leave nest at around 60 days.

RANGE Up to 46 races are recognised (though validity of some doubtful), breeding across Europe, Asia, Australia, North and South America and on many islands. Race *affinis* breeds sub-Saharan Africa; birds of Madagascar and Comoros are sometimes treated as a separate race, *hypermetra*.

STATUS Introduced to Mahé (race *affinis*) and now resident on Mahé, Praslin, Curieuse, Aride, Silhouette, North and probably other granitic islands. Occurred naturally but now extinct on Aldabra where Abbott found it common in 1892-3[4]. One collected on Aldabra in 1906, this being the last definite evidence of their presence, though the resident Scientific Officer reported seeing a bird, thought to be this species, on Picard, Aldabra in 1994 (I. Stirling pers. comm.). The Annual Report of the Department of Agriculture,

1949, stated three 'Cape Barn Owls' were introduced from East Africa in March of that year, kept first at the Botanical Gardens and fed solely on small rodents prior to release on Platte on 27 April. They were said to have settled well and '..gnawed nuts are by no means as conspicuous as before and in some areas are even absent. Moreover from information available it appears the owls are feeding solely on rats'. The 1950 Annual Report considered the experiment 'very promising' but warned '...before any large scale importation of owls is effected...caution should be exercised as it would be most unwise to upset the natural balance of the larger islands of the group'. The owls on Platte were reported to be doing well, but did not breed possibly because all three birds were the same sex (it was conjectured). On 31 December 1951, despite the warning given in the 1950 report, 15 owls were imported and liberated at Union Vale, Mahé. The Agricultural Board justified this on the grounds they 'felt that isolated liberations on small coral islands did not give a fair chance for the owls to establish themselves successfully'. Two further batches of six birds each were released at Le Niol on 29 July and 9 August 1952. Pellets of one bird that fed near Beau Vallon Bay Hotel were collected up to December 1952. At first they consisted entirely of cockroaches, but later 'began to contain the remains of rats' according to the 1952 Annual Report of the Department of Agriculture. By 1959 the bird was reported to be breeding and spreading and from pellets was claimed to be feeding mainly on rats, but also on Fairy Terns and tenrecs. By 1958, the owls had spread to North, Silhouette and Praslin. It was reported at this time that pellets contained a large proportion (about 80%) rat bones[5]. This view was later contested and according to Penny 'after much acrimonious debate and partly as a result of the evidence collected by the Bristol Expedition, the owl was finally convicted of killing birds rather than rats and a price was put on its head'[6]. Owls had reached Aride by 1966[7]. By 1996 the Aride population was at least 18, but active control measures subsequently reduced this figure. There has been a suggestion Barn Owl may have occurred as a natural coloniser in the granitic islands, Blackburn referring to 'hibou' as common on the smaller islands[8]. However it seems very unlikely and would almost certainly have been noted by Newton in 1865[9] and others had it been common at this time. Possibly Blackburn confused reports from Aldabra, where Barn Owl still survived at that time. On Platte, the introduced owls probably died out around the same time as introduction to Mahé in December 1951; E.S. Brown visiting in May 1952 searched unsuccessfully for birds which the inhabitants of the island told him had been seen up to six months earlier[10].

SIMILAR SPECIES No owl of comparable size recorded, but **Short-eared Owl** *Asio flammeus* (L37-39 cm, WS95-105 cm) is a potential vagrant. Palearctic Short-eared Owls regularly migrate across the Sahara, occasionally reaching Kenya, where records include two from ships off the coast; also observed at sea off west coast of India and an irregular visitor to Maldives. Plumage generally tawny-brown with blackish streaks on the crown, which are larger on the back. Scapulars more barred and rump appears paler due to lack of streaking. Black carpal-patch on upper and underwing. Pale underwing with black primary tips.

References: 1. Bullock *et al.* (1995a), 2. Gerlach (1996b), 3. Nicoll (1996), 4. Ridgway (1895), 5. Lionnet (1959), 6. Penny (1974), 7. Ridley and Percy (1966), 8. Blackburn (1883), 9. Newton (1867), 10. Matyot (2000).

EURASIAN SCOPS-OWL *Otus scops*   Plate 38

Other names: Scops (or Common Scops) Owl   **French:** Petit-duc scops   **Creole:** Skops Eropeen

DESCRIPTION Length 19-20 cm; wingspan 53-63 cm. A very small owl, with a large head, long wings and prominent ear-tufts. Slim, upright posture when perched. **Adult** is variable but generally grey-brown, greyer than Seychelles Scops-owl, fairly uniform with darker streaks and bars. Underparts paler between fine, grey vermiculations. Markings vary considerably and sometimes has rufous tones. In pale silver-grey forms facial disc is unstreaked but vermiculated dark grey with a darker border. A line of white or pale buff spots on the scapulars and broad, pale, buffish bars on primaries may be visible in good light. Underwing very pale buff and only faintly marked. Tail dark grey with fine vermiculations forming bands. Bright yellow eye with a dark grey surround. Legs thin and feathered, not bare. Bill blue-black, legs grey or dark brown. Sexes similar. **Immature** similar to adult but paler with brown wash and shorter ear-tufts.

VOICE Usually silent in winter quarters. Call a vibrant trill of about one second duration repeated at intervals of 2-4 seconds.

BEHAVIOUR Favours open areas with scattered trees. Nocturnal, roosting in trees by day, often against the trunk. A reluctant flier (may occasionally be caught in the hand) and when approached may stand erect with body feathers pressed in, ear-tufts raised, eyes reduced to slits and may rock from side to side. Most prey caught by a pounce from low perch, but may also hawk for insects. Diet includes insects, scorpions, occasionally lizards, frogs, small passerines and small mammals.

RANGE Many races breed across southern Europe east to Lake Baikal wintering northern tropical Africa to India. Regular visitor to East Africa November to March. Races recorded Kenya are the nominate *scops* (breeds Europe to western Russia), *pulchellus* (Russia to northern Asia; paler, greyer, more delicately marked, clearer white spotting on upperparts) and *turanicus* (Caspian Sea to Iran; even paler silver-grey with very fine shaft streaking and narrower, fewer pale bars on underparts, more distinct black streaks on upperparts, spotting on upperparts sometimes absent). Several other races of more sedentary nature occur, unlikely to reach Seychelles.

STATUS Vagrant to granitic islands; recorded October to December. Nominate has been recorded, but *pulchellus* and *turanicus* might also occur.

SIMILAR SPECIES Seychelles Scops-owl has only one colour form, generally darker and always more rufous on facial disc and underparts with longer wings and more rounded head (lacking obvious ear-tufts). Some reports of vagrant Scops-owls do not rule out the possibility of other, albeit much less likely, related species. Non-breeding birds are likely to remain silent, so voice, the best field character elsewhere, cannot be used to separate from other potential vagrants. Oriental Scops-owl *O. sunia* is unlikely, being only a partial migrant in the north and resident in the south of its range. African Scops-owl *O. senegalensis* is also very unlikely being a mainly sedentary species. There is perhaps a slight chance of **Striated (Pallid) Scops-owl** *O. brucei* (L20-21 cm, WS54-64 cm), a partial migrant breeding southern Turkey to western China and a rare visitor to northern India and Yemen. It is paler greyish often with a sandy wash above, especially on face, with longer tail, shorter primary projection, smaller, broader ear-tufts, lacks white spots or rufous tones on mantle/scapulars, buff not white braces and more finely streaked below without cross-barring. Central tail feathers have 2-4 bars, indistinct at tip (in Eurasian there are 5-7 bars, evenly spaced to the tail tip). Feathers of tarsus extend onto toes (in Eurasian, feathering is restricted to tarsus). Juveniles are more easily separated, underparts being completely barred (Eurasian juvenile similar to adult).

SEYCHELLES SCOPS-OWL *Otus insularis*   Plate 38

Other name: Seychelles Bare-legged Scops-owl   **French:** Petit-duc des Seychelles or Petit-duc scieur
Creole: Syer

DESCRIPTION Length 21-23 cm; wingspan 40 cm
A small, generally rufous-brown, forest owl known only from the mountains of Mahé. **Adult** is dark brown-chestnut above, mottled black and pale brown on the head and barred on the back. Tail brown, mottled and barred chestnut. Rufous underparts and facial disc, the underparts streaked black and barred whitish, undertail-coverts buffish. Small ear-tufts are not usually visible in the field, head appearing rounded, quite unlike the migratory Eurasian Scops-owl. Eyes are large and golden yellow. Powerful, thick yellow legs have no feathers. Claws are dark, becoming blacker towards the tip. Bill horn-coloured. **Juvenile** similar to adult, paler with less black streaking and heavily barred above and below.

VOICE Call a deep, slow, rasping, repetitive purr: 'rrrrr-rrrrr-rrrrr', often given from a perch deep within vegetation, the caller remaining hidden. When first described in 1880, it had already acquired the Creole name 'Syer' meaning woodcutter, a reference to the call, resembling the sound of wood being sawn. May begin calling a little before dusk calling most persistently at this time until about 8.30 pm and at dawn, but bursts of a few notes may continue throughout the night. Pairs will sometimes duet, call note becoming slightly faster and higher pitched after a time with more gurgling or quacking notes.

BEHAVIOUR Frequents forested areas, especially deep wooded river valleys, above 200 m. Has a preference for 'boulder-fields' in forest and has been observed to appear from beneath rocks when disturbed at roosts during the daytime. May once have occurred to sea level and is still occasionally reported from lowland regions with suitable habitat. An exclusively nocturnal owl only encountered by day if disturbed at a roost. Stomach contents of specimens have included lizards, insects and some vegetation. Diet also probably includes tree frogs, common in scops-owl habitat.

BREEDING In copulation a high-pitched whistle is uttered, commonly heard during wetter months of north-west monsoon from October to April[1]. Food supply is highest at this time, when as a result most other land-birds breed. Fledged young have been observed in several different months indicating perhaps a twice-yearly breeding cycle or perhaps that the breeding season is fairly extensive. A nest site was not discovered until May 1999, when Rodney Fanchette and Camille Hoareau discovered a nest in a hole of a bwa rouz tree. The nest was at a height of 7 m with a single white egg, but failed possibly due to predation.

ORIGINS On the basis of similarity of calls Marshall considered *O. insularis* and Moluccan Scops-owl *O. magicus* (which is also similar in plumage and size) to be conspecific[2]. There may be at least a superspecific relationship, which is remarkable if correct, as the first known avian link between the Indian Ocean and Australasian region[3] though voice similarity is not always firm proof of a close relationship between species. At one time, the species was placed in a genus of its own, *Gymnoscops* on the basis of unfeathered tarsus and feet.

STATUS Endemic to Seychelles, where it has only been recorded with certainty from Mahé. IUCN Critically Endangered. There have in the past been hearsay reports from the Vallée de Mai and elsewhere on Praslin and from Félicité but these appear very unlikely to be correct. A study in 1975-76 in the Mission area of Morne Seychellois National Park found 12 pairs, regularly spaced about 1 km apart and extrapolation over the known range gave a total population of 80 plus pairs[1], possibly as many as 160 pairs. In 1997, 80 to 160 territories were estimated with possibly a stable population over the previous 20 years[3]. There may have been some recovery in numbers due to secondary forest regeneration, but earlier much lower estimates of numbers have probably more to do with lack of observers, nocturnal habits and a habitat remote from most human habitation. Indeed it was declared extinct in 1958[5], partly because a specimen collected March 1940 was overlooked until its 'rediscovery' in 1960[6]. It is likely that Seychellois living in the Mission area, Fairview and elsewhere were always familiar with the strange call of the Scops-owl but no one asked their advice. There is also a surprising but convincing report from Alphonse where the manager from 1935 to 1940 and his family noted 3-4 birds present and heard their distinctive calls at night in bwa blan trees at the Settlement. Workers on the island believed the birds had been introduced by earlier German or French owners of the island, as were at least five other species (including another rare endemic, Seychelles Magpie-robin). Absent from the island 1940-1948, the manager reported the 'syer' had disappeared upon his return (D Gendron pers. comm.). Credit for confirmation of the species' survival into the second half of the 20th Century goes to P. Loustau-Lalanne who observed one in south Mahé in 1959. At the time, this was believed to be the first record since 1906 when Nicoll heard it and was advised by H. Thommasset it was sometimes seen when driven from its hiding place amongst rocks or hollow trees[7].

THREATS AND CONSERVATION Listed as IUCN critically endangered[8], the greatest threat is loss of habitat. However, its main stronghold lies within the Morne Seychellois National Park where it appears to be doing reasonably well. Some areas of the National Park have been developed, including for housing, felling for timber or tea cultivation. Should these activities be expanded, it could be a cause for grave concern. Other threats may include introduced predators including rats, cats, Barn Owls and even dogs in at least one case in 1979 (L. Chong-Seng pers. comm.). Uncontrolled and repeated tape-luring by bird-watchers or tour guides could be a threat to certain pairs.

SIMILAR SPECIES Eurasian Scops Owl is more variable in plumage often paler and greyer and always less rufous on facial disc and underparts. Ear-tufts may be lowered, giving appearance of sharp corners to top of head.

References: 1. Watson (1979), 2. Marshall (1978), 3. Benson (1984), 4. Rocamora (1997c), 5. Greenway (1958), 6. Benson (1960), 7. Nicoll (1909), 8. Collar et al. (1994).

BROWN FISH OWL *Ketupa zeylonensis*　　　　　Plate 38

French: Kétoupa brun　　**Creole:** Ketoupa

DESCRIPTION Length 54-57 cm; wingspan 145-150 cm. A large flat-headed owl with shaggy, fairly short ear-tufts, short tail and a large bill. No feathers on legs and feet. **Adult** is fairly uniform buff, upper cheeks darker, streaked blackish brown on the ear-tufts, crown, back, wing-coverts and underparts with a white patch on the throat. Facial disc not very pronounced. Eye bright yellow, bill pale horn and legs greyish. Sexes similar. **Immature** has upperparts more rufous with browner streaking and underparts pale brown with narrower streaks. Shows only a small white throat-patch.

VOICE A deep, moaning 'boom-o-boom' repeated at intervals.

BEHAVIOUR Frequents woodland often close to water. Usually found in pairs. Flight weak and slow, wings making quiet singing sound on down stroke. Hunts over water, dangling long legs, and often bathes. Catches fish, frogs and freshwater crabs, which are located from a low perch or by hunting close to surface of water. Also feeds on lizards, birds, rodents, crabs and large beetles.

RANGE Several very similar races resident Middle East, India and South-East Asia.

STATUS Vagrant; recorded November. Given sedentary habits and weak flight its occurrence in Seychelles is very surprising. There is one record only, of a bird photographed on Mahé, origin unknown. It may have be an Indian ship brought it in deliberately or accidentally, as has occurred with at least two other species (House Crow and Ring-necked Parakeet).

SIMILAR SPECIES None.

EURASIAN NIGHTJAR *Caprimulgus europaeus*　　　Plate 38

Other name: Common Nightjar　　**French:** Engoulevent d' Europe　　**Creole:** Sonmey Eropeen

DESCRIPTION Length 26-28 cm (including tail 10-11 cm); wingspan 57-64 cm. A fairly large, greyish brown long-winged, long-tailed nightjar. **Adult** is generally grey-brown, mottled darker. At close range, in good light, finely barred underparts, dark ear-coverts, a pale bar across lesser coverts and blackish shoulder are visible. Crown grey, streaked black, contrasting with rich brown cheeks and throat and with a white patch on lower throat. Upperparts are heavily marked with blackish blotches or spots, banded dark brown and buff on scapulars. In flight, male shows white tips on outer tail feathers and white spots on the three outermost primaries. Female similar to male but lacks white tips to outer tail feathers while white spots on primaries may be absent or buffish. Very large dark brown eye. Tiny dark bill; very short, weak orangey brown legs. **Immature** similar to adult female but paler and markings less distinct. No spots on tail or primaries.

VOICE A sustained chirring trill, though this is rarely heard in winter quarters. On passage, a rapid, repetitive, subdued 'chuck-chuck' similar to, but more rapid than a house gecko *Hemidactylus spp.* Alarm call 'chuck' or 'cheek-eek'. Usually silent in winter quarters, but may sometimes give a single sharp note on take off.

BEHAVIOUR Frequents open woodland, most races favouring mature woodland in winter quarters, though race *plumipes* favours open sandy country. A nocturnal insectivore, which by day roosts lengthways on a branch or on the ground, camouflaged by cryptic colouring. Hawks for insects amongst trees and may drink in flight like a swallow. Flight agile and buoyant with some gliding and sharp twists and turns in search of aerial prey. Diet mainly moths and beetles (90% of diet in Europe).

RANGE Several races breed across Europe to Lake Baikal and also in northwest Africa. Some winter West Africa, but most winter southeast Africa, including the entire Asian population. Many autumn ship records off Arabia, indicating some cut across the Indian Ocean and could possibly reach Seychelles, though northbound records in April for this region are rare. Only nominate *europaeus* (breeds north Europe to Russia, wintering central and southern Africa) is recorded Seychelles, but others could possibly occur including *unwini* (Hume's Eurasian Nightjar; breeds Iran to southwest Asia, wintering southeast Africa, East Africa, northwest India; very pale and greyish, with less streaking, a larger throat-patch and unbarred undertail-coverts) and *plumipes* (breeds central Asia, wintering East Africa to South Africa; very pale but ground colour cinnamon-buff, with large buff spots on scapulars and breast and unbarred central tail feathers).
STATUS Vagrant to Bird only, recorded December, but could possibly occur almost anywhere in Seychelles.
SIMILAR SPECIES Madagascar Nightjar is much smaller and confined to Aldabra. The nightjar-like **Wryneck** *Jynx torquilla* (L16-17 cm, WS25-27 cm) is a potential vagrant, regularly crossing the Sahara and reaching East Africa. It is considerably smaller, at a distance appearing uniform grey-brown, paler below but at closer range shows darker tail with brown-edged black bars, mainly whitish underparts barred black and a buff throat with fine barring. **Egyptian Nightjar** *C. aegyptius* (L24-26 cm, WS58-68 cm) is also a potential vagrant, (once personally observed hitching a ride for two days aboard a Seychelles-bound yacht). Nominate race of Egypt to Kazakhstan winters in eastern Sahel. It is a pale, sandy nightjar, dark primaries contrasting sharply with remainder of upperside.

MADAGASCAR NIGHTJAR *Caprimulgus madagascariensis*   Plate 38

French: Engoulevent malgache   **Creole:** Sonmey Malgas

DESCRIPTION Length 21 cm; wingspan 35 cm. A fairly large, typical cryptically-coloured nightjar confined to Aldabra. **Adult** has grey head and neck, pale brown cap, cheeks, chin and throat. An orange-brown line extends from under the throat to behind the ear. Upperparts are grey-brown mottled black, upperwing brown. Underparts are brown, paler on the belly and undertail-coverts, which are finely barred brown. Large dark brown eyes. Very short, greyish legs. **Male** has white in the middle of first four dark brown primary flight feathers giving a broad white wingbar and the top of the two outer pairs of tail feathers also white. **Female** similar to male but the wingbar and top of two outer pairs of tail feathers is less extensive white or buff. **Juvenile** similar to adult female, more cryptically coloured and may retain some down for a short time after fledging.
VOICE Highly vocal during breeding season, call is a rattling 'chook-ook-ok-ok-ok', variously compared to a ping pong ball bouncing on a table, or a knife thrown into a board. Single 'chook' notes also common. Probably calls exclusively from the ground.
BEHAVIOUR Found over most of Aldabra, but roosts almost always on the ground where cryptic colouring makes detection difficult. Will often permit approach to about 1 m, relying on camouflage, before opening bill to reveal deep red mouth and spreading tail as a threat. Silent flight, twisting and turning with long narrow wings and long tail giving falcon-like silhouette. Nocturnal, hawking silently for insects over open ground, taking larger species such as cicada and praying mantis.
BREEDING Breeding season probably August to January. Young have been recorded in September, while eggs have also been noted in November and December. Nest is a shallow scrape on the ground in which one or two eggs are laid. Eggs glossy white mottled brownish grey, barely visible against the ground colour.
ORIGINS Nominate *madagascariensis*, common throughout Madagascar, is of similar size with crown and scapulars darker and white tips of two outer pairs of tail feathers less extensive. This is certainly the original source of the Aldabran population, while the earlier origin is probably Asian, forming a superspecies with Indian Nightjar *C. asiaticus* which differs mainly in having a golden collar on the nape[1].
STATUS Endemic race *aldabrensis* recorded only from Aldabra, where it is fairly common.
THREATS AND CONSERVATION As a ground-nesting bird, some eggs and chicks may be taken by introduced rats and cats. However, it appears to be fairly common and the camouflage of eggs and birds alike may minimise losses. The protection of Aldabra as a nature reserve and World Heritage Site should secure the future of this endemic race.
SIMILAR SPECIES Eurasian Nightjar is similar but much larger.
Reference: 1. Benson (1984).

SEYCHELLES SWIFTLET *Aerodramus elaphrus*   Plate 39

Other name: Seychelles Cave Swiftlet   **French:** Salangane des Seychelles   **Creole:** Zirondel Sesel

DESCRIPTION Length 11 cm; wingspan 28 cm. The only resident swift. Fairly small, dark grey-brown above, paler below; usually seen in small groups hawking for insects. **Adult** appears uniform dark grey-brown above, very slightly paler on the crown and rump. Tail black-brown and slightly forked. Upperwing black-brown with a very slight gloss (often not visible in the field), the underwing slightly paler. Blunter, more rounded wing than the sickle-shaped wing of *Apus* swifts. Pale grey-brown on underparts, darker on undertail-coverts. Bill and legs black. **Juvenile** is similar to adult.

VOICE On the wing, sometimes utters a soft twitter. In breeding caves, uses echolocation to navigate in the darkness; this call is a sharp metallic 'click-nik-nik-nik-nik'.

BEHAVIOUR Found from coast to mountains, generally seen hawking for insects in small parties at 5-20 m above the ground, flying higher during the middle of the day and lower early morning, late afternoon and when there is a heavy cloud cover or rain. Some leave caves before sunrise, but most depart at dawn, returning from early afternoon onward, many after dark. Often flies over freshwater and close to the summit of mountain passes such as the St Louis and Fairview Estate areas of Mahé. Heavy outward movement in early evening, probably in response to appearance of aerial insects. Feeds on flying insects, mainly flying ants (more than 90% of the diet).

BREEDING Breeding may take place year round. Swiftlets may possibly pair for life. Bracket-shaped nest is built, often using casuarina needles or lichen filaments cemented with saliva. Nest is attached to a near-horizontal ceiling or reverse sloping wall. Clutch is one white egg. Two have been recorded, possibly due to two females laying in one nest, two chicks having never been recorded[1]. Eggs often lost (and may be replaced) but chick mortality very low. Incubation may last 25-30 days (but not known accurately), fledging 42 days[2]. Both parents feed young, perching on sides of nest at night. Once young have left the nest, another egg may appear within 14 days, though it is not known if this is the same pair laying again[2].

ORIGINS Historically, there has been controversy over the splitting of Mascarene Swiflet *A. francicus* and Seychelles Swiftlet, but more recent data strongly supports recognition as a distinct species. Watson *et al.* placed the Mascarene and Seychelles Swiftlets together as *Collocalia francica*[3]. Gaymer *et al.* agreed with this view, but considered Seychelles birds racially distinct as race *elaphra*[4]. Medway argued that the lack of knowledge of nesting habits of Seychelles birds justified referring to the species as *Collocalia elaphra*, nest structure being an important determining factor of species in some instances[5]. Procter, who located the first known nest site in 1972, considered it similar to that of Mascarene Swiftlet and suggested it should be classified as a race of this species[6]. Medway and Pye in their review of swiftlet systematics renamed it *Aerodramus francicus elaphra*[7]. However, Cheke summarised reasons to justify separation. These are: Seychelles birds are larger (119-124 mm versus 108-117 mm), heavier, have broad-based wings (Mascarene narrow at base), invariably lay one egg (Mascarene usually two), build nests tightly clumped sharing common brackets (Mascarene nests spread out), nesting on steeply-angled cave roofs (Mascarene nesting on vertical or horizontal surfaces) and feeding on flying ants (Mascarene feeding mainly on *Diptera*)[8]. Johnson and Clayton investigated the DNA sequence of swiftlets and proposed splitting the group between echolocating (*Aerodramus*) and non-echolocating (*Collocalia*) species. On this basis, they suggested Seychelles and Mascarene Swiflets may have been isolated from each other for half a million years and from other swiftlets for perhaps a million years[9]. *Collocalia* is restricted to Asia and Australasia.

STATUS Endemic to granitic islands where resident Mahé, Praslin and La Digue and formerly Félicité (total population *c*1,000 pairs or more). Recorded as a visitor to Aride. IUCN Vulnerable. Absent from Silhouette and elsewhere, possibly due to lack of suitable caves for nesting. There are only four known colonies, though others probably remain undiscovered, particularly within Morne Seychellois National Park. In 1997, the largest known colony at La Gogue held *c*2,263 birds (plus or minus 371)[10], Mount Cabris, Praslin held 79 nests (G. Rocamora and V. Laboudallon pers. comm.) and on La Digue two small colonies in one small area overlooking La Réunion, held 35-45 nests (P. Constance and D. Niole pers. comm.); the total population was estimated at 1,000 pairs and 2,500-3,000 individuals, possibly more[10].

THREATS AND CONSERVATION As one of the rarest of all swifts confined to just a few nesting sites, the species must be a conservation priority. Location of breeding sites, within steep-walled caves, is probably the most effective protection for this species, though some are prone to disturbance. In the 1970s there were colonies at Fond Azore and L'Amitie on Praslin, but by the 1990s the precise location of the former was unknown and the latter was abandoned following the destruction of the cave (V. Laboudallon pers. comm.). Consideration has been given to placing a metal grille over the entrance to the most easily accessible breeding caves on La Digue. Legal protection of all nesting sites is desirable. Introduced Barn Owls and cats may be a threat.

SIMILAR SPECIES None resident in Seychelles. The most likely vagrant swifts, **Common Swift** and **Pacific Swift**, are both considerably larger, the latter also being blacker with a large white rump.

References: 1. Watson (1979), 2. MacDonald (1978), 3. Watson *et al.* (1963), 4. Gaymer *et al.* (1969), 5. Medway (1966), 6. Procter (1972), 7. Medway and Pye (1977), 8. Cheke (1987), 9. Johnson and Clayton (1999), 10. Rocamora (1997c).

WHITE-THROATED NEEDLETAIL *Hirundapus caudacutus* Plate 39

Other names: Needle-tailed (or Spine-tailed) Swift, Northern Needletail, White-throated Spinetail
French: Martinet épineux   **Creole:** Zirondel Ledo Blan

DESCRIPTION Length 19-20 cm; wingspan 50-53 cm. A large swift with a heavy, cigar-shaped body, short, square or slightly rounded tail, characteristic long, broad-based, delta-shaped wings and a highly distinctive white area on the lower flanks and undertail-coverts. **Adult** appears wholly dark brown except for the prominent white throat and a diagnostic U-shape on rear underparts. In good light white marks on the inner webs of tertials may be visible. Also, a white patch on the forehead and the appearance of a brownish white 'bald patch' on mantle and back. In fresh plumage has a blue-green sheen on crown, tail and flight feathers, becoming purplish in worn plumage, or disappearing altogether, wings appearing black, sometimes with a pale patch on the nape and mantle. Square tail appears slightly rounded when spread and pointed when closed. Each tail feather has a spine-like protrusion beyond the webs of up to 0.5 cm, only visible at very close range. Bill black and legs pinkish with purple tinge. **First-winter** similar to adult but duller with very little gloss and a narrower U-shape on vent. May retain juvenile's grey-brown forehead and lore, dark brown tips to tail-coverts, some tipped black and less contrasting throat. Before moult into first-winter plumage, pale grey bases to body feathers give an untidy scaly effect.

VOICE Vagrants normally silent. Call a less piercing screech, more chattering than that of Common Swift.

BEHAVIOUR Has no strong habitat preference but often seen around cliffs and crags. More solitary than some swifts, but forms flocks on migration and in winter quarters. Seen in loose association with Seychelles Swiftlet on Mahé. Highly aerial, with powerful but fairly slow flight, often gliding with wings held slightly downward, hawking for insects.

RANGE Breeds central and eastern Asia; southern race *nudipes* mainly resident, but northern nominate race *caudacutus* is entirely migratory to Australia, vagrant Fiji and westward to Britain and Ireland. Other races occur South-East Asia and Taiwan and are much less likely to reach Seychelles.

STATUS Vagrant Bird and granitic islands; recorded October to November. Not recorded elsewhere in the Afro-Malagasy region.

SIMILAR SPECIES No swift of comparable size has been recorded in Seychelles. The even larger **Alpine Swift** *A. melba* (L20-23 cm, WS51-58 cm) is a potential vagrant, its range including East Africa and Madagascar. Readily distinguished by pale brown plumage with white belly and white throat (latter sometimes indistinct or absent), separated by broad breast-band. Madagascar race *willsi* is smaller and darker than the nominate.

COMMON SWIFT *Apus apus* Plate 39

Other names: Swift, Black (or Eurasian or Northern) Swift   **French:** Martinet noir   **Creole:** Zirondel Nwanr

DESCRIPTION Length 16-17 cm; wingspan 42-48 cm. A medium-sized, blackish brown swift with a short, forked tail. **Adult** appears entirely blackish brown with an indistinct greyish or whitish chin (sometimes not visible in the field) and paler brown upper surface to flight feathers. Worn plumage browner, darkest when fresh. Upperparts contrast with wing at the tertials and greater/median-coverts, generally uniform with lesser coverts. Underparts uniform, the underwing-coverts darker. Bill black; legs black sometimes with pinkish tinge. **Immature** similar to adult, but darker with scaly underparts and generally a more prominent throat-patch extending to forehead and contrasting with dark eye-patch. Also a shallower tail fork, the outer tail feathers more rounded. Flight feathers and tail feathers browner with narrow white margins. Broad white fringes on undertail-coverts; other feathers fringed white, densely on forehead and also distinct on the rump.

VOICE Highly vocal in the breeding season when birds scream in flight, but less vocal at other times.

BEHAVIOUR No habitat preference, attracted to anywhere where aerial insect prey are available. Gregarious, though vagrants may be solitary. Flight high, powerful and rapid with a twinkling beat. Glides with scythe-like wings held stiff and body appearing torpedo-shaped.

RANGE Breeds across Europe east to Mongolia and north China wintering sub-Saharan Africa to South Africa. Nominate western race *apus* reaches east to Tanzania while the Oriental race, *pekinensis* traverses Middle East and northeast Africa; browner with more prominent whiter throat-patch and paler forehead, wing-coverts distinctly pale-fringed and under optimal conditions, some contrast between dark outer primaries and browner inner primaries and secondaries visible. Worn adults and juveniles may be inseparable by race.

STATUS Near-annual at Aldabra and vagrant elsewhere. Both races have been collected at Aldabra. Most records September to December with a few February to March.

SIMILAR SPECIES Many reports of vagrant Common Swift fail to eliminate other possible vagrants notably African Black Swift *A. barbatus* and African Palm Swift *Cypsiurus parvus*, both of which occur in East Africa, Madagascar and Comoros. **African Black Swift** (16 cm) is a medium-sized, blackish swift with long, narrow wings, slightly forked tail and whitish throat. Black upper body contrasts with paler greater coverts and secondaries. Very similar to Common Swift and difficult to distinguish in field, though somewhat bulkier with broader less pointed wings; forehead always dark. In the hand the following characteristics may be distinguished: 1. The penultimate 2-5 tail feathers are shorter than the outermost compared to 6-11 shorter in Common. 2. The greater coverts and tertials are paler brown, which in the closed wing contrasts with blackish forewing, primaries and mantle. 3. More extensive, broader pale scalloping on underparts. 4. More bronze-green sheen on primaries. 5. Tenth primary equal to, or longer than, ninth (shorter in Common)[1]. Immature similar to adult, with more prominent throat-patch and many feathers fringed greyish; forehead dark, unlike Common. **African Palm Swift** (16 cm) is generally paler than any *Apus* swift. Crown is darker than ear-coverts giving a capped appearance. This is the most slender long-tailed swift, with nearly two-thirds of length being body and tail behind the wing. Outer wing and underwing appear

darker than body. Tail is usually held to a needle point, opening when banking. Has a strong association with tall palms, especially close to water. Flight low and rapid, streamers often held together masking indentation, while speed and vigour masks other plumage features, but has an unmistakable delicate, very slim flight silhouette. Abbott reported seeing a swift on Aldabra he thought to be of the genus *Collocalia* several times[2]. Benson and Penny suggested African Palm Swift was a possible alternative[3].
References: 1. Chantler (1993), 2. Ridgway (1895), 3. Benson and Penny (1971).

PACIFIC SWIFT *Apus pacificus*   Plate 39

Other names: Fork-tailed (or Pacific Fork-tailed) Swift, White-rumped (or Large White-rumped or Asian White-rumped) Swift
French: Martinet de Sibérie   **Creole:** Zirondel Azyatik

DESCRIPTION Length 18 cm; wingspan 48-54 cm. Slightly larger than Common Swift, this is a blackish, elegant swift with well-defined white rump which wraps around the tail base, and a deeply forked tail (deeper than Common). The largest *Apus* swift recorded in Seychelles. **Adult** is black above with large white rump-patch extending to rear flanks. Long wings are swept back with an obvious angle at the carpal joint. White throat-patch more prominent than Common, but does not contrast sharply with underparts; the head larger and more protruding, body similar while tail lacks 'lumpiness' of Common Swift at the point where it joins the body[1]. Tail long (longer than Common) and black. Outer tail feathers broader than other fork-tailed swifts, giving heavier appearance. Feathers on body and larger coverts of underwing-coverts show distinct grey fringes, giving a scaly appearance lacking in much smaller White-rumped Swift, but only visible at close range. Nevertheless, this is the heaviest marked *Apus* swift, the body often appearing slightly grey in bright sunlight. Bill and legs black. **Immature** similar to adult but may have white margins on upperparts, secondaries and inner primaries.
VOICE Similar to Common, but somewhat harsher. Less vocal than Common and generally silent in winter quarters.
BEHAVIOUR Occurs from lowlands to high mountain cliffs. Gregarious but vagrants usually solitary. Flight rapid, wings and tail held with slightly greater attenuation than Common. Hawks for insects at great height in clear weather, descending to lower levels when there are heavy clouds.
RANGE Several similar races, some sedentary or locally dispersive, breed central and eastern Asia. Long-distance migratory nominate race breeds Siberia to Kamchatka and Japan wintering south to Indonesia and Australia. Vagrant westward to Britain.
STATUS Vagrant Bird, granitic islands and Amirantes with most records October to November, but also May. Nominate is probably the only race likely to occur in Seychelles.
SIMILAR SPECIES Combination of size, white rump and deeply forked tail is diagnostic.
Reference: 1. Chantler (1993).

LITTLE SWIFT *Apus affinis*   Plate 39

Other name: House Swift   **French:** Martinet des maisons   **Creole:** Pti Zirondel

DESCRIPTION Length 12 cm; wingspan 34-35 cm. A small, stocky swift. All-black except for broad, white rump extending to flanks and a short square tail. Even overhead some white can be seen. **Adult** appears sooty black (occasionally with green or brown tones visible at close range), blacker on the back than on the wings. Forehead greyish white, throat and chin white and a grey-white line extends over the eye. Often shows a sharp contrast between head and back. Wings may appear blunt-tipped. White rump extends to flanks. Tail short, square, broad and rounded at corners often appearing broader than body; may have slight cleft but not a distinct fork, or may appear slightly rounded when spread. Tail can appear greyer than black mantle, sometimes translucent when spread, contrasting with dark body and central tail feathers. Bill black; legs pink-horn. **Immature** similar but duller, inner flight feathers tipped white.
VOICE Highly vocal, but less so out of breeding season. A shrill trill uttered in flight.
BEHAVIOUR Frequents towns with high buildings, cliffs and gorges. Gregarious, often associating with other swifts. Flight not as graceful as other swifts, much weaker, slower and more fluttering, wings held straight out or only slightly curved, with much gliding. Forages for insects at high altitude.
RANGE Several races breed over much of Africa where it is the commonest swift in towns. Nominate *affinis* breeds India to East Africa. Migratory in the north and south of its range, resident or with local movements elsewhere. Also breeds India and Sri Lanka. Racial differences slight.
STATUS Vagrant to granitic islands, recorded November to December, but could possibly occur anywhere in Seychelles.
SIMILAR SPECIES Pacific Swift is much larger, longer-winged, with a longer, forked tail. Other smaller swifts with white rumps which are potential vagrants include Mottled Spinetail *Telacanthura ussheri*, Madagascar Spinetail *Zoonavena grandidieri*, White-rumped Swift *Apus caffer* and Horus Swift *A. horus*. **Mottled Spinetail** (L12.5 cm) is endemic to Africa. There is a published record from Aldabra, pre-dating formation of SBRC, not confirmed as details did not survive[1]. It is smaller than Common Swift, but larger than Little. Generally black-

ish, with a prominent white rump. Tail square; tail spines not obvious, but tail can appear spiky, especially when spread. Long protruding head with indistinct pale throat (not white, as Little Swift). Wings shaped like a butter knife (due to large primaries and short secondaries). May show a dark blue gloss above, while underwing and undertail-coverts may be glossy black. Narrow white band across lower belly (unlike Little Swift). **Madagascar Spinetail** (L12 cm) is endemic to Madagascar (race *grandidieri*) and Comoros (race *mariae*), the latter slightly shorter-winged. It is a small, black-brown swift, with a heavily streaked whitish rump. Wing shape is typical of spinetails, like a butter knife. Tail black-brown and square (or very slightly forked at some angles) with spiky feathers, longest at centre (and longer in male). **White-rumped Swift** (L14-15 cm, WS33-37 cm) is increasing in numbers in East Africa, where it is probably the most common swift in rural areas. It is a slim, small swift with proportionately long, pointed wings and a deeply forked tail (though tends to keep tail closed for long periods giving spike-like appearance). Dark tail does not contrast with back. White throat and narrow, white, crescent-shaped rump, sometimes difficult to detect, especially in young birds. Narrow white tips to secondaries and often some inner primaries give a white trailing edge when in fresh plumage. **Horus Swift** (L13-15 cm) is endemic to Africa. It has been reported as vagrant to Agalega, which, if correct, indicates a strong possibility of reaching Seychelles. A small, blackish swift with forked tail and large, white rump-patch extending around body almost to legs, rectangular in shape (not curved as in White-rumped Swift). Rarely holds tail to a point, but fork may not be visible when turns. Larger, stockier and less elegant than White-rumped Swift, with broader, shorter, more deeply forked tail, white fringes to the secondaries and tertials and less contrast between the body and underwing. Outer tail feathers rounded, (pointed in Pacific Swift). **Reference:** 1. Frith (1974).

COMMON KINGFISHER *Alcedo atthis*   Not illustrated

Other names: Kingfisher, River (or European or Eurasian) Kingfisher
French: Martin-pêcheur d'Europe   **Creole:** Marten Peser Eropeen

DESCRIPTION Length 16-17 cm; wingspan 24-26 cm.
A small compact kingfisher, with long bill and large head, brilliant blue and green above, orangey below. **Adult** has electric blue upperparts with orange underparts, undertail-coverts and ear-coverts. Greenish blue on head, scapulars and wings. White spot to rear of head. Short, stubby tail. Small, red legs. Long dagger-like bill is mainly black. Sexes similar, female with reddish base to lower mandible. **Immature** is duller than adult, greener above with some dusky scaling on breast.
VOICE Mainly silent outside breeding season. May give short, sharp 'tzii' in flight, repeated two or three times.
BEHAVIOUR Frequents vicinity of slow-flowing streams or still pools, avoiding rapid-flowing water. Sometimes on coastlines in winter. Generally solitary. Flight usually rapid, wingbeats whirring, low over fresh water, sometimes suddenly changing direction. Hunts from a perch, diving to catch fish, aquatic insects, crustaceans, molluscs and amphibians. Occasionally hovers or flycatches for aerial insects and occasionally takes lizards.
RANGE Nominate *atthis*, the only race to occur in Africa, breeds North Africa, southern Europe to Asia. Winter visitor to Egypt (recorded south to Aswan), Yemen and Pakistan. Six other races recognised but only the nominate occurs in Africa. Most populations winter mainly within breeding range, but some Russian birds may move more than 3000 km[1]. Apart from the nominate race, the migratory race *bengalensis* of eastern Siberia and north China (brighter and smaller than nominate) might be a possibility in Seychelles being a long-distance migrant though its known wintering range is considerably to the east (Sumatra to Philippines). Vagrant to Maldives.
STATUS Uncertain; there is a record from Frégate of what was certainly a kingfisher species, probably this one, though other *Alcedo* species were not ruled out.
SIMILAR SPECIES It is difficult to see that any other kingfisher is a potential vagrant but even remote possibilities need to be considered to clinch a 'first' for Seychelles. Rufous ear-coverts distinguish from all but juvenile **Blue-eared Kingfisher** *A. meninting* (L17 cm), a potential albeit unlikely vagrant, being a resident of Sri Lanka to South-East Asia. It is much deeper blue on the upperparts, becoming purple blue on cheeks, nape and upper mantle and more chestnut on underparts; lacks greenish tones to crown, wings and scapulars. Other possibilities might include **Madagascar Kingfisher** *A. vintsioides* (L13-15 cm) of Madagascar and Comoros, given the proximity of its range to Seychelles; it is much smaller, purple-blue above, whitish orange below; bill black with dull red base. **Half-collared Kingfisher** *A. semitorquata* (L18-19 cm), is uncommon and local from Ethiopia to South Africa. It is very similar, with a white patch on neck and blue patch on side of breast.
Reference: 1. Fry *et al.* (1992).

BLUE-CHEEKED BEE-EATER *Merops persicus*   Plate 40

French: Guêpier de Perse   **Creole:** Manzer Mousdimyel Lazou Ble

DESCRIPTION Length 27-31 cm (including 4-7 cm tail-streamers); wingspan 46-49 cm. A large, slender, graceful, long-winged, bright green bee-eater, the crown concolorous with upperparts. **Adult** plumage is generally grass-green, relieved by a black face-mask, pale blue supercilium and cheek and a whitish fore-head. Blue cheek-patch, brighter than adjacent green feathers. Yellow chin merges to rufous throat. Underwing rufous, with narrow, dark trailing edge. Elongated green central tail feathers project up to 7 cm (slightly shorter in female). Bill black; legs dark grey. Eye deep red in male, orangey red in female. **First-winter** similar to adult but duller, with shorter tail-streamers dark brown eye, whitish green not bluish around face-mask and forehead greenish, yellowish at the bill. Pale yellow chin merges to rufous throat-patch, paler at the edges and extending to ear-coverts. Sometimes retains pale-edged feathers of juvenile, giving a scaly appearance. Indistinguishable from adult by midwinter.

VOICE A repeated 'dirrip'. Alarm call is 'dik-dik-dik'.

BEHAVIOUR Inhabits open woodland, usually near water. Often perches on dead branches and wires. Gregarious year round (most Seychelles records are of two or more individuals together). Swoops towards ground, or gains height rapidly with even wingbeats, gliding on outstretched wings then twisting sharply to catch bees, beetles, dragonflies, butterflies or other insect prey. Returns to perch to beat insect before eating though will occasionally consume in flight. Venomous insects rubbed against perch to remove venom.

RANGE Two races, of which the nominate *persicus* is the only one likely to reach Seychelles; this breeds Nile delta east to northwest Indian and Lake Balkhash, wintering mainly Africa from Ethiopia to South Africa (a few remaining in India and Pakistan).

STATUS Vagrant throughout Seychelles, with scattered records from granitic islands, Bird, Amirantes and Aldabra November to December and March to May.

SIMILAR SPECIES Madagascar Bee-eater has dull green plumage, brown crown, white forehead, chin and cheek-stripe and paler underwing. **European Bee-eater** easily distinguished by combination of chestnut cap, back and upperwing-coverts, turquoise underparts and yellow throat.

MADAGASCAR BEE-EATER *Merops superciliosus*   Plate 40

Other name: Olive Bee-eater   **French:** Guêpier de Madagascar   **Creole:** Manzer Mousdimyel Malgas

DESCRIPTION Length 30-33 cm including 7 cm tail-streamers; wingspan 43-46 cm. A large, slender, long-winged bee-eater, with brown or olive green crown and dull olive green plumage. **Adult** is generally dull green except for brown to olive green crown and black face-mask bordered by white above and below. Forehead is usually white (sometimes slightly yellowish, greenish or pale bluish). Supercilium usually white (sometimes blue or pale yellow), narrower and more restricted than Blue-cheeked. Cheek is usually white (but note occasionally Blue-cheeked may have white cheek). Chin duller yellow than Blue-cheeked, under-wing buff not rufous. Eye deep red, bill black and legs pinkish brown. **Juvenile** is duller than adult, lacking tail-streamers with dull greenish crown, some blue feathers on back, buff throat, pale blue breast merging to pale green belly and brown char.

VOICE Very similar to Blue-cheeked.

BEHAVIOUR Similar to Blue-cheeked, often favouring less arid habitat such as mangroves, streams and wetland margins. In Madagascar prey includes butterflies, grasshoppers, cicadas, wasps and beetles. In India prey mainly bees and dragonflies.

RANGE Nominate *superciliosus* breeds Madagascar, Comoros and East Africa. In Madagascar breeds September to December; birds present year round, but some may migrate to Africa. Another race, *alternans*, breeds west Angola and northwest Namibia.

STATUS Uncertain; some bee-eater reports are indeterminate between Madagascar and Blue-cheeked.

SIMILAR SPECIES Blue-cheeked Bee-eater is generally brighter grass-green including on crown, with darker underwing. **European Bee-eater** easily distinguished by combination of chestnut cap, back and upperwing-coverts, turquoise underparts and yellow throat outlined in black. **Blue-tailed Bee-eater** M. *philippinus* (L28-31 cm) is a potential vagrant; it breeds north Pakistan to Papua New Guinea, migrating south to southern India and Sri Lanka, vagrant to Maldives. It has blue or greenish blue rump and tail, a narrow blue line below the mask, green forehead and supercilium and yellow chin.

EUROPEAN BEE-EATER *Merops apiaster*   Opposite Plate 40

Other names: European Bee-eater, Bee-eater
French: Guêpier d'Europe   **Creole:** Manzer Mousdimyel Eropeen

DESCRIPTION Length 27-29 cm; wingspan 44-49 cm. A vividly coloured, elegant bee-eater with long pointed wings and pointed central tail feathers. Combination of chestnut cap and all-yellow throat diag-nostic. **Adult** of both sexes have unmistakable, vivid, multi-coloured plumage. Chestnut crown, nape and mantle merge to golden yellow scapulars and rump, which form a V shape in flight. Non-breeding is slightly duller, scapulars and rump greenish with some green feathers in chestnut areas of plumage. Throat yellow separated by black line from turquoise blue underparts. Underwing translucent, pale rufous with thick black trailing edge. Eye deep red. Long, curved black bill; legs blackish. **Male** is brighter than female, inner half of upperwing bright rufous (unlike any other bee-eater), outer half green. Scapulars of **female**

are greener and wings less chestnut. **Immature** is less colourful than the adult, with tail-streamers very short or absent, eye browner, a green tone to crown, nape and back similar to Blue-cheeked Bee-eater, from which it is separated by yellow throat and brown hindneck.

VOICE A liquid, melodious 'prruip', softer than Blue-cheeked Bee-eater, given year round.

BEHAVIOUR Frequents most open habitats in winter quarters, avoiding forest. Gregarious on migration but vagrants are likely to be solitary. Flight graceful, gliding on outstretched wings, sometimes making sudden changes in direction to catch insects. Hawks for aerial insects from a vantage point, prey including bees, dragonflies, butterflies and crickets. Occasionally takes insects, spiders, caterpillars and even worms from the ground.

RANGE Breeds southern Europe to Kazakhstan, wintering mainly sub-Saharan Africa. Also breeds South Africa. Monotypic.

STATUS Vagrant to Aldabra group reported December and April.

SIMILAR SPECIES No other bee-eater has the same combination of chestnut cap and all-yellow throat.

EUROPEAN ROLLER *Coracias garrulus*   Plate 40

Other names: Roller, European (or Northern or Blue) Roller, Blue Jay
French: Rollier d' Europe   **Creole:** Rolye Eropeen

DESCRIPTION Length 31-32 cm; wingspan 66-73 cm. The size and build of a small crow but with pale blue head and underparts, brown back and striking, vivid, ultramarine wings. **Adult** has head, neck and underparts pale azure-blue; back, scapulars and tertials chestnut brown, purple-blue on the lower back and rump. The tail is greyish brown in the centre with pale blue sides, a black spot in the corners and no tail-streamers. In flight, colours of wing are striking: flight feathers jet black above, ultramarine below, wing-coverts and bases of flight feathers pale blue. Autumn birds duller, many feathers brown-tipped and back more rufous. At close range, white at base of bill and a black line through the eye are visible. Eye brown, with grey eye-ring. Bill blackish; legs olive brown to yellow. Sexes similar. **Immature** similar to adult non-breeding but duller and whitish on face, brown on back and cheeks while throat and breast have a brownish hue, streaked with white; outer tail feathers lack black tips and are slightly shorter than other tail feathers. Bill pinkish at base.

VOICE Generally silent away from breeding grounds. Usual flight call a short, crow-like 'krak' often repeated.

BEHAVIOUR Frequents open woodland where hunts from a perch in a tree, on telegraph poles, walls, wires etc., usually above sparsely vegetated ground. Gregarious, often in hundreds of thousands in mass migration in East Africa, though Seychelles vagrants are usually solitary. Flight powerful and direct, with shallow, rapid wingbeats and occasional glides. Swoops down to ground from perch, ending with a glide; alights and eats small insects on the spot, while larger prey are taken back to look-out perch. Insects make up 97% of diet, but also takes frogs, lizards, snakes, small mammals, small birds, worms, spiders, millipedes, centipedes, scorpions and molluscs[1]. Regurgitates several small brown pellets daily. Always flies between different look-out perches, never hops, even when very close together, but sometimes hops clumsily on the ground chasing prey. Occasionally sunbathes.

RANGE Nominate *garrulus* breeds North Africa, southern Europe to southwest Siberia. Race *semenowi* breeds Iraq, Iran except where replaced by nominate in northwest Iran. Latter has paler head and rump, paler, greener underparts, browner back and narrower purple-blue band on forewing. Almost the entire world population winters sub-Saharan Africa, with a large concentration in East Africa of up to three million in eastern Kenya alone, outnumbering all other *Coracias* rollers 5-7 times[2]. Most of these are nominate, with *semenowi* passing through to winter southern Africa.

STATUS Vagrant; scattered records of unknown race throughout Seychelles from Aldabra group, Amirantes, granitic islands and Bird. Recorded October to March, with most reports October to November.

GENERAL This long-distance migrant makes a 10,000 km journey from the central Palearctic to central Africa at 67 km per day, returning at 110 km per day[3]. There is a spectacular mass movement along the East African coast in the first half of April with up to several hundred thousand passing some localities within a few hours[4,5].

SIMILAR SPECIES European Roller is one of the most easily identified vagrants to Seychelles, but descriptionless records have not been confirmed by SBRC due to other remote possibilities. Lack of tail-streamers distinguishes from adult **Lilac-breasted Roller** *C. cordata* (L32-35 cm, WS55-60 cm) and **Abyssinian Roller** *C. abyssinica* (L40-45 cm, WS 58-60 cm), potential vagrants from Africa. Immature Lilac-breasted Roller also lacks tail-streamers but has a tawny-rufous throat with broad, whitish streaks. Immature Abyssinian Roller also lacks elongated outer tail feathers, but has violet-blue not black on upperside of flight feathers, more extensive turquoise tail tip, brighter blue rump and broader white band on forehead. **Indian Roller** *C. benghalensis* (L31-35 cm, WS65-74 cm) of southern Asia, another potential vagrant, has a greenish blue cap, brown neck, throat and chin streaked white, and blue (not black) flight feathers; also a bright pale blue band near wingtips and terminal blue tail-band. All these potential vagrants are sedentary or perform only local movements (though Indian Roller has occurred as a vagrant to Maldives) but should be considered in confirming the identification of any roller seen in Seychelles.

References: 1. Fry *et al*. (1992), 2. Brown and Brown (1973), 3. Glutz and Bauer (1980), 4. Ash and Miskell (1980), 5. Feare (1983).

BROAD-BILLED ROLLER *Eurystomus glaucurus*   Plate 40

Other name: Cinnamon Roller   **French:** Rollier (or Rolle) violet   **Creole:** Rolye Malgas

DESCRIPTION Length 29-30 cm; wingspan 55-58 cm. A small, stocky, large-headed, long-winged roller with a short, broad, yellow bill. **Adult** is mainly cinnamon above and deep lilac below. Forewing cinnamon, the rest of the wing dark blue. Tail dark blue-grey with a shallow fork. Dark grey-blue undertail-coverts and vent in nominate *glaucurus*, the most likely race to occur (clear blue in other races). Short, stout, wide bright yellow hooked bill visible from long range. Legs olive brown. Sexes similar. **Juvenile** similar to adult but duller. Underparts below the breast dull blue and mottled greyish; upperparts rufous-brown. Forewing rufous-brown, the rest of the wing dull greenish blue. Forehead dark brown. Bill yellow, the culmen spotted brown.
VOICE A loud, harsh 'grak-grak-grak'.
BEHAVIOUR Frequents open woodland where perches alone or in pairs on a bare branch overlooking an open area. Generally solitary, but occasionally in twos or threes. Dashing flight appearing falcon-like but chunky and short-tailed, lacking the striking wing pattern of Northern Roller. May drink on the wing similar to a swallow. Swoops in a long glide to seize insect prey, which are eaten on the wing or upon return to the perch. Regularly used perches may be defended to a distance of 300 m[1]. Swarming winged ants and termites may form a large part of diet. May take up to ten insects in the bill per minute, feeding until nightfall, each roller taking 200-800 insects[1]. Fiercely territorial, chasing away other birds.
RANGE Breeds throughout Madagascar, migrating to mainly eastern Zaire February to November occurring on passage in East Africa. Also resident sub-Saharan Africa, though uncommon in South Africa, Namibia and Botswana. Nominate *glaucurus* of Madagascar is the largest of four recognised races (some authors recognise more). Other races are unlikely to reach Seychelles, though perhaps *suahelicus*, which ranges from 5°N south to Transvaal, should be considered; this race is smaller (wing 171-185 mm compared to 190-220 mm for nominate race) and uppertail-coverts is entirely blue, undertail-coverts pale blue. Other less likely races are *afer*, the smallest race, of West Africa to Sudan (central uppertail-coverts brown, lateral pale blue-green) and *aethiopicus* of Ethiopia (similar to *afer*, slightly larger and more brightly coloured).
STATUS Near-annual in Aldabra group and a vagrant elsewhere. Aldabran records are almost entirely October to December and where it has been possible to assign an origin, they have always been the Madagascan nominate race.
SIMILAR SPECIES None.
Reference: 1. Fry *et al*. (1992).

HOOPOE *Upupa epops*   Plate 40

Other names: Common (or Eurasian) Hoopoe   **French:** Huppe fasciée   **Creole:** Oupou

DESCRIPTION Length 26-28 cm; wingspan 42-46 cm A highly distinctive, pinkish brown bird with striking crest and broad, rounded black-and-white barred wings, lower back, rump and tail. **Adult** has pinkish brown body (often washed grey outside breeding season) and a prominent pinkish crest which is raised when excited or when lands and occasionally in flight. Race *africana* is more richly coloured than the nominate race. Crest feathers are tipped black (nominate race also has a white subterminal band). Striking broad, black-and-white barred wings (male *africana* having a large white patch due to white base of secondaries, both sexes with all-black primaries). Black tail with broad white band. Long, thin, downcurved bill, mainly black, pale horn at base. Legs slate-grey. Sexes similar, though male slightly larger and brighter; female has whitish chin, more streaked on the breast and sides of belly with rufous-grey lesser coverts. **Juvenile** is darker more brownish than adult, greyer on head and body, with a shorter crest and shorter less decurved bill.
VOICE A far-carrying 'poo-poo-poo' usually given from the ground with head lowered and neck distended. Usually silent outside breeding season.
BEHAVIOUR Frequents forest clearings and open areas. Generally solitary or in pairs. Flight slow and undulating, the black-and-white plumage of back and wings very obvious. Feeds on the ground in open areas, walking jerkily and probing for insects and insect larvae and may fly to a tree if disturbed. Some prey is captured on the surface but the skull and jaw muscles show adaptations that enable the bill to be opened with great force against resistance of the earth when probing for insect larvae.
RANGE Nominate *epops* breeds northwest Africa and western Palearctic, wintering mainly south of the Sahara, occurring commonly in East Africa October to March. Race *africana* (sometimes considered a separate species, African Hoopoe*)* breeds from the Equator throughout most of the southern half of Africa, where lowland populations are largely migratory. African Hoopoe is more richly coloured dark orange, lacks whitish subterminal band to crest, has entirely black primaries (nominate shows white-banded pri-

maries) while male has a large white patch in the secondaries, which are barred in female.
STATUS Vagrant to Aldabra, where both the nominate race and *africana* have been recorded in October.
SIMILAR SPECIES None recorded but **Madagascar Hoopoe** *U. marginata* (L32 cm) is a potential vagrant; it is larger (slightly larger than a Madagascar Turtle Dove, whereas Hoopoe is smaller) with barred primaries and secondaries and a totally different call (a resonant, repeated 'rrrooooo').

GREATER SHORT-TOED LARK *Calandrella brachydactyla*   Plate 42

Other names: (European) Short-toed Lark   **French:** Alouette calandrelle   **Creole:** Kalandrel

DESCRIPTION Length 13-14 cm; wingspan 25-30 cm. A small, pale, thickset lark, sandy buff with dark streaks above, whitish and unstreaked below and no crest. **Adult** has upperparts grey-brown (eastern races) to warm brown (nominate) streaked pale buff and blackish. Underparts buff-white and generally unstreaked. Uppertail-coverts tawny and unstreaked. Crown brownish (tinged rufous in the male) with fine blackish parallel lines extending to the mantle. No crest, but a broad whitish supercilium outlines cap, feathers of which may be raised when excited. Ear-coverts and lore dusky, throat and breast dingy white lacking buff tones or streaking. Indistinct black moustache interrupted by white throat-band, below this a dark patch on the side of the neck (sometimes streaked). Long tertials with no primary projection; dark bar and a pale bar on closed wing; long tertials, almost covering wingtip. Tail buff in centre, separated from white outer feathers by a black panel. Stout, pointed, yellowish, finch-like bill; legs flesh-brown. **Adult breeding** has more obvious dark patch on side of breast. **Immature** similar to adult, upperparts dark brown with buff edges and whitish tips to feathers, underparts white sometimes with dark brown or grey spots or streaks. In fresh autumn plumage, neck-patch less distinct, breast more buff.
VOICE A short, dry 'tchirrup' recalling a House Sparrow. Often imitates other birds.
BEHAVIOUR Frequents open sandy or stony areas, dry mudflats and fields. Gregarious outside breeding season, occurring in flocks of several thousand, though vagrants are likely to be solitary. Flight strong, low and undulating, body appearing torpedo-shaped. Walks with a jerky gait and makes short erratic runs. Feeds on the ground taking insects and seeds, occasionally flying insects.
RANGE Several races breed North Africa, southern Europe to northeast China. Eastern birds are paler and greyish. Western Palearctic birds migrate across the Sahara occasionally reaching East Africa. Nominate race of most of Europe to northwest Africa is warm brown and prominently streaked above. More likely in Seychelles is *longipennis* (Eastern Short-toed Lark), which breeds central Asia, Afghanistan, north India wintering from Senegal to Kenya (where it is a vagrant), crown, hindneck and rump sandier grey, narrow streaks on upperparts, breast washed brownish buff and a smaller bill. Races in winter are very difficult to separate.
STATUS Vagrant to granitic islands; recorded November to December.
SIMILAR SPECIES No other lark is likely to reach Seychelles.

SAND MARTIN *Riparia riparia*   Plate 41

Other names: Bank Swallow, Collared (Common, or European or Gorgeted) Sand Martin
French: Hirondelle de rivage   **Creole:** Zirondel Bren

DESCRIPTION Length 12 cm; wingspan 27-29 cm. Combination of well-defined, brown breast-band on white underparts, brown upperparts and squarish tail distinguish from any other swallow species occurring in Seychelles. **Adult** is brown or grey-brown above, white below with brown breast-band, broadest at sides; centre of breast is speckled brown. Underwing-coverts grey-brown, with pale fringes to lesser coverts. Secondaries and tail feathers have white outer edges and tips in fresh plumage. Tail lacks streamers and is slightly forked, sometimes appearing squarish. Eye dark brown; bill black; legs dark brown. Sexes similar. **Juvenile** has buff or creamy edges to feathers on the upperparts (conspicuous on inner wing-coverts and rump), which may wear off within a few months, and a buff wash to the throat and breast-band.
VOICE Contact call is a characteristic harsh rasp of one or two syllables. Also gives a low-pitched single note alarm call.
BEHAVIOUR Frequents freshwater marshes or open ground (such as grassy airstrips). Highly gregarious at all times though vagrants are likely to be solitary. Flight more direct than Barn Swallow, with rapid, shallow wingbeats and frequent glides. Feeds at an average height of 15 m, taking mainly flying insects, but will take insects off the surface of water and occasionally even land to feed on the ground.
RANGE Breeds across northern hemisphere, migrating to southern wintering grounds. European birds of nominate *riparia*, migrate to sub-Saharan Africa. Wintering grounds of race *ijimae*, which breeds Siberia, are poorly understood but includes Burma and South-East Asia; race *diluta* (sometimes considered a full species, Pale Sand Martin, overlapping with nominate without interbreeding) migrates to Pakistan and India, from Central Asia. Race *ijimae* has blackish brown upperparts, while *diluta* is distinctly paler grey-brown with paler, narrower, less distinct breast-band.
STATUS Vagrant throughout Seychelles. Single birds have been recorded from October to as late as June.

Nominate race is probably the most likely to reach Seychelles.

SIMILAR SPECIES None recorded, though potential vagrants should be considered. Only other hirundine with a clear breast-band is **Banded Martin** *R. cincta* (L16-17 cm, WS39-41 cm). It breeds throughout southern and central Africa, but is much larger, longer-winged, has diagnostic white streak over the eye and white underwing-coverts (Sand Martin underwing all-dark); migrates north within Africa in March and April following the rains. **Brown-throated Martin** *R. paludicola* (L12 cm, WS28 cm) of Africa and Asia is similar to Sand Martin though it performs only local movements; it has grey throat and breast and no breast-band. **Crag Martin** *Hirundo rupestris* (L14.5 cm, WS32-34.5 cm), which breeds in southern Europe and Asia, should also be considered; it is ashy brown, has no breast-band, speckled throat and white patches on square tail. Northern populations are mainly migratory, but rarely reach further than North Africa or India.

MASCARENE MARTIN *Phedina borbonica*   Plate 41

Other name: Madagascar Martin   **French:** Hirondelle des Mascareignes   **Creole:** Zirondel Malgas

DESCRIPTION Length 15 cm; wingspan 30 cm. A large, chunky swallow with short broad wings, heavy body and slightly forked tail. The only martin with boldly striped underparts likely to reach Seychelles. **Adult** has finely streaked dark grey-brown upperparts, lighter brown underparts with heavy dark brown streaks. Throat white with dark brown streaking. Sides of breast and flanks grey-brown. Blackish brown tail has very shallow fork. Underwing-coverts and axillaries grey-brown. Eye dark brown; bill and legs blackish. **Immature** similar to adult, with pale edges to secondaries.

VOICE Fairly silent, on the whole. Flight call is 'siri-liri, siri-liri', and when perched may utter a warbling call.

BEHAVIOUR Frequents woodland, mangroves, lake sides and rivers, from sea level to high mountains. May roost in buildings (one specimen now in the Natural History Museum at Tring, UK, was caught in a building on Desnoeufs[1]). Gregarious and sometimes seen in the company of other species, though vagrants likely to be solitary. Slow, fluttering, heavy flight with much gliding, feeding close to the ground or vegetation. Diet small aerial insects.

RANGE Race *madagascariensis* is locally common on Madagascar, breeding from August to November and migrating as far as East Africa. There is also a resident race *borbonica* on Mauritius and Réunion which has never been recorded away from those islands.

STATUS Vagrant, with single birds recorded throughout Seychelles. Most records are October to November.

SIMILAR SPECIES None.

Reference: 1. Feare (1977).

BARN SWALLOW *Hirundo rustica*   Plate 41

Other names: Swallow, House (or Chimney, or Common, or Rustic, or Eurasian or European) Swallow
French: Hirondelle rustique   **Creole:** Zirondel Eropeen

DESCRIPTION Length 17-19 cm; wingspan 32-34 cm. The most commonly observed hirundine in Seychelles. Steel-blue above, whitish below with rufous buff chin, throat and forehead, blue-black breast-band and a deeply forked tail. **Adult** has upperparts steel-blue with no pale rump; pale 'windows' on tail feathers may be visible in flight when long tail-streamers are visible. Well-defined dark blue breast-band, rufous or red on chin and front of forehead. Bill and legs black. Sexes very similar, though female has less glossy blue on upperparts and breast-band, generally whiter underparts and shorter tail-streamers. **Juvenile** (up to December) duller than adult, browner, shorter tail-streamers, much paler buff on the chin, throat and sometimes whiter underparts.

VOICE A fairly wide repertoire, including a loud 'vit, vit' and a melodious twitter. Also a variety of alarm calls which differ partly according to the degree of danger including a two-note 'tsee wit' and several muffled calls.

BEHAVIOUR Frequents farmland, human habitation and open areas. Gregarious, including on migration and in winter quarters. A strong flier, rarely gliding, usually seen at a height of 7-8 m hawking for insects. Also skims low over the ground or water to take non-flying insects, spiders and caterpillars.

RANGE The world's most widespread species of swallow, breeding across North America, Europe and Asia and migrating to southern wintering grounds. Nominate race *rustica* breeds throughout Europe, western Russia, North Africa, Middle East to eastern India. This is the most likely race in Seychelles. Less likely but perhaps possible is race *gutturalis* (breeds further east as far as Japan), which has an incomplete breast-band unlike the nominate race and is slightly smaller, or possibly *tytleri* (breeds central Siberia, northern Mongolia and winters India to South-East Asia), which has chestnut underparts and an incomplete breast-band.

STATUS Near-annual throughout Seychelles. Recorded every month from September to May, with the great majority of sightings in either November or March, rarely for more than a day or so, indicating stages

of southerly or northerly migration. Most sightings have been of single birds though small groups have been noted at Aldabra and, exceptionally, 50 birds in March 1999 following strong northwesterly winds[1]. **SIMILAR SPECIES** No similar species recorded in Seychelles, but even so, the majority of records submitted to SBRC have been accepted with the caveat that the description failed to rule out other possibilities[2]. One of these potential vagrants is **Lesser Striped Swallow** *H. abyssinica* (L15-17.5 cm), found over much of sub-Saharan Africa and occurring as a vagrant to Madagascar; it is, however, a distinctive swallow with a rufous head and rump, metallic blue-black back and white underparts boldly streaked black. Also worth consideration is **Red-rumped Swallow** *H. daurica* (L16-17 cm, WS32-34 cm), which ranges over much of southern Europe, Africa and Asia; distinguished by rufous rump (with faint streaking) and collar, lack of breast-band, streaked breast and whitish throat. In flight, black undertail-coverts and rufous underwing-coverts are diagnostic. There is a remote possibility of **Rufous-chested Swallow** *H. semirufa* (L18-19 cm), found over much of sub-Saharan Africa, distinguished by dark rufous throat and breast, buff underwing-coverts, rufous sides to neck and blue-black lore, cheek and ear-coverts. Also possible is **White-throated Swallow** *H. albigularis* (L14-17 cm), which is migratory throughout South Africa and has been recorded in East Africa. Chin and throat are white, forehead rufous, upperparts blue and underparts greyish below the blue breast-band.
References: 1. Betts (1999), 2. Skerrett (1996a).

COMMON HOUSE MARTIN *Delichon urbica*   Plate 41

Other names: House Martin, Northern (or European or Western) House Martin
French: Hirondelle de fenêtre   **Creole:** Zirondel Deryer Blan

DESCRIPTION Length 12.5 cm; wingspan 26-29 cm. Combination of conspicuous white rump, white underparts and glossy blue upperparts distinguish from any other hirundine likely to be seen in Seychelles. **Adult** has dark glossy blue upperparts, with black, moderately forked tail and well-defined white rump. Gloss to upperparts may only be obvious at close range, appearing black from a distance or in strong sunlight. Underparts white, flanks greyish white; in non-breeding plumage, when most likely to be seen, has greyish brown wash to cheek, chin, throat, flanks, shorter tail-coverts and rump. When opened, tail appears almost triangular. Dumpy body, obvious in side view. Eye blackish brown; bill small and black; legs pink, covered with white feathers to toes. Sexes similar though female has slightly greyer underparts, sometimes with grey spots on undertail-coverts and occasionally also on rump. **Immature** similar to adult but duller, pale grey on rump, with glossy blue confined to the mantle and a greyish black wash on sides of breast which may give the impression of a breast-band. Distinct white tips to tertials.
VOICE Less melodious than Barn Swallow, the usual call is a shrill 'prit' or a soft twitter. Several alarm calls of increasing pitch may be given.
BEHAVIOUR Associates with man at breeding sites, but vagrants to Seychelles unlikely to have any habitat preference. Winter habitat in tropical and southern Africa is poorly known: thousands of birds enter Africa each year and yet are hardly ever seen. They may mainly occupy high altitudes where they are difficult to observe. Gregarious, but vagrants likely to be solitary. Flight slower with more gliding than Barn Swallow, usually at a greater height (average of 21 m above the ground, lower in wet weather has been recorded at breeding grounds). Most prey is taken on the wing, with a marked preference for smaller insects than prey of Barn Swallow. Often 'towers' to catch insects, very occasionally perching to pick off insects, spiders and caterpillars from walls or trees.
RANGE Breeds throughout Europe, North Africa and Asia. Nominate *urbica*, breeding from Europe to Mongolia and the Himalayas, migrating to sub-Saharan Africa, is most likely to occur; has bred South Africa and Namibia. Less likely but not impossible is race *lagopoda* which breeds eastern Asia, wintering South-East Asia; has more extensive white on the lower back and a slightly less deeply forked tail.
STATUS Vagrant; recorded Farquhar and Aldabra groups November or March.
SIMILAR SPECIES The only other migratory hirundine with the same combination of blue-back, white rump and white underparts is **Asian House Martin** *D. dasypus* (L13.5 cm), which replaces Common House Martin in central and eastern Asia migrating through eastern China to South-East Asia, rarely reaching Indonesia. Unlikely in Seychelles but cannot be entirely ruled out. It may be distinguished by darker, less glossy upperparts, more distinct streaks on the rump, greyer underparts, darker underwing and less deeply forked tail (almost square-ended when spread).

RICHARD'S PIPIT *Anthus richardi*   Plate 42

French: Pipit de Richard   **Creole:** Pipit Richard

DESCRIPTION Length 18 cm; wingspan 29-33 cm. A large, thrush-like, long-tailed, long-legged pipit. Warm brown above with heavy blackish streaks, whitish or buff below with dense, short streaks on breast. Size and upright stance separate from all pipits recorded in Seychelles, though Tawny Pipit and Blyth's Pipit (both potential vagrants) should be considered. **Adult** has a bold face pattern: pale supercilium and lore; dark eye-stripe, moustachial-stripe and broad, blackish malar streak. Eye-stripe is more prominent behind than in front of eye (opposite of Tawny Pipit). Warm buff breast and flanks contrast with white centre to belly, streaked on upper breast, streaking sometimes forming a dark smudge on side of throat. Pale pinkish buff edges to wing-coverts form two wingbars; broad rusty margins to tertials. Warm brown streaked upperparts contrast somewhat with plainer nape and rump. Long, slender tail is dark brown, the outer tail feathers white. Legs pinkish and very long; hindclaw is long and curved. Long stout bill is blackish brown, yellowish at base of lower mandible. **Juvenile** plumage, retained to early winter, is more strongly streaked above than adult, breast spotted. Generally darker than adult, lore sometimes also appearing dark; worn birds may appear nearly black on crown and mantle. **First-winter** retains white fringed wing-coverts (fringed broadly and more diffusely buff in adult), contrasting with rest of wing.

VOICE A harsh, explosive 'rreep' recalling House Sparrow; harsher and louder than Tawny Pipit.

BEHAVIOUR Frequents wet grassland, lake margins and cultivation. Gregarious, in flocks of up to 100 outside the breeding season though vagrants are usually solitary. Flight strong and undulating. On the ground makes short dashes pausing to stand erect, occasionally dipping tail during pauses. Shy, retreating from observer with swift runs prior to long escape flight. Hunts on the ground for insects, arthropods and sometimes seeds, walking with a strutting gait. May occasionally leap or make short flights to catch aerial insects, sometimes hovering like a lark, tail spread.

RANGE Several races breed Siberia, Mongolia, China wintering southward. Nominate described above, others including *centralasiae* (larger and sandier above) and *dauricus* (more heavily streaked above). Formerly lumped with Australasian Pipit A. *novaeseelandiae*.

STATUS Uncertain, with unconfirmed reports from granitic islands

SIMILAR SPECIES Tree Pipit is smaller and less robust, less heavily marked with an unstreaked rump and a short hindclaw. Adult **Tawny Pipit**, a potential vagrant (see also under Tree Pipit), is smaller, less upright, with shorter bill, shorter legs and hindclaw, supercilium broad and square-ended to rear, ear-coverts more uniform, malar-stripe less prominent, generally lighter brown and less heavily streaked, creamy not warm buff on flanks. Juvenile Tawny Pipit is very similar to Richard's, but darker lore and eye-stripe contrast with broad, square-ended supercilium. **Blyth's Pipit** A. *godlewskii* (L17 cm, WS28-30 cm) is also a potential vagrant that breeds Mongolia and north China, wintering Indian subcontinent, Sri Lanka. It is a large pipit, very slightly smaller than Richard's, but less robust with shorter, more pointed bill, shorter legs, shorter less curved hindclaw, shorter wings and tail, though these characteristics are very difficult to evaluate in the field. Centres of median coverts compared to Richard's are darker and more squarely cut against pale tips and less triangular. In fresh plumage upperparts greyer, more heavily streaked and underparts more uniform than Richard's. Lore pale (distinguishing from Tawny Pipit which has dark loral-stripe). Bill finer than Richard's. Juvenile similar to adult but lacks distinctive median and greater coverts compared to Richard's and has buffer tone to plumage. Flight undulating, shallower and less powerful than Richard's. Less erect posture (most horizontal of all pipits) and less strutting gait than Richard's, tail, wagging more vigorous.

TREE PIPIT *Anthus trivialis*   Plate 42

Other names: Brown (or Eurasian or Witherby's) Tree Pipit   **French:** Pipit des arbres   **Creole:** Pipit Dibwa

DESCRIPTION Length 15 cm; wingspan 25-27 cm. A slender medium-sized pipit. Olive brown above, heavily streaked on the back but almost plain on the rump. By far the most likely pipit to be seen in Seychelles. **Adult** has upperparts olive brown streaked blackish brown, plain or very faintly streaked on rump. Tail dark brown with white outer tail feathers. Pale buff wash on flanks, breast and sides of white throat, white also on the belly. Heavy blackish streaks on breast, more finely streaked on the flanks. Pale eye-ring and bold yellowish or buff supercilium. Blackish moustachial-stripe and distinct buff submoustachial-stripe. Dark eye-stripe conspicuous behind the eye. Stout dark brown bill, pale flesh at the base of the lower mandible. Legs flesh, with short hindclaw. Fresh autumn birds are warmer buff on the throat, breast and flanks showing more contrast with white belly and narrow, pale edges to tertials. **Juvenile** similar to fresh adult, feather centres of upperparts blacker and more rounded with buff fringes. **First-winter** is indistinguishable from adult.

VOICE A short sharp 'tseep', sometimes drawn-out, or a softer 'zeep'. Usually silent in winter quarters.

BEHAVIOUR Frequents woodland margins, fields and grassy areas, often seen on margins of airstrips in Seychelles. Usually solitary or in pairs. Strong undulating flight. Walks on the ground with body held more upright than Red-throated Pipit. Engages in short runs, gently wagging tail during pauses. Readily perches in trees. Diet includes insects, spiders and seeds.

RANGE Nominate *trivialis* breeds Europe to Lake Baikal and migrates mainly to sub-Saharan Africa. Some also migrate to Indian subcontinent where they mix with races *schlueteri* (breeds Afghanistan to northwest China) and *haringtoni* (breeds northwest Himalayas). Differences between races are slight.

STATUS Near-annual in small numbers throughout Seychelles October to March with majority of records November to December and none February. This is the pipit most often encountered in Seychelles; indeed, it has been reported to SBRC more often than any other migrant passerine.

SIMILAR SPECIES Short hindclaw distinguishes from all pipits except **Olive-backed Pipit**, which has a greenish tone to plumage, is almost unstreaked on the mantle and is heavily streaked on the breast with a distinctive head pattern. **Red-throated Pipit** has heavier, more extensive streaking on flanks, darker, more heavily streaked upperparts including the rump, longer hindclaw, breeding birds with a brick-red throat, non-breeding with whiter underparts. Vagrants that should be considered include **Tawny Pipit** *A. campestris* (L16.5 cm, WS25-28 cm) which breeds North Africa, Europe to western Mongolia wintering south to Kenya. It is a slim, wagtail-like pipit with poorly marked sandy-brown upperparts and plain, pale, virtually unmarked underparts (paler and less patterned than any other pipit). Best identification feature is a line of bold, dark spots on the median-coverts contrasting sharply with rest of plumage. Conspicuous pale supercilium. Fairly long tail with white outer tail feathers. Long fine dark brown bill, pink at the base of lower mandible. Legs long, bright yellow without long hindclaw. Flight wagtail-like, but lacks power on take off and less undulating than other pipits. Often wags tail strongly. Posture the most horizontal of all pipits.

OLIVE-BACKED PIPIT *Anthus hodgsoni*   Plate 42

Other names: Indian (or Olive, or Hodgson's or Oriental) Tree Pipit, Spotted Pipit
French: Pipit à dos olive   **Creole:** Pipit Ledo Ver

DESCRIPTION Length 14.5 cm; wingspan 24-27 cm. Similar to Tree Pipit, but with more distinctive head pattern, less streaked above (can appear unstreaked) with a plain rump, more compact, finer bill and more elegant appearance. **Adult** in fresh plumage has upperparts unstreaked olive tinged greenish but wearing greyish (greyer than Tree Pipit) with contrasting green edges to flight feathers. Underparts thickly striped with vertical, evenly spread lines of black spots, often bolder than Tree Pipit, on whitish background, often warm buff on upper breast and throat. Lines join to form a black mark on side of throat. Head pattern is distinctive: crown less heavily streaked than Tree Pipit, usually with a bold black border above a thick supercilium which is buff in front of the eye, whitish behind edged black above, flaring upward near the nape (Tree Pipit supercilium usually entirely buff and less contrasting). Ear-coverts have white and black rear-spots (though white spot may be absent or indistinct). Bold white wingbar and less distinct buffish lower wingbar. White outer tail feathers and olive-yellow undertail. Bill small, blackish above and buff-horn below. Legs pinkish with a short hindclaw. **Immature** similar to adult.
VOICE Flight call and alarm call very similar to Tree Pipit, but fainter.
BEHAVIOUR Frequents forest margins, paths and wetland edges. Usually solitary or in pairs. Flight buoyant and undulating. An enthusiastic tail-wagger. Often shy, but approachable with care. Walks and runs, foraging on insects and seeds, walking confidently on branches, unlike other pipits.
RANGE Nominate *hodgsoni* breeds Himalayas to Honshu (Japan). Race *yunnanensis* breeds northern Eurasia to Hokkaido (Japan); has almost unstreaked upperparts. Winters Indian subcontinent, South-East Asia, vagrant west to Britain, south to Sri Lanka.
STATUS Uncertain; there are unconfirmed reports from the granitic islands.
SIMILAR SPECIES Tree Pipit generally lacks greenish tinge to upperparts and has finer streaking on flanks, buffish (not greenish olive) fringes to greater coverts, tertials and secondaries and a different head pattern.

RED-THROATED PIPIT *Anthus cervinus*   Plate 42

French: Pipit à gorge rousse   **Creole:** Pipit Lagorz Rouz

DESCRIPTION Length 15 cm; wingspan 25-27 cm. A medium-sized, sleek but robust pipit, heavily streaked on breast, flanks and rump/uppertail-coverts. **Male breeding** has distinctive brick-red throat and breast, upper breast sometimes with blackish streaks. **Male non-breeding** may retain some red on the throat. Upperparts, rump and uppertail-coverts pale brown, lacking olive tones of Tree Pipit, with blackish brown streaking. Face lacks strong pattern, but supercilium and chin deep pink-buff. Heavy streaks on flanks. Distinct white wingbars and white outer tail feathers. Tail a little shorter than Tree Pipit. Autumn birds can look distinctly cold with little or no buff on the underparts, but obvious creamy mantle braces. Bill dark brown, yellowish pink at base of lower mandible, finer than Tree Pipit. Legs yellowish or brownish flesh with fairly long, slightly curved hindclaw. **Female** similar to male, but with generally less intense colour on the breast, chin and supercilium which are buffish, sometimes with a pink tinge. **Immature** similar to adult female, lacking any pink cast to face and throat; throat may be pale buff often wearing to whiter below with a heavy black wedge on the side of the lower throat. Upperparts brown, heavily streaked black with pale edges to tertials and white mantle-stripes. Pale lore and white eye-ring.
VOICE Flight call is a high-pitched 'tzeez', more drawn-out than Tree Pipit, given when flushed or occasionally from the ground.
BEHAVIOUR Frequents water margins, coasts and damp cultivation. Gregarious, often in loose groups. Undulating flight, often fairly high. Walks on the ground with a horizontal posture generally creeping and probing for food, occasionally making short runs. Tends to hide behind cover, flushing at last moment when approached. Diet is insects, insect larvae, small worms, seeds and freshwater molluscs.
RANGE Breeds in high latitudes from northern Scandinavia to the Bering Strait wintering mainly Africa south to northeastern Tanzania. Vagrant to South Africa. Monotypic.
STATUS Vagrant to granitic islands, Amirantes and Bird; recorded every month October to March.

SIMILAR SPECIES Tree Pipit is more slender, has unstreaked (or very faintly streaked) rump, thicker bill, pale eye-ring, longer tail, shorter more curved hindclaw and less boldly streaked, warmer upperparts; more olive brown above, buff below (not brown above, white below).

YELLOW WAGTAIL *Motacilla flava*   Plate 43

French: Bergeronnette printanière   **Creole:** Bat Lake Zonn

DESCRIPTION Length 17 cm; wingspan 23-27 cm. A fairly small wagtail with relatively short tail. Mainly olive green or brownish above, pale yellow below, pale buffish wing edgings and usually a prominent pale supercilium. **Adult** generally has yellow underparts, in which the extent and intensity of yellow varies between individuals. Olive green tones are evident in the upperparts of most birds, unlike Grey or Citrine Wagtails. Short tail is black with white outer feathers. Bill horn; legs blackish. There are many races, **male breeding** having very different head patterns. Nominate *flava* (**Blue-headed Wagtail**) has blue-grey crown separated from darker grey face by a long white supercilium starting from the nostril; non-breeding has head, upperparts and rump greyish brown. Race *lutea* (**Yellow-headed Wagtail**) has entire head bright yellow in male breeding, greenish-crowned in other plumages. Race *beema* (**Sykes's Wagtail**) has paler grey head than *flava* and a broader supercilium, white chin, usually a white submoustachial-stripe and white cheek-patch. **Female** is paler and duller on underparts and upperparts, whiter on chin and throat with similar but less distinct head pattern. Female *flava* similar to non-breeding male but has yellowish supercilium, paler underparts and brown spots on the neck. Solitary females may be unassignable by race. **Male non-breeding** similar to female, underparts more uniform yellow and less buffy and head pattern less distinct; may be unidentifiable by race until spring. **First-winter** similar to non-breeding female, but whiter below, more distinct white wingbar and faint spotting on breast. Some first-winters may lack any yellow. First-year male may acquire some of adult head pattern by December.

VOICE A loud monosyllabic (sometimes disyllabic) 'shreep', also a hard 'trizz' uttered year round.

BEHAVIOUR Usually close to freshwater, but in winter also short-grass areas. Vagrants usually solitary. Bounding flight, gaining height with rapid wingbeats and curving downward, wings closed. Flight silhouette less attenuated than Citrine Wagtail, or White Wagtail, short tail obvious. On ground, stance less horizontal than other wagtails. Frequently wags tail. Feeds on ground mainly on insects with some spiders and berries also taken. Often follows cattle, taking insects disturbed. Sometimes hovers to pick prey from plants.

RANGE Breeds across Eurasia to Alaska and also North Africa. Migrates southward to sub-Saharan Africa, Indian subcontinent, South-East Asia and Indonesia. Race *flava* breeds central Europe to Urals; *lutea* breeds Lower Volga basin, Kazakhstan; *beema* breeds Lower Volga to Lake Baikal. There are Seychelles records of birds showing characteristics of each of these races and other races may occur including *leucocephala* (White-headed Wagtail) breeds central Asia, winters northwest India, vagrant to East Africa: male has white head; *thunbergi* (Grey-headed Wagtail) breeds Scandinavia to Siberia, winters sub-Saharan Africa, recorded Sri Lanka, Maldives: male has blackish or dark slate-grey forehead, forecrown, lore and ear-coverts merging to dark blue-grey hindcrown and a broad white stripe beneath lore and ear-coverts, lacking a supercilium; also *feldegg* (Black-headed Wagtail) breeds Balkans to Afghanistan, winters Kenya west to West Africa: male has entirely black head. Non-African races which might conceivably occur in Seychelles include *simillima* (Eastern Blue-headed Wagtail) which breeds Siberia to Philippines wintering South-East Asia), male similar head to Grey-headed Wagtail but with obvious white supercilium; also, *melanogrisea* (Eastern Black-headed Wagtail), which breeds central Asia, wintering south to Sri Lanka: head black, no eyebrow. Races breeding further south leave the wintering grounds first and those furthest north, despite the greater distance to be travelled, leave last. In East Africa and Zaire studies have shown *feldegg* leaves in February, *lutea* February to March, *flava* and *flavissima* March to April and *thunbergi* April to May. Wherever possible, an attempt should be made to identify any bird seen in Seychelles by race, particularly given the likelihood that some races will increasingly be treated as separate species.

STATUS Vagrant throughout Seychelles October to March, possibly near-annual at Aldabra.

SIMILAR SPECIES Grey Wagtail has entirely grey mantle and back, whitish throat (black in breeding male), darker wings and a much longer tail. **Citrine Wagtail** has upperparts slate grey, including rump and uppertail. Underparts yellow merging to whitish undertail-coverts, broad double broad white wingbars (broader than Yellow) and white margins to tertials. Immature lacks yellow, similar to Immature Yellow Wagtail, but has broader supercilium, with dark upper border, extending around ear-coverts to join white of throat and broader white wingbars.

CITRINE WAGTAIL *Motacilla citreola*

Plate 43

Other names: Yellow-hooded (or Yellow-headed) Wagtail
French: Bergeronette citrine **Creole:** Bat Lake Sitrin

DESCRIPTION Length 17 cm; wingspan 24-27 cm A small wagtail, grey above, yellow below with broad double white wingbars. **Male non-breeding** has yellow head and underparts with whitish undertail-coverts, a grey wash to sides of breast and flanks, greyish ear-coverts, crown and nape. May show trace of black half-collar with ash-grey mantle, rump and uppertail-coverts. Broad white wingbars and white edges to tertials evident in all plumages. Tail shorter than Grey but a little longer than Yellow Wagtail with similar white outer tail feathers. Long black-brown legs. Bill blackish. **Male breeding** has all-yellow head with black half-collar on lower nape and sides of neck. **Female** shows broad yellow supercilium that curves around and frames greyish ear-coverts, joining yellow throat and underparts, and has a greenish grey crown. Usually shows pale centre to ear-coverts and may show buff forehead (absent in Yellow). Some non-breeding females may be very pale buff-yellow on underparts, yellow sometimes absent below breast, but undertail always white. **First-winter is** dull ash grey above, faintly tinged brown and sometimes has pale brown tinge to forehead. Broad white supercilium, becoming yellow by November onward, including surround to ear-coverts and throat. Forehead is paler than crown and shows a pale lore. Sides of breast and flanks washed grey with a few weak spots on the breast. Yellow head appears by late winter. Bill all-dark, never pale at base.
VOICE Call a harsh, generally monosyllabic 'sreep', similar to Yellow Wagtail.
BEHAVIOUR Frequents margins of freshwater and grassy areas. Gregarious, but vagrants likely to be solitary. Flight stronger than Yellow, lacking stalling action. Feeds mainly on invertebrates, sometimes following cattle to take insects disturbed. Also picks insects and larvae from surface of water or by submerging head. Occasionally flycatches.
RANGE Breeds from central Palearctic to central Asia, wintering mainly India and South-East Asia. Vagrant to western Europe, Morocco, Djibouti and Ethiopia. In addition there are remarkable vagrant records to Australia and South Africa (probably due to reverse migrations), the only southern hemisphere records other than Seychelles. Several races: *werae* breeds Poland and Turkey eastward (slightly smaller, paler head and underparts, lighter grey above, less grey on flanks, narrower half-collar); race *calcarata* breeds Himalayas to China (deep yellow underparts including undertail-coverts); nominate race breeds Russia to Mongolia.
STATUS Vagrant; recorded granitic islands April. The occurrence of this species in Seychelles is surprising. There are very few records south of the Equator, the only others being a few from Australia and one from South Africa, all believed to be instances of reverse migration.
SIMILAR SPECIES Grey Wagtail is also grey above, yellow below but lower underparts brighter than upper (opposite of Citrine) and legs pinkish. Only one wingbar and has much longer tail. **Yellow Wagtail** has olive green tones in upperparts and rump, shorter tail, narrower wingbars and a stalling flight action.

GREY WAGTAIL *Motacilla cinerea*

Plate 43

French: Bergeronette des ruisseaux **Creole:** Bat Lake Sann

DESCRIPTION Length 18-19 cm; wingspan 25-27 cm A very long-tailed, graceful, slim wagtail. Plumage is grey above, with yellow vent and rump. Single white wingbar visible above and below. **Male non-breeding** has white or buffish white chin and throat and indistinct, buffish white supercilium. Grey upperparts contrast sharply with bright yellow undertail plumage that extends around the long thin tail onto yellow rump. No distinct covert-bars, but a bold white wingbar at base of flight feathers. Very long tail is black with white outer tail feathers. **Female** similar to male non-breeding but has a white throat, whiter sides to belly and an indistinct submoustachial-stripe (at all times). **Male breeding** has black throat, entirely bright yellow underparts and a distinct white supercilium and white submoustachial-stripe. Both sexes have brownish flesh legs (blackish in other wagtails). Bill black. **First-winter** grey above with yellowish green rump, buffish white below. Bill black (with pink base in juvenile).
VOICE A metallic, high-pitched, disyllabic 'tzitzi'.
BEHAVIOUR Favours fast-running clear streams, usually in wooded or grassy places. Also frequents town centres where perches on roof tops. May appear along rocky shorelines in winter. Usually solitary or in pairs. Flight is deeply undulating with more attenuated silhouette than other wagtails and wingbar obvious, appearing translucent from below against sunlight. Stance on the ground is more horizontal than other wagtails, incessantly wagging tail, sometimes raising tail when running. Runs, walks, pecks the ground or water surface taking mainly aquatic insects, occasionally chasing aerial prey like a flycatcher.
RANGE Nominate *cinerea*, the only race to occur in Africa, breeds northwest Africa and Europe, east to Iran wintering western Europe, northwest, western and eastern Africa. Race *melanope*, slightly smaller, especially in the tail, breeds from the Urals to central Asia. Race *robusta*, with darker slate grey upperparts, breeds eastern Asia to Japan. Other races occur which are unlikely to reach Seychelles.
STATUS Vagrant; recorded Bird, granitic islands and Amirantes October to November.

SIMILAR SPECIES Other wagtails have black not pink legs. **Yellow Wagtail** is smaller, has shorter tail, more upright stance and a greenish tone to the upperparts. **Citrine Wagtail** has ash-grey upperparts and rump, shorter tail and broader double wingbars. Yellow of breast is brighter than vent, which is usually whitish (opposite of Grey Wagtail).

WHITE WAGTAIL *Motacilla alba* Plate 43

French: Bergeronette grise   **Creole:** Bat Lake Blan

DESCRIPTION Length 18 cm; wingspan 25-30 cm. Adults readily distinguished by grey upperparts and rump, black crown and bib and the lack of any yellow on underparts. **Male non-breeding** has forehead, chin and upper throat white, crown black with some grey, while chin and throat are white. Underparts white with a black chest-band. Narrow double white wingbars and white underwing. Tail is grey and shorter than Grey Wagtail with two, (not three) white outer tail feathers. **Male breeding** has neater all-black crown and clear white forehead. **Female** is less clean-cut; crown grey (or at least very little black, unlike male), the lore and cheek mottled dusky, black mottling on forehead, white mottling on chin and duskier mantle, rump and flanks. Both sexes have bill and legs black. **Juvenile** has a whiter throat, dark grey-brown breast-band, grey crown with olive tinge and a yellowish wash on the head retained until December. **Male first-winter** is similar to adult female non-breeding except blacker on the crown. **Female first-winter** has dull grey crown tinged dusky.
VOICE A distinctive 'chi-sik' flight call.
BEHAVIOUR Frequents riverbanks, lakes, marshes, farmland and gardens. Usually solitary or in pairs, occasionally in flocks on migration. Flight deeply undulating; tail does not 'whip', unlike Grey Wagtail. Walks with a nimble gait and angled stance, the head moving back and forth, occasionally running. Wags tail frequently, flashing white outer tail feathers. May hop or leap to catch insect prey.
RANGE Breeds across whole of Eurasia and northwest Africa wintering Africa, Arabia, India and South-East Asia. Nominate *alba* breeds Europe to Asia Minor, wintering south to Kenya; this is possibly the race recorded in Seychelles, though several Asian races could also occur, notably *dukhunensis* (Siberian White Wagtail) breeds southern Russia to India, winters south Asia to northeast Africa and common on Socotra where no other race recorded: broad white tips to median and greater coverts, paler blue-grey mantle, scapulars and uppertail-coverts and paler wing.
STATUS Vagrant to Bird, granitic islands and Aldabra November to January and again in March.
SIMILAR SPECIES Adults distinguished by absence of yellow tones. Immature **Citrine Wagtail** is most similar, identifiable by very bold white wingbars, shorter tail, broader supercilium wrapping around ear-coverts and absence of breast-band.

RED-WHISKERED BULBUL *Pycnonotus jocosus* Plate 44

Other names: Red-eared (or Red-cheeked or Crested) Bulbul
French: Bulbul orphée   **Creole:** Merl Konde

DESCRIPTION Length 17-22 cm; wingspan 23 cm. Confined to Assumption, where it is the only bulbul. Easily identified by combination of brownish upperparts, white underparts, black crest and scarlet ear-patch. **Adult** has upperparts brown merging to black on crown to below the eye; a black band extends from the neck to the side of the breast. Prominent black crest, held erect, except in flight. Conspicuous scarlet patch behind eye and on undertail-coverts. Tail brown tipped white. Bill black; legs blackish brown. Sexes similar. **Juvenile** similar to adult but lacking scarlet ear-patch, vent pale orange-rufous, crest and nape brown.
VOICE Noisy, with a variety of lively calls, some harsh, some anxious. One of the most frequent calls is rendered 'pleased-to-meet-you' with the accent on 'meet'.
BEHAVIOUR Found in trees and scrub throughout Assumption, where often perches conspicuously. Gregarious, usually found in pairs or small flocks of ten or more birds. Flight strong, but slightly erratic. Diet includes fruit, berries, seeds, nectar and insects. Insects may be gleaned from vegetation or taken in short aerial pursuits.
BREEDING In courtship, male prostrates itself before the female on a perch, bowing head with tail spread and wings hanging. Wings are then quivered and a low, croaking call uttered prior to mating. May breed year round, probably with a peak during the rainy months of October to January when food is more plentiful. In India, the season is poorly defined, mainly March to July in the north and December to June in the south with a second period in September; there is occasional nesting in any month. Cup-shaped nest is built from grass, twigs and rootlets bound with spiders' webs and lined with softer material, deep in a tree or shrub. Both sexes share in nest building, which lasts about four days. Usually three pinkish eggs mottled brownish, purple or reddish are laid, but clutch may be 2-4. Incubation lasts about 12 days. Young are fed mainly on insects at first, fruit and berries later.
RANGE Several races, resident India, Andamans, Nepal east to Vietnam, Hong Kong and southern China. It has been introduced to Malaysia, Singapore, Sumatra, Java, Nicobar Islands, New South Wales,

California, Florida and Hawaii.

STATUS Introduced to Assumption, probably in 1976, from Mauritius. Race *emeria* was earlier introduced to Mauritius as a cage bird around 1892[1], from whence it escaped and is now common. In 1977, only six birds were seen on Assumption[2]. In 1986, the population was estimated at 200 pairs[3]. By 1997, it was estimated to have reached 1,000-1,500 birds (Rocamora pers. comm.). The presence of this introduced successful species so close to Aldabra could pose a threat. There are currently no introduced species on Aldabra.

SIMILAR SPECIES None

References: 1. Rountree *et al.* (1952), 2. Prys-Jones *et al.* (1981), 3. Roberts (1988).

MADAGASCAR BULBUL   *Hypsipetes madagascariensis*   Plate 44

Other names: Black (or Madagascar Black or Grey) Bulbul
French: Bulbul malgache   **Creole:** Merl Malgas

DESCRIPTION Length 21-22 cm; wingspan 35-37 cm. Restricted to Aldabra where it is common; greyish brown, paler below with a shaggy crest and bright orange bill. **Adult** has upperparts and rump grey-brown, tail somewhat darker and distinctly forked with a blackish shaggy crest on the forehead and crown. Paler grey-brown below washed with brown. Eye dark brown. Bright orange, long, slender bill. Brown, fairly short legs. Sexes similar, male averaging larger. **Juvenile** similar to adult but duller and with brown or dull orange bill and shorter tail. At close range chestnut margins to tail and flight feathers may be visible, though this plumage is soon replaced. Underparts grey, washed more strongly brown than adult.

VOICE A loud 'cheep' often two or more notes, the second and subsequent notes being lower in tone. Alarm call a harsh whinging 'chair' often repeated. Also a wide variety of cackling sounds with some softer notes, or a rapid repetitive series of notes.

BEHAVIOUR Found throughout Aldabra wherever trees or scrub are available, though less common in mangroves, especially on Malabar and absent from open sand dune areas of Grande Terre. Gregarious, usually in pairs or family parties, sometimes in groups of up to 20 birds especially near dawn and dusk when they often chase each other between trees. Flight is strong and direct. Feeds mainly in bushes and trees, occasionally on the ground. Diet includes insects, spiders, berries, fruit, seeds and flowers. Occasionally flycatches for aerial insects.

BREEDING Breeds mainly October to January. Nest is built at a height of 2-3 m. Clutch is one or two (in Madagascar and Comoros the clutch is often three), eggs whitish with brown speckles.

ORIGINS Undoubtedly from Madagascar. The region may have been colonised relatively recently from Asia, perhaps even simultaneously to Madagascar (represented by nominate *madagascariensis*), Iles Glorieuses (race *grotei*; extinct and probably invalid as indistinguishable from nominate), Comoros (race *parvirostris*) and Aldabra group[1]. In Asia there are 12 other races from Sri Lanka and India to China and Vietnam and Taiwan. Races *humii* of Sri Lanka and *ganeesa* of southwest India are different in voice from Madagascar birds and Asian birds may be specifically distinct, though plumage differences are very slight[2]. Race *rostratus* is recognised on account of the brownish wash to the upperparts and thighs

STATUS Endemic race *rostratus* found only on Aldabra (4,000-8,000 pairs[3]). Nineteenth-century records from Astove and Cosmoledo may have been this race, or some unrecorded, closely related race, now extinct[4].

THREATS AND CONSERVATION The future of the species should be secure given status of Aldabra as a nature reserve and World Heritage Site and the fact this is a fairly common species. Re-introduction to Astove and Cosmoledo may be a step towards restoring the avifauna of those atolls and restoring the former range of the species.

SIMILAR SPECIES None.

References: 1. Benson (1984), 2. Keith (1980), 3. Rocamora and Skerrett (1999), 4. Rivers (1878).

SEYCHELLES BULBUL   *Hypsipetes crassirostris*   Plate 44

Other names: Thick-billed (or Seychelles Black) Bulbul   **French:** Bulbul des Seychelles or Bulbul merle
Creole: Merl Sesel

DESCRIPTION Length 24-25 cm; wingspan 35-37 cm. A large well-built bulbul, brownish with a greenish hue. The only bulbul of the granitic islands. **Adult** is grey-brown above with an olive green hue, slightly paler below, especially on the belly, and a shaggy black crest on crown. Pale edge to tail is obvious in flight. Eye reddish brown. Bill and legs bright orange. Sexes similar, male slightly larger and heavier. **Juvenile** similar to adult but bill blackish and flight feathers chestnut. Eye yellowish at first, soon darkening. Legs dark brown.

VOICE Highly vocal, giving a variety of squawks, harsh chatters, clucks and whistles. One study recognised four basic types of call[1]. First a cluck, usually as an isolated call, sometimes in a series of two or more notes, apparently to maintain communication when moving through trees. Second, a raucous caw given at intervals of 5-10 seconds, usually by an isolated bird, apparently to maintain contact between separated group members. This sounds like a whinging, slightly drawn-out 'nyeer' (similar to Bugs Bunny prior to 'What's up doc?'). Third, a rapid chatter, often for more than 30 seconds, sometimes given by a single bird but more often by a bird in company or in a family group: 'ak-ak-ak-ok-ok-ok-ak'; this often draws a response from other groups and may have a territorial function. Fourth, a song comprising a series of clucks, squawks, whistles and chatters usually given by a solitary adult with wings held drooped, on a prominent perch; this may be associated with breeding. Immatures may give a soft begging call 'twee-twee-twee', during which the wings are drooped and quivered displaying chestnut flight feathers, while feathers on the upper back are erect, giving a hunched appearance.

BEHAVIOUR A woodland species found from the coast to the highest mountains, wherever there is suitable habitat. Highly gregarious within family groups, pairs or small parties noisily chasing each other between trees in a rapid, direct flight. Territories are established during the breeding season; may be mainly sedentary at other times. Some birds without territories, (often a group of up to four young birds), may move through the range of others until chased off. Territories occupy an area of approximately 200 m diameter with 318 territories per square kilometre and about 1,000 birds, excluding nomadic individuals[2]. They are aggressive towards most species within their habitats chasing off birds up to the size of Green-backed Heron, though Seychelles fruit bats displace bulbuls from fruit trees where both may feed. Catholic diet includes insects, fruit, flowers, eggs and possibly nectar. Insects are plucked from tree trunks or foliage and airborne prey taken in short aerial chases. Seen even to take venomous yellow wasps (P. Matyot pers. comm.).

BREEDING Breeding season coincides with onset of northwest monsoon, especially October to January, though breeding may also take place at other times of year. Nest site is generally a fork in a branch high up in a tree, generally over 10 m, sometimes as low as 2 m. Nest is circular, built mainly of grass, dead leaves, twigs, tree bark, moss and palm fibres. Both parents help with nest building in the early stages. Nest building takes 6-10 days[3]. Usual clutch size is two whitish eggs with russet spots, though in about 90% of family groups there is only one immature bird suggesting generally only one young survives beyond fledging or possibly some young (perhaps of one sex) disperse rapidly from natal territories[1]. Incubation lasts about 15 days, fledging a further 21 days. Immature birds associate with adults in loose social groups after fledging, collecting their own food and possibly learning feeding skills.

ORIGINS Probably derived from Comoros Bulbul *H. parvirostris*[4]. The original stock is almost certainly Asian.

STATUS Endemic to Seychelles and common on Mahé, Praslin, La Digue and Silhouette. Also occurs on a few neighbouring islands in small numbers.

THREATS AND CONSERVATION The species is thriving and like Seychelles Sunbird appears to have adapted well to the arrival of man. Nevertheless preference for mature woodland means its future is linked to the conservation of this habitat.

SIMILAR SPECIES None.

References: 1. Greig-Smith (1979a), 2. Cook (1978), 3. Loustau-Lalanne (1962), 4. Louette (1992).

RED-BACKED SHRIKE *Lanius collurio*   **Plate 45**

French: Pie-grièche écorcheur   **Creole:** Pigryes Ledo Rouz

DESCRIPTION Length 17 cm; wingspan 24-27 cm. A small, compact, short-billed shrike; adult male unmistakable, blue-grey and chestnut above; female and immature paler and more sparrow-like. **Male** is easily recognised by combination of bluish grey head with black mask, chestnut back, grey rump, pinkish wash on underparts and white edges to black tail. Shows no vermiculations on flanks. White chin and throat, and sometimes shows a very thin white supercilium merging to a pale patch on the forehead (not in race *kobylini*) and may show a small white primary flash. Central tail feathers all-black, other tail feathers white at base, outer web of outer pair white, forming an inverted black T on a white background, obvious in flight. Undertail pale grey with darker terminal band and white at tail tip. Bill hooked, thick, short and blackish. Legs black or dark brown. **Female** is dull rufous-brown above, rump tinged grey; creamy buff below barred with brown crescent shapes on breast and flanks. Warm brown ear-coverts; creamy white supercilium, spotted black. Tail dark brown with rufous tinge and white edges and tip, vent whitish, undertail grey. Bill similar to male, legs grey-brown. By **first winter**, when birds might reach Seychelles, the diagnostic strongly scalloped feathering of juvenile plumage is likely to be subdued, but still visible especially on rump and scapulars, otherwise resembling adult female. Plumage mainly rusty-brown above, pale whitish below. Tail often rufous but does not contrast with rest of upperparts (unlike first-winter Isabelline). A complete moult takes place in winter quarters so that all birds may be sexed by February. Bill paler than adult.

VOICE Commonest call a harsh 'chak-chak', used all year round. May also sing with sweeter notes in winter quarters from January onward.

BEHAVIOUR Frequents bushes and thickets, where hunts from a perch, pouncing to ground and returning. Flight rapid and agile undulating over long distances. Mainly solitary and often territorial outside breeding season. Diet mainly beetles, also other insects and sometimes small animals, birds and reptiles. Prey is often impaled on a branch.

RANGE Nominate *collurio* breeds Europe to Ukraine merging with the races *kobylini* (Georgian Red-backed Shrike) of Crimea to Iran and *pallidifrons* (Siberian Red-backed Shrike) of western Russia to Siberia. Racial differences slight and intermediates may occur; race *kobylini* has chestnut of mantle duller and darker and less extensive than the nominate, crown and hindneck paler, the grey of the nape extending to the back and the forehead never white; race *pallidifrons* is the palest race, forehead and supercilium whitish, very pale on the crown and neck). All three races are highly migratory and have been recorded in Kenya where *kobylini* is most numerous, while the nominate and *pallidifrons* have been recorded south to South Africa. Isabelline, Red-backed and Brown Shrikes form a superspecies and are said to hybridise where all three meet in the Altai Range and east Turkestan. A loop-migrant passing further east on spring migration compared to autumn.

STATUS Vagrant to Aldabra group in March, when pattern of migration suggests more likely than in autumn; race or races unknown.

SIMILAR SPECIES First-winter might be confused with first-winter **Woodchat Shrike** which is larger, paler (lacks rufous tones), has a more angular head; wing-patch and shoulder show traces of adult pattern. Isabelline Shrike *L. isabellinus* and Brown Shrike *L. cristatus* are potential vagrants which should be considered. **Isabelline Shrike** (L17.5 cm, WS25-28 cm) breeds Iran to central Asia, wintering south to India, northeast Africa to East Africa. It is similar to Red-backed Shrike in structure, except slightly larger with longer tail; paler and greyer, less barred above. Grey-brown upperparts contrast with rufous rump. Tail reddish rufous with ginger or orange at edges and a pale tip in fresh plumage. Wings grey-brown, blackish on primaries with white patch at base of primaries (indistinct or absent in female). Underparts creamy white. Crown of male varies from rufous to dull grey. Thin white supercilium contrasts sharply in rufous-crowned birds, less so in others. Blackish eye-patch from lore to upper ear-coverts (smaller and browner in female). Immature is uniformly pallid above, with a pale wing-panel and a white patch at the base of the primaries. Best distinguished from Red-backed Shrike by contrasting bright rufous tail and generally pale or unmarked upperparts; ear-coverts are never deep rufous (as in some Red-backed). **Brown Shrike** (L18 cm, WS26-28 cm) breeds Siberia, wintering south to India and Sri Lanka; occasionally reaches Maldives. It is a medium-sized shrike, darker brown above than Isabelline, with black face-mask, broad pale greyish supercilium extending to forehead and long, slender, graduated and rounded rufous tail. Heavily built and bull-headed with a large powerful bill. No pale patch on primaries (unlike Isabelline). Russet-brown upperparts and upperwing show little contrast with rufous-brown cap. Underparts white with pale buffish wash on breast, flanks and belly. Sexes similar but mask of female may be less distinct at front, supercilium tinged cream and may have some fine dark chevrons on underparts. Immature similar to adult with deeper buff wash and more heavily marked breast and flanks. Face-mask browner and incomplete on lore; supercilium shorter in front of eye not reaching forehead. Upperparts generally unmarked (strongly barred in immature Red-backed).

LESSER GREY SHRIKE *Lanius minor*　　　　　Plate 45

French: Pie-grièche à poitrine rose　**Creole:** Pigryes Pwatrin Roz

DESCRIPTION Length 20 cm; wingspan 32-34.5 cm. A white, grey and black shrike, underparts tinged pinkish in some birds, with fairly short tail (compared to other shrikes) and long, pointed wings. **Adult** has a black mask which extends to forehead in spring, black barring on forehead in autumn. Rest of head and upperparts are grey, slightly paler on the rump and uppertail-coverts. Underparts whitish washed salmon-pink in male in fresh plumage. Wings long and black with a large white wing-patch at the base of the primaries and secondaries. In fresh plumage scapulars edged white; shows distinct pale edges and tips to primaries. Tail black, outer feathers white and all but central pair of tail feathers tipped white; undertail white. Sexes similar, but female usually duller, mask greyer or browner, especially on forehead, underparts paler and less pink. Thick, stubby, blackish bill; legs blackish. **First-winter** is tinged brown above, mask brownish black and forehead grey, retaining some white scaling on wing-coverts.

VOICE Calls include a harsh piping chatter and a short 'chek'. Song a babbling chatter, often heard in winter quarters December onward.

BEHAVIOUR Frequents roadsides, open areas and scattered trees, where it generally perches high up in the open. Usually solitary in winter quarters. Long wings give fluent, rapid, direct flight, undulating over a distance. Sometimes glides or hovers. Sometimes hops on ground, appearing wheatear-like with an upright stance, tail often drooped. Diet almost entirely insects, occasionally snails, worms and lizards, hunted from an open perch. Prey may be taken in the air or on the ground, birds usually returning to their perch to feed.

RANGE Nominate *minor* breeds southern and eastern Europe, merging with race *turanicus* (sometimes not recognised) to western Asia; the latter, is slightly paler in the adult, first-winter more sandy brown. A loop migrant, with spring passage further east (in Zimbabwe, mainly west of 30°E in autumn and east of this in spring). Winters entirely southern and eastern Africa.

STATUS Vagrant to Aldabra group; recorded March, when pattern of migration suggests stronger possibility in spring than during autumn migration.

SIMILAR SPECIES Great Grey Shrike *L. excubitor* (L24-25 cm, WS30-35 cm), a confusion species over much of the Palearctic, is highly improbable in Seychelles being a partial migrant which almost never reaches as far south as North Africa. **Southern Grey Shrike** *L. meridionalis* (L24-25 cm, WS28-32 cm) of North Africa and southern Europe to Asia is unlikely, but has reached 5°N in Somalia. Both are different in structure being longer-tailed, longer-billed than Lesser Grey Shrike, lacking long primary projection and with less upright stance.

WOODCHAT SHRIKE *Lanius senator*   Plate 45

French: Pie-grièche à tête rousse   **Creole:** Pigryes Latet Rouz

DESCRIPTION Length 18 cm; wingspan 26-28 cm. A medium-sized, stocky shrike, adults basically black and white with diagnostic chestnut rear crown and nape. **Male** has upperparts black, underparts, rump and uppertail white. Wings are black with large oval white patch on the scapulars and a smaller patch at base of primaries, wing feathers fringed rufous in fresh plumage. Short, square-ended tail black, outer tail feathers and uppertail-coverts white. Eastern race *niloticus* breeds earlier and completes moult by autumn. This race, the only one likely in Seychelles, shows much more white in the uppertail than nominate race. Bill and legs blackish. **Female** similar but duller and browner on the back with dusky crescent-shaped bars on breast and flanks; whiter at base of bill, forehead and mask flecked chestnut. **First-winter** shows similar pattern to adult but is less tidy, and may retain some of the scalloping of juvenile on underparts, particularly on the upper breast. The mask is indistinct and the crown paler than adult, the forehead flecked brown-black and some buff in the white patches on scapulars and base of primaries. Pale tips to wing-coverts and tertials with dark subterminal markings. Distinguished from Red-backed Shrike by larger more angular, rufous head, absence of rufous tones on paler back and wing-coverts, more obvious eye-patch, brown wings with the crown, rump, wing-patch and shoulder showing traces of adult pattern. From Lesser Grey Shrike by rufous crown and white scapular-patch.
VOICE A sparrow-like chatter or a harsh 'kwik-kwik'. Song a musical warble with some harsher notes, sometimes heard in winter quarters.
BEHAVIOUR Frequents woodland margins, roadsides and clearings with scattered trees. Solitary and territorial in winter quarters. Flight rapid, direct and less undulating than other shrikes; very occasionally hovers and sometimes hops on ground. More shy than other shrikes. Hunts from a high open perch, mainly for insects taken in the air or on the ground.
RANGE Breeds northwest Africa, southern Europe to Iran, wintering mainly sub-Saharan Africa north of the Equator. Of three recognised races, only *niloticus* winters east of about 30°E; this race is remarkable in undergoing a complete moult prior to autumn migration.
STATUS Vagrant to granitic islands; recorded April.
SIMILAR SPECIES First-winter might be confused with first-winter **Red-backed Shrike**, which is darker above, with rufous tones on back/wing-coverts and an indistinct eye-patch. **Masked Shrike** *L. nubicus* (L17-18 cm, WS24-26.5 cm) is a potential vagrant that should be considered. It breeds southeastern Mediterranean to Iran, wintering sub-Saharan Africa to about 10°N; vagrant south to Kenya. It is a small, slim shrike with a long, graduated tail and fine bill. Adult male similar to Woodchat but with white forehead and black crown and nape. Female similar, black areas of plumage tinged grey. First-winter similar to adult female, retaining some pale barring of juvenile on wing-coverts and tertials; heavy barring on crown, back and sides of throat.

RUFOUS-TAILED ROCK THRUSH *Monticola saxatalis*   Plate 44

Other names: Rock Thrush, Common (or Rufous-tailed, or Chestnut-tailed, or White-backed or European) Rock Thrush   **French:** Monticole merle-de-roche   **Creole:** Merl Ros

DESCRIPTION Length 18.5-20 cm; wingspan 33-37 cm. A small, long-bodied, short-tailed, chat-like thrush with a long bill; long, tapering wings and black-centred orangey red tail. **Male non-breeding** is dull brown above, paler below, intensely mottled throughout with a blue wash on the greyish crown, some white on the back and rusty wash to breast and flanks. Mottled appearance is due to buffish fringes and black subterminal bands on most feathers. Wing-feathers are fringed brighter, the wing-coverts scaled. Long wings reach almost to tail tip. **Female** is intensely mottled, resembling male, but lacking blue tone and white back; centre of chin, throat and eye-ring creamy; upperparts mottled brown (sometimes with a trace of white on the back) and all ages have orangey red tail visible above and below, dark brown-horn bill and dark brown legs. **Male breeding** has grey-blue head, mantle and rump; also white central back and chestnut orange underparts and underwing; this plumage usually lost by September so unlikely in Seychelles unless a spring migrant may one day be recorded. **Immature** similar to adult but more buff, feathers of head and upperparts with dark brown tips and those of underparts tipped blackish. Male may have some adult colour especially to head and underparts. Female similar to adult but duller, wings browner.
VOICE Most frequent call is a moderate 'chak-chak'. Mostly silent in winter quarters, but sometimes gives a mellow subsong.
BEHAVIOUR At breeding grounds frequents dry rocky mountains with some trees for perches, vineyards

and farmland. Vagrants may turn up almost anywhere, often on coasts. Solitary and shy, diving for cover among rocks or leaping over ledges in upright stance recalling a wheatear. Flight, low; often quivers tail on landing. Will perch on walls, trees, wires, etc. appearing slim and upright, often with bill uptilted. Hops on the ground, also appearing chat-like, feeding mainly on insects (sometimes taking aerial prey), occasionally fruit.

RANGE Nominate *saxatilis* breeds southern Europe east to western China, wintering mainly sub-Saharan Africa and north India. Race *coloratus* breeds eastern Europe. Birds from the eastern limit of range may travel 7500 km or more to wintering grounds.

STATUS Vagrant to granitic islands and Farquhar; most likely October to November.

SIMILAR SPECIES Usual confusion species at breeding grounds, **Blue Rock Thrush** *M. solitarius* (L22 cm, WS33-37 cm) is extremely unlikely in Seychelles being resident or only a partial migrant, reaching about 8°N in Africa. No orange in tail. Male is dark blue above and below; female dark brown above, mottled and paler brown below.

SEYCHELLES MAGPIE-ROBIN *Copsychus sechellarum* Plate 46

French: Pie (or Pie chanteuse) des Seychelles **Creole:** Pisantez

DESCRIPTION Length 22-24 cm; wingspan 29 cm. An unmistakable, ground-feeding bird, entirely black except for white wing-patches. **Adult** plumage is all-black with a midnight-blue sheen when seen in sunlight and a large white patch on the upper half of the wing. Bill and legs black. **Juvenile** similar to adult but lacks sheen. White feathers of the wing-patch have ginger fringes. Resembles adult plumage at about nine months. Full adult plumage attained at around 45 weeks.

VOICE Male song is series of repeated, melodious, fluty warbles and harsher phrases, usually delivered from a tall tree, often a palm tree. Song may include mimicry of other species such as Madagascar Fody, Common Myna, Common Sandpiper and Whimbrel. Human whistling is also mimicked. Male sings at or before dawn and in late afternoon for up to two hours. Female sometimes produces a simpler song, occasionally in duet with male (perhaps prior to breeding). Subsong is a series of soft, throaty warbles and chuckling notes; these are most highly developed in adult male (though also by either sex and immature male on occasion), frequently from a low shady perch. Often produced from shelter in heavy rain. Usual alarm call is a series of loud ascending whistles 'whooeet-whooeet-whooeet'. A softer slower version is frequently produced immediately prior to roosting and may have territorial significance. In response to an intruder a characteristic three-note whistle is given 'wheeoo-whoo', the last note being distinctly lower in pitch. Usual response to disturbance in the vicinity of a nest or fledged chick is a fairly soft, slow, rising whistle, lacking the explosive quality of the full alarm call. Another call apparently associated with low-level anxiety is a descending four-note scale often ending in a short trill and most often produced by the female. Irritated birds in non-threatening situations produce a harsh churr. Birds attacking potential predators such as skinks and mynas in the vicinity of a nest, frequently produce an explosive, nasal 'cherr'. Vocalisation during courtship and aggressive displays resembling a harsher version of subsong with a distinct buzzing quality, though some more musical phrases may be included particularly in courtship. During aggressive encounters such calls are often interspersed with typical alarm whistles[1].

BEHAVIOUR A territorial ground-feeder frequenting bare earth or leaf litter in primary woodland or cultivation. May puff feathers out in territorial defence display. Favours deep shade with little or no undergrowth. On Frégate, sometimes follows giant tortoises, people, pigs or cows to feed on disturbed insects and invertebrates. A comparison of feeding success under natural conditions with foraging near people or animals, found success rate for the latter almost twice as great as the former[2]. Feeds under large trees, which provide, low, shaded perches and open ground with little or no herb layer. On Frégate, groves of sangdragon and mature fruit trees such as breadfruit are favoured. Flight is direct and undulating. Hops on the ground in a semi-erect posture, often cocking the tail. The most favoured prey is insect larvae and small invertebrates, notably cockroaches, millipedes, giant tenebrionid beetles *Pulposipes herculeanus* (endemic to Frégate), grasshoppers, crickets, spiders, scorpions, worms, etc. with some fruit or vegetable matter (including coconut, mango, cashew and pawpaw). Skinks, geckos, frogs, caecilians, baby mice and young snakes may also be taken. Small fish dropped by seabirds, seabird eggs and crabs are taken when available.

BREEDING May breed at any time of year given sufficient food available with peaks during times of highest rainfall around November to March. In courtship the male droops its wings and chases the female. Strongly territorial, the male proclaiming ownership with long bouts of singing from a prominent perch. A cavity nester and most favoured nest sites are rotted-out tree holes or the crowns of coconut palms. Nestboxes are also used where available. Both sexes help to build the nest, though the female carries out most of the work. Nest is unlined, built from dry grasses, small twigs and coconut fibre; man-made fibres are sometimes incorporated when available. Clutch size is always one. Incubation is carried out by the female only and may take 16-23 days[3]; male may take up a position guarding the nest from predators such as lizards during incubation. Fledging takes 16-22 days[3]. Both parents feed the chick until 5-12 weeks of age, though it can feed itself at four weeks[3]. Sometimes breeds co-operatively, helpers usually being earlier offspring of the dominant pair though sometimes apparently unrelated birds (mainly juveniles but occasionally adults) may assist. Breeding may take place at one year of age. May live to 14 years of age or more.

ORIGINS *Copsychus* is mainly an Asian genus though Madagascar Magpie-robin *C. albospecularis* is not very closely related. Seychelles Magpie-robin is probably derived from Indian Magpie-robin *C. saularis* though shows no sexual dimorphism (unlike either the Indian or Madagascar species) and is much larger than both. It is possible the Seychelles Magpie-robin represents an earlier colonisation from Asia than *albospecularis*[4].

STATUS Endemic to granitic islands and resident Frégate, Cousin, Cousine and Aride with a total population in 2000 of c90 birds. IUCN Critically Endangered. Former range included Mahé, Praslin, La Digue, Marianne, Aride, South-east Island, St Anne and probably other islands from which it was exterminated prior to any written record of its presence. By 1865, Newton was unable to find birds on Mahé and found them difficult to locate on Praslin[5]. Last record from Praslin was 1878[6] and it probably became extinct in the island around the turn of the century[3]. On Mahé, the last reference to the species was in an 1883 painting by Marianne North entitled 'Palm Forest, Mahé', which depicts a scene including a Magpie-robin[7]. Newton found it common on Marianne where it survived until the 1930s[8]. On Aride there was said to be a population of around 40 in 1868[9], where 24 were collected in one visit in the 1870s by a German collector, Lantz. Survived on Aride until the 1930s (P. Medor pers. comm.). At some time prior to August 1892 it was introduced to Alphonse[10], which by 1940 was said to be its stronghold[11]. It was first reported from Frégate in 1871 and the owner prevented collection of specimens on the grounds of rarity[12], though in 1873 the inhabitants were reported to snare birds[8]. By 1965 it survived only on Frégate where it was largely restricted to the coastal plateau. On Frégate, the first ever census in 1959 gave the population as ten pairs[13]. A short time later a figure of 30 birds was given[14], apparently declining to 15-20 birds by 1964[15]. In 1970 the population was estimated at 25[16]. By 1977-78, when a one-year study of the species was conducted, a maximum of 41 birds was present with groups breeding in 12-13 territories[17]. By 1990, this had fallen to 21 birds.

THREATS AND CONSERVATION Has been the subject of a BirdLife recovery programme since 1990, aimed at removing it from the list of threatened birds, or at least reducing the risk to no worse than the IUCN Vulnerable category and increasing the population to at least 300 birds on eight islands by 2010[3]. Conservation measures have included intensive habitat management, supplementary feeding, nestbox provision, protection of nest sites against predation by lizards, control of introduced Common Mynas and Barn Owls, translocation to other islands and the maximisation of genetic variation between populations. In its first ten years, this programme increased the total population more than fourfold from 21 at the end of 1990. Attempts to re-establish populations on islands other than Frégate began earlier in April 1978, when birds were transferred to Aride[18]. By October 1980, only one survived which lived on for a further eight years. Subsequent translocations in the 1990s also failed to bring any breeding success, disease problems and poor translocation techniques being implicated. Transfers to Cousin in 1994 and Cousine in 1995 have been more successful and these islands were probably at capacity by 1999. Lone birds have subsequently crossed from both Cousin and Cousine to Praslin. The main causes of decline up to 1980 are direct human persecution, habitat loss and possibly introduced mammalian and avian predators. At one time it was a practice to keep Magpie-robins as cage birds, indeed, the species was first described by Newton from one presented to Lady Barkly in Mauritius (wife of the Governor) in 1864, which lived for about two months. Honneger attributed their extinction on some islands to be their capture as cage birds, a habit encouraged by their rarity value[15]. Extreme tameness meant it fell easy prey to collectors and as the species became rarer the desire to collect specimens appeared to escalate. Newton noted it '...is the boldest and most familiar bird I ever saw. It will approach within a few feet and when sitting on the branch of a tree will allow itself to be knocked down with a stick as one was when I was there, (on Praslin) by a man whom we brought with us'. He further noted, 'the young birds are often taken from their nests but are seldom reared still more seldom live for any length of time'[15]. On some islands the use of pesticides may pose a threat. Since 1991 the use of conventional insecticides on Frégate has been discouraged in favour of compounds non-toxic to vertebrates. These compounds work by preventing the insects from reproducing and are delivered to them via mixing with food as bait. Cats may have contributed to the extinction of this species on some islands including a thriving albeit introduced population on Alphonse (G. Gendron pers. comm.). On Frégate, a project began in 1981 to eradicate cats, which had been introduced in 1951; 51 cats were trapped or poisoned, two others found dead and three assumed to be poisoned before the project came to a successful conclusion[19]. Nevertheless, the population failed to recover, being 21 in 1982, the same as in 1990 when an intensive recovery programme began. Cats may have caused the final extinction of the species on Aride, where numbers had already been greatly reduced by habitat destruction and collection of specimens. Cats were taken to Aride around 1918 to control mice and the birds are said to have practically disappeared from Aride because of cats[20]. In 1925 boys and dogs were employed to eradicate cats from Aride, but too late to save the Magpie-robin. Paul Medor, resident on Aride at this time, said hunting continued until cats were wiped out around 1932. Medor further reported a few Magpie-robins survived the cats, dying out around 1937 possibly due to poisons used to control mice, possibly from two neutered cats introduced around 1937, or possibly because birds of only one sex survived beyond 1932. Loss of habitat due to the felling of the original tall forest must also have played a part in the decline. Rats may also have contributed to the species decline on other islands though direct evidence is lacking. There were no rats on Frégate which contributed to an assumption that rats had played a significant role in island extinctions elsewhere. Brown rats were first detected on Frégate in 1995. Control measures began immediately, but rats ignored bait in favour of abundant natural food and some birds were poisoned before control was abandoned. A more radical eradication programme was implemented by BirdLife Seychelles in mid-2000 on Frégate and elsewhere. Rat eradication may render a number of islands suitable for translocation of birds in due course. Another threat is posed by Common Mynas, which compete for nesting sites and food. Barn Owls, which are absent on Frégate but survive on most other granite islands despite attempts to eradicate them, could also be a threat. One Magpie-robin found dead on Aride was possibly a victim of a Barn Owl but may have been killed by a Brown Noddy that shared the same roosting site. Control measures introduced by RSNC on Aride in 1996 greatly reduced the number of resident owls, though some may continue to invade from Praslin[21]. Cousin and Cousine are close enough to Praslin for non-resident owls to commute. However, passerines form a small part of owls' diet and the threat they create may not be significant. Natural predators include Wright's skink *Mabuya wrightii, which* takes both chicks and eggs, and possibly other lizards, notably Seychelles skink *Mabuya sechellensis* and bronze gecko *Ailuronyx seychellensis*. Cattle Egret, Green-backed Heron and Black-crowned Night Heron might possibly take eggs or nestlings. The main conservation actions required are supplementary feeding where useful, provision of nest boxes, myna control, predator eradication on potential new island sites, habitat restoration and management, protocols to prevent invasion by alien predators and translocation to suitable islands. Research priorities include disease problems, and disease risks during translocations, causes of mortality, genetic variation, natural dietary requirements and invertebrate availability at current and potential islands[22].

SIMILAR SPECIES None.

References: 1. McCulloch (1995), 2. Komdeur and Rands (1989), 3. McCulloch (1996), 4. Benson (1984), 5. Newton (1867), 6. Blackburn (1883), 7. North (1892), 8. Vesey-Fitzgerald (1936), 9. Hartlaub (1877), 10. Ridgway (1895), 11. Vesey-Fitzgerald (1940), 12. Pike (1872), 13. Crook (1960), 14. Loustau-Lalanne (1962), 15. Honneger (1966), 16. Procter (1970), 17. Watson (1978), 18. Hellawell (1979), 19. Todd (1982), 20. Ridley and Percy (1958), 21. Nicoll (1996), 22. Lucking and Lucking (1997).

COMMON REDSTART *Phoenicurus phoenicurus* Plate 46

Other names: Redstart, Eurasian (or White-fronted) Redstart
French: Rougequeue à front blanc   **Creole:** Lake Rouz Dife

DESCRIPTION Length 14 cm; wingspan 20.5-24 cm. A small, graceful chat distinguished at all times by rusty-red rump and tail with blackish central tail feathers. **Male** has black face, throat and upper breast contrasting with orangey red breast, paler on the rear flanks merging to whitish on the belly. White forehead extends as a short supercilium, obscured by grey feather tips in fresh autumn plumage. Crown, upperparts and scapulars bluish grey. Wings blackish brown with narrow grey-buff fringes. Bright orangey-red tail, with darker central feathers, obscured by pale feather fringes in autumn. Bill and legs black. **Female** is duller, grey-brown above, buffish below, whitish on the throat and belly with a pale eye-ring. Tail similar to male, a little duller. **Immature** similar to adult from August onwards but no obvious supercilium and back browner, lacking grey tones, orange breast much paler and mottled, greyish bars on throat and black bars on cheek.

VOICE Song, occasionally heard in winter quarters, is sweet and melancholy. Alarm call is 'twik' or 'whee-tik-tik'. Sometimes calls 'hweet', recalling Willow Warbler.

BEHAVIOUR Frequents woodland areas. Solitary and shy, sometimes in small parties on migration, but vagrants to Seychelles have been lone birds. Flight buoyant and slightly undulating. Restless, constantly quivering tail, making short flights and occasionally hovering to catch insects in the air or on the ground. Also sometimes eats fruit.

RANGE Race *phoenicurus* breeds northwest Africa, Europe to Lake Baikal and winters across the Afro-tropics north of the Equator. Race *samamisicus* (Ehrenberg's Redstart) breeds Crimea, Turkey to Iran, wintering Arabia, Sudan, Ethiopia and vagrant south to South Africa. Male of the latter has prominent white wing-panel, is slightly darker above, mantle mottled blackish and richer orange below; female sometimes has pale grey or buff wing-panel.

STATUS Vagrant to granitic islands October to November.

GENERAL First Seychelles record was of a female, possibly race *samamiscus*, trapped on Cousin on 22 October 1981 weighing 12.2 g. Retrapped on 9 November, it weighed 21 g, with fat visible on the breast, a weight gain of 72% in 18 days![1].

SIMILAR SPECIES Potential vagrants, which could be confused with immature or adult female Common Redstart unless seen well might be Rufous Bush Robin *Cercotrichas galactotes*, Thrush Nightingale *Luscinia luscinia*, Common Nightingale *L. megarhynchos* and White-throated Robin *Irania gutturalis*. **Rufous Bush Robin** (L15-17 cm, WS22-27 cm) breeds southern Europe and North Africa to Pakistan, wintering grounds including East Africa. It is a medium-sized, slim but robust, long-tailed chat with a rufous rump and tail. It is paler than Common Redstart with a white tail tip. **Thrush Nightingale** (L15-17 cm, WS24-26 cm) breeds northeast Europe to Russia wintering mainly south of 7°S in Africa; a passage migrant

in Kenya mainly October to December, but also March to April. It is stockier than female Common Redstart, lacks darker central tail feathers and is darker above. Similar to a Common Nightingale with darker, duller, less warm plumage, a reddish tail but brown rump. Dark malar-stripe borders white throat and chin (absent in Common Nightingale). **Common Nightingale** (L16.5 cm, WS23-26 cm) breeds north-west Africa, Europe to Kazakhstan, wintering mainly in the Afro-tropics, south to Kenya coast and north-east Tanzania. It is distinguished from Thrush Nightingale by longer, rufous rump and tail showing more marked contrast with warmer brown upperparts. Underparts are dull cream with buffish wash, paler on throat and vent and showing little or no mottling on breast (unlike Thrush Nightingale). **White-throated Robin** (L16.5 cm, WS27-30 cm) breeds southern Turkey to Turkestan, the entire population wintering in East Africa. Long black tail distinguishes from immature and adult female Common Redstart. Male easily identified by combination of slate grey upperparts, black tail, orange breast and black sides to face and throat contrasting with white throat and supercilium. Non-breeding male is paler with narrow pale fringes to orange and black areas. Female duller and browner above, still with contrasting blackish tail and orangey buff flanks and underwing-coverts, sometimes extending to breast. Immature is similar to adult female, with pale spots on primary coverts, the outer greater coverts and usually some tertials. Male usually tinged orange on breast.
Reference: 1. Phillips (1984).

WHINCHAT *Saxicola rubetra*   Plate 46

Other name: European Whinchat   **French:** Tarier des prés   **Creole:** Tektek Bren

DESCRIPTION Length 12.5 cm; wingspan 21-24 cm. A small, stocky, short-tailed chat. Identifiable at all times by combination of broad white or buff supercilium above dark cheek panel with pale lower border, white bar across inner wing-coverts and triangular white patches at sides of tail.**Male breeding** has blackish brown crown and cheeks bordered by a white supercilium and lower border and a warm orangey buff throat and chest. White wing-patches are evident in flight but sometimes concealed when at rest. Short, square blackish brown tail with white bases to sides. Blackish brown upperparts, strongly streaked, and paler on the rump which has smaller streaks. Brownish black bill; black legs. **Female** has paler brown or mottled cheeks, buff supercilium and cheek border, buffish tips to upperparts, with narrow buff bar across greater coverts and buff fringes to tertials and creamy-buff underparts, smaller white wing-patches and smaller tail triangles. **Non-breeding male** is similar to female but with more white in tail and larger white wing-patches. **Immature** similar to adult female, but warmer buff, lacking white areas in wing.
VOICE Usual call is a disyllabic 'tu-tik', sometimes 'tu-tik-tik'.
BEHAVIOUR Frequents open country, forest margins and cultivation. Sometimes territorial in winter grounds where it is generally solitary, but forms small groups on migration. Flight is low, straight and rapid. Flicks wings and bobs tail. Hops on ground, occasionally running. Perches on a low vantage point. Hunts insects by swooping from perch to catch in the air or on the ground. Diet also includes arthropods, earthworms, snails and seeds.
RANGE Breeds across Europe to central Russia, wintering mainly in sub-Saharan Africa. Generally considered monotypic.
STATUS Vagrant; recorded only on Bird in November.
SIMILAR SPECIES There is a slight chance that **Common Stonechat** *S. torquata* (L12.5 cm, WS18-21 cm) could occur as a vagrant and it should be considered. African and European races are resident or not highly migratory but Asian races are highly migratory wintering northeast Africa, Arabia, India and South-East Asia. There are also resident races close to Seychelles in Comoros and Madagascar. Longer-tailed than Whinchat with no supercilium, no white triangles on tail and shorter, rounded wings.

NORTHERN WHEATEAR *Oenanthe oenanthe*   Plate 46

Other names: Wheatear, Common (or Eurasian) Wheatear
French: Traquet motteux   **Creole:** Tektek Deryer Blan

DESCRIPTION Length 14.5-15.5 cm; wingspan 26-32 cm. A large, long-winged, short-tailed wheatear. Both sexes have a prominent white rump contrasting with a broad inverted black 'T' at end of tail. By far the most likely wheatear to be seen in Seychelles. **Male breeding** has blue-grey crown and upperparts, black mask, black wings, broad white supercilium and buff underparts (this plumage may be acquired January/February). **Female** has the same basic pattern of male, but grey-brown upperparts contrasting with blackish brown wings, a shorter supercilium and black mask replaced by brownish ear-coverts. **Adult non-breeding** birds are duller, particularly male; the female is less changed, but more uniform in tone. Bill and legs black in both sexes. In flight, shows darker upperwing and underwing compared to Isabelline Wheatear. **Immature** similar but duller than adult, broad pale buff edges to wing-feathers reduces contrast with back, the throat with some dark mottling and marked buff in front of eye. The male is usually distinguishable from female by October.
VOICE A hard 'chak' or a squeaky 'weet', sometimes combined.

BEHAVIOUR Frequents open areas with some rocks or low bushes for perches. Solitary, often territorial in winter grounds. Flight low, rapid, diving for cover or sweeping upward to perch and clearly revealing the tail pattern. On the ground usually makes short runs or hops, pausing to flick wings and tail, or bobs whole body or head. Usually has half upright stance, more erect when alarmed. Feeds by dashing or making short flights to catch insects. Diet also includes snails, very occasionally seeds and berries.

RANGE Four races (some recognise more) breed across Eurasia, northwest Africa, Alaska, northwest and northeast Canada, Greenland and Iceland. All winter mainly sub-Saharan Afro-tropics. Nominate *oenanthe* breeds northern Europe to Siberia and Alaska, wintering Sahel and East Africa. Race *libanotica* (southern Europe, Near East) similar but slightly paler than the nominate, with narrower black tail-band, the buff of underparts confined to upper breast/throat; female browner above than nominate. Race *seebohmi* (northwest Africa) has all-black bib and underwing in male (and sometimes female). Both sexes with black of tail tip less extensive than the nominate.

STATUS Vagrant throughout Seychelles, mainly to Aldabra group. Most records are December to March, occasionally as early as October.

SIMILAR SPECIES Isabelline Wheatear is larger, has paler, less contrasting wing, grey or white (not dark grey) underwing and paler ear-coverts; tends to show more upright stance, in which longer legs mean the tail is well off the ground. In a similar stance, Northern Wheatear tail reaches the ground. Pied Wheatear *O. pleschanka* and Desert Wheatear *O. deserti* are potential vagrants (albeit considerably less likely) which might be considered. **Pied Wheatear** (L14.5 cm, WS25.5-27.5 cm) breeds Black Sea to Mongolia, wintering East Africa and Yemen. It is a small, lightly built, long-tailed wheatear; darker than Northern Wheatear with black underwing. Also, has more extensive white on rump and black tail-band is narrower, occasionally broken. **Desert Wheatear** (L14-15 cm, WS24.5-29 cm) breeds Egypt to central Asia, wintering northeast Africa to India; upperparts tinged grey, breast deep buff. Race *oreophila* breeds Himalayas, central Asia, wintering Arabia, Socotra. All-black tail is diagnostic (a little white may be visible at the base). Smaller, stockier with longer tail and less pointed wings than Northern Wheatear.

ISABELLINE WHEATEAR *Oenanthe isabellina*   Plate 46

Other name: Isabelline Chat   **French:** Traquet isabelle   **Creole:** Tektek Izabel

DESCRIPTION Length 16.5 cm; wingspan 27-31 cm. A large, pale wheatear, lacking contrast above and below. White rump combined with absence of black on body distinguishes from most wheatears. **Adult** of both sexes generally uniform grey-buff, the tail with a suggestion of an inverted 'T' but with broad, dark brown (not black) terminal bar. Underwing pale cream or white (black in Desert Wheatear, dark grey in Northern Wheatear). Black alula contrasts with pale wing. Whitish supercilium is rather short but broad in front of eye, tapering and more buffish behind (Northern Wheatear supercilium is long at rear, not pointed and sometimes has greyish tinge). Lore greyish in autumn, blackish in spring and usually blacker in the male. Long, powerful black bill with a small hook at the end of culmen; long, black legs. **Immature** similar to adult after post-juvenile moult (which occurs prior to any likelihood of reaching Seychelles), but may retain duller brown flight feathers of juvenile. Buff fringes to wing-coverts and tertials. White in front of eye.

VOICE A high-pitched whistle and may also cheep like a chicken. Alarm call is 'chak'.

BEHAVIOUR Frequents open terrain and short grassy areas. Solitary and generally territorial. Strong, direct flight, in which tail appears short and broad. Usually a long escape flight. Upright stance with tail well off the ground (Northern Wheatear tail touches ground when similar stance adopted). An enthusiastic tail-wagger; 3-4 rapid jerky wags on the ground, less consistent when perched. Also bobs head emphatically. Feeds early and very late in the day using a dash-and-stab technique on the ground, or making short flights to catch aerial prey. Diet mainly insects but also some small seeds.

RANGE Breeds southeast Europe eastward to China, wintering India, Pakistan, Arabia and Africa. In East Africa occurs south to Tanzania September to April. Monotypic.

STATUS Vagrant to granitic islands November.

SIMILAR SPECIES Northern Wheatear has more contrast above and below, less upright stance, dark grey (not whitish) underwing and darker ear-coverts. Female **Desert Wheatear**, a potential vagrant (see under Northern Wheatear), has blacker wing and longer black tail which it wags even more vigorously; it is smaller, shorter-legged and has a less upright stance.

MADAGASCAR CISTICOLA *Cisticola cherina*   Plate 47

Other name: Malagasy Grass Warbler   **French:** Cisticole malgache   **Creole:** Timerl Tintina

DESCRIPTION Length 12 cm; wingspan 14-15 cm. A small, brownish warbler. The only warbler to be found on Astove and Cosmoledo, not recorded elsewhere in Seychelles. **Adult** has upperparts tawny-brown with heavy blackish brown streaks and rump light brown and unstreaked. Underparts whitish washed pale brown on the flanks, throat and chin white. White supercilium gives a capped appearance. Wings tawny-brown with dark brown flight feathers edged pale brown. Graduated tail, feathers of which are brown, tipped white with blackish subterminal bars except for central tail feathers, which are uniform brown. Eye pale brown. Bill horn, paler on lower mandible. Legs long and pinkish. Sexes similar but female slightly smaller, more boldly streaked on the head and upperparts and more brownish on the underparts. **Juvenile** has upperparts rusty with heavy brown streaking, a yellow wash to chin, throat and upper breast (especially in female) and suffused rusty on the flanks.

VOICE A loud explosive ticking that accelerates rapidly. Also, in flight, a sharp, pleasant 'tint-tint'. In courtship uses a similar but more drawn-out repeated note. Sometimes wingsnaps, producing a buzzing sound.

BEHAVIOUR Found in grass and low vegetation but also readily flies to trees, especially when calling. Solitary or in pairs. Flight low, fluttering, undulating and darting. Very active and agile, constantly on the move hunting for insects. Frequently cocks tail. Feeds on insects and small spiders.

BREEDING Season poorly documented, certainly breeds during northwest monsoon and possibly year round. Nest is domed with a side entrance near the top. It is built of grass bound by spiders' webs, low in a bush or in grasses, often less than half a metre above the ground. In Madagascar clutch is 3-5 greenish white eggs spotted chestnut, especially at larger end. Clutch size possibly averages smaller in Seychelles (three eggs recorded on Cosmoledo).

RANGE The only cisticola found in the Malagasy region, where it is resident on Madagascar and Iles Glorieuses in addition to Seychelles. Monotypic.

STATUS Resident Astove and Cosmoledo, where it is common. On Cosmoledo occurs on every island with vegetation. It has been speculated this may be a recent invader, the first real Seychelles record being in 1940[1]. However, Major Stirling, shipwrecked on Astove in 1836 observed a 'small bird like a wren' which nested in the bushes, and 'twitters very prettily like a linnet'[2]. This was almost certainly Madagascar Cisticola. A record from St Pierre of 'allouette'[3] has been interpreted as a cisticola, but this is doubtful, *zal-wet* being variously used in Creole for small waders. Also surprising would be an unconfirmed report from Rémire[4]. Apart from a tendency to smaller size on Cosmoledo, birds are indistinguishable from parent stock[5].

THREATS AND CONSERVATION Despite the degradation of Astove and Cosmoledo and the presence of rats, the species appears to be thriving. As a natural coloniser to these atolls, it will be of interest to note if it disperses further to Assumption or Aldabra. A genetic study would also be of interest to ascertain whether any significant differences to the parent stock have evolved.

SIMILAR SPECIES None resident in Seychelles, but somewhat similar to Sedge Warbler, which has longer wings, tawny rump and all-brown tail.

References: 1. Penny (1974), 2. Stirling (1843), 3. Bergne (1901), 4. Stoddart and Poore (1970c), 5. Benson (1970a).

ALDABRA WARBLER *Nesillas aldabranus*   Plate 1

Other names: Aldabra Tsikirity, Aldabra Brush Warbler
French: Fauvette (or Nésille) d'Aldabra   **Creole:** Timerl Dezil Aldabra

DESCRIPTION Length 17.5 cm; wingspan 18 cm. Once, the only warbler resident on Aldabra, but probably now extinct. Long-tailed, long-billed, dark brown above and whitish below. **Adult** brown above (female slightly greyer on crown) with rufous tinge to rump. Darker brown and edged rufous on primaries, secondaries and wing-coverts. White lore and supercilium. White on chin, throat and centre of breast, rest of underparts tinged buff. Tail very long. Eye brown. Bill horn, paler on lower mandible, long and slender with narrow gape. Legs grey. Sexes similar, male slightly darker above, lacking grey on crown, streaked dusky on chest (differences possibly due to individual variation in the few birds ever seen)[1]. **Immature** plumage is unknown; none has ever been seen.

VOICE Usually silent. Two types of call were distinguished, most common a brief 'chak' often repeated. Second call, used mostly by female and occasionally male, a loud nasal 'chir' always given after one or two 'chak' calls, giving a disyllabic or trisyllabic effect. Male sometimes sang from an open perch, beginning slowly, increasing in tempo and ending slowly[2].

BEHAVIOUR Favoured tall dense scrub (up to 5 m height) with considerable leaf litter, dense closed canopy and abundant pandanus and dracaena, the latter possibly particularly important for foraging[2]. Preferred habitat also characterised by absence of tortoises or goats. Other peculiarities of its habitat in northwest Aldabra are relatively high rainfall, relatively high species abundance in its flora, high humidity and still air (due to high rainfall and shelter from strong southeasterlies provided by tall mangrove and pandanus) and high invertebrate density[3]. Skulking, usually flying only very short distances. Foraged for insects, mainly below 1.5 m. Prey included spiders, ants, caterpillars, moths and other invertebrates up to 30 mm. Small items immediately swallowed, bigger items sometimes beaten against the ground or a branch. Three hunting techniques have been distinguished: picking (from the ground or leaves), probing (in leaf litter) and leap-snatching (mainly from leaves, involving a slight jump and a brief flight)[2].

BREEDING Season probably coincided with wet months of October to January. A female seen in December had a pronounced brood-patch[2]. Elsewhere in the region at Madagascar and Comoros Tsikirity Warbler *N. typica* breeds at this time. Territories of 0.75-1.5 ha were occupied year round. A nest with three eggs was collected in December 1967, found about 0.6 m above the ground in the leaf bases of a young pandanus. However, from the gonads of the parent, also collected, it was evident the clutch was incomplete[1]. Comparing to similar species in the region, clutch size probably normally two or three[1]. Eggs very pale purplish, spotted dark brown. Other empty nests were found in 1968 in pemphis at 1.5 m height and bwa mozet at 3.2 m[2].

ORIGINS Closely related to Madagascar Brush Warbler *N. typica*, of which it could possibly have been no more than a distinctive race. It is impossible to say whether colonisation took place directly from Madagascar or from Comoros. Earlier origin of *Nesillas* stock, endemic to the Malagasy region, is not known. A Pan-African origin appears most likely, but given the colonisation of the whole region by the Asian bulbul genus *Hypsipetes*, a possible Asian origin cannot be altogether ruled out[4].

STATUS Endemic to Aldabra, probably now extinct. Only ever recorded on a 10-ha coastal strip (2 km by 50 m) of the western end of Malabar, though once heard singing slightly outside this range at Anse Petit Grabeau. Once described as '...almost certainly the rarest, most restricted and most highly threatened species of bird in the world'[5] by 1994 it was listed as extinct[6]. It was only discovered in 1967 by Penny; a female with three eggs was collected on 14 December 1967 and a male collected 29 January 1968, skins and eggs now at The Natural History Museum, Tring, UK. A study in 1974-75 located only five birds: three males and two females[3]. No young have ever been seen. After November 1975 only males were ever seen. By February 1977 only two males were known to survive[2]. The last sighting of a lone male was in September 1983[6]. Extrapolation from these studies indicated a maximum population of 25 birds in a 9 km stretch considered to be suitable habitat[2]. This area may have been reduced to only 5 km following the penetration of goats and giant tortoises from about 1976, which would reduce the maximum population size to just 13. Roberts, resident on Aldabra in 1986, made specific searches each month from July to November, spending 11 days in the area of its previous occurrence using a tape recording of the bird's call, a bird squeaker and 80 m of nets. He found no sign of the species[7]. Many subsequent searches also found no sign of birds. Though it is impossible to say what caused the decline and probable extinction of the species, rats are prime suspects, being present on Malabar, while studies of other endemic landbirds of Aldabra show rats to be major predators. On the other hand, the location of nests in spiky pandanus may afford better than average protection from rats[2]. Cats may also have played a part, though this is unlikely to have been the decisive factor given their scarcity on Malabar and indeed there are no records of cats prior to 1967. Goats may also have contributed in reducing the available habitat through browsing on vegetation. Goats were eradicated on Malabar in 1994. Both goats and the endemic giant tortoise are absent from the dense vegetation of western Malabar. If it has disappeared, this may be a natural extinction, the environment on Aldabra offering little suitable habitat. It may have been due to a change in vegetation brought about by climatic change or the abundance of giant tortoises.

SIMILAR SPECIES None.

References: 1. Benson and Penny (1968), 2. Prys-Jones (1979), 3. Hambler *et al.* (1985), 4. Benson (1984), 5. Collar and Stuart (1985), 6. Collar *et al.* (1994), 7. Roberts (1987).

SEDGE WARBLER *Acrocephalus schoenobaenus*   Plate 47

Other name: European Sedge Warbler   **French:** Phragmite des joncs   **Creole:** Timerl Zon

DESCRIPTION Length 13 cm; wingspan 17-21 cm. A fairly small, bright, streaked warbler. Brown, boldly streaked upperparts (but rump tawny and unstreaked) and bold cream supercilium defining blackish streaked crown distinguishes from other *Acrocephalus* warblers likely in Seychelles. **Adult** has upperparts brown and heavily streaked with underparts creamy and a rufous tinge on the rear flanks. Brown, slightly rounded tail contrasts with more glowing, broad, tawny rump. Heavy black streaking on crown with a long, creamy supercilium somewhat pointed at the rear and a black eye-stripe. Eye brown; bill blackish, base of lower mandible yellowish; legs grey. **Juvenile** is similar to adult, more yellowish brown, often with a pale crown-stripe (usually absent in adult) and warmer creamy buff supercilium. Breast often has diffuse, round, brown spots. Flight feathers and tail feathers fresher in autumn than adult, tertials broadly edged buff. Legs brownish pink. Inseparable from adult after post-juvenile moult in winter quarters.

VOICE A short, loud 'tuk' or harsh 'churr'. Song is an accelerating harsh trill interspersed with soft whistles and clearer sweeter notes. May sing in winter quarters from January onward, though less intense.

BEHAVIOUR Frequents waterside scrub, reedbeds and dense damp vegetation. Solitary in winter quarters but may migrate in large flocks. Puts on large fat reserves prior to migration, permitting long, unbroken transits. Flitting, low-level flight, over short distances often with tail depressed and spread, suddenly diving for cover. Stance fairly horizontal. Regularly flicks tail. Not particularly shy, but often found in dense cover. Feeds mainly on insects.

RANGE Breeds Europe to Central Asia, entire population wintering sub-Saharan Africa south to South Africa. In Kenya seen mainly on passage November to December and more commonly March to April. Monotypic.

STATUS Vagrant to granitic islands, recorded November.

SIMILAR SPECIES Madagascar Cisticola, resident on Astove and Cosmoledo, has shorter wings, a brown (not tawny rump) and white tip to tail, with black subterminal band. There are many long-distance migrant

Acrocephalus and *Locustella* warblers unrecorded, but perhaps almost equally likely in Seychelles. These include the following: **Common Grasshopper Warbler** *L. naevia* (L12.5-13.5 cm, WS15-19 cm) breeding Europe to central Asia, wintering south to sub-Saharan Africa, vagrant Kenya, Sri Lanka; **River Warbler** *L. fluviatilis* (L13 cm, WS19-22 cm) breeding central and eastern Europe, wintering East Africa to northern South Africa; **Savi's Warbler** *L. luscinioides* (L14 cm, WS18-21 cm) breeding Europe to Kazakhstan, wintering sub-Saharan Africa, vagrant Kenya; **Eurasian Reed Warbler** *A. scirpaceus* (L13 cm, WS17-21 cm) breeding northwest Africa, Europe to central Asia, all populations migratory, mainly to sub-Saharan Africa south to Zambia; **Marsh Warbler** *A. palustris* (L13 cm, WS18-21 cm) breeding temperate middle latitudes of western Palearctic, wintering southeastern Africa mainly Zambia to South Africa; **Blyth's Reed Warbler** *A. dumetorum* (L13 cm, WS17-19 cm) breeding eastern Europe to central Asia, wintering Indian subcontinent, Sri Lanka; **Great Reed Warbler** *A. arundinaceus* (L19-20 cm, WS24-29 cm) breeding northwest Africa, Europe to central Asia, wintering south to South Africa); and **Basra Reed Warbler** *A. griseldis* (L17-18 cm, WS21-24.5 cm) breeding southern Iraq, wintering east and southeast Africa.

SEYCHELLES WARBLER *Acrocephalus sechellensis* Plate 47

Other name: Seychelles Brush Warbler **French:** Rousserolle des Seychelles **Creole:** Timerl Dezil Sesel

DESCRIPTION Length 13-14 cm; wingspan 17 cm. A small, brownish warbler with long legs and bill. The only resident warbler of granitic Seychelles found on Aride, Cousin and Cousine. **Adult** is very plain, having the upperparts dull greenish brown, underparts dingy white and otherwise no distinguishing features other than an indistinct pale supercilium and a red-brown eye. Long grey-blue legs and a long, slender bill, horn coloured with yellowish flesh base. **Juvenile** similar to adult but darker with some speckling on the breast. Birds can sometimes be aged at close range on eye colour, being grey-blue at first becoming grey-brown in the subadult.

VOICE Rich and melodious, has a clear, whistling song, similar to a human whistle. Alarm call a harsh chatter. Juveniles make rasping 'zhzh-zhzh-zhzh' notes.

BEHAVIOUR Establishes territories in woodland where dominant trees include in particular, pisonia and bwa torti. Usually solitary, in pairs or small family parties. Tame and approachable but often unobtrusive, as birds move through vegetation hunting for insects making only short, flitting flights between feeding stations. Feeds almost entirely on insects taken from the underside of leaves.

BREEDING May breed at any time of year with peaks of activity around June/July and January/February. In optimal habitat, has the smallest territory of its kind of any passerine[1]. Birds undergo a pre-breeding moult. Cup-shaped nest is constructed from grass stems and coconut fibres (sometimes man-made materials such as string, plastic and wood shavings are used), with a finer lining; usually situated at a fork in a tree or shrub. One or two eggs are laid, occasionally up to four. Eggs are ivory, with greenish brown speckles. Co-operative breeding (where young remain in their parents territory and help to raise subsequent broods) is commonly practised. This is usually explained by habitat saturation, but territory quality is also important[2]. When birds were transferred to Aride in 1988, territories vacated on Cousin were quickly filled, some within hours and over 90% of high quality territories were taken over by other birds from high quality territories, suggesting higher fitness of such birds. Studies on Cousin showed young remain in their natal territories longer when the territory quality is higher. By deferring, independent breeding birds may gain greater reproductive success later on, even if lower quality territory is immediately available[2]. Males born on high quality territories may 'bud off' a section of their parents' territory to live in and later acquire a female. Boundaries are then expanded at the expense of the parental territory, until on the death of the male parent, the young male acquires the entire territory. In this manner territories may be passed down through generations, remaining in the same family lineage[3]. By contrast, young males on low quality territories may become 'floaters' shortly after attaining sexual maturity at one year old. Only females incubate; helpers may share this duty, giving almost continuous protection from predators such as skinks and Seychelles Fody, which more than doubles hatching success. The presence of a helper significantly improves the reproductive success of the parents. Groups with one pair and one helper have the highest survival rate. Groups with one pair and two helpers are less successful, possibly due to joint nesting and reproductive competition or because of the increased pressure on resources in a territory where previous offspring remain resident[4,5,6]. Where helpers lay eggs (as is sometimes the case), survival rate declines. Otherwise, helpers do not play a significant role in raising the average clutch size. On Cousin, almost two thirds of territories contain helpers. Helpers share almost all aspects of nesting, rearing and defending young. Helpers are of close kin, including helpers to step-parents. Chicks hatch two months after an increase in rainfall, which is the same time lag required for an increase in insect abundance. Remarkably, sex ratio is known to depend strongly on the environment. In territories where food is scarce, nearly 80% of eggs will produce males (which will leave the territory as soon as possible), whereas where food is plentiful, nearly 90% will be females (which may be recruited as helpers). How this is done is unknown, though the mother's ovulation could be affected by diet[7]. On fledging, single chicks are a little heavier than adults, but lose weight quickly as parents reduce the feeding rate and they begin to search for their own food. Twins, which weigh less than adults on fledging, struggle to survive to independence. Triplets will usually all die before independence and quadruplets are likely to die in the nest within ten days.

Incubation lasts 14-19 days, fledging 16-20 days. Chicks may leave the nest even before they can fly[8]. Chicks remain dependent on their parents for four months after fledging and are fed mainly by the female. If parents breed again, their other offspring under one year of age do not become helpers. On Cousin, young birds remain in their natal territories for about two years and do not breed until almost four years of age. By contrast on Aride and Cousine, immediately after the transfer of parent birds from Cousin in 1988 and 1990 respectively, young left their natal territory at around four months, to breed at around eight months of age[9]. None of these young acted as helpers until around two years after transfer to Aride and one year after transfer to Cousine, by which time the best quality territories were occupied. Low quality territory was still available on both islands at this time so this confirmed both territory quality and habitat saturation are important factors in the co-operative breeding of Seychelles Warbler. One ringed bird on Cousin lived to 18 years of age.

ORIGINS Once tentatively associated with Swamp Warbler *A. gracilirostris*, a member of a group mainly of African origin, but also represented in Madagascar (Madagascar Swamp Warbler *A. newtoni*)[10]. However, Dowsett-Lemaire noted similarities in the song of Seychelles Warbler and African marshland species and concluded Seychelles Warbler evolved from stock close to Swamp Warbler. DNA studies by Schulze-Hagen confirm Seychelles Warbler, Swamp Warbler and Madagascar Swamp Warbler are all close relatives[11].

STATUS Endemic to Seychelles, resident on Cousin (*c*350 birds), Cousine (*c*150 birds) and Aride (*c*2,000 birds). IUCN Vulnerable. In 1870, the species was recorded on Marianne and Cousine[12] but disappeared from both. It has been suggested the Cousine record may possibly have been in error for Cousin which remained the last known refuge from the 1870s onward. However, given the geographical separation of Cousin and Marianne, the two islands where the species certainly occurred naturally, it seems probable it also occurred on other islands of the Praslin group. Also, one ringed on Cousin was recovered on Cousine in 1996 (P. Hitchins pers. comm.), which may indicate a certain amount of natural dispersal is possible. In 1938, it was considered rare on Cousin, its last refuge[13]. In 1953 it was regarded as 'in no danger of extinction'[14], though by 1959 a survey placed the population at only 26-29[15]. By 1982, there were nearly 320 birds and the Cousin population has since fluctuated around this level, suggesting this is the carrying capacity of the island[16]. Transferred to Aride in 1988 and Cousine in 1990 (see below) by the end of 1997, Aride's population exceeded 1,600 and Cousine's was estimated at 137.

THREATS AND CONSERVATION The species was exterminated from Marianne and possibly other islands due to habitat destruction. It was primarily to save the species from extinction that a conservation campaign was launched to purchase Cousin. The island was purchased in January 1968 in the name of the Society for the Promotion of Nature Reserves (now RSNC) and held in trust for ICBP (now BirdLife International) who managed the island until 1998 when it passed to BirdLife Seychelles. The regeneration of native woodland led to a large increase in numbers. It was recommended a second population should be established on another island in case any disaster befell Cousin[17]. ICBP, RSNC and the Seychelles Government approved a transfer to Aride in August 1988 and the following month 29 birds (16 males and 13 females), all adults of at least one year old, were transferred. One pair began nest building only a day after their transfer. In less than three weeks, seven territories had been established, four pairs were nest building and one completed nest contained two eggs. One year after the transfer, the Aride population had doubled to 59[18]. During this year, Aride birds showed more than three times more breeding activity than Cousin birds; the clutch size was almost double, chicks fledged up to six days earlier, the mean number of fledged chicks per pair was more than six times higher and mean reproductive success per pair was an amazing 16 times higher than on Cousin[9]. This ranks as one of the most successful rescue operations for any endangered species anywhere in the world. Given the greater size of Aride and the higher quality of territory, there may eventually be around 1,400 territories and 4,200 birds on Aride. Six were reported to have been introduced to Cousine around 1960 (a huge risk, representing 20% of the world population as estimated the previous year[15]). There were occasional subsequent sightings that may have been survivors or descendants, though it is also possible further birds arrived from Cousin by natural means. In July 1990, 29 birds were transferred to Cousine and have steadily increased in number[18].

SIMILAR SPECIES None resident in Seychelles.

References: 1. Diamond (1980), 2. Komdeur, (1992), 3. Komdeur and Gabrielsen (1993), 4. Komdeur (1991a and 1991b), 5. Komdeur (1994b), 6. Komdeur (1994c), 7. Komdeur (1996), 8. Komdeur (1988), 9. Komdeur *et al.* (1995), 10. Hall and Moreau (1970), 11. Dowsett-Lemaire (1994), 12. Oustalet (1878), 13. Vesey-Fitzgerald (1940), 14. Foster-Vesey-Fitzgerald (1953), 15. Crook (1960), 16. Komdeur (1994c), 17. Collar and Stuart (1985), 18. Komdeur (1994a).

ICTERINE WARBLER *Hippolais icterina*   Plate 48

French: Hypolaïs ictérine   **Creole:** Timerl Ikteren

DESCRIPTION Length 13.5 cm; wingspan 20.5-24 cm. A medium-sized, long-looking warbler with broad bill, blue legs, long wings, pale yellow panel in closed wing and a rather flat crown. Plumage basically yellow and green not brown, buff and grey-white. **Adult** in fresh plumage, greenish olive above, wings darker with diagnostic yellowish wing-panel and yellow underparts. In autumn due to wear, may be more greyish green above with a yellowish white or indistinct wing-panel. Rump and uppertail-coverts green-ish, tail olive brown indistinctly edged yellow. Shows a prominent yellow supercilium, narrow eye-ring and a pale lore with a greenish crown and ear-coverts. Note, there is no dark eye-stripe, unlike *Phylloscopus* warblers. Bill longish (but shorter than other *Hippolais*) and broad, dark brown on upper mandible, yellowish or orange on lower mandible; legs grey or bluish grey.

First-winter similar to adult, greyer or browner above, yellowish white below, flight feathers fresh in autumn (usually worn in adult) and yellowish buff wing-panel.

VOICE Song a long, sustained jumble of melodious notes and harsh chatters, often with some mimicry, usually given from high in a tree. Call note 'tek' often repeated with crown feathers raised. May sing February to March in winter quarters.

BEHAVIOUR Frequents forest margins, gardens and cultivation. Solitary and often territorial in winter quarters. Flight dashing, long-winged appearance obvious. Graceful gait contributes to alert look. Tail often slightly flicked and crown feathers may be raised when excited. Diet mainly insects, also some fruit.

RANGE Nominate race *icterina* breeds middle to upper latitudes of western Palearctic to western Russia; race *alaris* breeds Turkey to Iran; this race, not recognised by some authorities, is darker above with a slightly shorter wing. Both populations winter Africa, mainly south of the Equator.

STATUS Vagrant to granitic islands; most likely November to December.

SIMILAR SPECIES Some migrant *Phylloscopus* warblers may be tinged yellow, but are smaller with slen-der bills and legs and a dark streak at the lore. There have been unconfirmed reports of other *Hippolais* warblers. Potential vagrants can be separated by combination of size and plumage differences. **Olivaceous Warbler** *H. pallida* (L12-13.5 cm, WS18-21 cm) breeds North Africa, southern Europe to southwest Asia with European and Asian populations migratory. Race *elaeica* of southeast Europe to southwest Asia, winters mainly east and northeast Africa and is the most common *Hippolais* in Kenya September to April. It is a small *Hippolais*, appearing more rounded in shape than Icterine, legs longer with upperparts varying from grey-brown to pale brown, completely lacking any yellow or green tones and underparts very pale greyish white. Shorter wing projection than Icterine, wings reaching only to base of tail and no wing-panel. Face bland with plain pale lore, pale grey eye-ring and an indistinct supercilium, which may be emphasised by slight darkening along edge of crown but is restricted to in front of eye only, or only slightly projecting behind eye. **Upcher's Warbler** *H. languida* (L14 cm, WS20-23 cm) breeds lower to mid-latitudes of central Palearctic, the entire population wintering East Africa. It is a medium to large warbler, grey-brown above, dull white below (slightly buff on flanks) with more rounded head and longer tail than Olivaceous Warbler. Long wings, primary tips reaching beyond end of uppertail-coverts. Secondaries and tertials form a distinct wing-panel in fresh plumage. Narrow super-cilium in front of eye only or very slightly behind eye. Pale eye-ring contrasts with mottled face. Tail blackish, edged white (palest-edged tail of any *Hippolais* warbler); longer and wider than Olivaceous. Very exaggerated tail movements; fans, cocks and flicks tail downward displaying pale outer edges. **Olive Tree Warbler** *H. olivetorum* (L15 cm, WS24-26 cm) breeds east Mediterranean to Lebanon, the entire population wintering east and southeast Africa, mainly south of Tanzania. The largest grey-and-white warbler likely in Seychelles, with large, heavy dagger-like bill, long dark legs, large body and long wings. Brownish grey above, darker on wings with pale streaky wing-panel in fresh plumage, (usually lost by August). Dull white below sometimes with yellowish or greyish wash on throat and breast. Indistinct buff-white supercilium in front of eye only and a whitish eye-ring. Dusky ear-coverts and a spot in front of eye may give hooded appearance. Grey to blackish tail with indistinct grey-white outer tail feathers. Flicks tail, but less exaggerated than Upcher's.

WILLOW WARBLER *Phylloscopus trochilus*   Plate 48

French: Pouillot fitis   **Creole:** Timerl Ver

DESCRIPTION Length 10.5-11.5 cm; wingspan 16.5-22 cm. A small, slender warbler, with long primary projection; pale olive above, yellowish white below, pale yellow supercilium, usually pale legs and pale base to the bill. **Adult** is generally olive above, yellowish white below, but autumn birds are often brighter. Stronger eye-stripe and yellow supercilium, more clearly defined than Chiffchaff; also appears more slen-der and longer-winged. Often shows a greenish wing-panel (Chiffchaff wing more uniform). Bill usually brown with an obvious pale base to the lower mandible (Chiffchaff has all-dark bill). Legs generally pale; some have blackish legs but with dark orange feet. **First-winter** similar to adult, both adults and young being in fresh plumage by autumn, complicating ageing in the field. Throat, breast and flanks may be tinged buff.

VOICE Diagnostic: song is a liquid descending cadence, ending in a flourish 'suueet-suueetoo'. Call is an almost disyllabic 'hooeet'. Sometimes sings in winter quarters, especially September to November and March to May.

BEHAVIOUR Frequents woodland and low vegetation. Vagrants likely to be solitary. Dashing and agile with stronger flight than shorter-winged Chiffchaff. Occasionally flicks tail up and down but lacks persis-tent downward tail dip of feeding Chiffchaff. Hops with a horizontal posture. Diet mainly insects and spi-ders, also some berries, usually picked from twigs or foliage; occasionally takes aerial prey by flycatching or may hover for brief periods.

RANGE Nominate *trochilus* breeds western Europe to Poland wintering mainly West Africa but some reaching eastern and southern Africa. Race *acredula* breeds Scandinavia to eastern Europe and also regularly reaches eastern and southern Africa; often paler green above than the nominate with paler, more white below, though differences are hard to detect and there is some individual variation, some being very brown and white. Race *yakutensis* (Siberian Willow Warbler) breeds Siberia and again reaches eastern and southern Africa in winter; it is more grey-brown above, dull white below, greyer on breast with an olive tinge to rump, flight feathers and tail feathers and very little yellow coloration except on underwing-coverts and axillaries.

STATUS Vagrant to granitic islands and Aldabra November to December and February to March. Race unknown, but most in Kenya are *acredula* with much smaller numbers of *yakutensis*.

SIMILAR SPECIES Wood Warbler is larger, brighter, greener above, more yellow on breast and shows a sharper contrast between yellow breast and white belly with yellow-green edges to black-centred flight feathers of more pointed wings. **Chiffchaff** is more compact, generally dingier, darker and more olive green above, more olive-yellow (sometimes buffish) below and head pattern not as strong, with indistinct, pale supercilium, lore often dark, cheek plainer buff and noticeable eye-crescents.

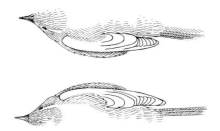

Primary projection of Chiffchaff (top), and Willow Warbler

Wing profiles of Chiffchaff (left), and Willow Warbler. Note position of tip of 9th primary and relative emarginations

CHIFFCHAFF *Phylloscopus collybita*　　　　**Not illustrated**

Other names: Common (or Eurasian) Chiffchaff, Brown Leaf Warbler
French: Pouillot véloce　**Creole:** Timerl Sifsaf

DESCRIPTION Length 10-11 cm; wingspan 15-21 cm. A rather dumpy, rounded warbler with relatively short wings, an indistinct supercilium and dark legs.
Adult is brownish green above, sometimes brighter on rump, whitish below with buff or yellowish tinge to throat and breast. Supercilium is short and indistinct and a dark eye-stripe splits the pale eye-ring at both front and rear of eye. Shorter primary projection than Willow Warbler, primaries extending beyond overlying tertials only 30-50% (75-100% in Willow). Wing plainer, lacking wing-panel often seen in Willow (though race *tristis*, Siberian Chiffchaff, often shows a pale narrow wingbar). Legs usually darker than Willow, dark brown to blackish, but colour variable (Willow legs usually orange but some have darker legs with orange feet). Bill is usually darker with little colour obvious at the base. **First-winter** is usually duller and browner than adult; moults slowly retaining wispy plumage of juvenile until about September. Compared to Willow, first-winter birds are much less yellow, more olive above and buffish below.
VOICE Song a diagnostic, repetitive 'chiff-chiff-chaff'. Call a more monosyllabic 'hweet' compared to Willow. Sings in winter quarters from January.
BEHAVIOUR Frequents woodland, plantations, gardens and mangroves. Vagrants likely to be solitary. Flight weaker than Willow. Habitually dips tail when feeding (unlike Willow). Posture horizontal, but tail dip and rounded shape give a half-upright appearance. Tame and more often encountered at edge of cover than Willow. Diet entirely insects picked from vegetation but sometimes hovers and flycatches (but less often than Willow).
RANGE Nominate *collybita* breeds western Europe east to Poland merging with *abietinus* (Scandinavian Chiffchaff; grey-brown above, paler below with more obvious supercilium), which breeds Scandinavia and western Russia and *tristis* (Siberian Chiffchaff), which breeds Russia from Pechora basin eastward. Siberian Chiffchaff may be identified by lack of yellow in underparts which are often warm buff on sides of breast, neck and ear-coverts and on supercilium, and lack of olive in upperparts which are grey or grey-brown, though rump/lower back often shows an olive tone. Bill and legs always black All three races are migratory, wintering from sub-Saharan tropical Africa, Arabia to northern India. Also possible might be race *lorenzii* (Caucasian Chiffchaff). Darker and warmer brown above it lacks any olive, including on

rump/lower back, supercilium often more well marked and paler than Siberian Chiffchaff.

STATUS Uncertain; some reports of **Willow Warbler** fail to eliminate Chiffchaff and have been accepted by SBRC as 'Willow Warbler/Chiffchaff'.

SIMILAR SPECIES Willow Warbler is more yellowish below (especially first-autumn birds), with longer, more pointed wings.

WOOD WARBLER *Phylloscopus sibilatrix*   Plate 48

French: Pouillot siffleur   **Creole:** Timerl Dibwa

DESCRIPTION Length 12 cm; wingspan 19.5-24 cm. A distinctive *Phylloscopus* warbler; very long primary projection, very bright yellowish green upperparts, white belly with upper breast and throat washed lemon-yellow and a bright yellow supercilium. **Adult** is much brighter than Willow Warbler or Chiffchaff (though fresh-plumaged Willow in autumn can be very yellow) with a strong head pattern: an obvious broad lemon-yellow supercilium and a dark greenish eye-stripe. Bright greenish upperparts with yellow ear-coverts throat and breast sharply demarcated from the clean white belly. Appears broader-chested and shorter-tailed than Willow, short tail accentuating the long wings, especially compared to Chiffchaff (Chiffchaff wing 53-68 mm, Willow 59-72 mm, Wood 70-81 mm[1]). Shows narrow yellow-green edges to flight feathers that are black in the centre. Bill blackish brown on upper mandible, pinkish yellow on lower mandible. Autumn birds are somewhat duller with worn flight feathers and tail feathers though tertials and some tail feathers often fresh. Legs yellowish brown, thicker than Willow Warbler. **First-winter** similar to adult, greyer above, paler yellow on throat and breast with fresh flight and tail feathers, the latter edged yellowish green (at least until October).

VOICE Song a diagnostic, accelerating trill. Call a piping 'pew'.

BEHAVIOUR A woodland bird though on migration could be found elsewhere. Vagrants likely to be solitary. Flight is fluent and agile, frequently hovering or flycatching; often glides through the canopy, wings half closed. Hops with a horizontal posture. Very active, but does not flick wings and tail. Diet mainly insects with some fruit and seeds.

RANGE Breeds Europe east to central Russia migrating to sub-Saharan Africa, mainly western and central. Uncommon in Kenya. Monotypic.

STATUS Vagrant to granitic islands and Aldabra November to December, but may be just as likely elsewhere in Seychelles.

SIMILAR SPECIES Willow Warbler is smaller, less green above, less yellow on breast and lacks yellow-green on flight feathers; tail longer but wings shorter, legs and bill more slender.

Reference: 1. Svensson (1992).

BLACKCAP *Sylvia atricapilla*   Plate 47

Other name: Black-capped Warbler   **French:** Fauvette à tête noire   **Creole:** Timerl Latet Nwanr

DESCRIPTION Length 13 cm; wingspan 20-23 cm. A fairly large, dark grey, robust warbler with long wings and legs, and sharply contrasting black or brown crown. **Male** has glossy black crown down to the eye, grey-brown upperparts; sides of head and underparts ash grey. Eyes hazel; bill black, paler at base; legs grey. **Female** has rusty-brown crown and browner underparts. **First-winter male** may have some brown feather tips in the crown (especially on the forehead) and a dull nape. Outer greater coverts edged brown contrasting with grey-green inner greater coverts. **First-winter female** has orangey brown crown and slight contrast between rufous-edged outer greater coverts and duller grey-brown edges of inner greater coverts. Eyes dark brown; legs brownish grey.

VOICE Song a rich varied warble. Commonest call a harsh 'tak-tak'. Sings in winter quarters, January onward, sometimes with a shortened, lower sub-song.

BEHAVIOUR In Africa favours a wide habitat range, from dry scrub to dense rainforest, mangroves, gardens and cultivation. Generally solitary in winter quarters. Flight straight and rapid. Hops and sometimes raises cap into a slight crest; often flicks wing and tail. Diet mainly insects in breeding season and plant material at other times.

RANGE Nominate *atricapilla* breeds Europe, northwest Africa, western Siberia and Turkey. Race *dammholzi* (South-eastern Blackcap) breeds eastern Turkey to north Iran; it is slightly paler and greyer above, whiter below than the nominate. Both races recorded Kenya (*dammholzi* most frequent) and both could conceivably reach Seychelles. Other races breed Corsica, Sardinia, Balearics, Azores, Cape Verde Islands, Madeira. Western birds are resident or partial migrants while eastern are wholly migratory, mainly to sub-Saharan Africa. Present Kenya and scarce Tanzania October to April.

STATUS Vagrant to granitic islands; recorded December.

SIMILAR SPECIES Potential vagrant *Sylvia* species include Garden Warbler *S. borin* and Barred Warbler *S. nisoria*, both long-distance trans-Saharan migrants. **Garden Warbler** (L14 cm, WS20-24.5 cm) is larger and plumper than Blackcap; uniform drab grey-brown plumage and stubby bill distinguishes from all other *Sylvia* warblers. **Barred Warbler** (L15.5 cm, WS23-27 cm) is the largest *Sylvia* warbler; size, dull wingbar and bright tips to tertials diagnostic. Male grey or grey-brown above, creamy below washed buff with barred crescent shapes (indistinct in winter). Female duller and browner above with less distinct barring below; white outer tail feathers (unlike Garden Warbler), less extensive in female.

COMMON WHITETHROAT *Sylvia communis* Plate 48

Other names: Greater Whitethroat, Whitethroat **French:** Fauvette grisette **Creole:** Timerl Lagorz Blan

DESCRIPTION Length 14 cm; wingspan 18.5-23 cm. A medium-sized, slim, long-tailed warbler; white throat contrasts with pale buff underparts; pale rufous wing-panel, white edges to tail and yellow-brown legs. **Adult** has crown and cheek of breeding male grey, often brown in winter as are those of female at all times. Both sexes have a white throat and a narrow white eye-ring, sometimes conspicuous. Upperparts greyish or brownish, underparts buffish white or greyish white, sometimes with a pinkish wash in the male. Distinct rufous edges to tertials and secondaries form a wing-panel, which distinguishes from any other warbler likely in Seychelles. Tail long, square at the end, dark brown with bold white edges. Eye pale brown; legs yellow-brown. Bill grey-brown, pinkish at base of lower mandible and along cutting edge. **First-winter** similar to adult female, browner above. In autumn outer tail feathers uniform light brown or possibly tipped and edged pure white, distinctly fresh compared to other tail feathers. Eye grey-olive or dark grey-brown (autumn adult paler varying from bright orange-brown to medium dull brown).
VOICE A rapid, scratchy warble. Sometimes sings in winter quarters December onward. Call note a repeated 'chek' or a harsh 'charr'.
BEHAVIOUR Favours open country with bushes, undergrowth and woodland margins. Vagrants likely to be solitary. Irregular wingbeats make flight appear erratic and a typical view is of a rather long-tailed brownish warbler flying jerkily into low vegetation while flicking tail to reveal obvious white outer tail feathers. Hops with a horizontal stance, tail often raised and rarely below horizontal. May raise crown feathers when excited. Diet mainly insects in breeding season and berries in winter quarters.
RANGE Race *icterops* (Southern Whitethroat) breeds Turkey to Iran, wintering East and southern Africa; this is the largest, greyest race with dark grey head (especially in male), pale whitish underparts, pale sandy rufous wing feathers and indistinct pale yellowish brown wing-panel. Race *volgensis* breeds eastern Europe to Altai range and has similar wintering grounds; it is less grey above, less white below and darker on flanks. Race *rubicola* breeds Turkmenistan to western China and probably winters East Africa; upperparts are more brown-grey. Nominate race of western Europe winters western to central Africa and is unlikely to reach Seychelles.
STATUS Vagrant; recorded only from Aldabra in October where either *icterops* or *volgensis* has been recorded based on measurements of an individual mist-netted, ringed and released.
SIMILAR SPECIES Lesser Whitethroat *S. curruca* (L12.5-13.5 cm, WS16.5-20.5 cm) is considerably less likely (wintering sub-Saharan Africa north of 10°N and Asia south to Sri Lanka), but should be considered a potential vagrant. It is smaller, more compact, shorter-tailed, shorter-winged with less peaked crown, lacks rufous wing-panel and has no obvious eye-ring, darker ear-coverts and darker legs.

SPOTTED FLYCATCHER *Muscicapa striata* Plate 49

French: Gobemouche gris **Creole:** Gobmous Pitakle

DESCRIPTION Length 14.5 cm; wingspan 23-25.5 cm. Almost uniform grey-brown above, paler dull sandy-brown or whitish on the forehead and forecrown and dull whitish below with brown streaking on sides of throat, neck and entire breast. The only migrant flycatcher recorded in Seychelles. **Adult** has upperparts uniform grey-brown, but primaries and tail a little browner. Short fine streaks on forehead and crown visible at close range. Underparts dull white with buffish wash on flanks, sides of breast and under-wing. Long dull brown streaking on lower throat, upper flanks and breast. Pale edgings, buff to whitish, on tips of wing-coverts and tertials form a narrow bar on the greater coverts and sharp lines on the tertials. Slight fork in tail. Whitish eye-ring; eye dark brown. Bill and legs blackish. **Immature** similar to adult, with a buff wash to face and forehead and plumage fresher in autumn. Inner greater coverts, tertials and tail-coverts are often tipped buffish.
VOICE Usually silent. Commonest calls are a thin 'tzee' and a rapid trisyllabic 'tzee-zuk-zuk'.
BEHAVIOUR Frequents woodland clearings and margins; also gardens. Flight rapid and slightly undulating. Usually solitary, perching in the lower branches of trees. Upright posture, but with head hunched into shoulders, tail slightly depressed occasionally flicking wings and tail. When feeding, sallies from a low perch to take aerial insects, usually returning to same perch. May also feed in foliage like a warbler or take prey from the ground. Occasionally takes berries, especially in autumn.
RANGE Several very similar races breed northwest Africa, Europe, east to beyond Lake Baikal. All races winter sub-Saharan Africa to South Africa. Nominate *striata* has been recorded East Africa south to South Africa. Race *neumanni* (Eastern Spotted Flycatcher) breeds west and southern Asia and has been recorded East Africa (where it is the most common race) to Namibia; paler grey above, whiter below. Race *sarudnyi* breeds northern Iran, Caucasus, Afghanistan and may also winter East Africa; paler, sandy brown above

with less distinct streaking. Other races possible in Seychelles include *inexpectata* (breeds Crimea), darker, browner above with heavier streaking and *mongola* (breeds Mongolia), similar to *sarudnyi* but greyer above with even less distinct streaking. In East Africa present September to April, but especially common December to February including on Kenya coast.

STATUS Annual or near-annual in Aldabra Group November to March, vagrant elsewhere in Seychelles.

SIMILAR SPECIES Potential vagrant small flycatchers might include **Brown Flycatcher** M. *dauurica* (L14 cm, WS19.5-21.5 cm), which breeds eastern Asia and winters southwestward as far as Sri Lanka. Head pattern is diagnostic: dull white lore, prominent white eye-ring and short submoustachial-stripe offset dark eye and dark malar-stripe and with throat conspicuously white, often the first indication of presence in low light. Eye appears large, emphasised by conspicuous eye-ring. Triangular horn-brown bill sometimes with flesh-yellow visible at base of lower mandible. Streaking, if visible is indistinct ashy brown on breast and flanks, absent on throat and crown.

SEYCHELLES PARADISE FLYCATCHER *Terpsiphone corvina* **Plate 49**

French: Gobemouche de paradis des Seychelles **Creole:** Vev

DESCRIPTION Length male 20-37 cm (tail 7-22 cm); female 17-20 cm; wingspan 23 cm. An unmistakable endemic bird exhibiting striking sexual dimorphism: the male all-black with long tail-streamers, the female chestnut-brown above, white below with a black head. **Male** has all-black plumage with a deep purple sheen and very long black tail-streamers. Pale blue eye-ring and strip of facial skin extending to pale blue bill are visible at close range. Legs blue-slate. **Female** has chestnut-brown upperparts, wing, rump and tail, creamy white underparts and an all-black head. Also has a pale blue eye-ring and bill, but lacks the bare facial skin of the male. **Immature** resembles adult female but is somewhat duller and browner. Bill blackish on upper mandible, pale lower mandible and legs blackish.

VOICE Male has a piping whistle often repeated, likened to a man calling a dog 'pli-pli-pli-pli'[1]. Female is usually silent, occasionally uttering a quiet twitter. Alarm call a harsh 'zweet'.

BEHAVIOUR Frequents mature woodland where territories are usually within Indian almond and takamaka. Territories are often adjacent to marshes, pools and streams in flat coastal areas. Birds flit between trees in an undulating flight and feed mainly in a 4-8-m height band by picking insects from leaves or flycatching in open areas between trees. One study found a strong vertical separation between the sexes in feeding, the male hunting in the lower layers, female higher up[2]. However, another could find no such separation, though they pointed out the difference in their results could be attributed to the former observations being close to nests where females do almost all incubation and are liable to stay close to nest height[3]. The sub-division of feeding height might reflect the sexual dimorphism of the species: the black male being camouflaged at lower levels and the counter-shaded females more successful higher up[2]. Most feeding is in Indian almond and takamaka, though also feeds elsewhere such as in vanilla and mangroves, indicating some degree of flexibility.

BREEDING Breeding may take place at any time of year but is concentrated in the wetter months of November to April when food is more plentiful, very rarely if ever breeding in the driest months of July to August. In courtship, male flutters up and down calling in front of female, displaying long tail feathers. Nest is built by both sexes, usually at the end of a branch, in Indian almond or takamaka trees, occasionally other mature trees such as mango and breadfruit. This location may reduce predation by skinks with the added bonus of protection from introduced rats and cats, but it also leaves the nest exposed to high winds or human destruction. Nest height is 2-10 m, the majority being in the range 3-5 m. There is a wide range, 12-48 days, between date of commencement of nest building and egg laying[2]. Clutch size is always one, the egg whitish with brownish red spots. Both sexes share incubation but the majority is done by the female. Incubation lasts 12-15 days, fledging a further 21 days.

ORIGINS Origin not known with certainty. Some writers link to Madagascar Paradise Flycatcher *T. mutata*[1,4]. There may be a link with African Paradise Flycatcher *T. viridis*[5], perhaps via Madagascar[6].

STATUS Endemic to Seychelles and confined to La Digue and Marianne, with occasional reports from Praslin. IUCN Critically Endangered. First described from specimens collected on Praslin in 1865, when it was also present on La Digue, Marianne and Félicité[7]. Last recorded on Marianne in 1936[8] until a single male was reported in January 1997[9] and three birds seen during a survey of one-fifth of the island in 1998[10]. On Félicité survived into the 1930s[11]. There have been occasional sightings on Félicité since then, but no evidence of breeding. It is probable that La Digue has been the last viable population of the species since around 1940. Contraction of range has been brought about largely by destruction of the original coastal lowland forest with which the species is closely associated. In 1969, it was thought the total population was not more than 50[12]. In 1977/78 the population was estimated to be 70-80 birds[2]. In 1988, it was estimated that the total population could be 90-100[3] and in 1995 at 150-200[13]. The apparent increase, despite a decline in habitat, might be due to changes in the water table and an increase in the extent of the marsh[14]. Another factor might be the cessation of wilful destruction of nests and killing of birds for 'sport' since an increase in awareness of their intrinsic value of birds and the recognition by residents of La Digue of the tourism value of the bird.

THREATS AND CONSERVATION The greatest threat is the loss of suitable habitat. The species is dependent on the conservation of mature Indian almond/takamaka coastal woodland and freshwater marshes that provide almost all feeding and nest sites. Woodland on the plateau decreased by 24% from 1978 to 1992[15]. Decline has continued at an alarming pace due to housing and tourism developments. Further loss of habitat around the turn of the 21st Century has come from a disease affecting takamaka trees. The establishment of a reserve was recommended in 1981[2]. In 1979 Christopher Cadbury rented a site at La Passe, which was managed jointly by RSNC and the Seychelles Government. The site was later acquired by the Government to be run as a reserve and achieved legal recognition in 1991 for an area of almost 8 ha. The Government subsequently purchased additional plots raising the total protected area to 18 ha by 2000. This protected area is only sufficient for a maximum of five pairs, too few to guarantee the survival of the species. A woodland conservation policy for La Digue has been recommended by several studies[3,13,16] including an extension to the protected area, limitation of housing development on the plateau and restoration and protection of suitable habitat. An additional threat comes from human disturbance. Nests situated more than 5 m from a path or road may be more than twice as successful as nests less than 5 m from them[2]. Another threat may come from the invasion of marshes by introduced water lettuce, first noted on La Digue in 1992; this rapidly covers open water, restricting breeding areas for the insect prey of flycatchers[17]. It has been proposed marshes should be skimmed to remove water lettuce[17]. The establishment of a second island population could be an important step to securing the survival of the species. Curieuse, which could hold 10-15 pairs,[13] has been suggested as one possible site. Aride, which once held flycatchers, has also been suggested and a wetland restoration project begun in 1996 stated the reintroduction of flycatchers was a long-term aspiration. Silhouette, though it has never been known to hold flycatchers, could support a substantial population.

SIMILAR SPECIES None.

References: 1. Penny (1974), 2. Watson (1981b), 3. Bullock *et al.* (1988), 4. Gaymer *et al.* (1969), 5. Hall and Moreau (1970), 6. Benson (1971), 7. Newton (1867), 8. Vesey-Fitzgerald (1936), 9. Ladouceur (1997), 10. Parr (1998), 11. Vesey-Fitzgerald (1940), 12. Fayon (1971), 13. Rocamora *et al.* (1995), 14. Rocamora (1997c), 15. Neufeld (1998), 16. Macdonald (1993), 17. Gerlach (1996b).

SOUIMANGA SUNBIRD *Nectarinia sovimanga*　　　**Plate 50**

French: Souimanga malgache　**Creole:** Kolibri Aldabra

DESCRIPTION Length 10 cm; wingspan 13 cm. The only sunbird and the most common landbird of Aldabra. A noisy, active, small, dumpy bird with a curved bill. **Male** has head and neck dark metallic green with a well-defined maroon chest-band and dark brown upperparts/wings and yellow pectoral-tufts. Upper abdomen is black and the lower abdomen dull white or yellowish white. Curved, black bill and black legs. **Male non-breeding** lacks the metallic sheen to the upperparts, being dull olive in tone; the throat, chin, chest-band and upper abdomen are less metallic, the lower abdomen olive-yellow. **Female** is much duller, dark grey-brown above, pale grey below. **Juvenile** similar to adult female, sometimes with black chin and throat and stronger olive wash to the upperparts.

VOICE Male gives a very fast, scratchy song, phrases frequently repeated. Calls 'chip' in flight with each flurry of wingbeats and also on take off. Alarm call a hoarse 'twir'.

BEHAVIOUR Found in almost all habitats and even on some small lagoon islets of Aldabra. Territorial but small groups gather at favoured feeding sites. Flight fast and undulating. Highly active, always on the move, flitting between feeding stations and trees or shrubs searching for insects, spiders and nectar.

BREEDING Breeds mainly August to March, with eggs laid throughout this period. Nest is usually situated in an open site at about 1-2 m. Sometimes it is below ground level in a sinkhole at a depth of up to 2 m; sink holes may offer some protection from predators and strong winds[1]. Nest is domed with a side entrance sheltered by an overhanging porch and material hanging untidily below the structure. It is usually suspended from a thin branch or other similar structure, even including wires and clotheslines in the vicinity of houses. Nest material includes grass stems, with some plant tendrils, coconut fibre, leaves, casuarina needles, twigs and a softer lining. The exterior is sometimes decorated with *Cymodocea ciliata*, a marine angiosperm blown onto land[2]. Nest building may take anything from 8-36 days. Nest construction and incubation is done entirely by the female. Clutch size is usually two, occasionally one. Eggs are dull white, mottled dark red. Male may feed the young, but most work is done by the female. Incubation lasts about 13-14 days, fledging a further 16-18 days.

ORIGINS Undoubtedly from Madagascar. Nominate *sovimanga* breeds in Madagascar and Iles Glorieuses. Another race, *apolis* breeds in southwestern Madagascar. The earlier origin of *sovimanga* has been attributed to Olive-backed Sunbird *Nectarinia jugularis* of South-East Asia[3]. Wing measurements show that compared to the nominate race of Madagascar, *aldabrensis* is smaller.

STATUS Endemic race *aldabrensis* is confined to Aldabra, where it is very common.

THREATS AND CONSERVATION Protection of Aldabra as a nature reserve and World Heritage Site should secure the future of this species. Does not appear to have been much affected by introduced predators. Location of nest gives some protection from cats, rats and also indigenous lizards.

SIMILAR SPECIES None on Aldabra.

References: 1. Stoddart and Wright (1967), 2. Benson and Penny (1971), 3. Delacour (1944).

ABBOTT'S SUNBIRD *Nectarinia (sovimanga) abbotti*   Plate 50

French: Souimanga d'Abbott   **Creole:** Kolibri Abbott

DESCRIPTION Length 11 cm**; wingspan**: 14 cm. Similar to Sovimanga Sunbird, but larger and with much darker underparts in both sexes, especially the male. The only sunbird of Assumption (nominate race *abbotti*) and Astove/Cosmoledo (race *buchenorum*). The most common landbird on these islands. **Male** has head and neck metallic green (absent in non-breeding), with well-defined maroon chest-band (broader than that of Souimanga Sunbird, particularly in race *buchenorum*), the rest of the underparts entirely dark brown (race *abbotti)* or blackish (race *buchenorum*). Blackish brown upperparts and wings with yellow pectoral-tufts visible when wings are lifted. Black curved bill and black legs. **Male non-breeding** lacks metallic sheen. **Female** is mainly dark olive brown above, tail black tipped greyish; scaly black throat pattern and only slightly paler dark olive grey on the underparts (contrast more obvious in race *abbotti*, while *buchenorum* is the darkest female sunbird of the Aldabra Group with upper breast concolorous with upperparts and only slightly paler below). **Juvenile** similar to adult female but uniform sooty olive grey on throat and upper breast and upperparts grey with olive tinge.
VOICE Male gives a complicated, variable musical trill, phrases frequently repeated. Calls 'chip' in flight with each flurry of wingbeats and also on take off. Alarm call a hoarse 'twir'.
BEHAVIOUR Found in almost all habitats throughout Assumption, Cosmoledo and Astove. Territorial but small groups gather at favoured feeding sites. Flight fast and undulating. Highly active, always on the move flitting between feeding stations and trees or shrubs searching for insects, spiders and nectar.
BREEDING Breeds mainly August to March. Nest is domed with a side entrance sheltered by an over-hanging porch, usually situated in an open site at about 1-2 m. It is usually suspended from a thin branch or other similar structure, material including grass stems, with some plant tendrils, coconut fibre, leaves, casuarina needles, twigs and a softer lining. Clutch size usually two, occasionally one. Eggs are dull white mottled dark red. Nest construction and incubation is done entirely by the female, the male sometimes helping to feed young. Incubation *c*13-14 days and fledging *c*16-18 days.
ORIGINS Undoubtedly similar in origins to the Aldabra race of Souimanga Sunbird, but it is not known whether all islands of the Aldabra group were invaded simultaneously from Madagascar, or perhaps Aldabra Atoll (where birds most closely resemble their relatives in Madagascar) was invaded at a later date. Wing measurement of race *abbotti* is greater than *N. s. aldabrensis*, but equal to *N. s. sovimanga*; wing measurement of race *buchenorum* is greater than that of both these races of Souimanga Sunbird[1]. Sinclair and Langrand first proposed separation of birds of Assumption, Cosmoledo and Astove as a separate species, a treatment we follow for consistency but with some reservation[2]. An alternative interpretation of relationships within the Aldabra group is that birds of Aldabra/Assumption form one species and those of Cosmoledo/Astove another. This makes a certain amount of sense in that these pairs of islands lie about 140km apart whereas only *c*20 km separates the islands within each pair with Assumption south-east of Aldabra and Astove southeast of Cosmoledo (the direction of prevailing winds)[3]. Certainly, the birds of Cosmoledo/Astove are most strikingly different in plumage and measurements from all other races of Souimanga Sunbird than can be said for birds of Aldabra and Assumption. If Cosmoledo/Astove birds are a separate species to Aldabra/Assumption birds they cannot be called Abbott's Sunbird as the race concerned is *buchenorum* and some other name such as Atoll Sunbird *N. buchenorum* might be appropriate.
STATUS Endemic to Seychelles, with two distinct isolated races: nominate *abbotti* on Assumption and *buchenorum* on Astove and Cosmoledo.
THREATS AND CONSERVATION Despite the almost total destruction of Assumption during the period of human exploitation, particularly when the island was stripped of vegetation to permit the mining of guano, the species remains very common, as it does on Cosmoledo and Astove where human exploitation was only marginally less ruthless. Nevertheless, the possibility that this may be one of the rarest sunbirds in the world merits a genetic study and consideration of measures to protect the whole of the Aldabra group. Further genetic work is required to confirm full species status.
SIMILAR SPECIES None on Assumption, Astove or Cosmoledo.
References: 1. Williams (1953), 2. Sinclair and Langrand (1998), 3. Skerrett (1999c).

SEYCHELLES SUNBIRD *Nectarinia dussumieri*   Plate 50

French: Souimanga des Seychelles   **Creole:** Kolibri Sesel

DESCRIPTION Length 11-12 cm; wingspan 15-16 cm. The only sunbird and most common endemic of the granitic islands; a small active, noisy, mainly grey bird with a curved bill. **Male breeding** is mostly dull slate grey with bright 'flame-coloured' orange or yellow pectoral-tufts and a dark blue iridescence on the head and throat; female is duller. Bill and legs black. **Male non-breeding** and **female** lack iridescence. Male larger than female. **Immature** similar to adult female.
VOICE Male song, given throughout the day though mainly in the morning, is a series of enthusiastic, harsh, high-pitched notes interspersed with occasional rasping calls and a quieter sub-song[1]. Female may also sing.
BEHAVIOUR Frequents gardens, clearings and open woodland. Highly active, constantly moving about

vegetation and flitting between food plants with undulating flight. Males may display facing each other while singing loudly, heads held back. Territorial and not generally gregarious, though small groups may feed together when a nectar source is particularly abundant. Feeds on nectar and insects, often chasing other sunbirds that may approach too closely. Nectar is taken from flowers with short corolla tubes by probing, while longer corolla tubes may be pierced. Usually perches to feed but sometimes hovers if a suitable perch is not available. Insects may comprise up to one-third of diet and be particularly important in woodland or at times when few suitable flowering plants are available. Ants, cockroaches, spiders and other prey are taken by probing vegetation, while aerial insects are sometimes taken by flycatching and occasionally by snatching from spider's webs.

BREEDING Breeding may take place at any time of year but is concentrated at the end of southeast and beginning of northwest monsoons around September to October. Nest site less conspicuous on Mahé and Silhouette than other islands, possibly due to presence of Seychelles Kestrel on these islands[1]. Pear-shaped nest about 16 cm deep with a diameter at its widest point of about 8.5 cm, is built suspended from a branch in a shrub or tree at anything from 1.5 to 19 m above the ground. Entrance, halfway up the structure, is protected by an overhanging porch. Older nests soon lose their shape, developing a trail of material hanging from the base up to 30 cm in length. Both sexes choose the nest site but all or near-ly all nest building is done by female, male rarely approaching the site at this time. Construction can take as little as three days, though where material robbers disturb a nest, female may continue to add to the structure for several weeks. Dead grass, leaves and stems form the principal building materials, with moss, palm fibre, broad leaves and spiders' webs also used; the chamber is lined with feathers or kapok. Some material may be gathered from old nests or robbed from active nests by hanging from the base and plucking material. This, together with robbing of material by Madagascar Fodies, may be the cause of the long tails hanging from the base of nests[1]. Nest is vigorously defended by both sexes, though more so by female at nest building stage and by male during incubation and fledging. Clutch size is invariably one egg, dull white with brown spots. Incubation takes 11-16 days, carried out solely by female, which sits with bill protruding from the nest hole in stints averaging 10 minutes. Most or all feeding of nestlings done by female. One study found the male made less than 4% of visits to the chick[2]. Young fledge in a further 20 days. Recently fledged young may be fed by both sexes, though female carries out more of this work too. Fledged young beg vigorously with wings quivering, but may be chased off by the parents with-in a few days of fledging. On Aride, during the early stages of colonisation (see STATUS), polygny was noted in this species for the first time, in which one male paired with two females in different home ter-ritories. Polygynous males defend both territories vigorously, delivering more food items (16% of all food, whereas monogamous Aride males not seen to contribute any food) and raising more chicks than monog-amous males[3].

ORIGINS Several derivations have been suggested, including Souimanga Sunbird, with characteristics similar to other Seychelles endemics such as increase in size (as in Seychelles Magpie-robin and Seychelles Bulbul) and decrease in sexual dimorphism (as in Seychelles Fody and Seychelles Magpie-robin). It is also possible that Seychelles Sunbird is related to Olive-backed Sunbird *N. jugularis* of South-East Asia to Australasia by direct colonisation from Asia[4]. Perhaps most convincing is the suggestion of a link with Humblot's Sunbird *N. humbloti* of Comoros[5]. Seychelles Sunbird and Humblot's Sunbird show several similarities of plumage including white tips to the tail (unlike Souimanga Sunbird) and orangey pectoral-tufts, similar male display and similar song[5].

STATUS Endemic to granitic islands, resident on all larger islands. Has re-invaded Aride by natural means, where there were sightings of individuals throughout the 1980s, including a ringed female from Cousin (20km distant) in 1987. However, there was no evidence of breeding until 1992. A census in 1993 put the population at 10-12 birds[2], reaching 70 by 1999[6]. At one time, Mahé birds were described as a different species, *Cyanomitra mahei*, from those found in the rest of the islands, *Cyanomitra dussumieri*, having lemon-yellow, not orange pectoral-tufts in the adult male. However, while the colour of pectoral-tufts varies considerably between individuals, it does not vary between islands; nor is there any significant size difference between birds of different islands[7].

THREATS AND CONSERVATION Possibly the only Seychelles endemic bird to have actually benefited from the arrival of man, feeding as it does on the nectar of introduced flowering trees and shrubs in gar-dens, hotel grounds etc. Also, it is too small to have ever been a food item and the nest site is usually inac-cessible to introduced mammalian predators.

SIMILAR SPECIES None within breeding range.

References: 1. Greig-Smith (1980), 2. Anderson (1994), 3. Lucking (1996a), 4. Benson (1967), 5. Louette (1992), 6. Bowler and Hunter (2000), 7. Hall (1953).

MADAGASCAR WHITE-EYE *Zosterops maderaspatana* Plate 50

French: Zostérops malgache **Creole:** Zwazo Linet Malgas

DESCRIPTION Length 10 cm; wingspan 14 cm. A small yellowish bird with a white eye-ring found only in the Aldabra group (but absent from Assumption) where it is the only species of white-eye. **Adult** has yellow forehead, chin and throat. White eye-ring is broken in front of reddish brown eye with a blackish lore. Upperparts and wings dark olive green, the tail sooty black. Underparts pale grey except for russet wash on flanks, yellow vent and yellow undertail-coverts. Bill black on upper mandible, lower mandible grey. Legs bluish grey. **Juvenile** similar to adult, eye greyish brown, the eye-ring tinged yellowish.

VOICE Constantly makes a soft disyllabic 'pee-u' call while moving throughout the vegetation in small groups. Lone birds and groups of just a few birds are usually silent. Sometimes gives a brief song or a harsher trill. Call of race *menaiensis* differs from *aldabrensis*, being much thinner (M Betts pers. comm.).

BEHAVIOUR Occurs wherever there are trees with apparently no particular preference, though less common in mangroves. Gregarious, usually in small feeding parties of up to ten birds, occasionally in larger groups up to about 30 birds. Feeding groups flit between trees with slow, fluttering flight. Sometimes engages in allopreening. Often hangs upside down to feed. Diet mainly insects, but may also take some seeds, fruit, flowers and possibly nectar.

BREEDING Breeds mid-September to March. Deep cup-shaped nest is constructed from casuarina needles, grassy stems and bark fibre with some softer material on the outside in dense vegetation, usually in the fork of a tree, at a height of about 2-4 m. Clutch is usually two pale green eggs (but up to four eggs reported).

ORIGINS Undoubtedly from Madagascar though the earlier origin of the species is not certain. Plumage is close to Oriental White-eye *Z. palpebrosa* of India, South-East Asia to Indonesia, rather than grey-bellied African races[1]. Other races occur Comoros (*comorensis* on Moheli and *anjouanensis* on Anjouan) and Europa (*voeltzkowi*). White-eyes have never been recorded from Assumption, but possibly once occurred and were wiped out during the wholesale destruction of the island before their presence had been recorded. Race *aldabrensis* of Aldabra is more yellow above than *menaiensis* of Cosmoledo but less yellow than birds of Iles Glorieuses and drier areas of Madagascar. Yellower birds may have reduced melanin linked to drier climates[2].

STATUS Endemic to Aldabra group. Race *aldabrensis* is resident throughout Aldabra atoll, *maderaspatana* on Astove and *menaiensis* on Menai, Cosmoledo. On Aldabra it is probably the most abundant landbird after the sunbird though on Astove and Cosmoledo it is not so abundant. Birds of Aldabra are more yellow, less green above than Astove and Cosmoledo birds and are also slightly smaller. The nominate race also occurs on Madagascar; it has black lore, is darker green above, darker grey sides to breast and slightly longer bill: 10-11 mm compared to 8-9 mm in *aldabrensis*. The validity of race *menaiensis* on Cosmoledo requires further investigation; it is a little paler than the nominate race but otherwise similar[3]. However, given the isolation of this tiny population, its different call and the propensity of white-eyes to subspeciate it may even be a valid species.

THREATS AND CONSERVATION The protection of Aldabra Atoll as a nature reserve and World Heritage Site should secure the future of this species. Introduced rats and cats may pose a threat, though the secretive location of the nest site may offer some protection from predators. Protection of Cosmoledo would be highly desirable to secure the future of the species (and possibly a separate race) on this atoll. Further genetic work is required to confirm the status of birds of the different islands of the Aldabra group.

SIMILAR SPECIES None.

References: 1. Benson (1984), 2. Benson and Penny (1971), 3. Benson (1970a).

SEYCHELLES CHESTNUT-FLANKED WHITE-EYE *Zosterops semiflava* Plate 1

French: Zostérops jaune des Seychelles **Creole:** Zwazo Linet Zonn

DESCRIPTION Length 11 cm; wingspan 15 cm. A small passerine, canary-yellow below with chestnut sides, now extinct. **Adult** had lemon-yellow underparts and chestnut flanks, dark grey-brown primaries. Brownish lore, crown to upperparts greenish yellow merging to yellowish on rump with a dark brown tail. Fairly long, blackish bill and pale grey-brown legs. **Juvenile** similar to adult.

VOICE Unknown.

BEHAVIOUR Largely unknown, though probably dependent on native forest, cleared by early settlers of Seychelles. Apparently gregarious with early reports of small flocks. Diet likely to be mainly insects, but also probably some seeds, fruit, flowers and possibly nectar.

BREEDING Unknown.

ORIGINS Formerly treated as a race of Mayotte White-eye *Z. mayottensis* of Mayotte, but this view has been challenged by Louette who placed the latter as a race of Madagascar White-eye *Zosterops maderaspatana*[1]. It would indeed be surprising if taxa on Mayotte and Seychelles were so close as to merit consideration as one biological species. Colour resemblance between *Z. semiflava* and *Z. m. mayottensis* may possibly be coincidental given similarities of white-eyes throughout their range and Seychelles Chestnut-flanked White-eye may possibly be of Asian origin (Louette pers. comm.).

STATUS Endemic species now extinct, probably due to habitat destruction when the original vegetation was replaced by a monoculture of coconuts. Only ever recorded with certainty from Marianne, although there are also unsubstantiated reports from Mahé, South-east Island, Praslin, La Digue and Silhouette. It may well have been this species noted in 1768 by the Marion Dufresne expedition on South-east Island and recorded as 'several little yellow birds like canaries'. First described by Newton who collected specimens from Marianne in 1865 noting hearsay reports from La Digue, Praslin and Silhouette (though the owner of Silhouette later told him it did not occur there) and he was 'fairly sure' he saw one on Mahé[2]. Abbott also took specimens in 1892[3]. The exact date of extinction is unknown, but was probably very early in the 20th Century.

References: 1. Louette (1985), 2. Newton (1867), 3. Ridgway (1895).

SEYCHELLES WHITE-EYE *Zosterops modestus*   **Plate 50**

Other names: Seychelles Grey (or Brown) White-eye
French: Zostérops des Seychelles   **Creole:** Zwazo Linet Sesel

DESCRIPTION Length 10-11 cm; wingspan 15 cm. A small greyish brown bird, paler on the underparts with a white eye-ring, obvious at close range. Confined to Mahé and Conception. **Adult** has head upperparts and tail dark olive grey to brown, paler on the rump that has a yellowish tinge, visible at close range. Underparts pale grey, dirty pale mustard yellow on the throat, the belly having a faint yellowish tinge and brownish on flanks. Broad white eye-ring has a tiny gap in front of the eye. Blackish grey bill and grey legs. Mahé birds tend to be on average larger than birds of Conception, with longer wings and tarsus[1]. **Juvenile** similar to adult; can only be distinguished in the first few weeks after fledging when downy feathers around the bill and on the belly remain visible and the bill is yellow at first, gradually changing to grey with age.

VOICE A low, soft, nasal contact call 'cheer-cheer' a little like a new-born chicken, becoming louder and more persistent when moving between trees and shrubs. There are two other calls given in alarm: a loud 'chewick' and a chattering trill[2,3]. Often most easily located by the loud contact-calls when foraging parties take off, given at all times of day but mainly early morning and late afternoon. Song is varied and pleasant most often heard at dawn, sometimes by day, during the breeding season from October to April[1,4].

BEHAVIOUR Frequents open woodland, gardens and areas with mature trees, mainly above 300 m on Mahé but at all levels on Conception. Usually seen in family groups of 2-5 birds on Mahé and in larger groups of up to 8 on Conception[1,4]. Outside the breeding season flocks of up to 10-12 birds have been reported[2,5]. When feeding, birds are constantly on the move, searching trees and shrubs for insects and rapidly moving between branches with leaps and weak, undulating flight. Sometimes may raise crown feathers and often flicks wings and tail. Does not appear to have a strong preference for native trees, often feeding in introduced fruit trees, clove, kalis, albizia, etc. However, several plants form an important part of their fruit diet including bwa dir, bwa siro and bwa kuiyer on Conception[1]. Also partial to berries of cinnamon and lantana, two common introduced plants[4].

BREEDING In courtship, female may beg and be fed by male. Eye-ring pecking by male, allopreening and huddling may also form a part of courtship, or have general social functions[2,3]. Breeding may take place at any time from beginning October to end April[1,4]. Cup-shaped nest is built mainly of grass, moss and spiders webs in dense vegetation. Usually two eggs are laid, pale greenish blue sometimes with very faint brown spots. Breeds co-operatively, similar to Seychelles Warbler, helpers contribute to nest-building and rearing duties. Despite the highly social nature of white-eyes, this has only been reported elsewhere in Réunion in Mascarene Grey White-eye *Z. borbonicus*. On Conception, Dr Rocamora and his team revealed a complex social breeding system with large nesting families of up to eight birds laying up to seven eggs in the same nest! Incubation takes 13-15 days and fledging 11-15 days[1].

ORIGINS Not known with certainty but may be related to Mauritius Olive White-eye *Z. chloronothos* and Réunion White-eye *Z. olivaceus*, which it resembles and which have similar wing-flicking action and staccato 'chipping' calls[2]. Its earlier origin is unknown.

STATUS Endemic to Seychelles and only ever found with certainty on Mahé and Conception. IUCN Critically Endangered. The only indication of its occurrence elsewhere is a specimen, which has now vanished and was claimed to have come from Marianne in 1879, and a report by Henri Dauban that white-eyes (of unknown identity) were abundant on Silhouette until the 1930s. Newton found it 'tolerably plentiful' in groves of clove trees[6], and in 1906, Nicoll found it 'fairly abundant' at Cascade[7]. It subsequently became extremely rare and in a survey of landbirds in 1936, it was reported 'rarely seen but not infrequent in mountainous parts of Mahé' by Vesey-Fitzgerald[8]. In 1960, Crook reported that it had recently become extinct[9]. However, it was 'rediscovered' around this same year by Loustau-Lalanne[10]. In 1975, it was thought to be abundant in the Mission to Casse Dents area[2], but during the 1980s apparently disappeared. Birds were also commonly seen at Fairview Estate during the 1980s, but later sightings became more infrequent and by 1994 it had apparently vanished. White-eyes still occur at La Misere, Upper Barbarons and Cascade. Work carried out in 1995/96 by the Division of Environment, BirdLife International, the Ministry of Tourism and Glasgow University suggested the world population, restricted to Mahé, numbered only c50 birds[4,5]. Then came the sensational discovery of a previously unknown population on

Conception, which had been rarely visited following the abandonment of its plantation around 1970, being a fairly inhospitable island to man, with steep slopes and no beaches or streams. Following public appeals for white-eye sightings through the media, L Chong-Seng made a report that a soldier, Harry Hoareau, had seen birds on Conception. A visit by C Awmarck, K Beaver L Chong-Seng, and R Youpa in February 1997 confirmed the presence of white-eyes on the island. A rough census by G Rocamora evaluated the population at a minimum of 30-50 territories and, later, 200 birds[4,11]. In 1999, accurate census work estimated a population of 244-336 birds[1]. Remarkably, it is the most common bird on Conception. In 1999, 34-40 birds were censused on Mahé where the population appears to have been stable since 1996 (Rocamora and François pers. comm.).

THREATS AND CONSERVATION Until the discovery of white-eyes on Conception, this species was thought to be the most endangered of the granitic island endemics[12,13]. Indeed, Collar and Stuart considered 'nothing positive can apparently be done for it and, on present evidence, it will possibly very shortly become extinct'[14]. Research conducted under the Seychelles White-eye Recovery Programme since 1998 has given a better understanding of the reasons for the species' decline. Comparison with Conception where birds survive in much higher densities suggests that nest predation is the main limiting factor on Mahé. On Conception there are no black rats (though the less arboreal Norway rat occurs), few Common Mynas and no Seychelles Bulbuls which are also nest predators (though Seychelles Kestrels, natural predators, are present)[1]. The relative abundance of indigenous fruiting trees may also help to explain differences in density between the two islands, as degraded habitat on Mahé may limit food supply[1,15]. As Conception is a dry island, fire may be a threat to the population, as may disease[4]. The protection of the island under national legislation is highly desirable. Other conservation options include translocation to predator-free islands or captive breeding. Small islands around Mahé such as Thérèse and St Anne have been found to be suitable if predators can be eradicated. Given significant genetic and morphological differences between Mahé birds and those of Conception it might be desirable to transfer birds from each population to separate islands and, possibly later, to try to mix and strengthen the stock transfer birds from both populations to a third island (G Rocamora and D Richardson pers. comm.).

SIMILAR SPECIES When seen well, cannot be mistaken for any other species, but at a distance could be confused with **Madagascar Fody** or **Seychelles Sunbird**. Optimists searching White-eye sites sometimes claim sightings of these last two species as Seychelles white-eye. The softer call of the Seychelles White-eye quickly identifies even distant birds not seen well.

References: 1. Rocamora and François (2000). 2. Feare (1975a), 3. Greig Smith 1979b), 4. Rocamora (1997b), 5. Mellanby *et al.* (1996), 6. Newton (1867), 7. Nicoll (1906), 8. Vesy-Fitzgerald (1940), 9. Crook (1960), 10. Loustau-Lalanne (1962), 11. Rocamora (1997a), 12. Gill (1970), 13. Gill (1971), 14. Collar and Stuart (1985), 15. Gerlach (1996b).

ALDABRA DRONGO *Dicrurus aldabranus* Plate 49

French: Drongo d'Aldabra   **Creole:** Moulanba

DESCRIPTION Length 24-26 cm; wingspan 33-35 cm. Confined to Aldabra; all-black with a long forked tail and a heavy black bill. **Adult** is entirely black, with a greenish blue sheen, duller on the flight feathers and tail feathers. Long nasal plumes are visible at close range. Tail long and forked, black in fresh plumage, browner when worn. Eye deep red; bill heavy, hooked and black and legs black. **Juvenile** similar in structure to adult, but greyish brown above and very pale brown, almost white below, streaked darker; shorter wings, shallower fork to tail and eye brown.

VOICE Call is a harsh chuckle. When chasing other birds utters a nasal 'titi-po fa fa'[1] often repeated; after repelling an intruder may land on a branch and call 'fa-wip' a few times.

BEHAVIOUR Arboreal, with a marked preference for casuarina woodland, mangroves or areas near to mangroves and dense scrub. Uncommon in areas of open scrub such as eastern Aldabra. Usually solitary or in pairs, occasionally groups of up to six birds. Flight heavily undulating. Diet includes insects and lizards caught by a 'perch and pounce' technique or by hawking for aerial insects. Often follows Madagascar Coucal, using speed and agility to steal prey. Such kleptoparasitism may enable drongos to occupy denser areas of scrub than suitable for 'perch and pounce' or aerial hunting methods alone[2]. It is fierce in defence of the nest site, sometimes even outside the breeding season, mobbing intruders including other drongos, Madagascar Kestrel and Pied Crow.

BREEDING In courtship, established pairs face each other, bills held a little above the horizontal, quivering wings and wagging tails while uttering high-pitched squeaky twitters[3]. Breeding season lasts approximately from mid-September to early January, much shorter than for most Aldabra birds, with most egg laying before the end of November. Neat cup-shaped nest is constructed from roots and casuarina leaves, bound externally with spiders' webs and lined with grass. Nest often in casuarina or in mangrove or other tree near to water, at height of 2-8 m. Usual clutch 2-3 eggs, pale creamy-pink with deep red spots at thicker end. Not measured until 2000, incubation lasts *c*18 days, fledging *c*14 days[4]. It is unusual for more than one chick to survive to fledging, possibly due to predation of eggs and young by Pied Crows, rats or other predators. Young remain with parents for many months after fledging, feeding mainly on the ground at first before mastering aerial agility of adults taking flying insects. Birds do not breed until at least two years of age, but may remain in the adult territory into the breeding season following their birth.

ORIGINS *Dicrurus* is both an African and an Asiatic genus. Aldabra Drongo's nearest relative may be Crested Drongo *D. forficatus* of Madagascar or Fork-tailed Drongo *D. adsimilis* of sub-Saharan Africa. Crested Drongo and Fork-tailed Drongo are much more similar in appearance to one another than either is to Aldabra Drongo, but on zoogeographical grounds Madagascar is the most likely origin of Aldabra birds[5]. The immature dress of the Aldabra Drongo is particularly distinctive being much paler than immature Crested Drongo which is generally black with some white feather fringes.

STATUS Endemic to Aldabra (c500 pairs[6]). IUCN Near-threatened. There is no evidence of it ever having occurred elsewhere.

THREATS AND CONSERVATION The protection of Aldabra as both a nature reserve and World Heritage Site should help secure the future of this species. The preference of drongos for dense scrub and mangroves may minimise the threat from the rats, present on Aldabra. However, predation may be high, potential predators including black rat, feral cat and Pied Crow[4,7].

SIMILAR SPECIES None.

References: 1. Penny (1974), 2. Prys-Jones and Diamond (1984), 3. Frith (1977), 4. Rocamora and Jules (2000), 5. Benson (1984), 6. Rocamora and Skerrett (1999), 7. Threadgold and Johnson (1999).

HOUSE CROW *Corvus splendens*   **Plate 51**

Other names: Indian House Crow, Indian Crow, Ceylon Crow, Colombo Crow
French: Corbeau Indien   **Creole:** Korbo Endyen

DESCRIPTION Length 41-43 cm; wingspan 76-85 cm. A fairly large crow, all-black except for greyish white patch from nape to mantle. **Adult** is black above and below, sometimes with deep purple gloss. Greyish white area from nape to upper back, often extending to the upper breast, but sometimes restricted to the nape. Nostrils are covered in long, stiff, black bristles. Bill powerful and black; long tibia, legs black. Sexes similar. **Juvenile** similar to adult but lacking gloss and more brownish black above. Greyish patch on neck is darker and does not extend to the upper breast.

VOICE Call is a deep but soft nasal 'kwah-kwah'. Highly vocal throughout the day.

BEHAVIOUR Commensal with man, favouring rubbish dumps where it scavenges for food. Nevertheless it is an alert, wary bird, not easily approached and liable to fly off a long distance if disturbed. Gregarious and social, roosting communally. Flight is slow and direct with shallow wingbeats. Walks on the ground or moves with a sidling hop, jerking wings. Omnivorous, feeding on seeds, fruits, buds, nectar, insects, lizards, rodents, crabs, small birds, eggs, nestlings, fish and all manner of household waste and scraps. In Seychelles, seen to destroy the nest of Seychelles Sunbird, illustrating the danger to Seychelles endemic avifauna.

BREEDING In courtship, commonly carries twigs (especially the female). Both sexes collect nest material, but most nest building is done by the female, the male presenting twigs for incorporation into an untidy platform. Nest has a softer inner lining of grass, coconut fibre or other similar material. Nest site is fairly high in tree-forks or (less often) on building ledges. Clutch is 3-6 pale bluish green eggs streaked and speckled brown. Incubation, performed solely by the female, lasts 16-17 days. Fledging 22-25 days. Hatching is asynchronous, so later-hatched chicks often starve. Breeding pairs often aggressive, sometimes striking human trespassers.

RANGE Southern Iran, Indian subcontinent, Sri Lanka, Laccadives, Maldives, southern China, Burma. Several races recognised. Expanding along coasts of Red Sea, East Africa south to Natal, Zanzibar, Oman, Australia, Malaysia, and Singapore. The dramatic spread of this species has been partly by ship, Indian sailors often encouraging birds to remain on board by feeding them, birds abandoning the vessel on reaching a foreign port. Some introductions have also been by deliberate transfer.

STATUS Introduced; five birds arrived at Mahé on an Indian cargo vessel in 1977[1]. Two were shot but a surviving pair bred successfully. A further nine birds were shot up to 1986, but despite this the population reached a peak of about 25 around this time, centred around Anse Etoile, Mahé, where birds nested in trees adjacent to a rubbish dump. Following the closure of this dump and the burial of exposed waste, birds dispersed to elsewhere in Mahé being most frequently reported from Machabée, Victoria and environs and from a dump at Roche Caiman (also now closed). Away from Mahé a single bird was sighted on Bird raiding Sooty Tern nests in 1985. On Aride a single bird was seen in 1992; nesting terns repelled it. There have also been isolated reports from Praslin, Moyenne, Cousin and Ile aux Vaches Marines. One pair was seen nest building on Silhouette in 1981, but only on Mahé has it bred with certainty. Continuous efforts to eliminate the species have prevented a population explosion. It is believed the species was successfully eradicated in 1994 when the last two birds present in Victoria were shot. However, re-invasion of Seychelles took place in 1998 with sightings of a single bird centred on the east coast of Mahé. One was shot in mid-1998, probably the same bird, at Providence dump. However, another single bird was seen at Machabee in mid-2000. Vigilance will be necessary to prevent the species from becoming established and uncontrollable, as has happened elsewhere.

SIMILAR SPECIES None in Seychelles.

Reference: 1. Ryall (1986).

PIED CROW *Corvus albus*   Plate 51

Other name: White-bellied Crow **French:** Corbo pie **Creole:** Korbo Blan e Nwanr

DESCRIPTION Length 45 cm; wingspan 98-110 cm. A medium-sized black-and-white crow found throughout the Aldabra group. **Adult** has glossy black head, throat, upperparts and wings with a broad white band across the hindneck and upper back extending to the belly. Eye dark brown. Powerful black bill, upper mandible feathered at the base; legs black. Sexes similar. **Juvenile** similar to adult but drabber black areas; some white feathers tipped black.

VOICE A loud raucous caw 'kaarr-kaarr-kaarr'.

BEHAVIOUR Found around human settlement and in areas with taller trees. Also frequently seen foraging on the beach. Social and gregarious, usually seen in pairs or small groups. Flight strong, relaxed and direct. Often soars to 200 m or more, possibly in display or as a means of locating distant food sources. There are many records of single birds or pairs crossing between islands of Aldabra via the channels or occasionally across the entire width of the lagoon. One bird colour ringed at the Research Station (northwest Aldabra) was seen at Cinq Cases (southeast Aldabra) 18 days later, a distance of 24 miles 'as the crow flies'[1]. Diet is omnivorous. May take eggs of terns and landbirds and turtle hatchlings. Large groups of crows may appear at a food source, such as fish laid out to dry in the sun, food thrown into a rubbish pit or a freshly dead turtle. Groups of up to 40 are reported appearing at such sites[2], a remarkably high number considering a population estimate of 80 birds for the whole of Aldabra.

BREEDING Breeds around November to February. A large nest of twigs is built high in a tree, the interior lined with softer material. Details from Seychelles are sketchy, though one nest was found to contain three eggs[2]. In Madagascar, the clutch size is 4-6 light green, mottled olive green, pale violet-grey and dark brown eggs[3].

RANGE Resident throughout Africa south of 20°N except for the most arid regions; also Madagascar, Comoros, Iles Glorieuses and Aldabra group. In the 19th Century, it was introduced to both Mauritius and Réunion, but disappeared. Monotypic.

STATUS Resident Aldabra, Cosmoledo, Assumption and Astove. Natural colonisation may have been due to birds being carried to islands because of their habit of soaring and riding on updrafts of air (as suggested by Frazier[2]). Continued inward gene flow by this means might also explain the lack of any apparent subspeciation. It has been suggested that none of the Indian Ocean islands was inhabited by Pied Crows prior to the arrival of man, on the basis of their strong association with human settlement. However, the earliest accounts mention the species present in the Aldabra group. Abbott in 1892 found it 'not common on either Aldabra and Assumption'[4]. It could be argued that it had been introduced to Aldabra at a date prior to 1892, but Major W. Stirling, shipwrecked on Astove in 1836, noted in his log '...crows with the white ring round their neck and white breast'[5]. In 1966, the Aldabra population was put at 'hundreds'[6]. However, in 1974, it was estimated at 60-70[7] in 1986 at about 80[1] and in 1998-2000 at 60-73[8]. It may be that food supply limits the population and indeed in 1999/2000, when there was a very poor season for turtle nestings, an increase in aggressive behaviour and the failure of breeding attempts by crows were both noted[8,9]. Assumption population was estimated at about 30 in 1977, 15-20 in 1978 and 10-15 in 1986[1]. Numbers on Cosmoledo are probably similar, with perhaps somewhat fewer on Astove.

THREATS AND CONSERVATION There has been some well-intentioned support for the control of crows in the belief they are a threat to other species. However, it has been the general view, expressed by Huxley in 1974 that no crows should be 'controlled' unless there is clear evidence of a permanent and damaging effect on the ecology of Aldabra[7].

SIMILAR SPECIES None.

References: 1. Roberts (1986), 2. Benson and Penny (1971), 3. Langrand (1990), 4. Ridgway (1895), 5. Stirling (1843), 6. Bourne (1966), 7. Huxley (1974), 8. Betts (2000b), 9. Betts (2000a).

EURASIAN GOLDEN ORIOLE *Oriolus oriolus*   Plate 45

Other names: (European) Golden Oriole **French:** Loriot d' Europe **Creole:** Loryol Eropeen

DESCRIPTION Length 24 cm; wingspan 44-47 cm. A fairly large passerine with long wings and long tail; the bright yellow-and-black male is unmistakable, female greener, duller and streaked. **Male** is readily distinguished by bright yellow plumage, with black wings, tail and lore, a yellow spot on the primary coverts and yellow corners to the tail. Crimson eye, heavy dark pink bill and grey legs. **Female** is greenish above with rump, uppertail-coverts and flanks tinged yellowish; rest of underparts very pale grey with becoming yellowish white on the belly, streaked dull brown, streaking heaviest on breast; wings dark greenish brown. Eye reddish, legs brownish pink, bill as male. **Immature** similar to adult female, but more heavily streaked. Male is less streaked with yellower underparts and more black in tail. Female has underparts whitish (more heavily streaked than adult female). Median coverts usually tipped yellowish, fresh primaries tipped and edged whitish. Progression to adult plumage lasts two years. Eye dark brown or blackish and bill dark brown, later becoming pinkish.

VOICE Song is a loud, clear whistle. Call a harsh 'kraar'.

BEHAVIOUR Frequents lowland forest, avoiding both dense forest and treeless areas. Vagrants generally solitary. Flight rapid, powerful and undulating, wings apparently set forward and long tail obvious. Rarely on the ground, feeding mainly in the tree canopy on insects and berries.

RANGE Nominate *oriolus* breeds Europe east to Caspian Sea, wintering sub-Saharan Africa. Race *kundoo* (Indian Golden Oriole), breeds Turkestan to central Asia, a partial migrant, which also winters India, Sri Lanka. It is slightly smaller than the nominate, the black loral-stripe of adult male continues around and behind the eye (usually no streak behind eye in nominate), has more yellow on sides of tail (only one-third, not two-thirds of outer feathers black) and a larger yellow patch on primary coverts.

STATUS Vagrant throughout Seychelles; recorded granitic islands September to November and Aldabra in March. Most are first-year or possibly female birds of unknown race, probably *oriolus*, though *kundoo* cannot be ruled out.

SIMILAR SPECIES None likely in Seychelles, except possibly **African Golden Oriole** *O. auratus* (L20 cm), a non-breeding visitor to East Africa March to September from southern Africa. Male has much more extensive black eye-patch and broad yellow edges to primary coverts and flight feathers. Female and immature birds are yellowish green above, yellowish streaked grey below, with a dark grey eye-patch. Immature most easily confused but greenish yellow bend of wing is not darker than scapulars and sides of neck (unlike Eurasian).

WATTLED STARLING *Creatophora cinerea*   Plate 51

French: Etourneau caronculé **Creole:** Marten Sann

DESCRIPTION Length 19-20 cm; wingspan 26-30 cm. A greyish starling with a white rump, black flight feathers and tail. **Adult non-breeding** has a pale grey or grey-brown body contrasting with whitish rump, short black tail and black flight feathers. Greater coverts are whitish in male, black in female. A bare bluish grey streak is visible each side of the throat. Bill pale pinkish, blackish at base, legs pink, grey or flesh-brown. **Female** similar, but primary coverts black or brown, wings brown in worn individuals. **Male breeding** loses head feathers exposing bright yellow skin from eye to hindcrown, black on forehead and throat; also acquires black hanging wattles on forehead/chin and smaller ones on centre of crown. Bill yellowish. **Immature** similar to adult non-breeding, more brownish above (particularly female), darker on wings and bare skin on sides of throat and behind eye brownish yellow or greenish yellow. Sometimes faintly streaked on underparts. Bill brownish.

VOICE High, squeaky notes or in flight, a harsh three-syllable call.

BEHAVIOUR Favours areas with scattered bushes or trees and cultivation. Gregarious and nomadic, moving in response to rains and the consequent availability of insect food, vagrants are likely to be solitary or in pairs. Flight direct, pointed wings and short tail recalling Common Starling *Sturnus vulgaris*. Sometimes follows animals, feeding on insects disturbed. Also sometimes takes berries and seeds.

RANGE Breeds in sub-Saharan Africa, moving widely outside the breeding season, with vagrant records within Africa and most notably from Madagascar. Monotypic.

STATUS Vagrant to Bird and Aldabra with records of long-staying birds in most months. Flocks of up to 14 birds recorded Aldabra.

SIMILAR SPECIES Common Myna is darker, browner with yellow bare parts and conspicuous white wing-patches. Immature **Rose-coloured Starling** has dark brown wing feathers with pale fringes and lacks white rump.

ROSE-COLOURED STARLING *Sturnus roseus*   Plate 51

Other names: Rosy Starling, Rose-coloured (or Rosy) Pastor
French: Etourneau roselin **Creole:** Marten Roz

DESCRIPTION Length 21.5 cm; wingspan 37-40 cm. Unmistakable in adult plumage with combination of black and rose-pink plumage and a shaggy crest. **Adult** has black hood, wings, tail and vent contrasting with striking pink rump, back, breast and belly. Purplish gloss on head, neck and upper breast. Shaggy crest on the nape. Undergoes a complete body moult in the autumn. Pale pink legs. Pale pink bill, black at base of lower mandible. Sexes similar, female duller and less contrasting with shorter crest. **Juvenile** lacks crest of adult, shows a pale supercilium and slightly darker ear-coverts and has upperparts more sandy brown than adult, palest on the rump. Wings are dark brown, with buffish white fringes. Underparts paler brown, throat and centre of belly almost white with brownish grey streaks on the breast. Bill yellowish and shorter than adult. This plumage may be retained as late as November moult, after which **first-winter male** is hard to separate from adult female, but differs from first-year female by pure pink belly, upperparts and rump and longer crest. Bill pale yellowish with a darker tip; legs pale pink. **First-year female** is much duller and browner with little or no pink. Both male and female have shorter crests than in adult of same sex.

VOICE A variety of loud twittering and chattering notes.

BEHAVIOUR Frequents open countryside and cultivation. Gregarious, though vagrants to Seychelles have been solitary. Flight rapid and direct with some gliding. Diet includes fruit, berries, seeds, nectar, lizards and insects.

RANGE Breeds eastern Europe to southern Siberia, wintering Iran, Indian subcontinent, Sri Lanka. Occurs erratically as a vagrant to western Europe (as far as Ireland, Iceland, Faeroes), North Africa and Andamans. There was an attempt at introduction to Mauritius in 1892, but birds disappeared immediately after release[1]. Monotypic.

STATUS Vagrant to granitic islands and Bird. Most records October to December, though one remained on Frégate for almost a full year.

SIMILAR SPECIES Wattled Starling has black wings and tail, unlike browner wings and tail of immature Rose-coloured Starling and a white rump.

Reference: 1. Meinertzhagen (1912).

COMMON MYNA *Acridotheres tristis*   Plate 51

Other names: Indian (or Brown or House) Myna (also often spelt 'Mynah'), Locust Starling
French: Martin triste   **Creole:** Marten Ordiner

DESCRIPTION Length 23 cm; wingspan 33-36.5 cm. A smart, dark brown bird with bright yellow bill and eye-patch and large white wing-patches conspicuous in flight. **Adult** has dark brown upperparts and underparts merging into glossy black head and upper breast. Yellow bare skin around the eye. Tail black tipped white. Vent and undertail-coverts white. Large white patches at the base of primaries show as conspicuous panels in flight. Axillaries and underwing-coverts white. Bright yellow bill, greenish at base of the lower mandible. Legs yellow. Sexes similar. Sometimes moult results in a variation known as the **King Myna** ('Lerwa Marten' in Creole), the head of which is completely bald and yellow, giving the appearance of a miniature vulture. According to local folklore, the yellow head is due to over-indulgence in mangoes, while the name is due to the supposed deference shown by other Mynas. **Juvenile** is similar to adult but paler and duller with paler brown throat and breast and dark grey-brown head, lacking gloss.

VOICE A wide variety of loud, piercing, high-pitched calls and whistles. Will mob dogs and cats with wild, angry raucous cries of 'chake-chake'. On take off gives a weak whistle. Sings within trees. In courtship and at other times, the male will fluff out its feathers, bob its head up and down and facing its mate utter a routine rendered 'keek-keek-keek, kok-kok-kok, churr-churr-churr'[1]. Will mimic other birds and in captivity copy human speech. May mimic Seychelles Bulbul, but call not as harsh and on Praslin may mimic whistle of Black Parrot, but call not as strong. Non-breeding birds gather sometimes in hundreds at dusk in favoured trees uttering a loud clamour until the flock eventually settles down to sleep.

BEHAVIOUR Ubiquitous, but particularly common around human habitation. In pairs or solitary by day, gathering in large flocks at dusk. Flight strong, direct and purposeful. Omnivorous, feeding in trees, on the ground and even foraging in the intertidal zone on both sandy and rocky shores like a wader. On the ground walks with big, jaunty strides. Wide diet includes fruit, seeds, insects, grubs, carrion (such as dead tenrecs and birds on roads), frogs, lizards, crabs, nectar and household scraps. May sometimes 'dive bomb' geckos, possibly to deliberately make them shed their tails, which the birds then eat. May follow cattle, feeding on flies. In India, it is an important pollinator of some species and disperser of others, but causes some damage to seed and fruit crops. In Seychelles, it is also a great disperser of seeds and is probably the cause of cinnamon being so common and widespread in the granitics; lantana, also extremely common in Seychelles, is favoured.

BREEDING Monogamous and may pair for life. Pairs frequently bow to each other while fluffing out feathers, especially of head and breast, wings drooped and tail spread. Pairs sometimes crouch down, necks extended bills open and touching or nearly touching while calling, sometimes followed by bowing. Courtship may also involve allopreening and presentation of nesting material. Breeds mainly September to March. Nest is an untidy cup-shaped collection of grass, roots and household waste such as bits of string, paper and kitchen foil. Nest site is usually deep in a tree or bush, or under the eaves of rooftops (especially palm-thatched roofs); will use natural tree cavities where available, competing aggressively with other species such as Seychelles Magpie-robin. Clutch is usually 2-4 turquoise eggs. Both sexes share nest duties. Incubation takes 14-18 days and fledging 22-34 days. Young follow parents, begging for food for several weeks after fledging.

RANGE Resident southern Afghanistan, Indian subcontinent east to South-East Asia. Several races, nominate *tristis* present throughout Indian subcontinent. Widely introduced including to Hong Kong, Australia, New Zealand, Réunion, Rodrigues, Mauritius, Comoros, Madagascar, Chagos, South Africa, Hawaii, Ascension, St Helena, Fiji, Solomon Islands, France, Russia, Georgia and Canary Islands. Unsuccessfully introduced to mainland USA and England but arrived in France around 1986 (probably escapes).

STATUS Introduced to Mahé from Mauritius; has spread throughout granitic islands and to Bird and Denis. Gaymer *et al.* suggested an introduction date around 1830[1], but Penny suggested late 18h Century[2], soon after first colonisation, when birds were sent from Mauritius to Seychelles by Mahé de Labourdonnais, where the nominate race had been introduced from India probably in 1762[3]. Now common on all the larger granitic islands except Aride where active control measures have been implemented. Has occasionally bred on Aride, but not on a regular basis. Numbers on Frégate have also been substantially reduced by trapping and shooting, to reduce competition with Seychelles Magpie-robin.

SIMILAR SPECIES None.

References: 1. Gaymer *et al.* (1969), 2. Penny (1974), 3. Diamond (1987).

ORTOLAN BUNTING *Emberiza hortulana*

Plate 52

Other name: Ortolan **French:** Bruant ortolan **Creole:** Ortolan

DESCRIPTION Length 16-17 cm; wingspan 23-29 cm A streaky brown, dumpy bird, it is the only bunting likely to reach Seychelles. Combination of pale yellow throat, dark malar-stripe, yellow eye-ring, pinkish bill and pinkish buff underparts diagnostic. **Male non-breeding** has head and upper breast greyish olive, with some fine dark spotting on the malar-stripe, crown and nape, pale yellow throat and olive brown malar-stripe. At close range, a narrow pale yellow eye-ring is visible. Upperparts reddish brown streaked black, rump and uppertail-coverts yellowish with darker streaks. Flanks and belly are tinged orange-brown. Tail blackish brown with white edges to outer tail feathers, obvious in flight. Upperwing dark brown edged lighter brown and grey with two indistinct buff wingbars, the underwing yellowish. In autumn, crown is more prominently streaked, the breast spotted or mottled similar to female. Colours duller with wear, olive tones sometimes absent. Bill brownish pink; legs flesh. **Male breeding** has olive grey head, broad breast-band and malar-stripe, the belly and flanks entirely orange-brown. **Female** similar to male but paler, crown tinged brown and distinctly streaked in spring (when male crown is unstreaked). Usually streaked heavily (sometimes only a few streaks) on the breast, which is greyish yellow with only a faint olive tinge; in winter, no breast-band separates yellow throat from pale rufous breast and belly. Brownish rump; pale buff vent and undertail-coverts. **Immature** similar to adult female non-breeding, but somewhat browner, with grey-brown head, pale yellow-buff eye-ring, dark malar-stripe, ear-coverts streaked greyish with a brownish lower edge and prominent white wingbars. Broad red-brown edge to outer tertials widen towards centre. Prominently streaked on mantle, upper flanks, breast and malar-stripe.

VOICE Male song, delivered from a bush or open perch, is a monotonous series of about seven notes, two pitched lower than the rest. Call a soft 'tsip' or piping 'tsoo'. When disturbed, calls 'twick'.

BEHAVIOUR Frequents open country, scrub, gardens and cultivation. Gregarious though vagrants to Seychelles have been single birds. Flight light and undulating. Short escape flight, often diving for ground cover. Feeds on the ground, the diet including insects, snails and seeds.

RANGE Breeds Europe east to Mongolia, wintering Iran, Arabia and sub-Saharan Africa. Vagrant to Kenya. Monotypic.

STATUS Vagrant to granitic islands around November.

SIMILAR SPECIES None likely in Seychelles though perhaps a slight chance of **Cretzchmar's Bunting** *E. caesia* (L16 cm, WS23-26.5 cm), which breeds southeast Europe to Israel wintering mainly northeast Africa. This has not been eliminated by some reports of Ortolan. Adults have a similar plumage pattern to Ortolan, but are easily distinguished by distinctive grey head, malar-stripe and breast pattern lacking olive tones, an orange, not yellow, throat and darker part body. Wings are dark brown edged lighter; more uniform than Ortolan. Juvenile difficult to separate from Ortolan but has whiter eye-ring, grey (not olive) tinge to crown, white (not yellow) tinge to throat, white (not yellow) underwing and warmer rufous ground colour to body with rufous tinge to rump and vent. Rufous edges to tertials widen towards centre. During first winter, throat becomes yellow-buff with rufous tinge; rump rufous, undertail-coverts and vent rufous buff.

Ortolan Bunting (two figures on left), and Cretzscmar's Bunting (two figures on right)

YELLOW-FRONTED CANARY *Serinus mozambicus* **Plate 52**

Other names: Mozambique (or Yellow-fronted) Serin, Yellow-eyed (or Icterine, or Green or Yellow-bellied) Canary, Green Singing Finch **French:** Serin du Mozambique **Creole:** Sren Zonn

DESCRIPTION Length 11.5 cm; wingspan 14 cm. A small, greenish brown and yellow finch found only on Assumption. **Male** is easily identified by bright yellow underparts and rump, with a slight greenish wash on sides of breast. Upperparts are more greenish brown with darker brown streaks, the tail black, narrowly edged yellow and tipped buffish. Bright yellow forehead extends behind the eye as a broad supercilium, grey on upper ear-coverts and cheeks, yellow on lower ear-coverts with a black line through the eye and blackish moustachial-stripe. Crown greyish, finely streaked blackish. Bill small, pointed, dark brown on upper mandible (mainly on the culmen), paler on lower mandible; legs dark brown. **Female** similar to male but duller and greyer particularly on upperparts; head pattern less striking, underparts duller yellow and rump yellow. **Juvenile** similar to adult female but with only a trace of pale yellow on underparts and face, white throat, some streaking on breast and flanks (more obvious in female) and a dull greenish yellow rump.
VOICE Call is 'tssp' sometimes repeated. Song is vigorous, monotonous and canary-like.
BEHAVIOUR Found in taller trees, mainly around human settlement where flits between branches with heavily undulating flight. Social and gregarious, usually seen in small groups. Diet mainly seeds, but also some flowers and buds.
BREEDING Cup-shaped nest is built from grass stems and roots, lined with softer material such as cobwebs, in a shrub or tree, built solely by the female. Clutch size is 3-4 eggs, blue with red streaking. Chicks are fed by the female on regurgitated seeds.
RANGE More than ten similar races that overlap and interbreed are found across sub-Saharan Africa except for the arid southwest and central West African rain forest. Introduced to Mauritius, Réunion, Rodrigues and Hawaii.
STATUS Introduced to Assumption (nominate *mozambicus*), where most easily located in tall casuarina trees between Settlement and coast. It was also introduced to Desroches, where specimens were collected in 1882[1] and 1892[2], but subsequently died out for unknown reasons. It may once have occurred elsewhere in the Amirantes. 20-30 birds were introduced to Assumption (contrary to Seychelles law) in April 1977[3]. In 1986, the population was estimated at 25-35[4] and numbers have possibly remained fairly low, around this level. The species was earlier introduced to Mauritius, probably from Cape of Good Hope 1756-63[5]. It may be a threat to the endemic avifauna of Aldabra that lies only a short distance to the north of Assumption and is entirely free of introduced birds.
SIMILAR SPECIES None on Assumption.
References: 1. Coppinger (1899), 2. Ridgway (1895), 3. Prys-Jones *et al.* (1981), 4. Roberts (1988), 5. Diamond (1987).

COMMON ROSEFINCH *Carpodacus erythrinus* **Plate 52**

Other names: Scarlet Rosefinch (or Grosbeak), Scarlet Finch, Hodgson's Rosefinch
French: Roselin cramoisi **Creole:** Sren Roz

DESCRIPTION Length 14.5-15 cm; wingspan 24-26.5 cm. A medium-sized, thickset finch with round head and heavy bill. **Male non-breeding** has red areas covered by brown feather tips and may appear almost uniform grey-brown above, and only very slightly paler greyish white below with faint streaking. Buffish brown on wing-coverts with pale edges to tertials. Dark eye stands out in featureless face with crown tinged brown or buff; faint, diffuse streaking on crown and upperparts. Plumage very worn September onward, moulting in winter quarters November onward. Two narrow, pale grey-white wingbars **Female** is uniform grey-brown above, with brown streaks on forehead and forecrown and also shows two pale indistinct wingbars; underparts pale buff or whitish with faint streaks on chin and throat becoming more prominent on breast and upper flanks. Both sexes have a short stout bill, adapted to cracking large seeds, with mandibles curving slightly to a pointed tip; bill grey-brown in male, paler in female; legs brown or brownish pink. **Male breeding** has bright red head and breast, with variable amounts of brown feathers showing through; upperparts and wings grey-brown, sometimes with a red wash or streaks and pinkish rump and double wingbars. **Juvenile** similar to adult female but entire plumage fresh and tinged olive, less grey-brown and more prominently streaked. Some birds have prominent pale buff wingbars. Base of bill pinkish (this colour sometimes retained on lower mandible of adults).
VOICE Song is a repetitive, slowly rising whistle. Non-breeding birds are usually silent.
BEHAVIOUR Frequents undergrowth and bushes, forest edges and cultivation, often near water. Gregarious in wintering grounds but vagrants likely to be solitary. Undulating flight. Hops on the ground, feeding on seeds, fruit and berries, occasionally nectar and insects.
RANGE At least five races, some authorities claim more, breed across Eurasia. Races are difficult to separate, particularly in non-breeding plumage. Northern races are entirely migratory, others partially so to India and South-East Asia. Vagrant to Faeroes, North Africa. Possibly most likely race to reach Seychelles is nominate *erythrinus*, a long-distance migrant breeding from the Baltic to central Siberia, wintering south

to India; this is the palest race.

STATUS Vagrant to granitic islands; recorded October.

SIMILAR SPECIES It is unlikely that any other rosefinch could reach Seychelles. Confusion may be possible with **Madagascar Fody** female, immature or male non-breeding, which lack streaking on underparts, with a triangular, less stout bill and has a lighter build. Female **House Sparrow** shows white wing-patches, no streaking on underparts and a less hefty bill.

COMMON WAXBILL *Estrilda astrild*   Plate 52

Other names: Waxbill, Barred (or Brown, or Red-bellied) Waxbill, Pheasant Finch
French: Astrild ondulé   **Creole:** Bengali

DESCRIPTION Length 11-13 cm; wingspan 12-14 cm. A small grassland finch with bright red bill and eye-stripe, usually seen in flocks. **Adult** has upperparts grey-brown with fine darker brown bars and underparts pale buff, whiter on chin and throat with fine brown bars on breast and flanks. Sometimes has a pink wash to lower breast and belly, or a reddish patch in centre of belly. Tail is dark brown, graduated towards the tip and appearing fairly long in flight. Triangular red bill. Eye brown; legs black. Sexes similar, though female is slightly paler.
Juvenile similar to adult but paler with very faint barring, very pale pink wash on underparts and a blackish bill. Bill colour changes to orange within a few months.
VOICE Call a short, hesitant 'chip-chip' repeated. In flight a rapid, quiet twitter. Song is a low, harsh rising 'tcher-tcher-preee', occasionally followed by 'dit' or a descending 'chewi-chee'[1].
BEHAVIOUR Favours long grass, roadsides and gardens. Highly gregarious, usually found in small active flocks which take off en masse. Feeds mainly on seeds, moving acrobatically between grass stems. Sometimes feeds on the ground and occasionally takes insects.
BREEDING Breeds in northwest monsoon, October to March. Nest is a large domed structure with a side entry corridor inclined slightly towards the base, built from grass stems in a bush or small tree at 3-4 m height. Clutch is 4-5 white eggs. Incubation last *c*13 days, fledging *c*20 days.
RANGE Many races distributed throughout most of Africa. Races not easily differentiated, with much overlap and interbreeding.
STATUS Generally considered introduced, though this view has been challenged. Resident Mahé, La Digue and Alphonse. Formerly occurred Desroches (reported in 1882[2]) and possibly elsewhere. Noted once on Assumption in 1977[3] and probably introduced there in 1976 or 1977 from Mauritius along with other species, but failed to establish and not reported subsequently. Generally assumed to be introduced due to lack of differences from African birds, weak flight (hence little chance of reaching Seychelles) and history of introductions elsewhere. However, G and R Gerlach drew attention to a 1789 report by Malavois who on returning to Mauritius from Seychelles noted 'the parrots and bengalis (the Creole name for waxbill) did great damage to the crops'. This report was only 19 years after the first settlement on St Anne from Mauritius, probably too short a period for an introduction to have reached epidemic proportions[4]. In addition, the species was unknown in Mauritius at this time (probably introduced *c*1800). However, the name bengali was also applied to other species in Mauritius from at least 1740[5] and may have been a convenient label readily transferred to any similar seed-eating bird. One version of their introduction is that it coincided with the liberation of slaves in 1833; planters found it difficult to recruit labour and former slaves grew their own rice at the La Passe marsh on La Digue, where one former slave owner deliberately introduced the birds[6].
SIMILAR SPECIES Easily separated from **Madagascar Fody** by red bill, smaller size and constant soft twittering.
References: 1.Clement *et al.* (1993) 2. Coppinger (1899), 3. Prys-Jones *et al.* (1981), 4. Gerlach and Gerlach (1994), 5. Cheke (1982), 6. Dauban (1979).

MADAGASCAR FODY *Foudia madagascariensis*   Plate 53

Other names: Red (or Madagascar Red, or Madagascan or Cardinal) Fody, Madagascar Weaver
French: Foudi de Madagascar or Foudi rouge   **Creole:** Kardinal, Sren, Tisren

DESCRIPTION Length 12-13 cm; wingspan 17-19 cm
By far the most common small passerine on most of the granitic islands and some outer islands. Breeding male unmistakable with bright red plumage and brown wings. Female and non-breeding male are pale brown, darker on the wings. **Male breeding** is bright scarlet above and below with some black streaks on the back and a black stripe through the eye; wings and tail are dark brown. Some breeding males fail to develop red on the belly, which remains brown or grey-brown but with red head and breast and red vent. Bill black. **Female** is generally brown or yellowish brown, darker and streaked on the upperparts, paler and more uniform on underparts; bill pale brown. By March/April, male begins to moult and appears scruffy. Some may retain a few red feathers at all times, but most resemble females by June. **Flavistic male** occurs, with golden yellow plumage, smaller than typical male but retaining black eye-stripe. Flavisim

occurs in the genus *Foudia* in this species and Aldabra Fody. Flavistic birds lack gonads of either sex but behave like males, sometimes even nest building and attempting to attract a mate. **Juvenile** similar to adult female.

VOICE Males utter a monotonous trill, apparently to lay claim to their territories. When chasing females they may call 'zzz-zzz-zzz'. Flocking birds call softly 'si-si-si-si'. During the southeast monsoon when birds are not breeding they may gather towards evening in huge collective roosts where they call loudly and monotonously, similar to, but not as loud and shrill as, roosts of Common Myna.

BEHAVIOUR Inhabits most habitats, particularly gardens, cultivation and woodland, but generally avoids dense forest, particularly at higher altitudes. Feeding flocks often seen around small shops and on the road to the New Port, attracted by leaking sacks of rice. Common around houses except where House Sparrow occurs; it appears this latter species is more aggressive and able to oust the Fody. Breeding males sometimes admire their own reflection, perching on car wing mirrors or the rear window of cars for long periods. Soon learns to come to houses for rice, which it may demand by perching on louvres, chirping loudly, or if ignored, flying through rooms. Highly gregarious, often in large flocks even during the breeding season. Unlike the insectivorous Seychelles Fody, it is mainly a seedeater. Crook found seeds taken in 84% of observations (only 25% for Seychelles Fody)[1]. About half the consumption of seeds are grass seeds which may be harvested by standing on the stalk and pulling the head downwards.

BREEDING Season extends from September to April with a peak in December. Male establishes breeding territory, driving off intruders, including females at first, possibly to show them the size and limits of their domain. On the coastal plateau territories may extend 10-30 m from the nest site while on higher ground it may be up to 75 m[1]. The nest is usually hung from the lowest tier of palm fronds of a coconut palm. Nest height is usually over 6 m (in contrast to Madagascar where it nests in bushes at 1-2 m). Strips from fronds are harvested by the male making a break with its bill and holding the broken end as it leaps downward, using its body weight to tear it off. An initial ring is constructed between two palm pinnae woven into either side. Materials include casuarina needles, grass and palm pinnae. Male does most construction work, female assisting later mainly with shaping and lining the interior. Kapok is used in lining and to plug gaps in the flimsy structure. Entrance often has a downward curving tubular porch. Nest building may be a long, drawn-out process, sometimes taking over 30 days, though this is highly variable. Sometimes after or near to completion, it is dismantled and the material used to construct another nest. This is done probably after failure to attract a female. The female rejects many nests: one study found more than three out of four constructions were rejected[1]. Male actively defends its territory, sometimes fighting off attempts by other males to supplant it, gripping claws or bills and falling to the ground. They then separate and may face each other with a threat posture in which the bill is thrust forward, neck drawn in, body crouched, wings raised and drooped, with red feathers fluffed out (especially on the back and rump). When territories are established, most birds are still in flocks. As a flock passes through a territory, the resident male singles out a female and alights beside her. If the female flies off, the male gives chase, calling 'zzz-zzz-zzz', but if she remains the male displays with a similar posture to that used in threat, puffing out its red feathers, the bill often lowered while leaning forward to emphasise the fluffed-out rump feathers. Female may leave flocks to check out territories, only to be chased out by the male. In time, the male's aggressiveness becomes less strong, and it may land beside female, tail depressed, wings quivering, head and breast feathers fluffed out. If the female remains, the male may raise its wings rigidly, then fly to the nest, followed by the female. Male may also invite female to the nest by chasing it just a short way, before returning to the nest with slow rigid wingbeats interspersed with gliding, wings held stiffly. Eventually, female does not flee and flies to the nest, pecking at the male when it attempts to drive it away. Male retreats, wings quivering. Eventually female crouches, raises its bill and calls 'tzip-tzip-tzip', inviting the male to approach, which it does with red feathers fluffed, wings spread and lowered and tail lowered; it then mounts female. Usually two or three (occasionally one or four) pale blue eggs are laid. Incubation lasts 12-14 days, fledging a further 11-14 days. Both parents feed the young on insects and seeds by regurgitation. After leaving the nest, young birds follow the parents and are fed mainly by the female, the male showing increasing intolerance with the begging young, which it may chase off for a short distance.

RANGE Endemic to Madagascar where it is common, occurring from sea level to 2300 m. Also widely introduced.

STATUS Introduced or a natural coloniser. Resident on all but the very smallest granitic islands. Also on Bird, Denis, Platte, Rémire, D'Arros, St Joseph Atoll, Desroches, Farquhar and Assumption. In 1866, Newton saw just one on Mahé and it has been suggested the species was probably introduced only shortly prior to this date[2]. By 1906, it was the most common landbird after Common Myna[3]. Nicoll reported at this time he was told on good authority 'two neighbours went to law concerning the ownership of a certain field which each claimed as his property. The loser, to be revenged on his adversary, brought from Madagascar a cage full of weaver birds which he liberated on his neighbor's land'[3]. The conventional view that this is an introduced species in the granitic islands was first challenged by G and R Gerlach, who drew attention to an observation by the Marion Dufresne expedition of 1768, two years prior to the first human settlement, of a bird seen on St Anne with a crimson head and breast, black bill and legs and a brown and coffee coloured body 'like a linnet.'[4] This was in September when the male Madagascar Fody would be moulting into breeding plumage. One explanation might be a surprising mistake by a team of observers instructed to '...especially give the greatest attention to the study and prospects of all the species of inland productions such as trees, bushes, plants, herbs, quadruped animals, birds, insects, freshwater fish, stones, soil, minerals. Nothing is unimportant'. The alternative explanation is Madagascar Fody were present in Seychelles at this time. Perhaps it was uncommon and being a bird of open terrain, did not greatly increase in numbers until land was cleared for housing and agriculture. G and R Gerlach suggested adaptive radiation may have been prevented by periodic colonisation from the parent stock. Introduced on Assumption from Mauritius in April 1977, the island manager releasing 20-30 birds[5]. In 1986, the population was estimated at 25-35[6]. It has probably remained around a similar level.

THREATS AND CONSERVATION If the species is native to Seychelles it is worthy of higher esteem than generally given to it. Nevertheless, it is extremely common and has benefited from man opening up habitat which it has invaded.

GENERAL Competition with this species has in the past been blamed for the demise of Seychelles Fody in much of its former range. However, each occupies a separate niche. Madagascar Fody generally avoids dense woodland, favouring open areas, gardens and the vicinity of houses. Seychelles Fody favours dense woodland, though it also adapts well to the vicinity of human habitation. Seychelles Fody is mainly insectivorous, while Madagascar Fody is mainly a seedeater. It is interesting that on D'Arros where just five Seychelles Fodies were introduced in 1965 that by 1995 they had become considerably more common than Madagascar Fody[7]. On the four islands where the two species coexist they do not interbreed, though a single female Seychelles Fody discovered on Aride in 1988[8] went on to breed with a Madagascar Fody producing at least two young[9]. The tongues of Madagascar Fody show a slight degree of brush development, typical of many nectar-feeding birds such as sunbirds and hummingbirds, a trait also seen in Aldabra Fody. It has been suggested that the appearance since 1990 of half-coloured males, in which the belly, vent and mantle remains brown, may be due to hybridisation. This plumage pattern resembles that of Aldabra Fody, Comoro Fody *F. eminentissima* and Forest Fody *F. omissa*. It has been speculated this may be due to the arrival of a single individual or possibly a small flock in the late 1980s[10].

SIMILAR SPECIES Seychelles Fody is more compact, darker brown and occurs only on Cousin, Cousine, Frégate and D'Arros. Female **House Sparrow** is larger, more patterned sandy brown above with white wing-patches.

References: 1. Crook (1961), 2. Newton (1867), 3. Nicoll (1909), 4. Gerlach and Gerlach (1994), 5. Prys-Jones *et al.* (1981), 6. Roberts (1988), 7. Skerrett (1995), 8. Bullock (1989), 9. Lucking (1997), 10. Gerlach (1996a).

ALDABRA FODY *Foudia (eminentissima) aldabrana* Plate 53

French: Foudi d'Aldabra **Creole:** Kardinal (Sren, Tisren) Aldabra

DESCRIPTION Length 15-16 cm; wingspan 21-22 cm. A small finch-like bird confined to Aldabra where it is common. Breeding male is scarlet on the head and breast, yellowish below. Female and non-breeding male are duller and sparrow-like. **Male breeding** has scarlet head and breast, fairly sharply divided from yellow on the belly and undertail-coverts. Back and wings are dark olive to grey-brown with a little red and are streaked blackish. The rump is red and the tail dark brown. Black patch around brown eye. Bill stout, triangular and blackish; legs pinkish. Sometimes the red plumage is replaced by orange/yellow; this **flavistic male** is probably sterile (see Madagascar Fody). **Female** is yellowish brown on head, upperparts and rump, dark brown on the tail, with light streaking on the crown and nape, heavier black streaking on the upperparts, the rump unstreaked. Brown eye-stripe and prominent yellow supercilium; underparts paler and more uniform greenish, greyer on the belly; undertail-coverts pale yellowish. Brown eye, pale pinkish brown bill at all times and pinkish legs. **Male non-breeding** loses bright red plumage by May/June, a few red feathers being retained through most of southeast monsoon by some birds, others being almost identical to adult female at this time; bill generally changes to pale brown to pinkish. Male is larger than female with hardly ever any overlap in wing length; also heavier but with some overlap[1]. **Juvenile** similar to adult female. Male may attain full breeding plumage at one year of age though some (probably born late in the season) may not achieve this until the following season.

VOICE The male display song, begins with two or three pairs of high-pitched yodels, the first note of each pair being an octave above the second. This is followed by a high-pitched trill and finally a rising series of notes likened to a bottle being filled, the whole expressed 'tsee-oo tsee-oo tsee-oo fsssssss look look look look look'[1]. The routine is performed with raised rump and breast feathers, wings drooped. Has a wide variety of other calls, including single high-pitched 'tweet' notes and variations of this. Male defending territory may puff out red feathers, droop wings and tail, loudly calling 'zeet-zeet-tweet' frequently repeated.

BEHAVIOUR Common in woodland, especially mangroves and casuarina and also around coconut palms. Generally absent from open areas such as sand dunes and from open woodland. There is a strong association with casuarina woodland, which provides nesting material, nest sites and food. Around the settlement it is also associated with man, Nicoll noting 'it seems to take the place of the English Sparrow'[2]. Highly gregarious, especially outside the breeding season when flocks of 30 or more birds may be seen, adult males leaving these flocks around mid-July to establish their territories. Non-breeding birds may roost communally with several hundred birds in a roost. Feeds on seeds, flowers, berries and insects. Also where available household scraps and rice. Insects are picked from branches and vegetation, birds often hanging upside down to search under leaves, or even climbing trees like a treecreeper, searching under loose bark. The tips of twigs may be snapped off and the freshly broken tip searched for insects[3]. Noted skewering praying mantis in the manner of a shrike[4]. Casuarina seeds are popular, especially after heavy rain when seeds may be easier to extract. Grass seeds may be taken by a variety of methods, including perching near seeding heads and pulling the head to the perch while holding it in the feet, or possibly pulling the stem to the ground using the bill, then standing on it to feed[3].

BREEDING Season extends from October to mid-April. Male establishes a territory of typically about 650 square metres[3]. In courtship male puffs out the feathers of the back, breast, crown and especially the rump. Courtship is very similar to Madagascar Fody. At first male chases off females as if they were rivals (possibly to teach potential mates territorial limits). Later the male displays, drooping wings and raising red feathers as in the threat display given to a rival mate. Female retreats, but as nest building commences, may move off just a short way, male following and again displaying. These short pursuits eventually develop into a glide flight: a slow flight, with shallow stiff wingbeats followed by a short glide. This is a nest-invitation display. When a female no longer retreats, after much courtship, the male may display with wings held vertical (not beaten as in Madagascar Fody). Nest building may commence as early as September, though eggs are rarely if ever laid before November, continuing as late as April. Nests are built from less than 1 m up to 17 m[3]. There is a marked preference for coconut groves, though this may be a recent development as coconut palms are probably introduced to Aldabra. Elsewhere, in mixed scrub, most nests are in the densely-leafed bush *Acalypha claoxyloides*. Male commences the construction of the domed nest. Female assists with the finer points of completion, including the lining. If a male has failed to attract a mate to a nest that is almost complete, he may choose a new site and commence building often by using the material from the old nest. Nest has a side entrance and usually no porch. It is suspended from a twig or pinnae of a palm frond, these supports woven into the sides of the nest. Material includes casuarina needles, grass stems, lepeka tendrils and roots with some exterior decoration of lichens, casuarina needles and *Cymodocea ciliata* (a marine angiosperm). It is lined with softer material including grass stems and cotton. Eggs are laid about three days after the completion of nest building. Clutch size is two or three pale blue-green eggs (more often three). Incubation is carried out solely by the female and lasts about 15 days. Fledging lasts a further 15-18 days. Both parents feed the young, though for the first few days it may be done solely by the female. Food for nestlings is regurgitated. More than 80% of eggs may be predated, mainly by rats and sometimes by Pied Crow[3]. About half of nestlings are predated by rats and crows[3].

ORIGINS The genus *Foudia* is confined to the western Indian Ocean, probably having earlier origins in Africa. All fodies are sexually dimorphic in breeding plumages, which are generally of a similar pattern. Birds show considerable diversity in bill shape and size. The type may have evolved in Comoros and then spread to Madagascar[5]. If so, Aldabran birds may have originated in Comoros. The Aldabran species has a larger bill than birds of both Madagascar and Comoros. It is larger than the Forest Fody *F. omissa* of Madagascar, which has grey (not yellow) belly and undertail-coverts. Further genetic research is required to confirm full species' status[6].

STATUS Endemic, confined to Aldabra (1,000-3,000 pairs[7]) where it is found throughout the atoll, wherever there is sufficient vegetation, including larger lagoon islets. There may have been closely related forms, now extinct, on Astove, Cosmoledo and Assumption. Rivers recorded 'cardinal' on Astove and by inference on Cosmoledo (he stated that landbirds of Astove 'are also found on Cosmoledo')[8]. However, Stirling shipwrecked on Astove in 1836, did not mention it in his fairly detailed observations[9].

THREATS AND CONSERVATION Though the species remains fairly common, rats take an enormous toll on eggs and nestlings. It has also been suggested that Madagascar Fody, introduced to neighbouring Assumption, might pose a threat through possible hybridisation and competition should birds ever succeed in crossing to Aldabra. The danger of hybridisation has been highlighted on Aride, where a female Seychelles Fody arrived on the island, paired with Madagascar Fody and raised at least two hybrid offspring[10].

SIMILAR SPECIES None on Aldabra.

References: 1. Benson and Penny (1971), 2. Nicoll (1909), 3. Frith (1976), 4. Chapman (1997), 5. Moreau (1960), 6. Sinclair and Langrand (1998), 7. Rocamora and Skerrett (1999), 8. Rivers (1878), 9. Stirling (1843), 10. Lucking (1997).

SEYCHELLES FODY *Foudia sechellarum*   **Plate 53**

French: Foudi des Seychelles   **Creole:** Toktok

DESCRIPTION Length 13 cm; wingspan 17-18 cm. A small generally drab, dark brown, dumpy weaver, confined to Cousin, Cousine, Frégate and D'Arros. **Adult** of both sexes are dark brown, above and below, somewhat darker on back and wings. **Male breeding** acquires yellow crown, face and chin, which gently merges to sooty brown then dark brown on throat and rear crown. The extent of this yellow patch is variable. Often shows a small white patch in the wing. Bill black. **Female** and **male non-breeding** are generally dark brown above and below, darker on the back and wings; bill and legs brown. **Juvenile** similar to adult female.

VOICE Alarm calls include a 'tchrr' and at the nest site, 'tok-tok-tok'.

BEHAVIOUR Probably originally a forest species, but with a catholic diet it has few habitat constraints. Social, often in groups of ten or more, but not as highly gregarious as Madagascar Fody. Flight is direct, compactness obvious and appearing almost tailless. Constantly flicks wings. Usual social unit is the family group, not the flock. Insects may be picked from vegetation or tree trunks, or taken by flycatching in short aerial pursuits. There is little overlap between Seychelles Fody and Madagascar Fody in feeding. The former is mainly insectivorous, the latter mainly seed-eating. Indeed, where there is an overlap such as at a copra drier, the superior hunting methods of Seychelles Fody could give it a decisive advantage[1]. When the two species encounter one another, it is almost always Madagascar Fody that withdraws. It is also interesting that on D'Arros where five Seychelles Fodies were introduced in 1965, by 1995 numbers exceeded Madagascar Fody[2]. Sometimes takes eggs of seabirds.

BREEDING Adults undergo a full pre-nuptial moult, which takes about three months, between February and May. Breeding takes place mainly May to September. In courtship male and female call to each other continuously 'tseep-tseep-se-se-se' in turns. Male may land near female, wings quivering and drooped, tail turned downwards, breast and back feathers fluffed out, finally stretching forward in an erect posture while beating its wings above its back. A short trill is then uttered, often several times. Pairs engage in short chases, male giving its wing-beating display, often followed by copulation. Territories are established and defended for about 50 m from the nest. Territories are fairly large and sometimes overlap with two or more Madagascar Fody territories. Nest is built in a tree attached at or near the end of the branch, or in palms attached to the underside of leaf pinnae. Nest material includes twigs, casuarina needles, plant tendrils and palm fibres which are loosely entwined in a bulky, untidy, domed structure often with a slight porch and lined with softer material such a kapok. Construction may last 14 days, most building being done by the male, the female assisting particularly with the lining. Clutch size is one or (more often) two[1], the second egg being laid after a 24-hour interval. Incubation lasts 14 days and is carried out entirely by the female; fledging lasts c18 days. Both parents feed the young in the nest, mainly by regurgitation for small insects, larger insects being offered in the bill. Post-fledging parental care lasts up to four months, until late December. During this period, juveniles are fed almost exclusively by the female. After breeding, pairs continue to defend the nest area.

ORIGINS Being more of a forest species than Madagascar Fody, Seychelles Fody is probably more closely related to Forest Fody *F. omissa* of Madagascar. As with Aldabran Fody it may have been derived either from Madagascar or the Comoros. Given the greater degree of variation in the breeding male of Seychelles Fody compared to parent stock it is probable the granitic islands were colonised much earlier than Aldabra[3]. The genus probably has its earlier origins in Africa.

STATUS Endemic to the granitic islands of Seychelles where it is found on Cousin (c1,000 birds), Cousine (c500 birds) and Frégate (c1,000 birds); there is also an introduced population on D'Arros (c300). IUCN Vulnerable. Five birds were introduced to D'Arros in August 1965 by the Bristol University Seychelles Expedition. Birds survived at least until 1968[4], but there were no further reports until 1995 when it was found to be thriving, with at least 100 pairs[2]. Seychelles Fody formerly occurred on Marianne and possibly La Digue, Aride and other islands in this vicinity. It has been suggested two (Tring) specimens in The Natural History Museum, labelled Praslin and dated 1907, may have come from Cousin, but given that the former range appeared to include most of the islands around Praslin it seems virtually certain it also occurred on Praslin itself. Single birds have been recorded on Bird Island[5] and on Aride (twice during 1988-1995). In 1959, populations were estimated at 250-300 on Frégate, 105 on Cousin and 80 on Cousine[6]. In 1964/65 the population was estimated as stable on Frégate, somewhat higher on Cousine at 100-150 birds and dramatically higher on Cousin at 400-600 birds[4]. The increase to present numbers has probably been due to a general cessation of persecution of birds and in the cases of Cousin and Cousine due to protection of birds and habitat. There is some evidence of subspeciation on different islands; bills of Cousin birds are deeper at the base than those of Frégate[1].

THREATS AND CONSERVATION Collar and Stuart suggested 'rats clearly emerge as the major and perhaps only cause of local extinction'[7]. However, on D'Arros there are rats as well as cats, Madagascar Fody and House Sparrow (possible competitors, also introduced). Despite all this competition, the Seychelles Fody has quickly established itself on the island following its introduction in 1965. Elsewhere, Common Myna may possibly be a nest predator. Introduction to other islands was recommended in 1960[6] with Aride suggested by Collar and Stuart[7]. A genetic study is desirable to examine the degree of divergence on different islands, which may have conservation implications.

GENERAL Crook compared the social organisation of Seychelles Fody to that of Madagascar Fody. Five

Seychelles Fodies (including two males in breeding plumage) and five Madagascar Fodies were kept in a metre-cube aviary. At night, all five Seychelles Fody slept in contact, whereas Madagascar Fody slept two or three inches apart and would peck fiercely at any Seychelles Fody that snuggled up to them by mistake in the fading light. In daylight, where the two species encountered one another, Madagascar Fody nearly always withdrew. Likewise, no Madagascar Fody would approach the food bowl if the Seychelles Fody were feeding. Madagascar Fody were quick to establish a recognisable peck order, whereas Seychelles Fody did not[1].

SIMILAR SPECIES Dark brown plumage and dumpy shape readily distinguish from **Madagascar Fody** where the two occur alongside each other.

References: 1. Crook (1961), 2. Skerrett (1995), 3. Benson (1984), 4. Gaymer *et al.* (1969), 5. Diamond and Feare (1980), 6. Crook (1960), 7. Collar and Stuart (1985).

HOUSE SPARROW *Passer domesticus*   Plate 52

Other names: English Sparrow, European Sparrow   **French:** Moineau domestique   **Creole:** Mwano

DESCRIPTION Length 14-15 cm; wingspan 21-25.5 cm. One of the most widespread, common and familiar birds in the world, but with a restricted introduced range in Seychelles. **Male breeding** has a distinctive head pattern: grey forehead and nape, with crown and sides of nape chestnut, black lore and black chin extending as a bib to upper breast contrasting with whitish cheek. Upperparts warm brown, boldly streaked black, lower back and rump grey-brown (greyer in worn plumage). Tail blackish brown with buffish edges. Underparts grey, sometimes washed buffish. Bill black, legs dark brown. **Male non-breeding** has the bib reduced in extent and flecked grey, chestnut sides of nape obscured by grey feather tips; edges and tips of underparts paler; bill brownish sometimes with a yellowish base. **Female** has a less distinctive head pattern: crown, nape and upper back and rump grey, with long pale buffish supercilium beginning above or very slightly in front of eye to behind eye. Upperparts sandy brown with blackish streaks and tail brown with paler edges; two indistinct white wingbars; underparts pale brown, whitish in centre of belly; yellowish horn bill and brown legs. **Juvenile** similar to adult female but scruffier, with more buff brown on upperparts, dingy brown on underparts. Throat and chin greyish in male, whitish in female. Bill usually paler than adult. Adult plumage assumed within first year.

VOICE A monotonous 'chirrup', perhaps the world's most familiar bird call.

BEHAVIOUR Frequents areas around human habitation. On Desroches, where it exists alongside Madagascar Fody, it has ousted the fody from the vicinity of buildings; on islands with no sparrows, fodies are very common in such areas. Highly social and gregarious. Flight rapid and direct, wings whirring. Diet mainly seeds including cultivated crops. Will also take household scraps, fruits, berries, flower buds, molluscs and insects (which are sometimes taken in flight). Young are fed mainly on insects and their larvae.

BREEDING May breed at any time of year, but mainly during northwest monsoon October to April. Nest is a large, untidy domed structure built by both sexes, often situated in a building cavity or inside a copra shed. Grass stems make up the bulk of the nest with a softer lining of finer vegetation and feathers. Clutch is 3-6 greenish white eggs speckled and streaked grey-brown. Incubation, carried out almost entirely by the female, lasts 14 days, fledging a further 15 days.

RANGE Breeds across most of Europe, Asia and Africa wherever there is human habitation. It has been introduced successfully to North and South America, Australia, New Zealand, South Africa, Mozambique, Kenya, Somalia, Sudan and many other countries. In the Indian Ocean, it was introduced to Mauritius, Rodrigues, Réunion, Comoros and to Chagos (via Mauritius).

STATUS Introduced to Amirantes (race *indicus*, smaller than nominate race, with whiter cheek and underparts) and established on D'Arros, St Joseph Atoll (breeds Ressource, Var and St Joseph), Rémire, Desroches, Desnoeufs, Marie-Louise, Bijoutier, St François and Alphonse. Also recorded in the granitics. The origin of the Amirantes population is unknown. Interestingly, the Asian race of Brook's house gecko *Hemidactylus brookii brookii* exists in Seychelles only on Desroches and it is likely this and the sparrow were accidental ship-assisted introductions. At one time, some plantation owners in the outer islands had connections with India and supply boats would visit directly rather than via Mahé, as is the case today. Alternatively sparrows may have been introduced via Mauritius, the source of other introductions to Seychelles. The granitic islands escaped introduction, but this situation may not last indefinitely. There have been occasional, unconfirmed reports from the granitics, most notably a flock of about 20 sparrows seen behind the Post Office, Victoria in 1965[1]. If correct it is strange the world's most successful avian coloniser failed to establish itself in the granitic islands earlier with such numbers present. A record on Aride (race *indicus*) in July-August 1998 was the first reliable report from the granitics subsequent to 1965 and this was followed by further reports of single birds on Mahé. It is possible these latest arrivals have also been ship-assisted with regular connections between Mahé and Mombasa, just a few days sailing time, a possible source.

SIMILAR SPECIES None apart from possible confusion with female, immature or non-breeding **Madagascar Fody**, which is smaller, not as pale grey-buff, lacks sandy brown in upperparts and white patch in wing and has more slender legs.

Reference: 1. Penny (1974).

BIBLIOGRAPHY

Ali, S. and Dillon Ripley, S. 1964-74. *Handbook of the Birds of India and Pakistan.* Vol 1, 1964, Vol 2, 1969, Vol 3, 1969, Vol 4, 1970, Vol 5, 1972, Vol 6, 1971, Vol 7, 1972, Vol 8, 1973, Vol 9, 1973, Vol 10, 1974. Bombay: Oxford University Press.

Anderson, C. 1994. Seychelles sunbirds (*Nectarinia dussumieri*) on Aride Island. *Phelsuma* 2: 67-70.

Ash, J.S. and Miskell, J.E. 1980. A mass migration of Rollers *Coracias garrulus* in Somalia. *Bull. B.O.C.* 100: 216-218.

Ash, J.S. and Shafeeg, A. 1994. Birds of the Maldive Islands, Indian Ocean. *Forktail* 10: 3-32.

Ashmole, N.P. 1965. Adaptive radiation in the breeding regime of a tropical seabird. *Proc. Natn. Acad. Sci. USA* 53: 311-318.

Augeri D. and Pierce, S. 1995. *Aldabra 1995 Science and Conservation Program. Summary of bimonthly reports #1-#4 and update.* SIF Internal Report.

Ayrton, V. 1994. *Aride Island 1994 seabird report.* RSNC Internal Report.

Ayrton, V. 1995. *1994 Roseate Tern breeding season-Aride Island.* RSNC Internal Report.

Bailey, R.S 1968. The pelagic distribution of seabirds in the western Indian Ocean. *Ibis* 110: 493-519.

Bailey, R.S. and Bourne, W.R.P. 1963. Some records of petrels handled in the northern Indian Ocean. *J. Bombay Nat. Hist. Soc.* 60: 256-258.

Barré, N. Barau A. and Jouanin, C. 1996. *Oiseaux de la Réunion.* Paris: Les Éditions du Pacifique.

Bathe, H. and Bathe, G. 1982. *Territory size and habitat requirement of the Seychelles Brush Warbler Acrocephalus (Bebrornis) sechellensis.* Cousin Island Research Station, Tech. Rep. 18. ICBP Internal Report.

Baty, S.C.E. 1895. *Report on the Aldabra Group of Islands.* Mahé: Crown Lands Dept.

Beaman, M. and Madge, S. 1998. The Handbook of Bird Identification for Europe and the Western Palearctic. London: A & C Black.

Beamish, T. 1981. *Birds of Seychelles.* Hong Kong: Government Printer.

Benedict, B. 1957. The Immigrant Birds of Mauritius. *Avicult. Mag.* 63: 155-157.

Benson, C.W 1960. Les origines de l'avifaune de l'archipel des Comores. *Mem. Inst. Scient. Madagascar.* A14: 173-204.

Benson, C.W 1967. The birds of Aldabra and their status. *Atoll Res. Bull.* 118: 63-111.

Benson, C.W. 1970a. Land (including shore) birds of Cosmoledo. *Atoll Res. Bull.* 136: 67-81.

Benson, C.W. 1970b. Land (including shore) birds of Astove. *Atoll Res. Bull.* 136: 115-120.

Benson, C.W. 1970c. An introduction of *Streptopelia picturata* into the Amirantes. *Atoll Res. Bull.* 136: 195-196.

Benson, C.W. 1970d. The Cambridge collection from the Malagasy region. *Bull. B.O.C.* 90: 168-172.

Benson, C.W. 1971. Notes on *Terpsiphone* and *Coracina* spp. in the Malagasy Region. *Bull. B.O.C.* 91: 56-60, 61-64.

Benson, C.W. 1984. Origins of Seychelles landbirds. In Stoddart (ed.) *Biogeography and ecology of the Seychelles Islands*: 469-486. The Hague: Dr W. Junk Publishers.

Benson, C.W., Beamish, H.H., Jouanin, C., Salvan, J. and Watson, G.E. 1975. The birds of the Iles Glorieuses. *Atoll Res. Bull.,* 176: 1-34.

Benson, C.W. and Penny, M.J. 1968. A new species of warbler from the Aldabra Atoll. *Bull. B.O.C* 88: 102-108.

Benson, C.W. and Penny, M.J. 1971. The landbirds of Aldabra. *Phil. Trans. R. Soc. Lond.* B 260: 417-527.

Bergeson, M. 1996a. News from Aldabra. *Birdwatch* 20: 6-7.

Bergeson, M 1996b. Flamingo news from Aldabra. *Birdwatch* 20: 16-17.

Bergeson, M. and Rainbolt, R. 1995. Flamingos nesting. *Birdwatch* 15: 7-10.

Bergne, H. A'C. 1901. *Rough notes on a voyage to the Aldabra Group.* Manuscript.

Bernadin de St Pierre, J.H. 1773. *Voyagé À L'isle de Bourbon, au Cap de Bonne-Esperance par un Officier du Roi.* Neuchâtel: Societé Typographique.

Betts, F.N. 1940. The birds of the Seychelles II. The sea birds – more particularly those of Aride Island. *Ibis* 14 4: 489-504.

Betts, M. 1998. *Aride Island Nature Reserve, Seychelles. Annual Report 1997.* RSNC Internal Report.

Betts, M. 1999. News from Aldabra *Birdwatch* 31: 12-16.

Betts, M. 2000a. News from Aldabra *Birdwatch* 34: 4-6.

Betts, M. 2000b. *Research Officer's Annual Report June 1999-July 2000.* SIF Internal Report.

Birch, D. 2000. The feeding ecology of Greater Frigatebirds *Fregata minor* and Lesser Frigatebirds *Fregata ariel* on Aride Island. In Bowler and Hunter. 2000. *Aride Island Nature Reserve, Seychelles. Annual Report 1999.* RSNC Internal Report.

Blackburn, C.A. 1883. Quelques courtes descriptions des oiseaux de Praslin. *Trans. Roy. Soc. Arts Sci. Maurice* 12: 125-130.

Bourne, W.R.P. 1960. The petrels of the Indian Ocean. *Sea Swallow* 13: 9-22.

Bourne, W.R.P. 1966. Observations on islands in the Indian Ocean. *Sea Swallow* 18: 40-43.

Bourne, W.R.P. 1971. The birds of the Chagos Group. *Atoll Res. Bull.* 149: 175-207.

Bourne, W.R.P. 1976. On subfossil bones of Abbott's Booby *Sula abbotti* from the Mascarene Islands, with a note on the proportions and distribution of the Sulidae. *Ibis* 118: 119-123.

Bowler, J. and Hunter, J. 1999. *Aride Island Nature Reserve, Seychelles. Annual Report 1998.* RSNC Internal Report.

Bowler, J. and Hunter, J. 2000. *Aride Island Nature Reserve, Seychelles. Annual Report 1999.* RSNC Internal Report.

Bretagnolle, V. and Attie, C. 1995. Coloration and biometrics of fledgling Audubon's Shearwaters *Puffinus lherminieri* from Réunion Island, Indian Ocean. *Bull. BOC* 116: 194-197

Bretagnolle, V., Attie, C. and Mougeot, F. (in press). The Audubon's Shearwater on Réunion Island, Indian Ocean : behaviour, habitat, census, distribution and breeding biology. *Ibis.*

Britten, P.L. and Osborne, T.O. 1976. The race of *Sterna bergii* in Kenya. *Bull. B.O.C.* 6: 132-134.

Brooke, M de L. 1985. The annual cycle of the Toc-toc *Foudia sechellarum* on Cousin Island, Seychelles. *Ibis* 127: 7-15.

Brown, L.H. and Brown, B.E. 1973. The relative numbers of migrant and resident rollers in eastern Kenya. *Bull. B.O.C.* 93: 126-130.

Bullock, I.D. 1988. *The Roseate Tern Sterna dougallii in Seychelles.* ICBP Internal Report.

Bullock, I.D. 1989. *Aride Island Nature Reserve Scientific Report April 1987-April 1989.* RSNC Internal Report.

Bullock, I.D. 1995. Tree-climbing shearwaters. *Birdwatch* 14: 15-17.

Bullock, I.D., Cadbury, J.C., Vogtlin, J. and Carty, P. 1995. *Barn Owls on Aride Island, Seychelles.* RSNC Internal Report.

Bullock, I.D., Komdeur, J.G., Komdeur, M.D., Laboudallon, V. and Lewis, G.F. 1988. *The Seychelles Black Paradise Flycatcher (Terpsiphone corvina).* RSNC/ICBP Internal Report.

Bullock, I.D. Laboudallon, V. and Cedras, A. 1989. *The Black Parrot Coracopsis nigra barklyi on Curieuse Island, Seychelles. A summary of recent sightings.* ICBP Internal Report.

Burger, A.E. 2000a. *A census of Red-footed Boobies, Greater Frigatebirds and Lesser Frigatebirds On Aldabra Atoll in March 2000.* Report to BirdLife Seychelles, The Second Dutch Trust Fund and Seychelles Islands Foundation

Burger, A.E. 2000b. *Census of Audubon's Shearwater, White-tailed Tropicbird and White Tern on Cousin Island, February 2000.* BirdLife Seychelles Internal Report.

Burger, A.E. 2000c. *Diving Depths of Shearwaters.* BirdLife Seychelles Internal Report.

Burger, A.E. and Betts, M. 2000. *Populations of Red-footed Boobies, Greater Frigatebirds and Lesser Frigatebirds breeding on Aldabra Atoll, Indian Ocean, in 2000.* Unpublished manuscript.

Burger, A.E., and Lawrence, A.D. 1999a. *Seychelles Seabird Monitoring Project 1999-2000.* BirdLife Seychelles. Internal Report to BirdLife Seychelles, the Dutch Trust Fund and IUCN Nairobi.

Burger, A.E. and Lawrence, A.D. 1999b. *A census of seabirds on Frégate Island, Seychelles in August 1999.* BirdLife Seychelles Internal Report.

Burger, A.E., Lawrence, A.D. and Davis, L.M. 1999. *Census and distribution of Bridled Terns on Cousin Island in 1999.* BirdLife Seychelles. Internal Report.

Burger, A.E. and Lawrence, A.D. 2000. *Census of Wedge-tailed and Audubon's Shearwaters on Cousin Island, Seychelles using call playback.* Unpublished manuscript.

Carey, G. and Olsson, U. 1995. Field identification of Common, Wilson's, Pintail and Swinhoe's Snipes. *Birding World.* Vol 8. No 5.

Carty, P. and Carty, H. 1996. *Aride Island Nature Reserve Seychelles. Scientific Report for 1995.* RSNC Internal Report.

Castle, G.E. and Mileto, R. 1991. *Aride Island Scientific Report 1989-1991.* RSNC Internal Report.

Chantler, P. 1993. Identification of Western Palearctic swifts. *Dutch Birding* 15: 97-135.

Chantler, P. and Driessens, G. 1995. *Swifts. A Guide to the Swifts and Treeswifts of the World.* Robertsbridge: Pica Press.

Chapman, R. 1997. News from Aldabra. *Birdwatch* 23: 5-6.

Charpentier de Cossigny, J.F. 1764. *Mémoire sur i'Ile de France.* In Crépin, P. 1922. *Charpentier de Cossigny, Fonctionnaire Colonial, d'apres ses écrits et ceux de quelques-uns de ses contemporains.* Thesis, Universitié de Paris.

Cheke, A. 1982. *Les noms des oiseaux dans les iles francophones de l'Ocean Indien.* Essai ornithologique. Institut International D'Ethnosciences.

Cheke, A.S. 1987. The ecology of the smaller land-birds of Mauritius. In Diamond, A.W. *Studies of Mascarene islandbirds.* Cambridge: Cambridge University Press.

Clement, P., Harris, A. and Davis, J. 1993. *Finches and Sparrows.* London: A & C Black.

Clement, P. and Norfolk, T. 1995. Southern and eastern Great Grey Shrikes in northwest Europe. *Birding World* Vol: 8 No: 8.

Collar, N.J., Crosby, M.J. and Stattersfield, A.J. 1994. *Birds to watch 2. The world list of threatened birds.* Cambridge: BirdLife International.

Collar, N.J. and Stuart, S.N. 1985. *Threatened Birds of Africa and Related Islands. The ICBP/IUCN Red Data Book, Part 1.* 3rd edition. Cambridge: ICBP.

Colston, P. and Burton, P. 1988. *A Field Guide to the Waders of Britain and Europe with North Africa and the Middle East.* London: Hodder & Stoughton.

Cook, M. 1978. *Group territory and dispersion in the Seychelles bulbul (Hypsipetes crassirostris crassirostris).* Aberdeen University Expedition to the Seychelles 1977 Report.

Coppinger, R.W. 1899. *Cruise of the "Alert".* Fourth edition. London: Swan Sonnenschein & Co.

Copsey, J.A. and Kraaijeveld, K. 1998. A short study of the wedge-tailed shearwater (*Puffinus pacificus*) on Cousine Island, Seychelles. *Phelsuma 6: 53-57.*

Cowles, G.S. 1987. The fossil record. In Diamond, A.W. *Studies of Mascarene islandbirds.* Cambridge: Cambridge University Press.

Cramp, S. (ed.) 1985. *The Birds of the Western Palearctic.* Vol 4, 1988 Vol 5, 1992. Vol 6. Oxford: Oxford University Press.

Cramp, S. and Perrins C.M., 1993. *The Birds of the Western Palearctic.* Vol 7, 1994. Vol 8, 1994. Vol 9. Oxford: Oxford University Press.

Cramp, S. and Simmons, K.E.L. (eds.). 1977. *The Birds of the Western Palearctic.* Vol 1, 1979. Vol 2, 1982. Vol 3. Oxford: Oxford University Press.

Crook, J.H. 1960. The present status of certain rare landbirds of the Seychelles islands. *Seychelles Government Bulletin.*

Crook, J.H. 1961. The fodies (*Ploceinae*) of the Seychelles Islands. *Ibis* 103a: 517-548.

Curry-Lindahl, K. 1981. *Bird Migration in Africa. Movements Between Six Continents.* Vol 1and 2. London: Academic Press.

Cuthbert, R.J. and Denny, M.J.H. 1995. *Population estimate of the Seychelles Brush Warbler (Acrocephalus sechellensis) on Aride Island, Seychelles – May 1995.* RSNC Internal Report.

Dauban, H. 1979. The Cateau Vert is not a Seychelles bird. *Letter to the 'Nation' 23rd February 1979.*

Davis, S. 2000. The breeding biology of White-tailed Tropicbirds *Phaethon lepturus* on Aride Island, Seychelles, 1997-1999. In Bowler and Hunter. 2000. *Aride Island Nature Reserve, Seychelles.* Annual Report 1999. RSNC Internal Report.

Day, D. 1981. *The Doomsday Book of Animals.* London: Ebury Press.

Delacour, J. 1944. A revision of the family *Nectariniidae* (Sunbirds). *Zoologica,* N.Y. 29: 17-38.

Del Hoyo, J., Elliott, A. and Sargatal, J. (eds.). 1992-1999. *Handbook of the Birds of the World.* Vol 1, 1992, Vol 2, 1994, Vol 3, 1996, Vol 4, 1997, Vol 5, 1999.

Devillers, P., Ouellet, H., Benito-Espinal, E., Beudels, R., Cruon, R., David, N., Erard, C., Gosselin, M. and Seutin, G. 1993. *Noms francais des oiseaux du monde.* Quebec: Editions Multimondes/Bayonne: Editions Chabaud.

Diamond, A.W. 1971. The ecology of the seabirds of Aldabra. *Phil. Trans. Roy. Soc. Lond.* B 260: 561-571.

Diamond, A.W. 1974. The Red-footed Booby on Aldabra Atoll, Indian Ocean. *Ardea* 62: 196-218.

Diamond, A.W. 1975a. Biology and behaviour of Frigatebirds *Fregata* spp. on Aldabra Atoll. *Ibis* 117: 302-323.

Diamond, A.W. 1975b. *Cousin Island Nature Reserve management plant 1975-1979.* ICBP Internal Report.

Diamond, A.W. 1975c. The biology of Tropicbirds at Aldabra Atoll, Indian Ocean. *Auk* Vol 92: 16-39.

Diamond, A.W. 1976. Subannual breeding and moult cycles in the Bridled Tern *Sterna anaethetus* in the Seychelles. *Ibis* 118: 414-419.

Diamond, A.W. 1979. Dynamic ecology of Aldabran seabird communities. *Phil. Trans. R. Soc. Lond.* B 286: 231-240.

Diamond, A.W. 1980. Seasonality, population structure and breeding ecology of the Seychelles Brush Warbler *Acrocephalus sechellensis. Proc. IV Pan-Afr. Orn. Congr.:* 253-266.

Diamond, A.W. (ed.). 1987. *Studies of Mascarene Island Birds.* Cambridge: Cambridge University Press.

Diamond, A.W. and Feare, C.J. 1980. Past and present biogeography of central Seychelles birds. *Proc. IV Pan-Afr. Orn. Congr.:* 89-97.

Diamond, E.P. 1981. An early report of the flora and fauna of the Aldabra Group. *Atoll Res. Bull.* 255: 1-10.

Diamond, A.W and Prys-Jones, R.P. 1986. The biology of terns nesting at Aldabra Atoll, Indian Ocean, with particular reference to breeding seasonality. *J. Zool. Lond (A)* 210: 527-549.

D'Offay, D. and Lionnet, G. 1982. *Diksyonner kreol-franse.* Hamburg: Helmut Buske Verlag.

Dowsett, R.J. and Forbes-Watson, A.D. 1993. *Checklist of the Birds of the Afrotropical and Malagasy Regions. Volume 1: Species Limits and Distribution.* Liège: Turaco Press.

Dowsett-Lemaire, F. 1979. The imitative range of the song of the marsh warbler, *Acrocephalus palustris,* with special reference to imitations of African birds. *Ibis* 121: 453-68.

Dowsett-Lemaire, F. 1994. The song of the Seychelles Warbler *Acrocephalus sechellensis* and its African relatives. *Ibis* 136: 489-491.

Dupont, R.P. 1907. Report on a visit of investigation to St Pierre, Astove, Cosmoledo, Assumption and the Aldabra Group. *Seychelles Government Printer.*

Dupont, R.P. 1916. *Aldabra: its guano deposits and other resources.* Seychelles National Archives: Typesript; Aldabra Group, Correspondence relating 1889-1943.

Dupont, R.P. 1929. *Report on a visit of investigation to the principal outlying islands of the Seychelles archipelago.* Dept. of Agriculture files, Mahé: Typescript.

Evans, P.G.H. 1979. Status and Conservation of the Seychelles Black Parrot. *Biological Conservation* 16: 233-240.

Farquhar, Commander S. St J. 1900. On two nesting places of gannets and terns in the South Indian Ocean. *Ibis* (7) 6: 63-67.

Fauvel, A.A. 1909. *Unpublished Documents on the History of the Seychelles Anterior to 1810.* Victoria: Govt. Printer.

Fayon, M. 1971. The plight of the Paradise Flycatcher. *J. Seychelles Soc.* 7: 8-11.

Feare, C.J. 1975a. Observations on the Seychelles White-eye *Zosterops modesta. Auk* Vol 92, No 3: 615-618.

Feare, C.J. 1975b. Post-fledging parental care in Crested and Sooty Terns. *Condor* 77: 368-370.

Feare, C.J. 1975c. Scavenging and kleptoparasitism in feeding methods of Seychelles Cattle Egrets *Bubulcus ibis. Ibis* 117: 388.

Feare, C.J. 1976a. Desertion and abnormal development in a colony of Sooty Terns infested by virus-infected ticks. *Ibis* 118: 112-115.

Feare, C.J. 1976b. Seabird ecology and tick distribution in the Western Indian Ocean. *National Geographic Society Research Reports.*

Feare, C.J. 1976c. The breeding of the Sooty Tern *Sterna fuscata* in the Seychelles and the effects of experimental removal of its eggs. *J. Zool., Lond.* 179: 317-360

Feare, C.J. 1976d. The exploitation of Sooty Tern eggs in the Seychelles. *Biological Conservation* 10: 169-181.

Feare, C.J. 1977. *Phedina borbonica madagascariensis* in the Amirantes. *Bull. B.O.C.* 97: 87-89.

Feare, C.J. 1978a. The decline of booby (*Sulidae*) populations in the western Indian Ocean. *Biological Conservation* 14: 295-305.

Feare, C.J. 1978b. Ringability and edibility. *BTO News.* 93: 6.

Feare, C.J. 1979a. Ecology of Bird Island. *Atoll Res. Bull.* 226: 1-29.

Feare, C.J. 1979b. Ecological observations on African Banks, Amirantes. *Atoll Res. Bull.* 227: 1-7.

Feare, C.J. 1981. Breeding schedules and feeding strategies of Seychelles seabirds. *Ostrich* 52: 179-185.

Feare, C.J. 1983. Mass spring migration of European Rollers *Coracias garrulus* in eastern Tanzania. *Bull. B.O.C.* 103: 39-40.

Feare. 1989. Underwater booby-trap. *BBC Wildlife Magazine.* March 1989.

Feare, C.J. 1999. *The sustainable exploitation of sooty tern eggs in the Seychelles.* Seventh Annual Report, June 1999. Unpublished Report to the Division of Environment, Seychelles.

Feare, C.J. and Bourne, W.R.P. 1977. The occurrence of "Portlandica" Little Terns and absence of Damara Terns and British Storm Petrels in the Indian Ocean. *Ostrich* 49: 64-66.

Feare, C.J. and Gill, E. 1993. *The sustainable exploitation of sooty tern eggs in the Seychelles.* Report of a visit to Seychelles 7-21 August 1993.

Feare, C.J. and Gill, E. 1994. *The sustainable exploitation of sooty tern eggs in the Seychelles.* Second Annual Report June-July 1994.

Feare, C.J. and Gill, E. 1995a. *The sustainable exploitation of sooty tern eggs in the Seychelles.* Third Annual Report June-July 1995.

Feare, C.J. and Gill, E. 1995b. The turtle doves of Bird Island, Seychelles. *Bull. B.O.C.* 115(4): 206-210.

Feare, C.J. and Gill, E. 1996. *The sustainable exploitation of sooty tern eggs in the Seychelles.* Fourth Annual Report June 1996.

Feare, C.J., Gill, E.L., Carty, P., Carty, H.E. and Ayrton, V.J. 1997. Habitat use by Sooty Terns *Sterna fuscata* and implications for colony management. *Biological Conservation* 81: 69-76.

Feare, C.J., Temple, A.A. and Procter, J. 1974. The status, distribution and diet of the Seychelles Kestrel *Falco araea. Ibis* 116: 548-551.

Feare, C.J. and Watson, J. 1984. Occurrence of migrant birds in the Seychelles. In Stoddart (ed.) *Biogeography and Ecology of the Seychelles Islands:* 469-486. The Hague: Dr W. Junk Publishers.

Forsman, D. 1999. *The Raptors of Europe and the Middle East. A Handbook of Field Identification.* London: T & AD Poyser.

Foster-Vesey-Fitzgerald, D. 1953. Wild life in Seychelles. *Oryx* 2: 28-32.

Frith, C.B. 1974. New observations of migrants and vagrants for Aldabra, Farquhar and Astove Atolls, Indian Ocean. *Bull. B.O.C.* 94: 12-19.

Frith, C.B. 1975a. Predation upon hatchlings and eggs of the green turtle, *Chelonia mydas,* on Aldabra Atoll, Indian Ocean. *Atoll Res. Bull.* 176: 11-12.

Frith, C.B. 1975b. Field Observations on *Centropus toulou insularis* on Aldabra Atoll. *Ostrich* 46: 51-257.

Frith, C.B. 1976. A Twelve-month field study of the Aldabran Fody *Foudia eminentissima aldabrana. Ibis* 118(2): 115-178.

Frith, C.B. 1977. Life History Notes on some Aldabran landbirds. *Atoll Res. Bull.* 201: 1-17.

Froberville, E. de. 1848. *Rodrigues, Galega, Les Sechelles, Les Almirantes, etc.* In M.A.P. d'Avezac (ed.). *Iles de l'Afrique 3me partie, II.* Paris: Firmin Didot Freres, Editeurs, 65-114.

Fry, C. Hilary 1994. Kingfishers and Bee-eaters. *Birding in Southern Africa.* Vol 46 No 4.

Fry, C.H., Fry, K. and Harris, A. 1992. *Kingfishers, Bee-eaters and Rollers.* London: A & C Black.

Fry, C.H., Keith S. and Urban E.K. 1988. *The Birds of Africa.* Vol 3. London: Academic Press.

Fryer, J.C.F. 1908. *Diary.* Manuscript.

Fryer, J.C.F. 1911. The structure and formation of Aldabra-with notes on their flora and fauna. *Trans. Linn. Soc. Lond.* (Z.) 14: 397-442.

Gadow, H. and Gardiner, J.S. 1907. Aves, with some notes on the distribution of the land-birds of the Seychelles. *Trans. Linn. Soc. Lond.* (2) 12 (Percy Sladen Exp. Rpts. 1): 103-110.

Gardiner, J.S. and Cooper, C.F. 1907. Description of the expedition. *Trans. Linn. Soc. Lond.* (2) 12 (Percy Sladen Exp. Rpts. 1): 111-175.

Gaymer, R. 1967. Observations on the birds of Aldabra in 1964 and 1965. *Atoll Res. Bull.* 118: 113-125.

Gaymer, R., Blackman, R.A.A., Dawson, P.G., Penny, M. and Penny, C.M. 1969. The endemic birds of Seychelles. *Ibis* 111: 157-176.

Gerlach, R. 1995. Seychelles kestrels taking Madagascar fodies. *Birdwatch* 15: 31.

Gerlach, J. 1996a. Recent changes in Madagascar Fodies. *Birdwatch* 18: 26 -27.

Gerlach, J. 1996b. New threats to Seychelles birds. *Birdwatch* 20: 18-24.

Gerlach, J. 1997a. *Seychelles Red Data Book 1997.* Victoria: Nature Protection Trust of Seychelles.

Gerlach, J. 1997b. Water lettuce (*Pistia stratiotes*) – a new invader. *Phelsuma* 5: 75-76

Gerlach, J. 1998. Malagasy Turtle Doves – More Historical Muddles? *Birdwatch* 28: 17-19.

Gerlach, J. 1999. Further speculations on turtle doves. *Birdwatch* 32: 21-23.

Gerlach, J. and Canning, K.L. 1996. *Seychelles Terrapin Conservation Project*. NPTS Internal Report.

Gerlach, G. and R. 1994. Introductions or Natural Colonists? Historical confusion in the case of *Foudia madagascariensis* and *Astrilda astrild*. *Phelsuma* 2: 64-66.

Gerlach, R. and Gerlach, J. 1997. Recent natural colonisation of the granitic islands by three bird species. *Phelsuma* 5: 63-67.

Gerlach, J. and Skerrett, A. (in press). The status of the yellow bittern *Ixobrychus sinensis* in Seychelles.

Gibson-Hill 1950. Ornithological notes from the Raffles Museum. 9. Notes on Abbott's Booby. *Bull. Raffles Mus.* 23: 65-76.

Gill, F.B. 1970. The taxonomy of the Mascarene Olive White-eye *Zosterops olivacea* (L.). *Bull. B. O. C.* 90: 81-82.

Gill, F.B. 1971. Ecology and evolution of the sympatric Mascarene white-eyes *Zosterops borbonica* and *Zosterops olivacea*. *Auk* 88: 35-60.

Gillham, M.E. 1977a. Observations on vegetation of Blue-faced Booby colonies on Cosmoledo Atoll, western Indian Ocean. *Atoll Res. Bull.* 199: 1-11.

Gillham, M.E. 1977b. Vegetation of sea and shore-bird colonies on Aldabra Atoll. *Atoll Res. Bull.* 200: 1-19.

Glutz von Blotzheim, U.N. and Bauer, K.M. 1980. *Handbuch der Vögel Mitteleuropas*. Vol 9. Wiesbaden: Akademische Verlagsgesellschaft.

Goodwin, D. 1967. *Pigeons and Doves of the World*. London: British Museum (Natural History).

Grant, P.J. 1986. *Gulls: A Guide to Identification*. Second edition. Calton: T & AD Poyser.

Grantlett, S. and Harrap, S. 1992. Identification forum: South Polar Skua. *Birding World* Vol 5, No 7.

Greenway, 1958. *Extinct and Vanishing Birds of the World*. New York: American Committee for International Wildlife Protection.

Greig-Smith, P.W. 1979a. Notes on the biology of the Seychelles Bulbul. *Ostrich* 50: 45-58.

Greig-Smith, P.W. 1979b. Observations of nesting and group behaviour of Seychelles White-eyes *Zosterops modesta*. *Ibis* 121: 344-348.

Greig-Smith, P.W. 1980. Foraging, Seasonality and Nesting of Seychelles Sunbirds *Nectarinia dussumieri*. *Ibis* 121: 307-321.

Gretton, A. 1991. *The Ecology and Conservation of the Slender-billed Curlew* (Numenius tenuirostris). Cambridge: ICBP Monograph 6.

Hall, B.P. 1953. On the status of the Seychelles Sunbirds *Cyanomitra dussumieri* and *Cyanomitra mahei*. *Ibis* 95: 545-546.

Hall, B.P. and Moreau, R.E. 1970. *An Atlas of Speciation in African Passerine Birds*. London: British Museum (Natural History).

Hambler, C., Hambler, K. and Newing, J.M. 1985. Some observations on *Nesillas aldabranus* the endangered Brush Warbler of Aldabra Atoll, with hypotheses on its distribution. *Atoll Res. Bull.* 290.

Hambler, C., Newing, J. and Hambler, K. 1993. Population monitoring for the flightless rail *Dryolimnas cuvieri aldabranus*. *Bird Conservation International* 3: 307-318.

Hancock, J. and Kushlan, J. 1984. *The Herons Handbook*. London: Croom Helm.

Harper, P.C. and Kinsky, F.C. 1978. *Southern Albatrosses and Petrels*. Wellington: Price Milburn/Victoria University Press.

Harris, A., Shirihai, H. and Christie D. 1996. *The Macmillan Birder's Guide to European and Middle Eastern Birds*. London: Macmillan.

Harris, A., Tucker, L. and Vinicombe, K. 1989. *The Macmillan Field Guide to Bird Identification*. London: Macmillan.

Harrison, P. 1985. *Seabirds: An Identification Guide*. Revised edition. London: Christopher Helm.

Hartlaub, C.J.G. 1877. *Die Vögel Madagascars und der benachbaren Inselgruppen*. Schmidt, Halle.

Hayman, P, Marchant, J. and Prater, T. 1986. *Shorebirds: An identification guide to the waders of the world*. London: Croom Helm.

Haysom, S. 1995. *Tourism and the breeding success of Fairy Terns (Gygis alba monte) and White-tailed Tropicbirds (Phaethon lepturus) on Aride Island Nature Reserve, Seychelles*. RSNC Internal Report.

Hellawell, C.E. 1979. *The Seychelles Magpie Robin on Aride*. ICBP Internal Report.

Herzig, H. and Carty, P. 1995. News from Aride Island. *Birdwatch* 15: 20-21.

Hirschfeld, E. 1990. White-cheeked Tern Identification. *Birding World* Vol 3, No 7.

Hirschfeld, E. 1992. Identification of Rufous Turtle Dove. *Birding World* Vol 5, No 2

Hirschfield, E., Roselaar, C.S. and Shirihai, H. 2000. Identification, taxonomy and distribution of Greater and Lesser Sandplovers. *British Birds* 93 (4): 162-189.

Hockey, P. 1993. Identification forum: jizz identification of sand plovers. *Birding World* Vol 6 No 9.

Hollom, P.A.D., Porter, R.F., Christensen, S. and Willis, I. 1988. *Birds of the Middle East and North Africa*. Calton: T & AD Poyser.

Honneger, R.E. 1966. Ornithologische Beobachtungen von den Seychellen. *Natur u. Mus.* 96: 481-488.

Huffstadt, A. and Prast, W.E. 1992. *Comparative studies on the breeding biology of two isolated Seychelles Warbler populations*. University of Amsterdam and ICBP.

Huxley, C.R. 1974. *Aldabra Station Report*. Unpublished.

Huxley, C.R. and Wilkinson, R. 1977. Vocalizations of the Aldabra white-throated rail *Dryolimnas cuvieri aldabranus*. *Proc. R. Soc. Lond. B*. 197: 315-331.

Huxley, C.R. 1979. The tortoise and the rail. *Phil. Trans. R. Soc. Lond. B*. 286: 225-230.

Johnson, L. 1994. News from Coëtivy. *Birdwatch* 9: 8.

Johnson, K.P. and Clayton, D.H 1999. Swiftlets on islands: genetics and phylogeny of the Seychelles and Mascarene swiftlets. *Phelsuma* 7: 9-13.

Keith, S. 1980. Origins of the avifauna of the Malagasy Region. *Proc. IV Pan-Afr. Orn. Congr.* 99-108.

Keith, G.S., Urban, E.K. and Fry, C.H. 1992. *The Birds of Africa*. Vol 4. London: Academic Press.

Kennedy, W.R. 1901. *Hurrah for the Life of a Sailor!* Edinburgh and London. William Blackwood and Sons.

Komdeur, J. 1988. The ecology and cooperative breeding of the Seychelles Brush Warbler *Acrocephalus sechellensis* with some conclusions for its future conservation. *Proc. VII Pan-Afr. Orn. Congr.* 16 pp.

Komdeur, J. 1991a. *Cooperative breeding in the Seychelles Warbler*. University of Cambridge: PhD Thesis.

Komdeur, J. 1991b. Influence of territory quality and habitat saturation on dispersal options in the Seychelles Brush Warbler (*Acrocephalus sechellensis*); an experimental test of the habitat saturation hypothesis for cooperative breeding. *Proc. 20th. Int. Ornithol. Congr.*, 1,325-1,332.

Komdeur, J. 1992. Importance of habitat saturation and territory quality for evolution of cooperative breeding in the Seychelles warbler. *Nature* 358: 493-495.

Komdeur, J. 1994a. Conserving the Seychelles Warbler *Acrocephallus sechellensis* by translocation from Cousin Island to the islands of Aride and Cousine. *Biological Conservation* 67: 143-152.

Komdeur, J. 1994b. Experimental evidence for helping and hindering by previous offspring in the cooperative-breeding Seychelles warbler *Acrocephalus sechellensis*. *Behav. Ecol. Sociobiol.* 34: 175-186.

Komdeur, J. 1994c. The effect of kinship on helping in the cooperative breeding Seychelles Warbler (*Acrocephalus sechellensis*). *Proc. R. Soc. Lond. B* 256: 47-52.

Komdeur, J. 1996. Facultative sex ratio bias in the offspring of Seychelles warblers. *Proc. R. Soc. Lond. B* 263: 661-666.

Komdeur, J. and Gabrielsen, L. 1993. *Report on a visit to the islands of Cousin and Aride (Seychelles) in order to collect blood samples from and update the record of colour-ringed Seychelles Warblers (August 1993)*. Unpublished.

Komdeur, J., Huffstadt, A., Prast, W., Castle G., Mileto, R. and Wattel, J. 1995. Transfer experiments of Seychelles warblers to new islands: changes in dispersal and helping behaviour. *Anim. Behav.* 49: 695-708.

Komdeur, J. and Rands, M. 1989. *The Conservation of the Seychelles Warbler (Acrocephalus sechellensis) on Cousin Island, Seychelles.* ICBP Internal Report.

Ladouceur, F. 1997. *Sighting of Paradise Flycatcher (La Vev) on Marianne Island.* Ministry of Environment Internal Report.

Langrand, O. 1990. *Guide to the Birds of Madagascar.* New York: Yale University Press.

Langrand, O. and Sinclair, J.C. 1994. Additions and Supplements to the Madagascar avifauna. *Ostrich* 65: 302-310.

Le Corre, M. 1999. Plumage polymorphism of red-footed boobies (*Sula sula*) in the western Indian Ocean: an indicator of bio-geographic isolation. *J. Zool., Lond.* 249: 411-415

Le Corre, M. and Jouventin, P. 1999. Geographical variation in the White-tailed Tropicbird *Phaethon lepturus*, with the description of a new subspecies endemic to Europa Island, southern Mozambique Channel. *Ibis* 141: 233-239.

Lewington, I., Alström, P. and Colston, P. 1991. *A Field Guide to the Rare Birds of Britain and Europe.* London: HarperCollins.

Lionnet, G. 1959. A Review of the Biological Control of Agricultural Pests in the Seychelles. *East Af. Agr. Journal* Vol XXIV, No 4.

Lockwood, W.B. 1993. *The Oxford Dictionary of British Bird Names.* Oxford: Oxford University Press.

Lodge, W. 1991. *Birds' Alternative Names. A World Checklist.* London: Blandford.

Long. J. L. 1981. *Introduced Birds of the World.* London: David and Charles/A.W. Read Pty. Ltd.

Louette, M. 1985. Double invasions of birds on the Comoro islands. *Proc. VI Pan-Afr. Orn. Congr.*: 77-86.

Louette, M. 1992. The origin and evolution of Comoro landbirds. *Proc. VII Pan-Afr. Orn. Congr.*: 207-215

Loustau-Lalanne, P. 1962. Land birds of the granitic islands of the Seychelles. *Seychelles Soc. occ. Publ.* 1.

Loustau-Lalanne, P. 1963. Sea and shore birds of the Seychelles. *Seychelles Soc. occ. Publ.* 2.

Lowe, K. W. and Richards, G. C. 1991. Morphological variation in the Sacred Ibis *Threskiornis aethiopicus* Superspecies Complex. *EMU* Vol 91: 41-45.

Lucking, R.S. 1995. Cinnamon Bittern *Ixobrychus cinnamomeus* in Seychelles, first for the Afro-Malagasy Region. *Bull. ABC* 2(2): 107-108.

Lucking, R.S. 1996a. Polygyny in the Seychelles Sunbird *Nectarinia dussumieri*. *Bull. B.O.C.* 116 (3): 178-179.

Lucking, R.S. 1996b. Eastern race Bar-tailed Godwit on Frégate. *Birdwatch* 17: 9-11.

Lucking, R.S. 1997. Hybridization between Madagascan Red Fody *Foudia madagascariensis* and Seychelles Fody *Foudia sechellarum* on Aride Island, Seychelles. *Bird Conservation International* 7: 1-6.

Lucking, R.S. and Lucking, V. 1997. Monitoring and conservation management of the Seychelles Magpie-robin *Copsychus sechellarum*. In Rocamora, G. 1997c. *Rare and threatened species, sites and habitats monitoring programme in Seychelles. Project G1 EMPS. Final Report. Volume 1 Monitoring methodologies and priority actions.* Ministry of Environment/European Union/BirdLife International.

Lucking, R.S. and Lucking, V. 1998. Inter-island translocations of the Seychelles Magpie-robin. *Bird Conservation International* 8: 109-112.

Macdonald, R.A. 1978. *The biology of the Seychelles cave swiftlet Aerodramus (francicus) elaphrus* Aberdeen University Expedition to the Seychelles 1977 Report.

Macdonald, K. 1993. *Project run down for La Digue CNP.* CNP Internal Report.

MacFarland, C.G. and Reeder, W.G. 1974. Cleaning symbiosis involving Galápagos tortoises and two species of Darwin's finches. *Z. Tierpsychol.* 34: 464-483.

Mackworth-Praed, C.W. and Grant, C.H.B. 1962. *Birds of the Southern Third of Africa.* London: Longmans, Green and Co., Ltd.

Madge, S. 1982. April records of White-cheeked Terns in Sinai. *Dutch Birding* 4: 104-105.

Madge, S. and Burn, H. 1986. *Wildfowl: An Identification Guide to the Ducks, Geese and Swans of the World.* London: Christopher Helm.

Marshall, J. 1978. Systematics of smaller Asian nightbirds based on voice. *Orn. Monogr.* 25: i-viii,1-54.

Matyot, P. 1996. News from Silhouette. *Birdwatch* 18: 16-18.

Matyot, P. 2000. Ornithological notes on Ile Platte from the diaries of E.S. Brown. *Birdwatch* 33: 15-16.

Maul, A.M. 1996. *Aride Island seabird report May-August 1995.* RSNC Internal Report.

Maul, A.M. 1997. *The Roseate Tern Sterna dougalli on Aride Island, Seychelles: 1996 breeding season.* RSNC Internal Report.

McCulloch 1995. *Vocalisations of the Seychelles Magpie Robin.* Unpublished.

McCulloch, N. 1996. *The Seychelles Magpie Robin Recovery Plan.* BirdLife International/RSPB Internal Report.

Mearns, B. and R. 1988. *Biographies for Birdwatchers.* London: Academic Press.

Medway, Lord 1966. Field characters as a guide to the specific relations of swiftlets. *Proc. Linn. Soc. Lond.* 177, 2: 151-172.

Medway, Lord and Pye, J.D. 1977. Echolocation and the systematics of swiftlets. in Evolutionary Ecology B. Stonehouse (ed.) 225-238.

Mellanby, R., Mee, A., Cresswell, W., Irwin, M., Jensen, M., McKean, M., Milne, L., Shepherd, E. and Bright, S. 1996. *Glasgow University Expedition to the Seychelles 1996.* Glasgow University Report.

Meinertzhagen, R. 1912. On the Birds of Mauritius. *Ibis* 82-108.

Moreau, R.E. 1960. The ploceine weavers of the Indian Ocean islands. *J. Orn., Lpz.* 101: 29-49.

Morris, P. and Hawkins, F. 1998. *Birds of Madagascar, A Photographic Guide.* London: Pica Press.

Mortimer, J.A. 1984. Rediscovery of the Turtle Dove *Streptopelia picturata* on Cosmoledo Atoll in the Seychelles. *Ibis* 126: 81-82.

Mortimer, J.A. and Constance, A. 2000. Observations on the birds of Cosmoledo Atoll, Seychelles. *Bull. B.O.C.* 120: 46-57.

Mullarney, K., Svensson, L., Zetterström, D. and Grant, P.J. 1999. *Collins Bird Guide.* London: HarperCollins.

Nelson, J.B. 1967. Etho-ecological adaptations in the Great Frigate Bird. *Nature* 214: 318.

Nelson, J.B. 1971. The biology of Abbott's Booby *Sula abbotti*. *Ibis* 113: 429-467.

Nelson, J.B. 1974. The distribution of Abbott's Booby *Sula abbotti*. *Ibis* 116: 368-369.

Neufeld, D. 1998. Nest site use and changes in habitat of the Seychelles Black Paradise Flycatcher. *Biological Conservation* 84: 103-105.

Newton, E. 1867. On the land-birds of the Seychelles archipelago. *Ibis* (2)3: 335-360.

Newton, A. and Newton, E. 1876. On the *Psittaci* of the Mascarene Islands. *Ibis* 23: 281-289.

Nicoll, M.A.C. 1996. *The biology and control of the Barn Owl on Aride Island, Seychelles.* RSNC Internal Report.

Nicoll, M.J. 1906. On the birds collected and observed during the voyage of the 'Valhalla', R.Y.S. from November 1905 to May 1906. *Ibis* (8) 6: 666-712.

Nicoll, M.J. 1909. *Three Voyages of a Naturalist, Being an Account of many little known Islands in Three Oceans visited by the "Valhalla." R.Y.S.* London: Witherby.

Normaja, J. 1994. Plumage Variation in River Warblers. *Birding World* Vol 7, No 5.

North, M. 1892. *Recollections of a Happy Life.* London: Macmillan.

Olsen, K.M. and Larsson, H. 1995. *Terns of Europe and North America.* London: Helm.

Olsen, K.M. and Larsson, H. 1997. *Skuas and Jaegers.* East Sussex: Pica Press.

Oustalet, E. 1878. Etude sur la faune ornithologique des iles Seychelles. *Bull. Soc. philomath. Paris* ser. 7 2: 161-206.

Parker, I.S.C. 1970. Some ornithological observations from the western Indian Ocean. *Atoll Res. Bull.* 136: 211-220.

Parr, S. 1998. Recolonisation of Marianne Island by Seychelles Paradise-Flycatcher after 60 Years. *Bull. ABC* 5(2): 90-91.

Peake, J.F. 1971. The evolution of terrestrial faunas in the western Indian Ocean. *Phil. Trans. Roy. Soc. Lond.* B 260: 581-610.

Penny, M. 1967. A new sanctuary in the Seychelles. *Oryx* 9: 214-216.

Penny, M. 1968. Endemic birds of the Seychelles. *Oryx* 9: 267-275.

Penny, M. 1974. *The Birds of Seychelles and the Outlying Islands.* London: Collins.

Penny, M.J. and Diamond, A.V. 1971. The White-throated Rail *Dryolimnas cuvieri* on Aldabra. *Phil. Trans. Roy. Soc. Lond.* B 260: 529-548.

Perrins, C. 1994. *L'Encyclopedie mondiale des Oiseaux.* Editions Bordas.

Phillips, N.J. 1982. *Forty-sixth report of the Scientific Administrator 1st April-30th June 1982, Cousin Island, Seychelles.* ICBP Internal Report.

Phillips, N.J. 1984. Migrant species new to Seychelles. *Bull. B.O.C.* 104: 9-10.

Phillips, N.J. 1987. The breeding biology of White-tailed Tropicbirds *Phaethon lepturus* at Cousin Island, Seychelles. *Ibis* 129: 10-24.

Phillips, N.J. and the Seychelles Bird Records Committee. 1997. Migrant landbirds in Seychelles. *Phelsuma* 5: 13-26.

Pike, N. 1872. A visit to the Seychelles Islands. *Trans. R. Soc. Arts Sci. Maurit.* ser. B 6: 83-142.

Pingré, G. c1763. *Voyage à L'île Rodrigue.* Paris: Bibliothèque Ste. Geneviève.

Pocklington, R., Willis, P.R. and Palmieri, M. 1972. Birds seen at sea and on island in the Cargados Carajos shoals. *Atoll Res. Bull.* 158: 1-8.

Prior, J. 1820. *Voyage in the Indian Seas in the Nisus Frigate during the years 1810 and 1811.* London: Sir Richard Phillips and Co.

Procter, J. 1970. *Conservation in the Seychelles.* Victoria: Seychelles Government Printer.

Procter, J. 1972. The nest and identity of the Seychelles swiftlet *Collocalia. Ibis* 114: 272-273.

Procter, J. 1974. *A management plan for Aride Island, Seychelles.* SPNC Internal Report.

Prys-Jones, R.P. 1979. The ecology and conservation of the Aldabran brush warbler *Nesillas aldabranus. Phil. Trans. R. Soc. Lond.* B. 286: 211-224.

Prys-Jones, R.P. 1984. The occurrence of migrant and vagrant terns at Aldabra Atoll, Indian Ocean. *Bull. B.O.C.* 104(2): 73-75.

Prys-Jones, R.P. 1993. *The conservation of avian biodiversity on a unique atoll ecosystem.* Unpublished proposal document.

Prys-Jones, R.P. and Diamond, AA.W. 1984. Ecology of the land birds on the granitic and coralline islands of the Seychelles with particular reference to Cousin Island and Aldabra Atoll. In Stoddart (ed.) *Biogeography and ecology of the Seychelles Islands:* 529-558. The Hague: Dr W. Junk Publishers.

Prys-Jones, R.P. and Peet, C. 1980. Breeding periodicity, nesting success and nest site selection among Red-tailed Tropicbirds *Phaethon rubricauda* and White-tailed Tropicbirds *P. lepturus* on Aldabra Atoll. *Ibis* 122: 76-81.

Prys-Jones, R.P., Prys-Jones, M.P. and Lawley, J.C. 1981. The birds of Assumption Island, Indian Ocean: past and future. *Atoll Res. Bull.* 248: 1-16.

Ramos, J.A. 1998. Roseate Terns on Aride: summary of 1997 breeding season and research activities. In Betts, M. 1998. *Aride Island Nature Reserve, Seychelles. Annual Report 1997.* RSNC Internal Report.

Ramos, J.A. 2000. The 1999 Roseate Tern breeding season on Aride. In Bowler and Hunter. 2000. *Aride Island Nature Reserve, Seychelles. Annual Report 1999.* RSNC Internal Report.

Rands, M. 1989. Saving the Seychelles brush warbler. *Oryx* 23: 3-4.

Reville, B.J. 1980. *Spatial and temporal aspects of breeding in the Frigatebirds Fregata minor and F. ariel.* University of Aberdeen Ph.D thesis.

Reville, B.J. 1983. Numbers of nesting frigatebirds, *Fregata minor* and *F. ariel,* on Aldabra Atoll Nature Reserve, Seychelles. *Biological Conservation* 27: 59-76.

Reville, B.J. 1988. Effects of spacing and synchrony on breeding success in the Great Frigatebird (*Fregata minor*). *Auk* 105: 252-259.

Reville, B.J. 1991. Nest spacing and breeding success in the Lesser Frigatebird (*Fregata ariel*). *The Condor* 93: 555-562.

Ridgway, R. 1893. Descriptions of some new birds collected on the islands of Aldabra and Assumption, northeast of Madagascar by Dr W.L. Abbott. *Proc. U.S. Natn. Mus.*16: 597-600.

Ridgway, R. 1895. On the birds collected by Dr W.L. Abbott in the Seychelles, Amirantes, Gloriosa, Assumption, Aldabra and adjacent islands, with notes on habits etc., by the collector. *Proc.U.S. Natn. Mus.*18: 509-546.

Ridley, M.W. and Percy, R. 1958. The exploitation of sea birds in Seychelles. *Col. Res. Stud.* 25.

Ridley, M.W. and Percy, R. 1966. *Report on the exploitation of sea birds eggs in Seychelles 1966.* Victoria: Seychelles Government Printer.

Rivers, F. 1878. *Letter to the Chief Civil Commissioner, Seychelles, 11th December 1878, on a visit to Aldabra on board the "Flower of Jarrow".* Seychelles National Archives: Letter book (outward), 1878-1880, B37.

Robenarimangason, H and Réne de Roland, L.A. 1998. Notes on nesting Madagascar Kestrels *Falco newtoni* on Masoala Peninsula, NE Madagascar. *Working Group on Birds in the Madagascar Region* 8 (1): 20-21.

Roberts, P. 1986. *Notes on Ringing and Censusing of Pied Crows in 1986.* Unpublished.

Roberts, P. 1987. Is the Aldabra brush warbler extinct? *Oryx* 21: 209-210.

Roberts, P. 1988. Introduced birds on Assumption Island – a threat to Aldabra. *Oryx* 22: 15-17.

Roberts, T.J. 1991. *The Birds of Pakistan.* Vol 1. Oxford: Oxford University Press.

Roberts, T.J. 1992. *The Birds of Pakistan.* Vol 2. Oxford: Oxford University Press.

Rocamora, G. 1997a. The Seychelles Grey White-eye – a species facing extinction. *Bull. ABC* 4(1): 13.

Rocamora, G. 1997b. Seychelles Grey White-eye. *World Birdwatch.* June 1997: 20-21.

Rocamora, G. 1997c. *Rare and threatened species, sites and habitats monitoring programme in Seychelles. Project G1 EMPS. Final Report. Volume 1 Monitoring methodologies and priority actions.* Ministry of Environment/European Union/BirdLife International.

Rocamora, G. and François, J 2000. Seychelles White-eye Recovery Programme. Phase 1 'Save the Seychelles Grey White-eye'. Final report. Ministry of Environment and Transport/Dutch Trust Fund/International Union for the Conservation of Nature.

Rocamora, G. and Jules, T. 2000.*Aldabra Drongo study. Brief analysis and conclusions from results obtained during the 1999-2000 breeding season.* SIF/DoE Internal Report.

Rocamora, G., François, J., Nolin, R. and Constance, P. 1999. Zwazo Linet Info No 2. Ministry of Environment and Transport Newsletter.

Rocamora, G., Remie, S., Constance, P. and Niol, D. 1995. *Results of the Seychelles Black Paradise Flycatcher Census, La Digue November 1995.* Report from the Conservation and National Parks Service, Seychelles.

Rocamora, G. and Skerrett, A. 1999. *First Inventory of Important Bird Areas of the Republic of Seychelles.* Report for BirdLife International.

Rountree, F.R.G., Guérin, R., Pelte, S. and Vinson, J. 1952. Catalogue of the Birds of Mauritius. *Bull. Mauritius Inst.* 3: 155-217.

Ryall, C. 1986. Killer crows stalk the Seychelles. *New Scientist* 2 October 1986.

Schoeberl, B. 1994. Goat hunters view of Aldabra. *Birdwatch* 12: 5-8.

Schreiber, R.W. and Ashmole, N.P. 1970. Sea-bird breeding season on Christmas Island, Pacific Ocean. *Ibis* 112: 363-394.

Scoones, T., Hambler, C., Woodroffe, R., Linfield, M. and Spillett, D. 1988. *Aldabra '88.* Oxford University Expedition Final Report

Seabrook, W.A. 1990. The impact of the feral cat (*Felis catus*) on the native fauna of Aldabra Atoll, Seychelles. *Rev. Ecol. (Terre et Vie)* 45: 135-145.

Shirihai, H. 1996. *The Birds of Israel.* London: Academic Press.

Shirihai, H. and Christie, D.A. 1996. A new taxon of small shearwater from the Indian Ocean. *Bull B.O.C.* 116(3): 180-186.

Shirihai, H., Roselaar, C.S. (Rees), Helbig, A.J., Barthel, P.H. and Van Loon, A.J. 1995. Identification and taxonomy of large *Acrocephalus* warblers. *Dutch Birding* 17 No 6.

Shirihai, H., Sinclair, I. and Colston, P. 1995. A new species of *Puffinus* shearwater from the western Indian Ocean. *Bull. B.O.C.* 115(2): 75-87.

Sibley, F.C. and Clapp, R.B. 1967. Distribution and dispersal of Central Pacific Lesser Frigatebirds *Fregata ariel*. *Ibis* 109: 328-337.

Sinclair, I. and Langrand, O. 1998. *Birds of the Indian Ocean Islands.* Cape Town: Struik.

Skead, C.J. 1967. *The Sunbirds of Southern Africa.* Cape Town: Trustees of the South African Bird Book Fund.

Skerrett, A. 1994a. The introduced birds of Assumption. *Birdwatch* 10: 4-8.

Skerrett, A. 1994b. A photographic visit to Aldabra. *Birdwatch* 10: 8-12.

Skerrett, A. 1994c. Wild birds and the law. *Birdwatch* 11: 3-8.

Skerrett, A. 1995. Birds of the Amirantes. *Birdwatch* 15: 10-20.

Skerrett, A. 1996a. The first report of the Seychelles Bird Records Committee. *Bull. ABC* 3(1): 45-50.

Skerrett, A. 1996b. Ornithological observations south of the Amirantes. *Birdwatch* 18: 3-8.

Skerrett, A. 1996c. Seychelles Bird Records Committee. *Birdwatch* 20: 13-15.

Skerrett, A. 1997. Seychelles Bird Records Committee. *Birdwatch* 24: 16-20.

Skerrett, A. 1999a. Birds of Cosmoledo. More questions. *Birdwatch* 29: 4-6.

Skerrett, A. 1999b. Cosmoledo yields up a few of its secrets. *Birdwatch*: 31: 4-12.

Skerrett, A. 1999c. Does the Abbott's Sunbird exist? *Birdwatch* 31: 25-27.

Skerrett, A. 1999d. Ferruginous Duck *Aythya nyroca*: the first record for Seychelles. *Bull. ABC* 6 (2): 148.

Skerrett, J and A. 1996. *Aride Island Nature Reserve Newsletter.* Issue 19 First half 1996.

Skerrett, A. and Matyot, P. 2000. *Zwazo Sesel: Seychelles bird names and their meanings.* Unpublished Document.

Skerrett, A. and the SBRC. 1998. Seychelles Bird Records Committee. *Birdwatch* 26: 14-20.

Snow, D.W. (ed.) 1978. *An Atlas of Speciation in African Non-passerine Birds.* London: British Museum (Natural History).

Staub, R. 1970. Geography and ecology of Tromelin Island. *Atoll Res. Bull.* 136: 197-209.

Staub, R. and Guého, J. 1968. The Cargados Carajos shoals of St Brandon: resources, avifauna and vegetation. *Proc. R. Soc. Arts Sci. Mauritius* 3: 7-46.

Staub, F. 1973. Birds of Rodriguez Island. *Proc. Roy. Soc. Arts and Sci. Mauritius* 4, part 1: 17-59.

Staub, R. 1976. Birds of the Macarenes and Saint Brandon. *Mauritius: Org. Normale des Enterpr. L'Tée.*

Stevenson, J. 1992. *A Brief visit to caves at La Gogue, N. Mahé.* Unpublished.

Stirling, Major W. 1843. *Narrative of the Wreck of the Ship Tiger, of Liverpool...on the Desert Island of Astova etc.* Exeter: W. Roberts.

Stoddart, D.R. 1971. White-throated Rail *Dryolimnas cuvieri* on Astove Atoll. *Bull. B.O.C.* 91: 145-147.

Stoddart, D.R. 1977. Identity of pelicans on St Joseph Atoll, Amirantes. *Bull. B.O.C.* 97(3): 94-95.

Stoddart, D.R. 1981. Abbott's Booby on Assumption. *Atoll Res. Bull.* 255: 27-32.

Stoddart, D.R. 1984. Breeding seabirds of the Seychelles and adjacent islands. In Stoddart, D.R. (ed.) *Biogeography and ecology of the Seychelles Islands*: 575-592. The Hague: Dr. W. Junk Publishers.

Stoddart, D.R. and Benson, C.W. 1970. An old record of a Blue Pigeon *Alectroenas* species and seabirds on Farquhar and Providence. *Atoll Res. Bull.* 136: 35-36.

Stoddart, D.R., Benson, C.W. and Peake, J.F. 1970. Ecological change and effects of phosphate mining on Assumption Island. *Atoll Res. Bull.* 136: 121-145.

Stoddart, D.R. and Poore, M.E.D. 1970a. Geography and Ecology of Farquhar Atoll. *Atoll Res. Bull.* 136: 7-26.

Stoddart, D.R. and Poore, M.E.D. 1970b. Geography and Ecology of Desroches. *Atoll Res. Bull.* 136: 155-165.

Stoddart, D.R. and Poore, M.E.D. 1970c. Geography and Ecology of Remire. *Atoll Res. Bull.* 136: 171-181.

Stoddart, D.R. and Poore, M.E.D. 1970d. Geography and Ecology of African Banks. *Atoll Res. Bull.* 136: 187-191.

Stoddart, D.R. and Wright, C.A. 1967. Geography and ecology of Aldabra Atoll. *Atoll Res. Bull* 118: 11-52.

Svensson, L. 1992. *Identification Guide to European Passerines.* Fourth edition. Stockholm: Privately Printed.

Tafforet c1726. *Relation de l'isle Rodrigue.* Paris Archives: Manuscript.

Threadgold R. and Johnson S. 1999. *Aldabra Drongo Observation Study.* SIF Internal Report.

Todd, D.M. 1982. *Seychelles Magpie Robin: cat eradication on Fregate.* ICBP Emergency Project Report.

Travis, W. 1959. *Beyond the Reefs.* London: Arrow Books.

Turner, A. and Rose, C. 1989. *A Handbook to the Swallows and Martins of the World.* London: Christopher Helm.

Tyzack, S. and Volcere, O. 1984. *The Roseate Tern Sterna dougallii arideensis on Aride Island, Seychelles: notes on 1984 season.* Unpublished.

Unienville, M.C.A.M. d', Baron 1838. *Statistique de l'Ile Maurice et ses Dependances, suivie d'une notice historique sur cette colonie et d'une essai sur l'Ile Madagascar.* Paris: Gustac Barba, Libraire 3 volumes.

Urban, E.K., Fry, C.H. and Keith, S. 1986. *The Birds of Africa.* Vol 2. Vol 5, 1997. London: Academic Press.

Van der Elst, R. and Prys-Jones, R. 1987. Mass killing by rats of roosting common noddies. *Oryx* 21: 4.

Vesey-Fitzgerald, L.D.E.F. 1936. *Birds of the Seychelles and Other Islands Included Within that Colony.* Victoria: Seychelles Government Printer.

Vesey-Fitzgerald, D. 1940. The birds of the Seychelles 1: The endemic birds. *Ibis* (14) 4: 480-489.

Vesey-Fitzgerald, D. 1941. Further contributions to the ornithology of the Seychelles Islands. *Ibis* (14) 5: 518-531.

Voeltzkow, A. 1897. Einleitung: Madagaskar, Juan de Nova, Aldabra. *Abhand. Senckenb naturf. Gesellsch.* 21: 1-76.

Voous, K.H. 1965. White-faced storm petrels in the Indian Ocean: corrections and additions. *Ardea* 53: 237.

Voous, K.H. 1966. Prions in the tropical Indian Ocean. *Ardea* 54: 89.

Wanless, R. 2000. Aldabra Rail Translocation. *Birdwatch* 33: 7-8.

Warman, S. 1977. *Aride Island Nature Reserve, Seychelles.* First Report to SPNC.

Warman, S.R. 1979. The Roseate Tern *Sterna dougallii arideensis* on Aride Island, Seychelles. *Bull. B.O.C.* 99(4): 124-128.

Warman, S. and Todd, D. 1979. *Aride Island report to SPNC. Part 3 Vertebrate fauna.* SPNC Internal Report.

Warman, S. and Todd, D. 1984. A biological survey of Aride Island Nature Reserve, Seychelles. *Biological Conservation* 28: 51-71.

Watson, G.E., Zusi, R.L. and Storer, R.E. 1963. *Preliminary field guide to the birds of the Indian Ocean.* Smithsonian Institution.

BIBLIOGRAPHY

Watson, J. 1978. 1978 *The Seychelles Magpie Robin* (Copsychus sechellarum). World Wildlife Fund Project 1590: Endangered landbirds, Seychelles. Final report 2 (a). Unpublished.

Watson, J. 1979. Clutch sizes of Seychelles' endemic landbirds. *Bull. B.O C.* 99(3): 102-105.

Watson, J. 1980. Distribution and nesting of the Yellow Bittern in Seychelles. *Ostrich* 51: 120-122.

Watson, J. 1981a. *Population, Ecology, Food and Conservation of the Seychelles Kestrel*. PhD Thesis Aberdeen University.

Watson, J. 1981b. *The Seychelles Black Paradise Flycatcher on La Digue 1977-8*. WWF Project 1590. Final Report 1(b).

Watson, J. 1992. Nesting ecology of the Seychelles Kestrel *Falco araea* on Mahé, Seychelles. *Ibis* 134: 259-267.

Wells, M.G. 1998. *World Bird Species Checklist: with Alternative English and Scientific Names*. Herts: Worldlist.

White, C.M.N. 1951. Systematic notes on African birds. *Ibis* 93: 460-465.

Williams, J.G. 1953. Revision of *Cinnyris sovimanga*: with description of a new race. *Ibis* 95: 501-504.

Wilson, J. (undated) *Ecology of Desnoeufs Island, Amirantes Islands*. Unpublished report to the Department of Agriculture and Land Use, Seychelles. Typescript.

Wilson, J.R. 1982. *An investigation into the effects of tourists on the breeding of Fairy Terns along paths on Cousin Island, Seychelles*. Cousin Island Research Station Technical Report No 20.

Wilson, J.R. and Chong-Seng, L. 1979. *Report on the exploitation of Sooty Tern eggs on Desnoeufs in 1979 and the implementation of controls on cropping*. Unpublished.

Woodell, R. 1976a. Variation in juvenile plumage of *Centropus toulou toulou* (Müller) and *Centropus toulou insularis* Ridgway. *Bull. B.O.C.* 96(2): 72-75.

Woodell, R. 1976b. Notes on the Aldabran Coucal *Centropus toulou insularis*. *Ibis* 118: 263-268.

Wright, G. and Passmore, K. 1999. *Breeding Seabird Census – July 1999: Lesser Noddy, Brown Noddy, Fairy Tern and White-tailed Tropicbird*. Cousine Island Internal Report.

Wyse. E. (managing ed.) 1998. *The Guinness Book of Records 1998*. Enfield: Guinness Publishing.

Zimmerman, D.A., Turner, D.A. and Pearson D.J., 1996. *Birds of Kenya and Northern Tanzania*. London: A & C Black.

Birding Journals

Nature Protection Trust of Seychelles publishes a quarterly journal, *Birdwatch*. For details of subscriptions contact NPTS, PO Box 207, Seychelles.

Sound Recordings.

An excellent CD is widely available in Seychelles entitled *Sounds of Seychelles*, produced by Gérard Rocamora and Aurora Sole. It is a collection of the sounds of the granitic islands with 51 species represented including 33 bird species.

INDEX OF ENGLISH NAMES

Albatross, Black-browed 140
  Wandering 139
Bee-eater, Blue-cheeked **40**, 262
  European **40**, 263
  Madagascar **40**, 263
Bittern, Cinnamon **12**, 172
  Eurasian **12**, 172
  Yellow **12**, 170
Blackcap **47**, 289
Booby, Abbott's **1**, 154
  Brown **7**, 157
  Masked **7**, 155
  Red-footed **7**, 156
Bulbul, Madagascar **44**, 274
  Red-whiskered **44**, 273
  Seychelles **44**, 274
Bunting, Ortolan **52**, 302
Canary, Yellow-fronted **52**, 303
Chiffchaff 288
Cisticola, Madagascar **47**, 282
Cormorant, Great **2**, 159
  Long-tailed **2**, 159
Corncrake **17**, 192
Coucal Madagascar **37**, 253
Crake, Spotted **17**, 192
  Striped **17**, 192
Crow, House **51**, 298
  Pied **51**, 299
Cuckoo, Asian Lesser **37**, 252
  Common **37**, 251
  Great Spotted **37**, 250
  Jacobin **37**, 250
Curlew, Eurasian **23**, 209
  Little **23**, 210
  Slender-billed 208
  Stone **18**, 197
Darter, African **2**, 158
Dove, Barred Ground **36**, 245
  European Turtle **35**, 242
  Madagascar Turtle **35**, 243
  Oriental Turtle **35**, 243
Drongo, Aldabra **49**, 297
Duck, Ferruginous **13**, 178
  White-faced Whistling **13**, 175
Eagle, Booted **14**, 182
Egret Cattle **10**, 164
  Dimorphic **10**, 166
  Great White **10**, 165
  Intermediate **10**, 166
  Little **10**, 166
Falcon, Amur **15**, 186
  Eleonora's **16**, 187
  Peregrine **16**, 188
  Red-footed **15**, 185
  Sooty **16**, 188
Flamingo, Greater **9**, 172
Flycatcher, Seychelles Paradise **49**, 291
  Spotted **49**, 290
Fody, Aldabra **53**, 306
  Madagascar **53**, 304
  Seychelles **53**, 308
Francolin, Grey **17**, 189

Frigatebird, Great **8**, 160
  Lesser **8**, 162
Gallinule, Allen's **17**, 194
Garganey **13**, 177
Godwit Bar-tailed **23**, 207
  Black-tailed **23**, 207
Grebe, Black-necked **2**, 139
Greenshank, Common **24**, 211
Gull, Black-headed **29**, **30**, 227
  Brown-headed 226
  Great Black-headed **29**, **30**, 225
  Grey-headed 226
  Heuglin's **29**, **30**, 224
  Lesser Black-backed **29**, **30**, 224
Harrier, Western Marsh **14**, 181
Heron, Black-crowned Night **12**, 170
  Green-backed **11**, 169
  Grey **9**, 163
  Purple **9**, 164
  Squacco **11**, 167
Hobby, Eurasian **16**, 186
Honey-buzzard, Western **14**, 179
Hoopoe **40**, 265
Ibis, Madagascar Sacred **9**, 175
  Sacred **9**, 174
Kestrel, Lesser **15**, 183
  Madagascar **15**, 184
  Seychelles **15**, 184
Kingfisher, Common 262
Kite, Black **14**, 180
  Yellow-billed **14**, 181
Knot, Great **21**, 204
  Red 204
Lark, Greater Short-toed **42**, 266
Lovebird, Grey-headed 247
Magpie-robin, Seychelles **46**, 278
Mallard **13**, 176
Martin, Common House **41**, 268
  Mascarene **41**, 267
  Sand **41**, 266
Moorhen, Common **17**, 193
Myna, Common **51**, 301
Needletail, White-throated **39**, 259
Nightjar, Eurasian **38**, 257
  Madagascar **38**, 258
Noddy, Brown **34**, 239
  Lesser **34**, 240
Oriole, Eurasian Golden **45**, 299
Osprey **14**, 179
Owl, Barn **38**, 254
  Brown Fish **38**, 257
Oystercatcher, Eurasian **18**, 196
Parakeet, Ring-necked **36**, 249
  Seychelles **1**, 249
Parrot, Seychelles Black **36**, 248
Pelican, Pink-backed **1**, 160
Petrel, Barau's **3**, 143
  Bulwer's **3**, 145
  Cape **3**, 142
  Jouanin's **3**, 145
  Mascarene **3**, 142
  Southern Giant **3**, 141

Phalarope, Red-necked **25**, 215
Pigeon, Comoro Blue **36**, 245
  Feral 241
  Seychelles Blue **36**, 246
Pintail, Northern **13**, 177
Pipit, Olive-backed **42**, 270
  Red-throated **42**, 270
  Richard's **42**, 268
  Tree **42**, 269
Plover, Caspian **20**, 200
  Common Ringed **21**, 201
  Crab **18**, 195
  Grey **21**, 203
  Little Ringed **21**, 202
  Pacific Golden **21**, 202
  Oriental **20**, 200
Pond-heron, Indian **11**, 168
  Madagascar **11**, 168
Pratincole, Black-winged **19**, 198
  Common **19**, 197
  Oriental **19**, 198
Prion, Antarctic 143
  Slender-billed 144
Quail, Common **17**, 189
Rail, Aldabra **17**, 190
Redshank, Common **24**, 210
Redstart, Common **46**, 280
Roller, Broad-billed **40**, 265
  European **40**, 263
Rosefinch, Common **52**, 303
Ruff **27**, 220
Sanderling **26**, 215
Sandpiper, Broad-billed **27**, 219
  Buff-breasted **27**, 219
  Common **25**, 213
  Curlew **27**, 218
  Green **24**, 211
  Marsh **24**, 210
  Pectoral **27**, 217
  Sharp-tailed **27**, 218
  Terek **25**, 212
  Wood **24**, 212
Sandplover, Greater **20**, 199
  Lesser **20**, 199
Scops-owl, Eurasian **38**, 255
  Seychelles **38**, 256
Shearwater, Audubon's **4**, 148
  Flesh-footed **4**, 146
  Mascarene 150
  Wedge-tailed **4**, 147
Shelduck, Ruddy **13**, 176
Shoveler, Northern **13**, 178
Shrike Lesser Grey **45**, 276
  Red-backed **45**, 275
  Woodchat **45**, 277
Skua, Antarctic **28**, 221
  Arctic **28**, 222
  Long-tailed **28**, 223
  Pomarine **28**, 222
  South Polar **28**, 220
Snipe, Common **22**, 205
  Great **22**, 206
  Pintail **22**, 205

Swinhoe's **22**, 206
Sparrow, House **52**, 309
Starling, Rose-coloured **51**, 300
  Wattled **51**, 300
Stilt, Black-winged **18**, 196
Stint, Little **26**, 216
  Long-toed **26**, 217
  Red-necked **26**, 216
  Temminck's **26**, 217
Stork, White **9**, 174
Storm-petrel, Black-bellied **5**, 151
  Matsudaira's **5**, 151
  Swinhoe's **5**, 152
  White-faced **5**, 151
  Wilson's **5**, 150
Sunbird, Abbott's **50**, 293
  Seychelles **50**, 293
  Souimanga **50**, 292
Swallow, Barn **41**, 267
Swamphen, Purple **1**, 195
Swift, Common **39**, 260
  Little **39**, 261
  Pacific **39**, 261

Swiftlet, Seychelles **39**, 258
Tattler, Grey-tailed **25**, 214
Teal, Common **13**, 176
Tern, Black-naped **34**, 234
  Bridled **33**, 235
  Caspian **31**, 229
  Common **32**, 232
  Fairy **34**, 241
  Greater Crested **31**, 230
  Gull-billed **31**, 229
  Lesser Crested **31**, 231
  Little **33**, 238
  Roseate **32**, 232
  Sandwich **31**, 231
  Saunders' **33**, 238
  Sooty **33**, 236
  Whiskered **32**, 227
  White-cheeked **32**, 234
  White-winged **32**, 228
Thrush, Rufous-tailed Rock **44**, 277
Tropicbird, Red-billed **6**, 152
  Red-tailed **6**, 153
  White-tailed **6**, 153

Turnstone, Ruddy **25**, 214
Wagtail, Citrine **43**, 272
  Grey **43**, 272
  White **43**, 273
  Yellow **43**, 271
Warbler, Aldabra **1**, 283
  Icterine **48**, 286
  Sedge **47**, 284
  Seychelles **47**, 285
  Wood **48**, 289
  Willow **48**, 287
Waterhen, White-breasted **17**, 193
Waxbill, Common **52**, 304
Wheatear, Isabelline **46**, 282
  Northern **46**, 281
Whimbrel **23**, 208
Whinchat **46**, 281
White-eye Madagascar **50**, 295
  Seychelles **50**, 296
  Seychelles Chestnut-flanked **1**, 295
Whitethroat, Common **48**, 290

INDEX OF SCIENTIFIC NAMES

Acridotheres tristis **51**, 301
Acrocephalus sechellensis **47**, 285
  schoenobaenus **47**, 284
Actitis hypoleucos **25**, 213
Aenigmatolimnas marginalis **17**, 192
Aerodramus elaphrus **39**, 258
Agapornis canus 247
Alcedo atthis 262
Alectroenas pulcherrima **36**, 246
  sganzini **36**, 245
Amaurornis phoenicurus **17**, 193
Anas acuta **13**, 177
  clypeata **13**, 178
  crecca **13**, 176
  platyrhynchos **13**, 176
  querquedula **13**, 177
Anhinga rufa **2**, 158
Anous stolidus **34**, 239
  tenuirostris **34**, 240
Anthus cervinus **42**, 270
  hodgsoni **42**, 270
  richardi **42**, 268
  trivialis **42**, 269
Apus affinis **39**, 261
  apus **39**, 260
  pacificus **39**, 261
Ardea cinerea **9**, 163
  purpurea **9**, 164
Ardeola grayii **11**, 168
  idae **11**, 168
  ralloides **11**, 167
Arenaria interpres **25**, 214
Aythya nyroca **13**, 178
Botaurus stellaris **12**, 172
Bubulcus ibis **10**, 164

Bulweria bulwerii **3**, 145
  fallax **3**, 145
Burhinus oedicnemus **18**, 197
Butorides striatus **11**, 169
Calandrella brachydactyla **42**, 266
Calidris acuminata **27**, 218
  alba **26**, 215
  canutus 204
  ferruginea **27**, 218
  melanotos **27**, 217
  minuta **26**, 216
  ruficollis **26**, 216
  subminuta **26**, 217
  temminckii **26**, 217
  tenuirostris **21**, 204
Caprimulgus europaeus **38**, 257
  madagascariensis **38**, 258
Carpodacus erythrinus **52**, 303
Catharacta antarctica **28**, 221
  maccormicki **28**, 220
Centropus toulou **37**, 253
Charadrius asiaticus **20**, 200
  dubius **21**, 202
  hiaticula **21**, 201
  leschenaultii **20**, 199
  mongolus **20**, 199
  veredus **20**, 200
Chlidonias hybridus **32**, 227
  leucopterus **32**, 228
Ciconia ciconia **9**, 174
Circus aeruginosus **14**, 181
Cisticola cherina **47**, 282
Clamator glandarius **37**, 250
Columba livia 241
Copsychus sechellarum **46**, 278
Coracias garrulus **40**, 264

Coracopsis (nigra) barklyi **36**, 248
Corvus albus **51**, 299
  splendens **51**, 298
Coturnix coturnix **17**, 189
Creatophora cinerea **51**, 300
Crex crex **17**, 192
Cuculus canorus **37**, 251
  poliocephalus **37**, 252
Daption capense **3**, 142
Delichon urbica **41**, 268
Dendrocygna viduata **13**, 175
Dicrurus aldabranus **49**, 297
Diomedea exulans 139
  melanophris 140
Dromas ardeola **18**, 195
Dryolimnas (cuvieri) aldabranus **17**, 190
Egretta alba **10**, 165
  dimorpha **10**, 166
  garzetta **10**, 166
  intermedia **10**, 166
Emberiza hortulana **52**, 302
Estrilda astrild **52**, 304
Eurystomus glaucurus **40**, 265
Falco amurensis **15**, 186
  araea **15**, 184
  concolor **16**, 188
  eleonorae **16**, 187
  naumanni **15**, 183
  newtoni **15**, 184
  peregrinus **16**, 188
  subbuteo **16**, 186
  vespertinus **15**, 185
Foudia (eminentissima) aldabrana **53**, 306
  madagascariensis **53**, 304

sechellarum **53**, 308
Francolinus pondicerianus **17**, 189
Fregata ariel **8**, 162
  minor **8**, 160
Fregetta tropica **5**, 151
Gallinago gallinago **22**, 205
  media **22**, 206
  megala **22**, 206
  stenura **22**, 205
Gallinula chloropus **17**, 193
Gelochelidon nilotica **31**, 229
Geopelia striata **36**, 245
Glareola maldivarum **19**, 198
  nordmanni **19**, 198
  pratincola **19**, 197
Gygis alba **34**, 241
Haematopus ostralegus **18**, 196
Heteroscelus brevipes **25**, 214
Hieraaetus pennatus **14**, 182
Himantopus himantopus **18**, 196
Hippolais icterina **48**, 286
Hirundapus caudacutus **39**, 259
Hirundo rustica **41**, 267
Hydroprogne caspia **31**, 229
Hypsipetes crassirostris **44**, 274
  madagascariensis **44**, 274
Ixobrychus cinnamomeus **12**, 172
  sinensis **12**, 170
Ketupa zeylonensis **38**, 257
Lanius collurio **45**, 275
  minor **45**, 276
  senator **45**, 277
Larus brunnicephalus 226
  cirrocephalus 226
  fuscus **29**, **30**, 224
  heuglini **29**, **30**, 224
  ichthyaetus **29**, **30**, 225
  ridibundus **29**, **30**, 227
Limicola falcinellus **27**, 219
Limosa lapponica **23**, 207
  limosa **23**, 207
Macronectes giganteus **3**, 141
Merops apiaster **40**, 263
  persicus **40**, 262
  superciliosus **40**, 263
Milvus aegyptius **14**, 181
  migrans **14**, 180
Monticola saxatilis **44**, 277
Motacilla alba **43**, 273
  cinerea **43**, 272

citreola **43**, 272
  flava **43**, 271
Muscicapa striata **49**, 290
Nectarinia dussumieri **50**, 293
  sovimanga **50**, 292
  (sovimanga) abbotti **50**, 293
Nesillas aldabranus **1**, 283
Numenius arquata **23**, 209
  minutus **23**, 210
  phaeopus **23**, 208
  tenuirostris 208
Nycticorax nycticorax **12**, 170
Oceanites oceanicus **5**, 150
Oceanodroma matsudairae **5**, 151
  monorhis **5**, 152
Oenanthe isabellina **46**, 282
  oenanthe **46**, 281
Oriolus oriolus **45**, 299
Otus insularis **38**, 256
  scops **38**, 255
Oxylophus jacobinus **37**, 250
Pachyptila belcheri 144
  desolata 143
Pandion haliaetus **14**, 179
Passer domesticus **52**, 309
Pelagodroma marina **5**, 151
Pelecanus rufescens **1**, 160
Pernis apivorus **14**, 179
Phaethon aethereus **6**, 152
  lepturus **6**, 153
  rubricauda **6**, 153
Phalacrocorax africanus **2**, 159
  carbo **2**, 159
Phalaropus lobatus **25**, 215
Phedina borbonica **41**, 267
Philomachus pugnax **27**, 220
Phoenicopterus ruber **9**, 172
Phoenicurus phoenicurus **46**, 280
Phylloscopus collybita 288
  sibilatrix **48**, 289
  trochilus **48**, 287
Pluvialis fulva **21**, 202
  squatarola **21**, 203
Podiceps nigricollis **2**, 139
Porphyrio porphyrio **1**, 195
Porphyrula alleni **17**, 194
Porzana porzana **17**, 192
Psittacula (eupatria) wardi **1**, 249
  krameri **36**, 249
Pterodroma aterrima **3**, 142

baraui **3**, 143
Puffinus atrodorsalis 150
  carneipes **4**, 146
  lherminieri **4**, 148
  pacificus **4**, 147
Pycnonotus jocosus **44**, 273
Riparia riparia **41**, 266
Saxicola rubetra **46**, 281
Serinus mozambicus **52**, 303
Stercorarius longicaudus **28**, 223
  parasiticus **28**, 222
  pomarinus **28**, 222
Sterna albifrons **33**, 238
  anaethetus **33**, 235
  dougallii **32**, 232
  fuscata **33**, 236
  hirundo **32**, 232
  repressa **32**, 234
  saundersi **33**, 238
  sumatrana **34**, 234
Streptopelia orientalis **35**, 243
  picturata **35**, 243
  turtur **35**, 242
Sturnus roseus **51**, 300
Sula abbotti **1**, 154
  dactylatra **7**, 155
  leucogaster **7**, 157
  sula **7**, 156
Sylvia atricapilla **47**, 289
  communis **48**, 290
Tadorna ferruginea **13**, 176
Terpsiphone corvina **49**, 291
Thalasseus bengalensis **31**, 231
  bergii **31**, 230
  sandvicensis **31**, 231
Threskiornis aethiopicus **9**, 174
  bernieri **9**, 175
Tringa glareola **24**, 212
  nebularia **24**, 211
  ochropus **24**, 211
  stagnatilis **24**, 210
  totanus **24**, 210
Tryngites subruficollis **27**, 219
Tyto alba **38**, 254
Upupa epops **40**, 265
Xenus cinereus **25**, 212
Zosterops maderaspatana **50**, 295
  modestus **50**, 296
  semiflava **1**, 295